Advanced High Strength Sheet Steels

Nina Fonstein

Advanced High Strength Sheet Steels

Physical Metallurgy, Design, Processing, and Properties

 Springer

Nina Fonstein
Scientific Advisor, Global R&D
ArcelorMittal
East Chicago, IN, USA

ISBN 978-3-319-37301-0   ISBN 978-3-319-19165-2   (eBook)
DOI 10.1007/978-3-319-19165-2

Springer Cham Heidelberg New York Dordrecht London
© Springer International Publishing Switzerland 2015
Softcover reprint of the hardcover 1st edition 2015

Printed on acid-free paper

Springer International Publishing AG Switzerland is part of Springer Science+Business Media
(www.springer.com)

Foreword

A book coauthored by N.M. Fonstein and S.A. Golovanenko under the title "Dual-phase low-alloy steels" was published in 1986 in Russia. Deep and wide research conducted for various applications of dual-phase steels with ferrite–martensite structure, including intensive trials of cold-rolled sheet steels for cars and long products for fasteners, laid the foundation for this book. It reflected the results of numerous studies by graduate students and engineers of laboratory of formable steels, which was led by Dr. Fonstein, and the analysis of world literature, spanning research methods of structure investigation of multiphase steels, the relationship of parameters of the structure and mechanical properties, and the fundamentals of heat treatment and characteristics that are important for the application of these steels such as formability, weldability, fatigue, etc. It was the first and so far, it seems, the only monograph devoted to dual-phase steel. However, the extensive presented information, including industrial development, was limited to the strength levels between 500 and 700 MPa.

Later on, when Nina Fonstein and a couple of her former graduate students joined the research center of ArcelorMittal, where Nina for over 10 years led the development of steels for the automotive industry, they had to develop dual-phase steels of higher and higher strength, 780, 980, 1180 MPa, TRIP steels, new martensitic steels, complex phase steels, to explore ways of various structural approaches to meet the requirements for the third-generation steels.

This book is once again the world's first attempt to outline the fundamental features of structure and mechanical properties of all modern Advanced High-Strength Steels families. It is important not only that, in contrast to numerous works of multiple conferences on the subject, this book provides a comparative analysis and synthesis of existing literature and author's data, but also that results are summarized from the position of the researcher with the dual responsibilities for not only fundamental understanding of physical metallurgy but also for the industrial implementation of developed grades.

The volume and quality of the data interpreted in this book came about not only because of Nina's individual ability, which itself is quite remarkable, but also due to major contributions of the postgraduate students and young engineers of Research Center at ArcelorMittal whom Dr. N. Fonstein continues to lead. Information presented in this book is transcending geographical boundaries and fully capitalizing of the global nature of ArcelorMittal research organization with 12 centers around the world and close cooperation with Academia. It should be noted, however, that the major contribution came from North America and European Centers.

I am thoroughly convinced that the published book will be a useful reference and guiding tool for engineers engaged in the development, production, and use of AHSS, as well as students of Universities and graduate students involved in the creation of new automotive products.

VP, Head of Global R&D, ArcelorMittal, Europe   Greg Ludkovsky

Acknowledgments

This book became possible thanks to the outstanding contributions of my graduate students and employees of the I.P. Bardin Central Research Institute of Ferrous Metallurgy in Moscow, in particular, A. Bortsov, T. Efimova, and A. Efimov, and intensive investigations of ArcelorMittal Global Research Center in East Chicago and especially by Olga Girina, Oleg Yakubovsky, Hyun-Jo Jun, and Rongjie Song, as well as due to creative atmosphere of the Center and the constant support of its leaders Greg Ludkovsky, Debanshu Bhattacharya, and Narayan Pottore.

Sincere thanks to John Speer for fruitful discussion of the book contents and Sebastien Allain for helpful advice during the preparation of a chapter on TWIP steels.

And last but not the least—special thanks to my husband and a supporter for my entire life for his help and patience.

Contents

About the Author

Dr. Nina Fonstein is currently a scientific advisor of automotive product development group, Global R&D, ArcelorMittal, in East Chicago labs, USA, being for more than 10 years the manager of that group, leading investigations, development, and commercialization of various grades of AHSS. Before joining ArcelorMittal in 1999, she had been working in the I.P. Bardin Central Research Institute of Ferrous Metallurgy in Moscow, Russia, as a senior research associate and later as a head of the laboratory of formable steels working on dual-phase steels since 1979 and on TRIP steels since 1988. She graduated from the Moscow Institute of Steel and Alloys and holds Ph.D. in solid state physics from I.P. Bardin Central Research Institute of Ferrous Metallurgy (1968) and Dr. Science degree in physical metallurgy from Moscow Institute of Steel and Alloys (1986) as well as the title of Professor from the Supreme Attestation Commission of Russian Federation (1990). She is the author of the book "Dual-phase low alloyed steels" (in Russian) and more than 100 publications in English. She has supervised 16 PhD students in Russia and is supervising PhD students—employees of ArcelorMittal.

List of Abbreviations

AHSS	Advanced high strength steels
B	Bainite
Bcc	Body centered cubic lattice
Bct	Body centered tetragonal lattice
BH	Bake hardening
BIW	Body in white
CAL	Continuous annealing line
CCT	Continuous cooling transformation
CE	Carbon equivalent
CFB	Carbide free bainite
CGL	Continuous galvanizing line
CP	Complex phase steels
CR	Cold rolled
DP	Dual phase steels
EBSD	Electron back scattering diffraction
F	Ferrite
Fcc	Face centered cubic lattice
GJC	Gas jet cooling temperature
HE	Hole expansion
HER	Hole expanding ratio
HR	Hot rolled
HSLA	High strength low alloyed steels
IBT	Isothermal bainitic transformation
IQ	Image quality
LOM	Light optical microscopy
MP	Multi-phase steels
P	Pearlite
PAGS	Prior austenite grain size
PHS	Press hardening steels
PM	Partial martensitic steels (DP steels with %martensite ≥ 50)
Q&P	Quenching and partitioning

Q&T	Quenching and tempering
RA	Retained austenite
RA	Reduction of area
RD	Rolling direction
ReX	Recrystallization
RT	Room temperature
SEM	Scanning electron microscopy
SS	Stress–strain curve
TBF	Trip steels with bainiticferrite
TD	Transverse direction
TE	Total elongation
TEM	Transmission electron microscopy
TRIP	Transformation induced plasticity
TS	Tensile strength
TTT	Time temperature transformation
UE	Uniform elongation
UFG	Ultra fine grains
ULC	Ultra low carbon
UTS	Ultimate tensile strength
WH	Work hardening
wt%	Weight percent
XRD	X ray diffraction
YPE	Yield point elongation
YS	Yield strength

List of Symbols

ΔG	Mechanical driving force
A_{c1}	Temperature at which austenite formation starts during heating
Ac3	Temperature at which austenite formation is completed during heating
A_{e1}	Equilibrium temperature for the lower boundary of the $\alpha + \gamma$ range
Ae3	Equilibrium temperature for the upper boundary of the $\alpha + \gamma$ range
Ar1	Temperature at which austenite disappears during cooling
Ar3	Temperature at which austenite starts to transform during cooling
B_S	Bainite start temperature
C_M	Martensite lattice parameter
C_{tot}	Total carbon content [wt%] in steel
C_α	Ferrite C content [wt%] in the final microstructure
$C\gamma$	Austenite C content [wt%]
f_B	Bainitic ferrite fraction in the final microstructure
f_M	Martensite volume fraction
f_α	Ferrite fraction in the final microstructure
f_γ	Austenite volume fraction
M_d	Temperature above which austenite is stable and no deformation-induced transformation to martensite takes place
M_F	End of martensitic transformation
M_S	Martensite start temperature
M_S^σ	Temperature at which the mode of transformation of austenite to martensite changes from stress-assisted to strain-induced
n	Strain hardening coefficient
v	Velocity
V_M	Martensite volume fraction in percentage in the final microstructure
$V_{\gamma ret}$	Retained Austenite volume fraction in percentage in the final microstructure
YS	Yield stress
α	α-phase, ferrite
α'	Martensite
γ	Austenite

γ_{res}	Residual austenite in the process of transformation
ε	True strain
ρ	Dislocation density
σ	True stress

Chapter 1
Evolution of Strength of Automotive Steels to Meet Customer Challenges

Contents

1.1 Ancient History

Almost for a century, the sheet steel materials for passenger vehicles did not undergo big changes. Exposed parts as well as structural and functional elements were made of mild steels that were required to be as soft as possible to ensure formability, repairability, etc., including ability to be weldable, paintable, and so forth. Yield strength of steels for parts, most critical for forming operations, was limited by 180–200 MPa, tensile strength accordingly by 330–350 MPa while some outer panels reached 1.5–2 mm in thickness.

It was in 1975, when the first bell rang as the worldwide oil crisis burst out. The goal to decrease fuel consumption by lowering the weight of automobiles immediately projected onto increase in steel strength that could allow thinner gauges of stamped parts. Although the crisis did not last long, this period, however, had a revolutionary impact on sheet steel development.

It was the time when the weight of publications devoted to high-strength steels for automotive applications was heavier than really commercially produced higher strength steels. However, just then the batch-annealed microalloyed HSLA with yield strength (YS) above 280–300 MPa at tensile strength (TS) of about 450 MPa was developed (Meyer et al. 1975). The first mentioning of a unique combination of strength and ductility of then unknown dual-phase steels (Hayami and Furukawa 1975) immediately attracted attention of leading metallurgists and was followed by

© Springer International Publishing Switzerland 2015

N. Fonstein, *Advanced High Strength Sheet Steels*,

DOI 10.1007/978-3-319-19165-2_1

numerous symposia and conferences focused on those steels (Dual-Phase 1978; Structure and Properties 1979; Fundamentals 1981). Industrial trials in Japan and in the USA [in the latter case, by the way, at galvanizing lines using hot-rolled coils (Bucher et al. 1979)] confirmed the expectation for tensile strength of 500–600 MPa at 25–30 % elongation. A few parts were stamped but without any commercial follow-up continuation at that time.

Nothing much happened even after continuous annealing or galvanizing lines quite suitable to produce dual-phase microstructure had been introduced in the steel industry. Excellent new equipment started to be utilized for manufacturing steels of exposed quality, the IF steels developed shortly thereafter solved problems with potential aging of continuously annealed steels, so that automotive customers enjoyed high formability of a new group of very formable steels that, however, possessed even lower strength than traditional mild Al-killed steels.

There was no big motivation for carmakers yet to apply necessary efforts

- To develop and to tune tooling of higher wear resistance,
- To stamp steels of higher strength, overcoming their often unpredictable spring-back and consequent distortion of part geometry,
- To adjust parameters of welding, which required higher current, pressure, or time, therefore jeopardizing cost and productivity,
- To cut/punch/shear steels of higher hardness, etc.

Low interest from auto industry chilled the excitement of steel producers. Even the most enthusiastic researchers of dual-phase steels such as R. Davis, the head of Ford Material Department (Davies and Magee 1981; Davies 1978), gave up. In the early 90s, the idea of DP steels seemed to have been buried.

As noted by GM at Auto/Steel Partnership annual meeting in 2011, in three decades from 1970 to 2000, only the so-called conventional HSS including high-strength IF steels, BH-IF steels, and mostly batch-annealed or sometimes continuously annealed HSLA with TS ~ 450–550 MPa were developed (Hall 2011).

Still a few pioneers including Inland steel, USA (Gupta and Chang 1985; Southwick et al. 1985), grounding on their previous experience of processing water-quenched martensite began steadily supplying the developed dual-phase steel with very high (for that time) strength (965 MPa). This steel is still being produced and successfully used for bumpers, door beams, and similar roll-formed parts. Luckily, a competition between carmakers made them seek a higher strength for critical safety parts, while batch-annealed or continuously annealed HSLA did not offer TS higher than 550 MPa.

In the middle of 1990s, Honda, USA, requested bare and coated DP590 grades to be supplied domestically. This request was based on the accumulated data regarding supply of high-strength DP steels for cold forming in the USA, as well as on successful trials and gradually growing applications and commercial supply of DP steels in Japan. This marked the start of reviving dual-phase steels and their subsequent escalation.

Among other motivations, a competition between steel and low-density metals such as Al and Mg was growing, as well as the requirements for passenger safety, fuel economy, and environmental friendliness continued to toughen, which resulted in extensive new developments of steels grades with higher strength and good formability (Fonstein et al. 2013).

1.2 Reviving Dual-Phase Steels

ThyssenKrupp presented interesting illustrations of the evolution of automotive steels that, in fact, reflect the worldwide history of high strength progress in steel industry (Sebald 2008). As shown in Fig. 1.1a, by 1990, mostly mild steels, high-strength IF, and microalloyed HSLA steels were available.

However, when in 1994 Audi introduced its A8 vehicle to the market, the first model based on aluminum space frame technology, the dominant position of steels as materials for vehicle body began to shatter (Samek and Krizan 2012). The weight of A8 all-aluminum body-in-white (BIW) was considerably lighter than that of competing conventional steel car bodies.

In 1994, the world steel industry reacted to the lightweight challenge by launching an international consortium of sheet steel producers that comprised 35 companies from 18 countries. This consortium initiated the so-called ULSAB project (Ultra-Light Steel Auto Body) to explore opportunities for weight saving using automotive components made of higher strength steels. Engineering efforts resulted in the development of steel "body-in-white" with a 90 % of high-strength steels (HSS). This car body had 25 % less mass at 14 % less cost than the benchmarked sedans. Additionally, the torsion and bending stiffness were improved by 80 % and 52 %, respectively.

The latest evolution of ULSAB concept, ULSAB—AVC (Advanced Vehicle Concept), proposes 100 % of HSS with over 80 % of advanced high-strength steel (AHSS) grades (Berger et al. 2009).

This project triggered an extensive development of novel steel grades with predominantly higher strength and satisfactory formability.

Existing challenges coming from carmakers and competitive materials gave rise to intensive trials and implementation of DP steels. By 1995 DP steels were commercialized not only in Japan but in the USA and Europe as well (Fig. 1.1b).

Dual-phase steels demonstrated very high combinations of tensile strength and elongation, high strain and bake hardening, as well as higher fatigue resistance. Since by that time the demand in coated, corrosion-protected steels grew as well, it was important to make DP steels coatable without specific problems, and their production on hot-galvanizing lines was implemented by many companies. As the experience of stamping and further processing of higher strength steel was accumulating, the demand for even higher strength of DP steels emerged. DP590/600 grade was followed by DP780 and DP980 grades (Bhattacharya et al. 2003; Pottore et al. 2006).

Fig. 1.1 Changes in cold-rolled automotive steels: (**a**) by 1990; (**b**) 1995; (**c**) 2000 years (Sebald 2008)—modified

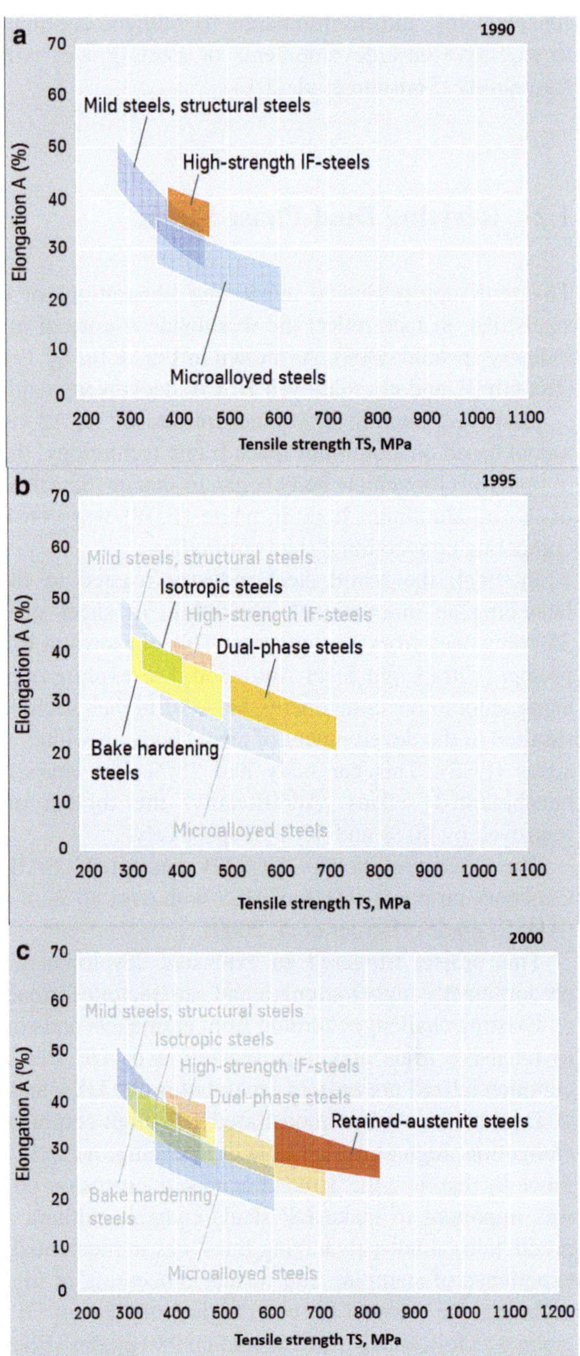

1.3 Development of AHSS

Growing complexity of safety part design together with the necessity to ensure prescribed stiffness of smaller wall thickness required special combinations of strength and formability parameters. Those requirements were summarized by the term "Advanced High-Strength Steels" that generally implies yield strength above 280/300 MPa and tensile strength above 590/600 MPa at improved formability.

Whereas the main category of AHSS is the dual-phase (DP) steels, automotive customer requirements brought about the development of new special microstructures of high-strength sheet steels to meet specific requests for particular safety conditions. First worth mentioning are TRIP steels with **T**ransformation-**I**nduced **P**lasticity featured by extremely high strain hardening rate up to the limit of uniform elongation that can help ensuring extremely high energy absorption at collisions.

As shown in Fig. 1.1c, the first attempts to commercially produce TRIP steels with retained austenite took place already by 2000. In Japan, their birth was due to NSC (Akisue and Hada 1995), and in Europe, the TRIP steels were pioneered by Thyssen (Engl et al. 1998).

TRIP steels demonstrated enhanced stretchability and higher absorbed energy and fatigue limit compared to conventional and DP steels with the same yield strength (Takahashi 2003).

The trend of wider utilization of high-strength steels for weight saving resulted in more stringent requirements for certain specific features of their formability, in particular, for higher flangeability. It was shown that a homogeneous microstructure with a minimal difference in strength between microstructure constituents is essential to obtain a high hole-expanding ratio (Senuma 2001).

No correlation of elongation with hole expansion has been established. Numerous studies point at the advantages of substituting, at least partially, martensite constituent by bainite that improves flangeability (hole expansion). These findings gave life to the so-called complex phase (CP) steels with ferrite–bainite–(martensite) microstructure.

In this direction, several different microstructure approaches were elaborated including the so-called ferrite–bainite (FB) steels, complex phase (CP) steels with dominant bainitic matrix, strengthening of ferrite phase in DP steels by Si or by precipitation hardening to decrease the strength difference between phases aiming to promote higher stretch flangeability (SF).

The demand for maximum possible resistance to intrusion led to development and growing usage of martensitic steels featured by the highest YS to TS ratio.

There are subcategories of classification for the above large groups of steel grades. Some standards distinguish between DP and PM steels: both types possess with ferrite–martensite microstructure, but PM (partially martensitic) steels are supposed to have more than 50 % martensite. TRIP steels, in turn, include TRIP-assisted DP steels containing not only ferrite, bainite, and retained austenite but also some fraction of martensite. Instead of the "CP steels," the term used mostly for cold-rolled substrates with preferably bainitic matrix, as-rolled steels with a similar microstructure are called multi-phase (MP) steels.

Martensitic steels should be classified depending on the method of their manufacturing: annealed, as-rolled, or made by quenching in the press; the latter group is termed press-hardened steels (PHS).

The level of the strength of AHSS determines the potential weight saving of parts, while a specific steel microstructure provides the combinations of strength and ductility, strain hardening, and/or other features of their mechanical behavior.

In particular, DP and TRIP steels as materials with high potential for energy absorption are used for parts that can undergo dynamic loading during car collisions.

Modified DP/PM steels with increased yield-to-tensile strengths ratio, as well as martensitic steels, are used for applications that require high stiffness, anti-intrusion, and elements of safety that should withstand high loads to protect the driver and the passengers.

Wider application of AHSS implies even more characteristics of formability depending on the forming operations. High elongation and strain hardening are critical for stretchability. However, with increasing strength, the edge cut cracking, poor bending, or insufficient flangeability that correlates with hole expansion becomes of great importance.

In the new Millennium, AHSS started to play a crucial role in making vehicles lighter while securing their safety. In addition to high strength of steels, improving ductility to make them suitable for wider applications became progressively important. Steel industry responded by the most rapid growth in number of new AHSS grades. Achieving new combinations of properties required new microstructure/processing approaches. All steel strengthening mechanisms, including solid solution strengthening, grain refinement, strengthening by controlled transformations, precipitation hardening, and strain aging, are employed alone or in combinations, also along with simultaneous additions of appropriate alloying and microalloying elements and application of thermomechanical processing.

During the first decade of 2000s, high Mn TWIP austenitic steels with unique combination of TS ~ 1000 MPa and TE ~ 50–60 % were developed (Allain et al. 2002; Kwon 2008). So far, these have become the highest combination of strength and ductility benchmark and are regarded as the new "second" generation of AHSS. Manufacturability problems, cost, and high sensitivity to hydrogen embrittlement imposed serious restrictions on commercial application of these steels. The progress with these steels was quite slow, and even after finding the ways to improve manufacturability and suppress their susceptibility to delayed fracture, they are still in the trial stage or beginning of commercialization.

Understanding of the advantages of high-strength and ultra-high-strength steels based on carbide-free bainite (CBF) matrix with retained austenite initiated a series of new developments (Sugimoto et al. 2000, 2002).

Establishing the relationship between high ductility and the amount and stability of retained austenite motivated the development of a new processing technology termed Q&P (quenching and partitioning) when mixture of martensite/bainite and retained austenite are formed (Speer et al. 2003)

A very important feature of all AHSS, both already developed and under development, is their so-called baking hardenability (BH), that is, the increase in YS after forming the part, i.e., without any impact on the initial formability. BH is typically specified as ~40 MPa but often reaches 60–90 MPa.

1.4 Nowadays

The current trends in AHSS development are to a great extent motivated by new environment of the end users, as carmakers are facing drastic changes in regulations that dictate new rules for both safety and fuel consumption.

For example, adopted in the last decade, the Euro-NCAP and the US-NCAP (New Car Assessment Program) standard tests for safety in new passenger vehicles impose much more severe requirements to car behavior at frontal and side impact, specify roof strength, etc. Evolution of safety requirements is illustrated, in particular, in Fig. 1.2.

The global presence of leading carmakers such as Ford, GM, Toyota, Honda, and Hyundai make them ready to all upcoming stringent safety and emission regulations in different countries.

As a result, BIW of leading cars such as Hyundai i40 contains only 35.6 % of traditional low-strength materials, 23.2 % high strength, 28.5 % AHSS, 6.1 % UHSS, and 3.1 % PHS ("Review of EuroCarBody 2011").

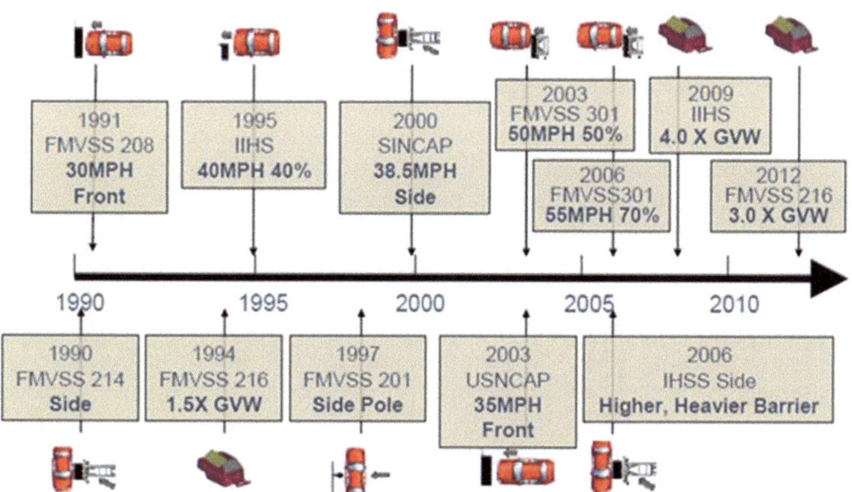

Fig. 1.2 Evolution of safety requirements (Hall 2011)

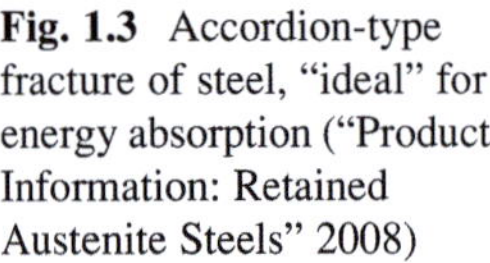

Fig. 1.3 Accordion-type fracture of steel, "ideal" for energy absorption ("Product Information: Retained Austenite Steels" 2008)

According to the Honda data, the usage of HSS in their MDX models increased from 13 % in 2001 to 56 % in 2007 (Mallen et al. 2007).

Due to their different functions, the driver/passengers' protective materials should have quite different features whether they are applied for engine compartment, energy management zones, or passenger compartment. It is possible to sacrifice the engine compartment and trunk to make them work as buffers, so that the deformation at collision absorbs the impact energy. For these materials, the highest combination of tensile strength and elongation is a key regardless of the yield strength. High strain hardening is beneficial because the expected absorbed energy is roughly proportional to the area under the stress–strain curve. Here, the "accordion-type" fracture of steel illustrated in Fig. 1.3 is very desirable. The best steel grades used in these cases are DP steels and TRIP steels with superior elongation.

On the contrary, the passenger compartment should protect the driver and the passenger from any part intrusions. This means that the safety elements, while ensuring stiffness of this section, should have the highest resistance to deformation, i.e., the highest possible yield strength. The most successful here are the parts made of as-annealed martensite or press-formed martensite.

In parallel, very strict requirements for fuel consumption were developed.

Fuel economy has a few equally important aspects. First, the transportation sector was reported to be responsible for over half of the world's oil consumption. For carmakers, the fuel consumption is one of critical indicators in competition, and its role is growing when the oil prices increase. As shown in Fig. 1.4, fuel consumption together with engine performance determines another important parameter, that is, the CO_2 emissions, which will be strictly controlled in nearest future (Horwarth and Cannon 2012).

A number of governments developed severe restrictions on permissible levels of emissions that should be implemented shortly, aiming at reducing gas emission by

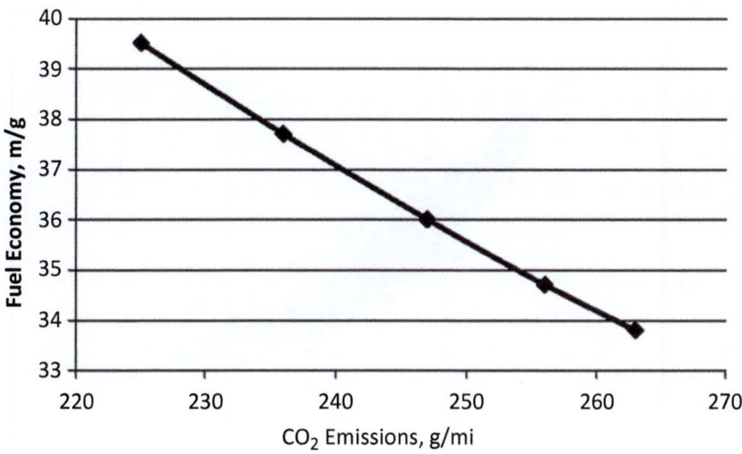

Fig. 1.4 Relationship between fuel economy and CO_2 emission (Horwarth and Cannon 2012)

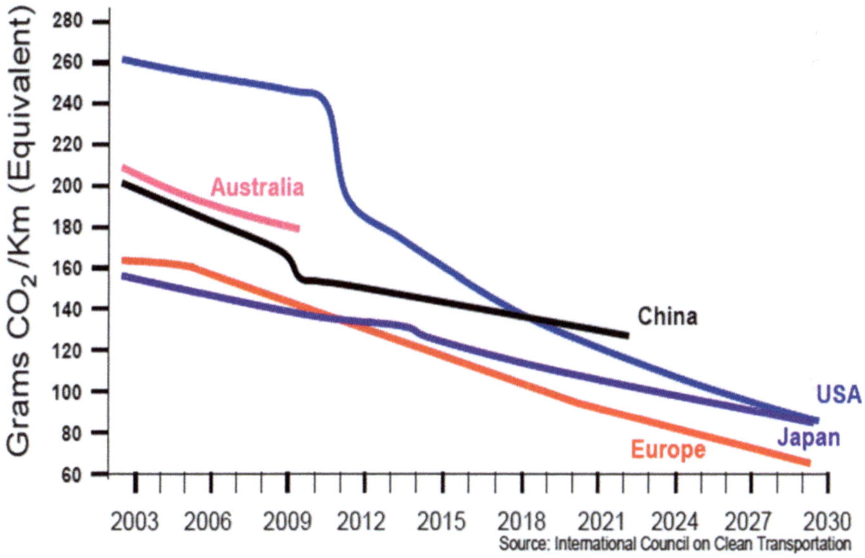

Fig. 1.5 Automotive CO_2 emission regulation in different countries (Shaw et al. 2010)

60–80 % in the decades to come compared to the level of 2005 (Galan et al. 2012). In particular, the Corporate Average Fuel Economy (CAFE) are the regulations enforced in the USA, first enacted by the U.S. Congress in 1975 and intended to improve the average fuel economy of cars and light trucks sold in the USA. As shown in Fig. 1.5, the current restrictions on emission levels are quite different in different countries but tend to converge by 2030.

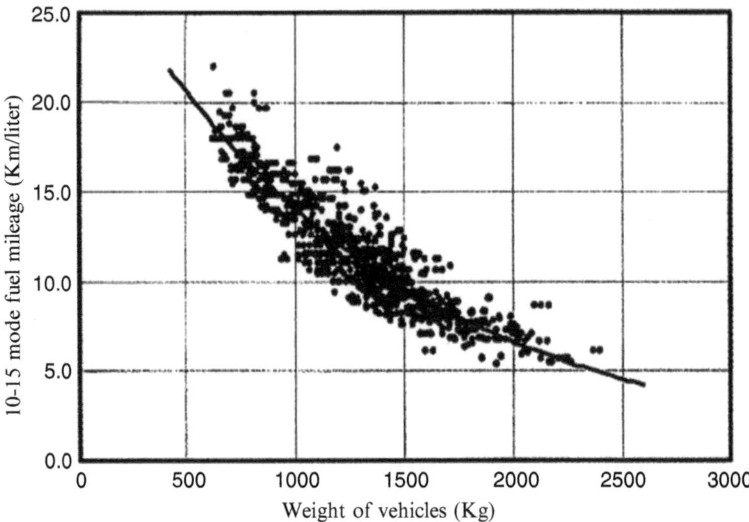

Fig. 1.6 Relationship between the fuel mileage (km/l) and automobile weight (Senuma 2001)

On July 29, 2011, the agreement with 13 large automakers was announced to increase fuel efficiency to 54.5 miles per gallon for cars and light-duty trucks by model of 2025 year. This agreement was joined by Ford, GM, Chrysler, BMW, Honda, Hyundai, Jaguar/Land Rover, Kia, Mazda, Mitsubishi, Nissan, Toyota, and Volvo (together they account for over 90 % of all vehicles sold in the USA), as well as by the United Auto Workers (UAW) and the State of California, who were all participants in the deal. The agreement will result in application of the new CAFE regulations to every model that will be on the US road in 2025, i.e., to all vehicles that are to be finalized on August 28, 2017 (counting average life time of a car as 8 years).

New norms of fuel consumption unambiguously mean new requirements for reducing car weight which is the main factor defining fuel consumption, as presented in Fig. 1.6.

As follows from a number of estimates, a reduction of total vehicle weight by ~10 % can correspond to 6–8 % improvement in fuel economy. However, as shown in Fig. 1.7, higher consumer expectations led to a steady gain in car weight due to improvements in vehicle safety, bigger space, better performance, reliability, passenger comfort, and overall vehicle quality. Thus the goal is not so much to contribute to a decrease in weight of cars but to minimize this trend of weight increase.

To satisfy new safety requirements to elevate crash worthiness at lower fuel consumption, the only alternative to additional reinforcement or/and to increase in thickness (weight) of the safety critical parts is the development of high- and ultra-high-strength steels with adequate formability.

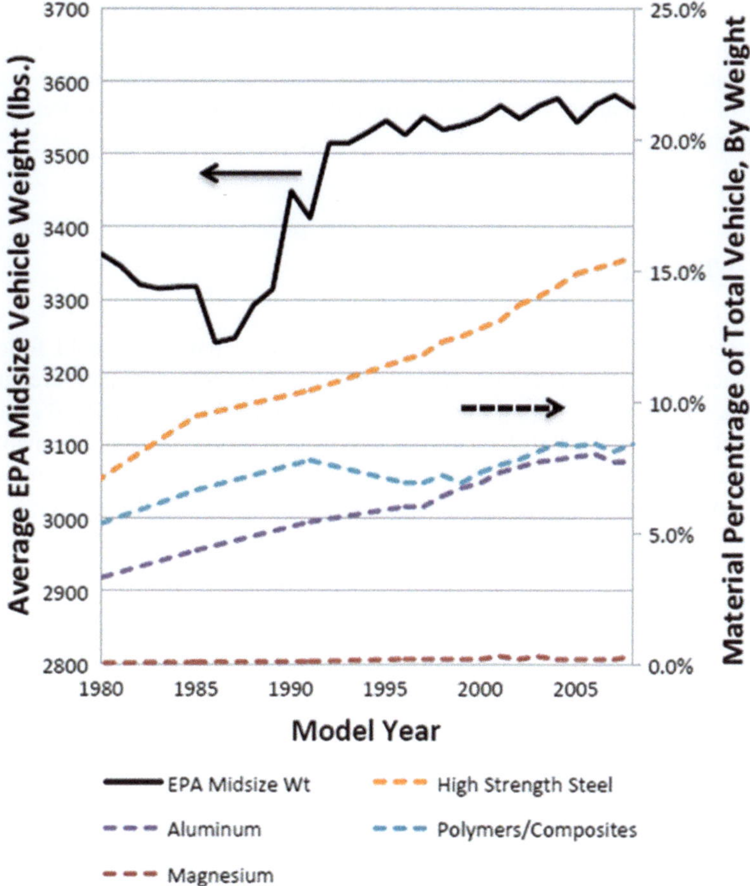

Fig. 1.7 Changes in car weight and usage of different materials (Data: US Department of Energy, 2014—public)

Crucial role of intrusions in the "passenger cage" during collision determines the growing application of ultra-high-strength martensitic steels with TS $\geq$ 1500 MPa. Some parts are roll-formed from as-annealed martensitic grades. New grades of martensitic steels with TS $>$ 2000 MPa are under development. Growing consumption of steels that quench to martensite in pressing (press hardenable steels, PHS) is determined by important advantages of hot stamping to help keeping the necessary stiffness of thinner parts due to simplified processing to obtain the desired geometry. PHS steels with TS $>$ 1500 MPa are commercial and new targets include TS $>$ 1800 and $>$2000 MPa.

Progress in hot-rolled steels follows to trends in changes of properties of cold-rolled steels. As-rolled dual-phase steels with TS $>$ 590 MPa demonstrating advantages in fatigue are widely used in manufacturing car wheels. Hot-rolled steels of the same strength with ferrite–bainite microstructure are featured by significantly better flangeability and are used in some parts of suspension. Requirements for higher strength of hot-rolled steels, such as TS > 800 and 1000 MPa, resulted in the development of so-called multi-phase steels (MP), in which substantial strengthening of ferrite–bainite matrix was achieved due to hardening by extremely fine precipitates.

1.5 "Third Generation"

The overall trend in car manufacturing, which steel industry should follow, is to replace an existing HSS or AHSS grade with the grade of higher strength but with the same level of formability. For example, steels with TS-590/600 should be replaced by steels with 980 MPa tensile strength that should be highly formable and have the same elongation of 21 % as steels with TS = 590 MPa. Likewise, the 780 grades should be replaced with steels that have TS > 1180 MPa at TE = 14 %, same as current DP780. Successful and timely development of highly formable AHSS and UHSS is the only way to sustain steel as the first choice for auto makers or at least to keep a significant share of steel in future cars.

These requirements, as well as, for example, a combination of TS >1470 MPa with TE > 20 %, cannot be achieved using existing types of dominant microstructure, DP or TRIP and require entirely new approaches.

There are numerous "banana" diagrams reflecting the evolution of existing high- and ultra-high-strength formable steel. The diagram in Fig. 1.8 modified from D. Matlock (Matlock and Speer 2006) depicts the "area" of new steels of the so-called "third Generation" that possess essentially high strength at significantly higher elongation approaching the breakthrough values of high Mn TWIP steels. The demand for these steels is growing both in volume and in the significance for

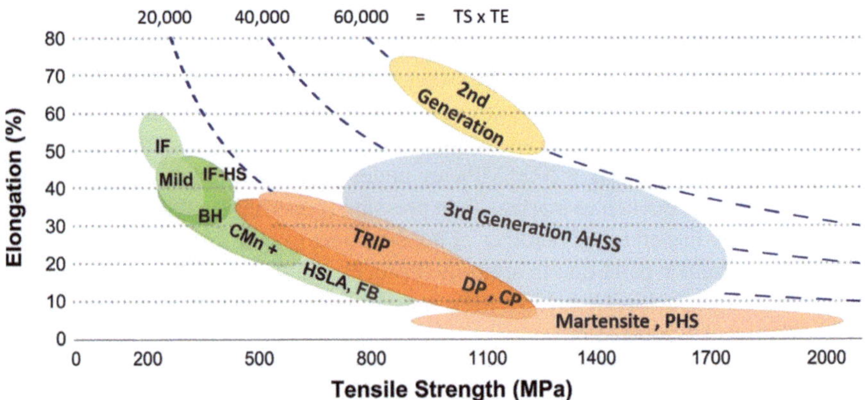

Fig. 1.8 Elongation—tensile strength balance diagram for existing variety of formable steels and prospective "third-generation" grades—modified

sustainability of steel industry. Carmakers expect them in new models to be launched already in 2015–2017 years.

It is important to emphasize that the combinations of strength and formability expected from the third-Generation AHSS assume as strength criterion not only the tensile but the yield strength as well. The concept of formability includes not only the total but also the uniform elongation, as well as the hole expansion. There are a few prospective candidates that demonstrate extremely high combinations of those parameters and attract as much attention from the industry as they have received from researchers.

1.5.1 Carbide-Free Bainitic Steel

Bainitic matrix concept is intensively studied in view of manufacturing high-strength high-formable (HSHF) steels with microstructure consisting of carbide-free lath bainite and inter-lath retained austenite. They are also called TRIP steels with bainitic ferrite (TBF) (Sugimoto et al. 2002). Besides the expected high formability, including hole expansion, there is a possibility for improving strength by producing ultrafine grain size of the bainitic ferrite plates and for enhancing ductility by the transformation-induced plasticity effect.

1.5.2 Medium Mn Steels

One of the promising types of steels that possess the combinations of properties attributable to the "third-Generation" are steels with relatively high but still medium Mn content (4–10 %) (Merwin 2006; Lee et al. 2008; Jun et al. 2011a, b). These steels with very high level of retained austenite and ductility are not commercialized yet but have already passed a trial at Baosteel.

Apart from alloying with Mn, certain amount of Si/Al and microalloying elements are also added in chemistry that has relatively low carbon matrix. Various microstructures can be generated using different heat treatments, and steels with a large range of strength levels can be produced (Arlazarov et al. 2012).

1.5.3 Q&P

According to the forecast by D. Matlock, the third-generation steels should have the microstructure of martensite and austenite mixture. This type of microstructure can be produced using the developed Q&P treatment (Speer et al. 2003). It was shown (Jun and Fonstein 2008; De Moor et al. 2012) that Q&P-processed steels are featured not only with high combination of TS and TE but also have higher

YS/TS ratio and higher hole expansion than the steels of the same strength with other types of microstructure (DP, TRIP, Q&T). Successful development of these steels is carried out by many companies despite the fact that upgrades of the annealing equipment are often necessary.

Perhaps, this chapter should have rather been entitled "Revolutionary trends in automotive steels," as it describes the multifold increase in strength of automotive steels and the acceleration of the pace of introducing the necessary changes. Looking at the retrospect of the progress made by steel industry over the last two decades, there are all reasons to believe that obstacles on the way of manufacturability and application of the third-generation steel will be overcome.

Chapters that follow will present the main effects of processing and chemical composition of each class of AHSS on the formation of microstructure that defines their mechanical behavior.

References

Akisue, O., and T. Hada. 1995. "Past Development and Future Outlooks of Automotive Steel Sheets." *Nippon Steel Technical REport*, no. 64 (January): 1–6.

Allain, S., and et al. 2002. "Characterization of the mechanical Twinning Microstructure in a High Manganese Content Austenitic Steel." In *TRIP-Aided High Strength Ferrous Alloys*, 75–78. Gent: GRIPS.

Arlazarov, A., A. Hazzotte, O. Bouaziz, and et. al. 2012. "Characterization of Microstructure Formation and Mechanical Behavior of an Advanced Medium Mn Steel." *Material Science and Technology, MS&T*, 1124–31.

Berger, l., M. Lesemann, C. Sahr, and et al. 2009. "Superlight-CAR – the Multi-Materials Car Body." In .

Bhattacharya, D., N. Fonstein, O Girina, I. Gupta, and O. Yakubovsky. 2003. "A Family of 590 MPa Advanced High Strength Steels with Various Microstructures." In *45th MWSP Conf.*, 173–86.

Bucher, J.H., E.G. Hamburg, and J.F. Butler. 1979. "Property Characterization of VAN-QN Dual-Phase Steels." In *Structure and Properties of Dual-Phase Steels*, 346–59. New Orleans.

Davies, R.G. 1978. "The Deformation Behavior of a Vanadium-Strengthened Dual Phase Steel." *Metallurgical Transactions A* 9 (1): 41–52. doi:10.1007/BF02647169.

Davies, R.G., and C.L. Magee. 1981. "Physical Metallurgy of Automotive High-Strength Steels." In *Conference of Structure and Properties of Dual-Phase Steels, Symposium at the AIME Annual Meeting*, 1–19. New Orleans, LA, USA: Metall Soc of AIME, Warrendale, Pa, USA.

De Moor, E., D. K. Matlock, J Speer, and et al. 2012. *Comparison of Hole Expansion Properties of Quench & Partitioned, Quenched & Tempered and Austempered Steels*. SAE International 2012-01-0530.

Dual-Phase. 1978. *Dual-Phase and Cold-Pressing Vanadium Steels*. Berlin: VANITEC.

Engl, B, L Kessler, F.-J. Lenze, and T.W. Schaumann. 1998. "Recent Experience with the Application of TRIP and Other Advanced Multiphase Steels." In *IBEC 98*, 1–8.

Fonstein, N., H. J. Jun, O. Yakubovsky, R Song, and N. Pottore. 2013. "Evolution of Advanced High Strength Steels (AHSS) to Meet Automotive Challenges." In . Vail, CO, USA.

Fundamentals. 1981. *Fundamentals of Dual-Phase Steels*.

Galan, J., L. Samek, P. Verleysen, and et al. 2012. "Advanced High Strength Steels for Automotive Industry." *Revista De Metalurgia* 48 (2): 118–31.

Gupta, I., and P.-H. Chang. 1985. "Effect of Compositional and Processing Parameters on the Variability of the Tensile Strength of Continuously Annealed Water Quenched Dual Phase Steels." In *Technology of Continuously Annealed Cold-Rolled Sheet Steel*, 263–76. Detroit, MI, USA: The Metal. Society of AIME.

Hall, Judy. 2011. "Evolution of Advanced High Strength Steels in Automotive Applications." In . Detroit: Auto/Steel Partnership.

Hayami, S., and T. Furukawa. 1975. "A Family of High-Strength Cold-Rolled Steels." In *Microalloying-1975*, 311–21. Union Carbide Corporation'.

Horwarth, C.D., and M. Cannon. 2012. "Future Material Opportunities and Direction for Lightweighting Automotive Body Structures." In . Southfield, Michigan.

Jun, H. J., and N. Fonstein. 2008. "Microstructure and Tensile Properties of TRIP -Aided CR Sheet Steels: TRIP-Dual and Q&P." In 155–61. Orlando, Florida.

Jun, H.J., O. Yakubovsky, and N. Fonstein. 2011a. "Effect of Initial Microstructure and Parameters of Annealing of 4% and 6.7& Mn Steels on the Evolution of Microstructure and Mechanical Properties." In Houston, TX.

Jun, H. J., O. Yakubovsky, and N. M. Fonstein. 2011b. "On Stability of Retained Austenite in Medium Mn TRIP Steels." In *The 1st International Conference on High Manganese Steels*, paper A5. Seoul, South Korea.

Kwon, Ohjoon. 2008. "Next Generation Automotive Steels at POSCO." In.

Lee, S, K. Lee, and B.C. DeCooman. 2008. "Ultra Fine Grained 6wt% Manganese TRIP Steel." *Materials Science Forum* 654-656: 286–89.

Mallen, R.Z., S. Tarr, and J. Dykeman. 2007. "Recent Applications of High Strength Steels in North American Honda Production." In *Great Design in Steel*. Detroit.

Matlock, D., and J Speer. 2006. "Design Consideration for the next Generation of Advanced High Strength Sheet Steels." In Korea.

Merwin, M.J. 2006. "Low-Carbon Manganese TRIP Steels." *Materials Science Forum* 539–543: 4327.

Meyer, L., F. Heisterkamp, and W. Mueschenborn. 1975. "Columbium, Titanium and Vanadium in Normalized, Thermo-mechanically Treated and Cold-Rolled Steels." In *Internat. Symposium on High Strength Low-Alloy Steels*, 153–67. Pittsburgh.

Pottore, N., I. Gupta, and R. Pradhan. 2006. "Effect of Composition and Processing in Cold-Rolled Dual-Phase Steels for Automotive Applications." In *MS&T'06*, 653–63. Cincinnati, Ohio.

"Product Information: Retained Austenite Steels." 2008. ThyssenKrupp Steel Europe.

"Review of EuroCarBdy 2011." 2011. In Bad Nauheim.

Samek, L., and Krizan. 2012. "Steel-Material of Choice for Automotive Lightweight Applications." In 1–6. Brno.

Sebald, Ronald. 2008. "Development Trends in Advanced High-Strength Steels." ThyssenKrupp Auto Day, 2008.

Senuma, Takehide. 2001. "Physical Metallurgy of Modern High Strength Steel Sheets." *ISIJ International* 41 (6): 520–32.

Shaw, J., H Singh, and A Farahani. 2010. "Reinventing Automotive Steel." In *Great Designs in Steel*.

Southwick, P.D., and et al. 1985. "The Development of High Strength Steels on Inland's #3 Continuous Annealing Line (CAL)." In *Technology of Continuously Annealed Cold-Rolled Sheet Steel*, 277–98. Detroit: The Metal. Society of AIME.

Speer, J.G., D. K. Matlock, B. C. De Cooman, and J. G. Schroth. 2003. "Carbon Partitioning into Austenite after Martensite Transformation." *Acta Materialia* 51 (9): 2611–22.

Structure and Properties. 1979. *Structure and Properties of Dual-Phase Steels.*

Sugimoto, K.-I., A. Kanda, R. Kikuchi, and et. al. 2002. "Ductility and Formability of Newly Developed High Strength Low Alloy TRIP-Aided Sheet Steels with Annealed Martensitic Matrix." *ISIJ International* 42 (8): 910–915.

Sugimoto, K.-I., J. Sakaguchi, T. Ida, and T. Kashima. 2000. "Stretch-Flangeability of a High-Strength TRIP Type Bainitic Sheet Steel." *ISIJ International* 40 (9): 920–026.

Takahashi, Manabu. 2003. "Development of High Strength Steels for Automobiles." *Nippon Steel Technical Report* 88 (July): 2–7.

Chapter 2
Main Features of Heat Treatment from Intercritical Region

Contents

Heat treatment from two-phase $\alpha + \gamma$ temperature range, more often referred to as intercritical region, was originally designed for dual-phase steels, but currently, it is the main manufacturing tool for most of cold-rolled AHSS including some of the third generation. The process implies controlled cooling from two-phase range

© Springer International Publishing Switzerland 2015   17
N. Fonstein, *Advanced High Strength Sheet Steels*,
DOI 10.1007/978-3-319-19165-2_2

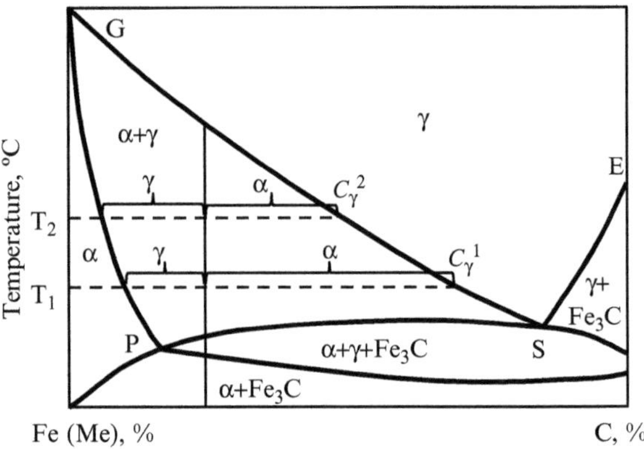

Fig. 2.1 Pseudo-binary Fe(Me)–C diagram—(original)

between the critical A_{c1} and A_{c3} temperatures, i.e., between the onset of austenitization and its completion, respectively.

This kind of heat treatment allows obtaining a combination of microstructure constituents, which is almost impossible for steel of the same composition cooled after full austenitization. Annealing in the intercritical region and phase transformations during subsequent cooling are determined by three main features. First of all, it is a substantial enrichment of austenite phase with carbon, as shown schematically in Fig. 2.1, accompanied by a certain partitioning of austenite forming alloying elements as well. Second, the transformation upon cooling occurs under the condition of coexisting austenite and ferrite phases and thus at the presence of austenite–ferrite boundaries that significantly facilitate the formation of ferrite from austenite. Last, but not least, is that due to the short holding time at the annealing temperature, typical for continuous annealing/coating lines, austenite in the two-phase mixture is characterized by a considerable heterogeneity of its chemical composition.

Mechanical characteristics of the microstructure consisting of two and more phases depend on ratio of their fractions, morphology, and properties of individual microstructure constituents obtained under given annealing conditions. To produce dual- or multi-phase steel with a desired combination of phases in their desired state, it is necessary to account for the effects of various heat-treatment parameters such as heating rate, annealing temperature, cooling profile, tempering conditions, etc. Special attention should be paid to the holding time of intercritical annealing which influences the isothermal growth of austenite and its chemical homogeneity. It is also important to consider the initial microstructure of steel prior to annealing that can affect the austenitization kinetics and the morphology of the resultant ferrite–austenite mixture before cooling.

The pseudo-binary diagram in Fig. 2.1 applies to the equilibrium conditions so that it can be used only to tentatively estimate the carbon content and the amount of austenite that *could be* approached at the end of holding time at selected intercritical temperature. Apart from insufficient time to achieve equilibrium, the temperature that determines the desired phase ratio can be shifted under the influences of heating rate, initial microstructure, concentrations of alloying elements, and their distribution.

Most studies of phase transformations that produce dual-phase microstructure are focused on the processes that follow the austenitization. However, the transformations occurring upon heating and the microstructure immediately after heating, i.e., fraction of austenite, its morphology (size, shape), distribution, and chemical composition of austenite grains, strongly impact phase transformations during subsequent cooling, final microstructure, and therefore the mechanical properties of steel.

2.1   Main Characteristics of Continuous Annealing/Coating Lines

Parameters of existing continuous lines define the capabilities and restrictions to obtain the desired microstructures and properties of AHSS. These lines differ in heating rates and soaking times and heating and cooling profiles, both of which can have a few steps, and in the available options, to ensure auto-tempering/tempering/ "overaging" of resultant ferrite–martensite/bainite mixture.

There are two major types of continuous annealing lines for bare cold-rolled steels. Their thermal cycles are presented in Fig. 2.2. Lines with water quenching (Fig. 2.2a) can, in turn, differ in capability to water quench immediately from annealing/soaking temperature or after some initial cooling. As to the type of lines, shown in Fig. 2.2b, the accelerated cooling can be provided by water-air mist or by

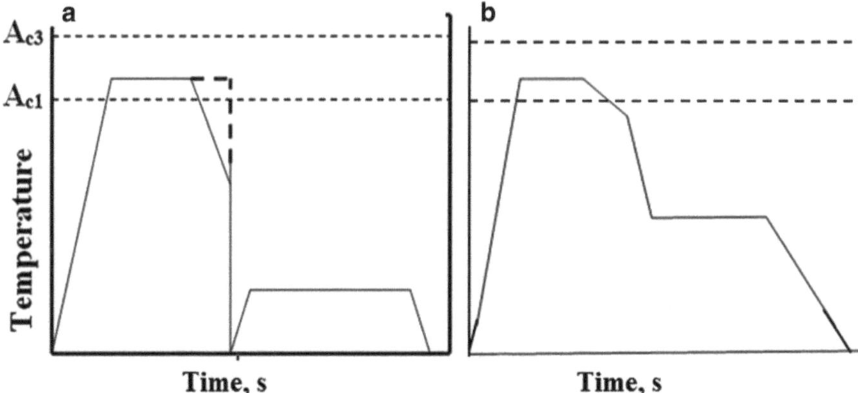

Fig. 2.2 Main thermal cycles of continuous annealing lines: (**a**) line equipped with water quenching, (**b**) line with accelerated cooling and direct transition of strip to overaging/tempering zones—(original)

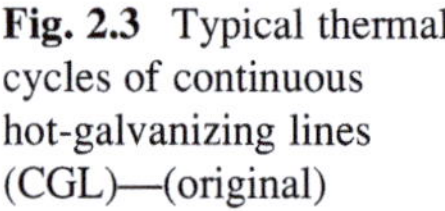

Fig. 2.3 Typical thermal cycles of continuous hot-galvanizing lines (CGL)—(original)

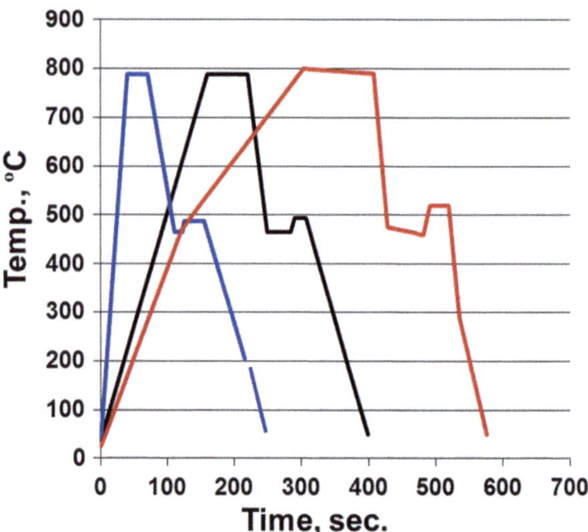

so-called quenching-on-rolls. Heating is provided by radiant tubes, so that the heating rate is about 10 °C|s. Timing parameters are defined by the length of the corresponding sections and the line speed with the latter depending on steel strip thickness. On average, the holding at annealing/soaking temperatures is ranging from 100 to 170 s, tempering time in water quenching lines is from 120 to 200 s for 1 and 2 mm strip thickness, respectively. Holding in annealing lines with accelerated cooling ranges from 100 to 165 s, cooling rate is about 80 and 40 °C/s, overaging time is 360–600 s for 1 and 2 mm thick strips, respectively.

There are much more differences in thermal cycles pertaining to coating lines, some of which are presented in Fig. 2.3. They can have different heating rates: 10 °C/s if radiant tubes (RTF) are used, or up to 50 °C/s when initial heating is boosted by direct fire furnaces (DFF).

Different CGL substantially differ by holding time at annealing temperatures, cooling profiles, as well as by cooling capacity. Often the initial cooling down to 650–700 °C is very slow (2–5 °C/s) but becomes faster (15–30 °C/s) at the next step. Being built for coating of mild low-carbon steels, many CGL are equipped with overaging (equalizing) zones in front of the Zn pot with ~460 °C. In the case of Fe–Zn coating (galvannealing), the Zn-coated strip is heated to 480–520 °C for various duration (up to 20–30 s) depending on the line design. The rate of final cooling to below 80 °C before temper rolling depends on the distance to the temper mill.

In the last decade, a few lines intended for AHSS production were built. Their main features presented in Fig. 2.4 include the capability of achieving high cooling rates from annealing temperatures; some of the lines do not have the overaging zone or provide the capability for cooling below Zn temperature followed by induction heating of the strip.

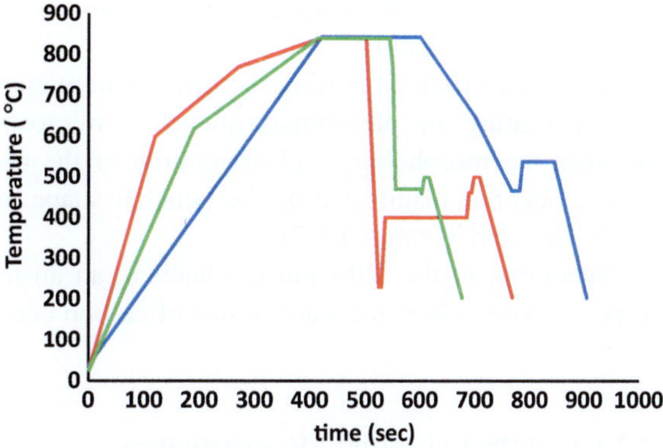

Fig. 2.4 Diagrams of a modern galvanizing lines (original)

2.2   Austenite Formation During Heating and Soaking of Low-Carbon Steel in the Intercritical Temperature Range

Various aspects of the austenitization including heating and holding within the "intercritical" A_{c1}–A_{c3} temperature range were addressed and described in such fundamental publications as (Garcia and DeArdo 1979; Speich et al. 1981; Azizi-Alizamini et al. 2011; Dyachenko 1982; Wycliffe et al. 1981; Golovanenko and Fonstein 1986). The main factors important for the practice of intercritical heat treatment include specific mechanisms of nucleation and growth of austenite, the existence and duration of the incubation time of transformation, the microstructure "inheritance," the nonuniformity of carbon content in austenite, as well as partitioning of carbon and alloying elements between α and γ phases.

This section summarizes theoretical and experimental data that should be taken into account for obtaining the desired fraction and morphology of the austenite phase.

2.2.1   Nucleation of Austenite

When steel is cooled from the intercritical temperature region, the products of austenite transformation emerge within the parent γ-phase. Thus, the final dual-phase microstructure is governed by the preferential sites of austenite nucleation and by the geometry of austenite grains.

Formation of austenite is considered to be a diffusion-controlled transformation that includes fluctuations and carbon redistribution in the α-phase, followed by $\alpha \rightarrow \gamma$ transformation. It was also suggested that the initial low-carbon portion of austenite can possibly be formed by a diffusionless $\alpha \rightarrow \gamma$ transformation, and then carbon is redistributed by diffusion. A diffusionless $\alpha \rightarrow \gamma$ transformation can take place, in particular, in the case when diffusion is suppressed during, for example, extremely rapid heating.

Important proof supporting the diffusionless nucleation of austenite was found during heating of preliminary-quenched microstructure. Certain relationships between the morphology and orientation of the α'-phase and emerged austenite were evident as manifested by the lamellar shape of austenite portions (Sadovsky 1973; Koo and Thomas 1977).

According to the diffusion mechanism, an austenite nucleus should primarily appear at sites where the fluctuations of carbon concentration are high.

2.2.1.1 Effect of Initial Microstructure

The diffusive nucleation of austenite in heating of steels with initial ferrite–pearlite microstructure occurs predominantly at the interphase boundaries between pearlite colonies and ferrite grains.

In steel with spheroidized carbides in ferrite matrix, austenite nucleates primarily at carbide particles located at ferrite grain boundaries. These nucleation sites are both more abundant and are distributed more uniformly. It was suggested that carbides located within ferrite grains do not contribute to austenite nucleation but dissolve later in ferrite matrix with carbon diffusing to the growing γ-phase (Garcia and DeArdo 1981; Hornbogen and Becker 1981).

When a preliminary-quenched steel with martensite microstructure is heated into the intercritical region, austenite first nucleates at the prior austenite grain boundary triple junctions, then at the surfaces of prior austenite grains, and finally at martensite plate boundaries. This produces a Widmannstatten-like microstructure with acicular shape of austenite that may have transformed to martensite upon cooling retaining acicular fiber-type morphology (Sadovsky 1973; Hornbogen and Becker 1981; Koo et al. 1979).

The conditions facilitating the inheritance of a prior microstructure, i.e., oriented phase nucleation during repeated heating of steel with initially martensitic microstructure, were discussed in detail elsewhere (Dyachenko 1982; Singh and Wayman 1987; Lenel and Honeycombe 1984).

Preliminary cold working increases the density of defects that can serve as heterogeneous austenite nucleation sites leading to a larger amount of uniformly distributed γ-phase nuclei.

Phase transformation kinetics during heating to intercritical region is controlled by parallel processes of transformation of parent phase into austenite and carbide dissolution. In accordance with Speich et al., nucleation of austenite at the ferrite–pearlite interface is the first stage of austenite formation with very fast growth of austenite in pearlite (Speich et al. 1981).

Comparison of austenitization kinetics during isothermal heating of hot-rolled and cold-rolled (with 25 and 50 % reduction) 0.11C–1.58Mn–0.4Si steel revealed that the time of complete "dissolution" of pearlite substantially decreased after cold deformation due to the higher driving force and/or lower activation energy of transformation (El-Sesy et al. 1990).

Higher cold rolling reduction generates more dislocations and more fragmented carbides increasing the number of sites for austenite nucleation and shortening carbon diffusion paths. In studying of 0.13C–1.54Mn–0.22Si–0.016Nb steel subjected to various cold reductions, a finer martensite distribution was found after quenching of samples that had undergone heavier cold reduction which is indicative of finer austenite grains in heating (Granbom 2008).

2.2.1.2   Effect of Heating Rate

The effect of initial microstructure on austenite nucleation strongly depends on the heating rate, i.e., on factors controlling the rate of rearrangement of lattice defects and the extents of recovery or recrystallization before austenitization. Diffusional processes during heating before $\alpha \rightarrow \gamma$ transformation such as tempering of martensite in preliminary-quenched steel, recrystallization of ferrite, spheroidization of lamellar cementite, etc., have crucial effects on nucleation and subsequent growth of austenite.

In fact, slow heating into the intercritical temperature range of either hot-rolled or normalized steel allows for spheroidization of pearlite before $\alpha \rightarrow \gamma$ transformation or complete tempering of martensite. Cold working additionally increases the spheroidization rate during heating. Hence, in slow heating, the nucleation of γ-phase in such steels can occur in similar fashion independently of original spheroidized microstructure.

According to Garcia and DeArdo (1981), the speroidized cementite particles in cold-rolled ferrite do not act as austenite nucleation sites. However, with more rapid heating, the nucleation of austenite at the interfaces between spheroidized cementite and ferrite and, as a consequence, the presence of austenite (martensite after quenching) within ferrite grains were observed (Shirasawa and Thomson 1987).

The work carried out at University of British Columbia using 0.16C–1.38Mn–0.24Si steel (Cho 2000) compared initial microstructures of hot-rolled ferrite–pearlite (F + P), bainite, and F + P cold rolled at 20 and 50 % reduction. In continuous heating experiments with the rates ranging from 0.3 to 300 °C/s, the austenite volume fractions were measured along with the transformation start (T_s) and finish (T_f) temperatures. The initially bainitic microstructure demonstrated the lowest (almost zero) sensitivity to heating rate, while hot-rolled F + P microstructure appeared to be the most sensitive when T_s and T_f increased with heating rate by 20 and 50 °C, respectively. Cold-rolled F + P microstructure resulted in smaller changes in both T_s and T_f temperatures.

Huang et al. performed a careful study of the effects of heating rate on the formation of austenite during continuous heating and isothermal holding within the intercritical temperature range (Huang et al. 2004) of typical DP (0.06C–1.86Mn–0.155Mo) and TRIP (0.178C–1.55Mn–1.74Si) steels. Heating rates of 1, 10, and

100 °C/s were employed. In continuous heating, an increase in heating rate shifts the onset of transformation to higher temperatures thus reducing the initial amount of austenite at a given temperature. At the same time, a significant acceleration of austenite growth during isothermal holding was observed after reheating to a given temperature at higher rate. Similar behaviors were found for both initially hot-rolled and cold-rolled microstructures; this fact was explained by the authors by the dominant role of nucleation process that is sensitive to the heating rate.

After fast transformation of pearlite colonies into austenite in hot-rolled steels, austenite can further nucleate at ferrite grain boundaries competing with growth of austenite formerly nucleated from pearlite. This is the key of effects of heating rate on transformation kinetics. Thus, slower heating rate favors substantial growth of austenite nucleated at pearlite sites, while faster heating rates promote additional nucleation at ferrite grain boundaries.

In cold-rolled steels, additional factors come into play. The distribution and shape of pearlite colonies are modified by cold deformation. In slow heating, the recrystallization is complete before the nucleation of austenite and therefore the reasoning outlined above for initial hot-rolled microstructure can be applied to austenitization kinetics in this case as well, albeit the morphology of resultant austenite must be different due to modified shape of pearlite during cold working. With higher heating rates, the completion of recrystallization is attained at higher temperatures so that the recrystallization can occur concurrently with the formation of austenite.

According to the data due to Huang et al., ferrite is still recrystallizing when austenite has already nucleated within pearlite colonies and starts to grow there. The authors observed only a few austenite particles located at ferrite grain boundaries, so the austenite nucleated in pearlite can then grow without significant competition from grain boundary austenite. As a result, a block-like austenite was observed formed parallel to rolling direction with the highest transformation rate.

In studying of 0.08C–1.9Mn–0.01Nb steel additionally alloyed with Mo and Cr, the increase in heating rate from 10 to 50 °C/s was found to produce larger volume fraction of austenite within the entire intercritical temperature range regardless of whether the amount of austenite was controlled by nucleation during continuous heating or some austenite growth during isothermal holding was involved (Mohanty et al. 2011).

2.2.2 Growth of Austenite

Speich et al. point out three stages of isothermal austenitization during heating of steels with ferrite–pearlite microstructure (Fig. 2.5): (1) rapid growth of γ-phase within the pearlite regions until the latter are fully consumed (in 0.06C–1.5Mn steel, the duration of this stage was estimated to be about 15 s at 740 °C and 0.2 s at 780 °C, but it required 30 s to 2 min during annealing of steels with 0.12 or 0.2 % C at 740 °C); (2) slow growth of austenite into former ferrite regions; the rate of this

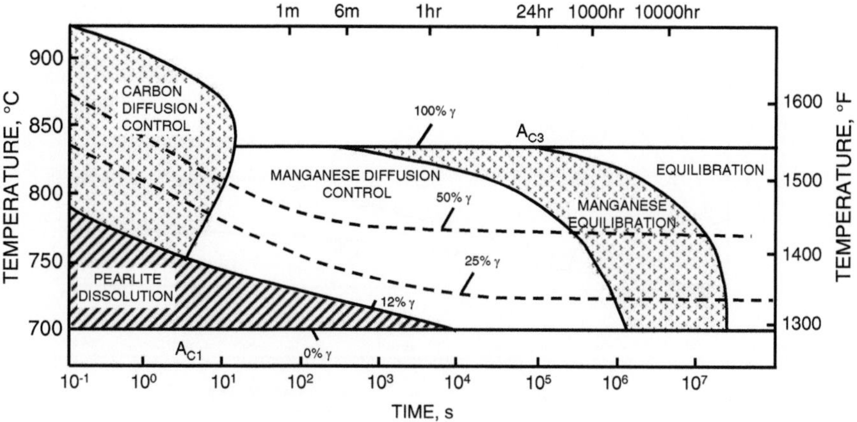

Fig. 2.5 Diagram for 0.12C–1.5Mn steel illustrating austenitization stages depending on temperature and time (Speich et al. 1981)

stage at elevated temperatures (~850 °C) is controlled by carbon diffusion in austenite (with the process lasting 2–9 s) or by manganese diffusion at lower temperatures (between 740 and 780 °C), which can take 4–24 h; (3) achievement of equilibrium at a rate controlled by the diffusion of alloying elements (in this case manganese) in austenite, the rate being approximately three orders of magnitude slower than in ferrite (the time for this stage has been assessed theoretically as 2000–4000 h) (Speich et al. 1981).

During stage 1 of austenitization, the initial increase in the amount of austenite was reported of being entirely due to transformation of pearlite with no change in the amount of ferrite (Souza et al. 1982). Some amount of pearlite, however, can retain at stage 2 of austenite growth. At this stage, the nucleation of new austenite preferentially at ferrite–pearlite interphase boundaries and the growth of earlier emerged austenite grains into ferrite occur simultaneously.

A detailed study showed that in medium-carbon (0.36–0.47 % C) steels, the pearlite-to-austenite transformation can, in turn, proceed in either one or two steps depending on the heating rate. At very low heating rate (0.05 °C/s), ferrite and cementite plates transform simultaneously. At higher heating rate (20 °C/s), a two-step process was observed: ferrite within a pearlite grain transforms into austenite first and then cementite lamellae start to dissolve (Savran et al. 2007).

Kinetics of austenite growth at various carbon contents (0.06, 0.12, and 0.20 %) in steel and temperatures and holding times in the intercritical temperature range were studied in detail by Speich et al. The isothermal austenitization curves for these steels are summarized in Fig. 2.6. As can be seen, approaching the "equilibrium amount of austenite" was observed only at low temperatures of intercritical region and only for steels with 0.06 and 0.12 % C and required the time more than 15 min.

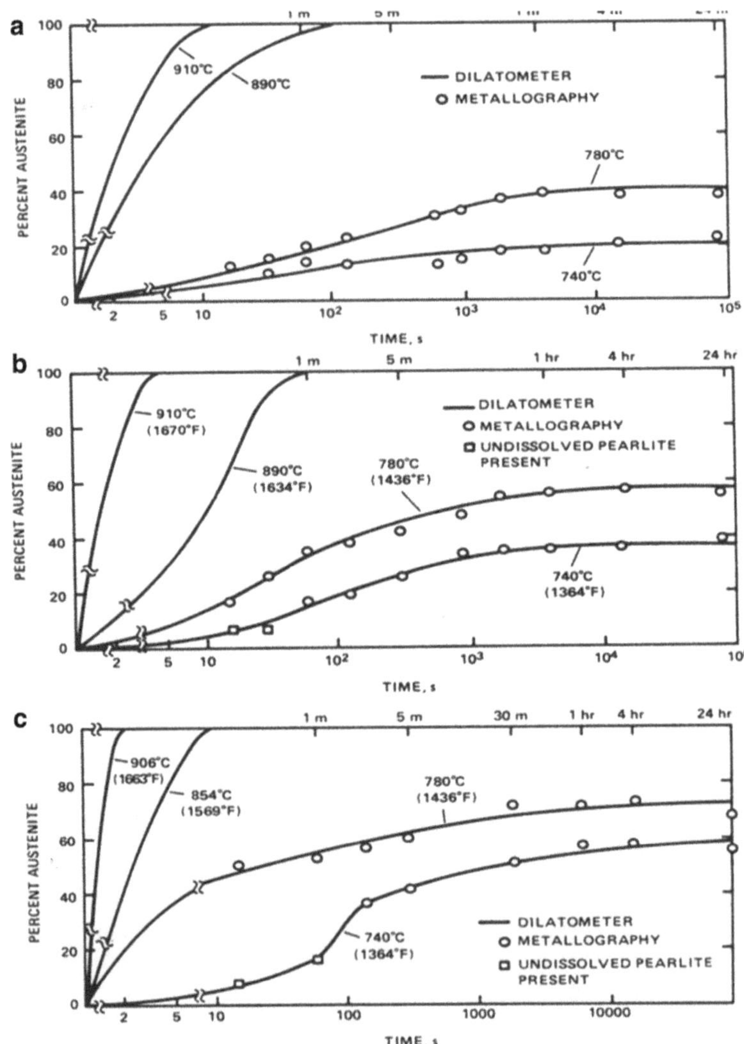

Fig. 2.6 Kinetics of isothermal austenitization at different temperatures : (**a**) steel 0.06C–1.5Mn, (**b**) steel 0.12C–1.5Mn; (**c**) steel 0.20C–1.5Mn (Speich et al. 1981)

According to other reports, the time to reach the "austenite plateau" ranges from 10 min in normalized 0.14C–1.6Mn–0.4Si steel austenitized at 740 °C (Souza et al. 1982) to 400 min in steels with (0.01–0.22) % C and 1.5 % Mn containing spheroidized carbides and held at 725 °C (Garcia and Deardo 1981).

The isothermal austenitization curves exhibit some kinks indicative of a certain decrease in the austenitization rate or a delayed transformation. Similar kinks are observed also during continuous heating (Pawlovski 2011) and can be interpreted in terms of different stages of austenitization: (a) a range of rapid formation of

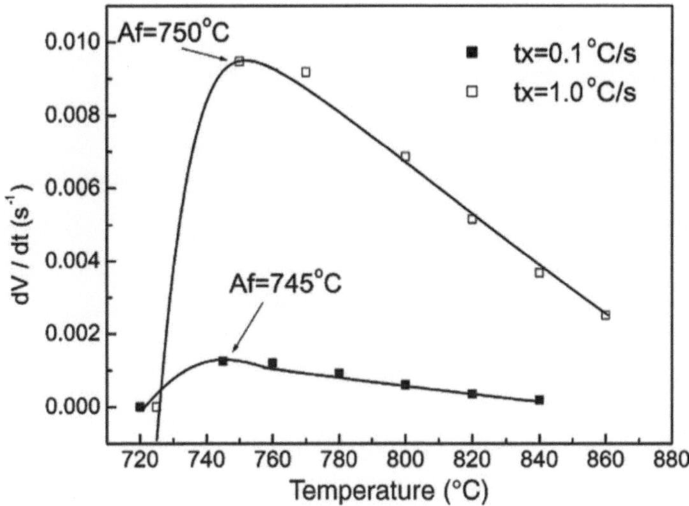

Fig. 2.7 The rate of austenite formation as a function of temperature for two heating rates (Oliveira et al. 2007)

austenite from pearlite, (b) relatively slow further transformation as a result of exhaustion of the preferential γ-phase nucleation sites, and (c) a rapid austenitization in ferrite regions due to enhanced carbon diffusion at higher temperatures.

Maximum austenitization rate is always observed in the beginning of isothermal holding and then it progressively decreases. In accordance to the data obtained using the 0.15C–1.42Mn–0.37Si–0.03Nb steel, the austenitization rate reaches maximum near the finishing temperature of pearlite "dissolution" as depicted in Fig. 2.7 (Oliveira et al. 2007).

The presence of a delayed stage of transformation has been noted also in the study of the heating of 0.08C–1.6Mn–0.4Si steel containing spheroidized carbides. The austenitization rate first increased in the temperature range of 750–775 °C and then decreased. Similar transition points were explained by the lack of nuclei when all of the original carbides (that are the preferential nucleation sites for austenite) had dissolved and the rate of austenite growth into the ferrite matrix became controlled by the slow rate of carbon diffusion across the α–γ interface (Surovtsev et al. 1986).

An interesting interpretation of the inflexion points in the austenitization curves observed at very short holding periods (10–100 s) during heating of 0.04C–1.8Mn–2.2Si steel was proposed by Yi et al. (1985). In this steel, the plateau was always detected at ~25 % of γ-phase, which was substantially higher than the pearlite volume fraction in the pre-annealed samples. It is known that the rate of carbon diffusion along grain boundaries is much higher than that of bulk diffusion. This leads to preferential growth of austenite along ferrite grain boundaries. Thus, in authors' opinion, the rate of isothermal austenitization sharply decreases when the

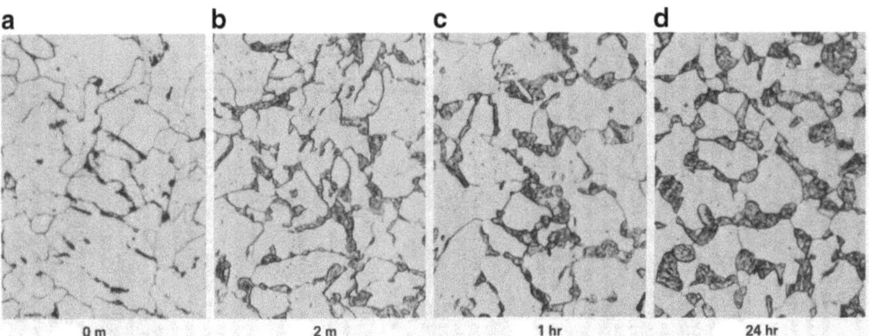

Fig. 2.8 Austenite growth in 0.06C–1.5Mn steel at 740 °C (Speich 1981)

original ferrite grains are completely separated by austenite so that an austenite growth starts to be controlled by bulk diffusion rather than by grain boundary diffusion.

Since the diffusion coefficients along and across grain boundaries are different, the growth of austenite along the original ferrite grain boundaries with holding time is faster than its growth into ferrite matrix. Eventually, as the amount of γ-phase increases, all the original ferrite boundaries become covered with austenite (increased austenite continuity), while the growth rate of austenite into ferrite grains remains essentially constant after initial period of 30–60 s. This can result in martensite skeleton after rapid cooling of ferrite–austenite mixture of such a type, as shown in Fig. 2.8.

There are significant inconsistencies in the experimental data on the kinetics of isothermal austenitization in the $\alpha + \gamma$ region, primarily with respect to existence of the incubation period and its duration. The latter varies from zero or several seconds, in most publications, to several minutes (Dyachenko 1982; Shtansky et al. 1999). The data on the time for attaining the "saturation" stage are also controversial varying from a few seconds to several minutes in most studies, although some researchers reported even the absence of saturation after isothermal holding for an hour or longer.

In addition to the variations in steel composition, initial microstructure, and heating temperatures used in various studies, the differences in the heating rate should be also taken into account to explain those inconsistencies.

2.2.2.1 Effect of Steel Chemical Composition

The effect of steel composition on the process of austenitization encompasses a number of factors: the changes in activation energy of $\alpha \rightarrow \gamma$ transformation, carbon diffusivity, thermodynamic activity of carbon in austenite, carbon concentration gradients between α and γ phases, as well as the values of critical A_e, A_{c1}, and A_{c3} temperatures described, for example, by following equations (Andrews 1965):

$$A_{e1} = 723 - 10.7\text{Mn} - 16.9\text{Ni} + 29.1\text{Si} + 16.9\text{Cr} + 290\text{As} + 6.38\text{W} \qquad (2.1)$$

$$A_{e3} = 910 - 203\sqrt{\text{C}} - 15.2\text{Ni} + 44.7\text{Si} + 104\text{V} + 31.5\text{Mo} + 104\text{V}$$
$$+ \, 13.1\text{W} - 30\text{Mn} + 11\text{Cr} + 20\text{Cu} - 700\text{P} - 120\text{As} - 400\text{Ti} \qquad (2.2)$$

An important feature of the early austenitization stages is the concurrent occurrence of nucleation and growth, which poses problems in distinguishing the effects of individual alloying elements that can also depend on the experimental conditions.

For example, the increase in manganese content raises the activation energy of $\alpha \rightarrow \gamma$ transformation apparently due to manganese segregation at the surface of carbide particles, thus reducing the rate of carbide dissolution and modifying the nucleation of austenite.

At short holding times, the dissolution of cementite occurs at a rate that is several times higher than that of special carbides formed by chromium, molybdenum, vanadium, and niobium. This means that prior to dissolution of special carbides, the effective carbon content in austenite can be lower than that calculated from simple equation as

$$C_\gamma = \frac{C_{st}}{V_\gamma} \qquad (2.3)$$

where C_{st}—total carbon content in the steel and V_γ—a fraction of austenite.

At short soaking at temperatures close to A_{c1}, a "three-phase" microstructure, i.e., a mixture of carbides, ferrite, and low-carbon austenite (martensite after quenching) can be observed as shown in Fig. 2.9.

Short soaking times in the industrial production of AHSS often leads to nonuniform concentration of carbon and alloying elements in austenite (Stevenson et al. 1979; Speich and Miller 1981) that leads, in particular, to the enrichment of the periphery of γ-phase grains with manganese. The formation of such a fringe of manganese, that reduces thermodynamic activity of carbon, should be accompanied by a local increase in carbon content and its lower concentration in the core. As a joint result, it leads to the formation of a high-manganese and carbon rim around the austenite particle. Since this rim has a higher hardenability than the central core, upon slow cooling a martensite ring envelopes the central ferrite–pearlite core (Speich et al. 1981).

In contrast, the growth of γ-phase around high-carbon nuclei (at the former site of carbide particles) progresses with a gradual decrease in carbon content towards α/γ interface as happens at insufficient hardenability of steel annealed at relatively high intercritical temperature. The core of austenite formed at lower annealing temperatures during heating has relatively high carbon and Mn content as a result of mostly pearlite-to-austenite transformation. The austenite formed later at higher annealing temperatures involves ferrite-to-austenite transformation and consequently has lower concentration of carbon and Mn. This austenite transforms to pearlite at cooling as depicted in Fig. 2.10 (Girina and Bhattacharya 2003).

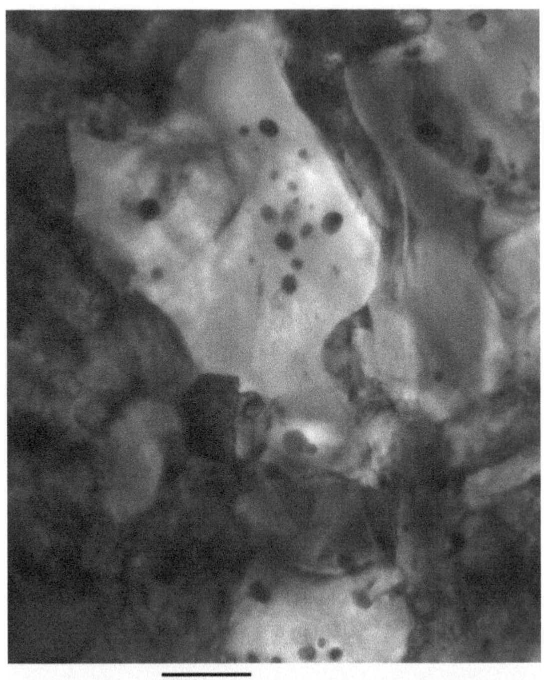

Fig. 2.9 Coexisting of carbides, martensite, and ferrite in a steel quenched from low part of intercritical range;—TEM—original

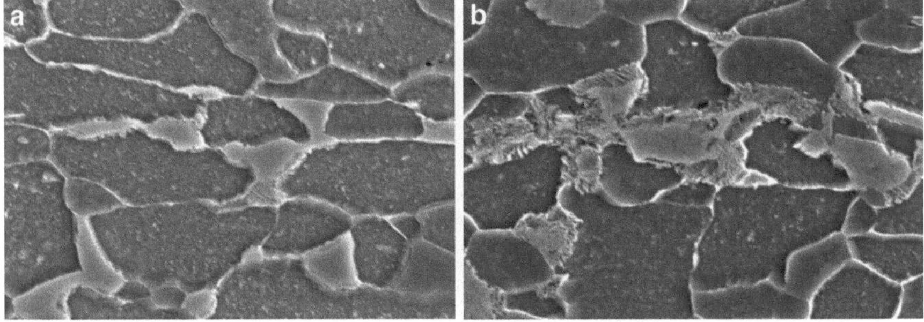

Fig. 2.10 Microstructure of steel with 0.1C–0.5Mn after quenching from (**a**) 740 °C and (**b**) 760 °C (Girina and Bhattacharya 2003)

Some elements can affect the observed changes in austenite growth indirectly, in particular, by retardation of recrystallization or by altering the initial microstructure. For example, in studying the formation of austenite during continuous heating of cold-rolled samples in dilatometer at 10 °C/s rate, it was found that an increase in

Nb content to 0.03 % has insignificant effect on A_{c1} and A_{c3} temperatures so that, within the accuracy of experiments, the A_{c1} and A_{c3} temperatures were measured equal to 715 and 840 °C, respectively, for all studied steels. However, at a given temperature, a larger amount of austenite in steel with 0.03 % Nb than in "no Nb" steel was detected (Girina et al. 2008). Evidently, several factors contributed to the acceleration of austenitization by additions of Nb. First, Nb refines the initial hot-rolled microstructure before cold rolling, creating more interface boundaries and other defects that could facilitate the nucleation of austenite. Secondly, Nb is known to retard the recrystallization, which leads to overlay of austenitization and recrystallization and hence to accelerated growth of austenite (Maalekian et al. 2012).

The rate of increase in the amount of austenite with annealing temperatures in the two-phase region is also a function of the steel composition that affects the width of the intercritical A_{c1}–A_{c3} temperature range.

2.2.2.2   Effect of Heating Temperature

The heating temperature is the key factor controlling the volume fraction of austenite and therefore its average carbon content [in accordance with Eq. (2.3)] that has a crucial effect on the stability of the austenite and the final microstructure formed during subsequent cooling.

The velocity of austenite phase boundary, v, i.e., the phase growth rate, is described by the following equation:

$$v = D\frac{dC}{dx}\left(\frac{1}{\Delta C^{\gamma \leftrightarrow \alpha}} + \frac{1}{\Delta C^{c \leftrightarrow \alpha}}\right) \tag{2.4}$$

The value of v is proportional to the diffusivity of carbon in austenite (D) and to the gradient of carbon concentration dc/dx within the austenite grain; also v is inversely proportional to the diffusion path x as well as to the differences between the carbon concentrations on austenite–ferrite $\Delta C^{\gamma \leftrightarrow \alpha}$ and carbide–austenite interfaces $\Delta C^{c \leftrightarrow \gamma}$ (Wycliffe et al. 1981)

With increasing temperature, changes in all of the above parameters contribute to an abrupt increase in v: the carbon diffusion rate increases, while the concentration gradients across austenite–ferrite and austenite–carbide boundaries (lines GS and SE in Fig. 2.1, respectively) decrease. Since the decrease in carbon concentration gradient is more pronounced at the austenite–ferrite than at the austenite–carbide boundaries, the growth rate of austenite towards ferrite is higher than into carbide. As a result, as it was mentioned above, in steels with more stable carbides or with their bigger amounts, an increase in heating temperature can lead to greater number of undissolved carbides within the austenite (martensite after cooling) forming a microstructure similar to the one shown in Fig. 2.11.

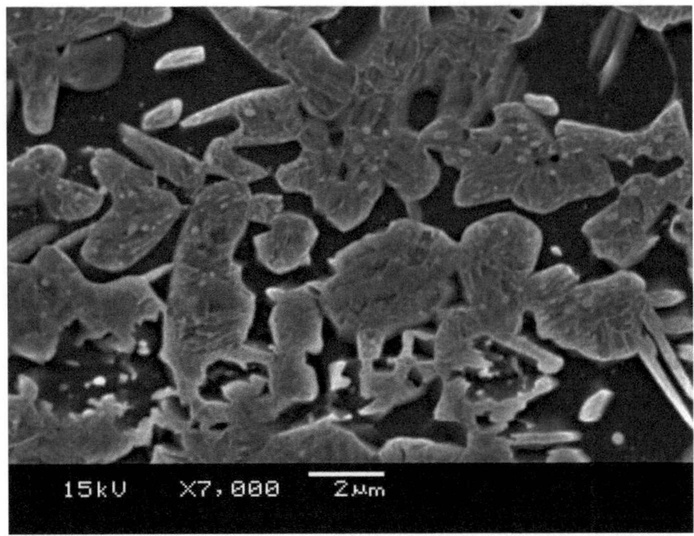

Fig. 2.11 Carbides, undissolved in former austenite—original

With initial ferrite–pearlite microstructure, the "dissolution" rate of pearlite increases in three orders of magnitude when annealing temperature is raised from 730 to 780 °C.

In general, both the rate of austenitization and the amount of austenite increase with intercritical heating temperature. For example, in 0.06C–1.2Mn–0.5Cr–0.003B steel, a 5 min holding produced 20 % of austenite at 750 °C, 35 % at 780 °C, 50 % at 800 °C, and 80 % at 830 °C. The formation of equal amounts of austenite (e.g., 20 %) at 750 and 800 °C required 300 and 24 s, respectively. The dilatometric data showed that during isothermal holding at 740 °C (close to the $A_{c1} = 735$ °C), this steel exhibited incubation period of 90 s during which there was no evident growth in the austenite volume. No incubation period was observed at all other temperatures (Fonstein 1985).

2.2.2.3 Effect of Initial Microstructure

In addition to the earlier-discussed effect on the nucleation of austenite, the initial microstructure also controls the austenitization kinetics and the morphology of austenite–ferrite mixture, which is later inherited by the final microstructure of steel after cooling.

In terms of austenitization, the effect of the original microstructure manifests itself as a shift in the critical temperatures of phase transformations and in the rate of austenite growth at increasing holding temperature or time. In the order of increasing rate of austenitization, Dyachenko categorized the initial microstructures

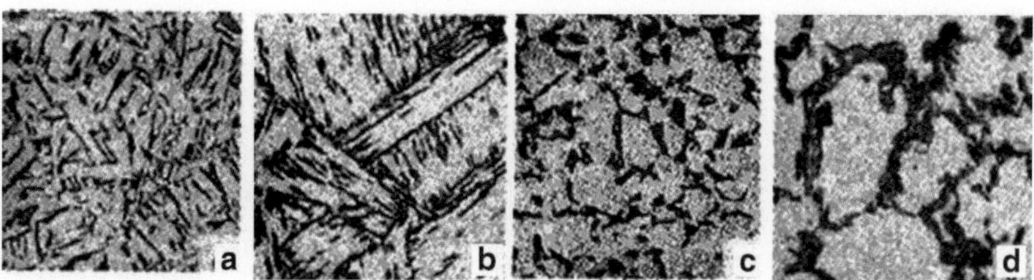

Fig. 2.12 Effect of the initial microstructure on morphology of growing austenite (martensite after water quenching): (**a**) initial martensite microstructure after quenching from 950 °C; (**b**) initial martensite structure after quenching from 1150 °C, (**c**) initial ferrite–pearlite microstructure, (**d**) ferrite plus initially spheroidized carbides (original)

as follows: ferrite with spheroidized carbides, ferrite with lamellar pearlite, as-quenched, and cold-deformed microstructures (Dyachenko 1982).

As shown in Fig. 2.12, the austenite, produced by heating martensite into the intercritical temperature region, reflects the acicular orientation of the initial martensite, and the greater the similarity is, the higher has the original heating temperature been during pre-quenching. The inheritance of the initial acicular martensite structure was found enhanced by faster heating and in the presence of silicon (Koo and Thomas 1977).

Higher austenitization rate was found during reheating of initially martensitic microstructure compared to that for bainitic microstructure reheated under identical conditions. Microstructure containing spheroidized carbides demonstrated the biggest shift in A_{c1} and A_{c3} temperatures to higher values, as well as the slowest rate of austenitization. Since austenite usually nucleates at carbide/ferrite interfaces, the amount of γ-phase and its growth rate increase with increasing specific area of ferrite–carbide interfaces, i.e., with refinement of carbide particle and/or ferrite grains (Law and Edmonds 1980).

It is both of fundamental and practical importance to reveal whether the initial microstructure has similar effects on austenitization after cold rolling. The existing data are controversial. For example, Shirasawa and Thompson employed special heat treatment of hot bands of homogenized 0.07C–1.5Mn–1Si steel to obtain speroidized carbides, ferrite–pearlite, bainite, and fully martensite microstructures prior to cold rolling with 50 % reduction (Shirasawa and Thomson 1987). After annealing of cold-rolled specimens at 800 °C for 1 min (to get ~40 % austenite), cooling at 10 °C/s to 750 °C followed by water quenching and tempering at 200 °C, the authors found negligible differences in tensile properties regardless of initial microstructure.

These results differ from the data obtained by the author using steel of different composition with various initial microstructures that simulated the microstructure after hot rolling under different conditions.

It was shown that the difference in amount of austenite became really negligible at long holding times at temperatures close to A_{c3}. However, when holding times less than 5 min were used, the initial (pre-deformation) microstructure appeared to

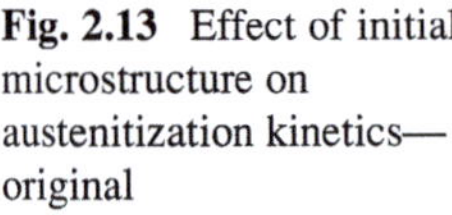

Fig. 2.13 Effect of initial microstructure on austenitization kinetics—original

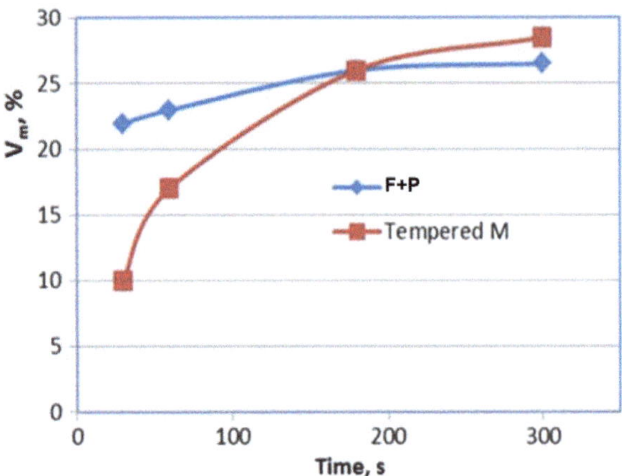

have an important effect on the austenitization rate. With short holding times (30–60 s), the maximum amount of austenite corresponded to the initial microstructure containing coarse-lamellar pearlite that simulated high coiling temperatures and slow cooling of a hot-rolled strip. The minimum amount of austenite was produced from the microstructure containing martensite tempered at high temperature that simulated low coiling temperature involving quenching and subsequent long tempering during coil cooling. These results are presented in Fig. 2.13 and are in good agreement with data obtained in a later work (Mohanty and Girina 2012), where effect of coiling temperature on austenitization kinetics was studied.

Banded ferrite–pearlite microstructures after high temperature coiling are usually featured by manganese segregation in pearlite bands. Therefore, the local A_{c1} in high-manganese regions (pearlite) can be lower than in surrounding volumes of ferrite that should additionally accelerate pearlite–austenite transformation. The following growth of the forming austenite into ferrite should be slower in a coarsely banded microstructure than in a refined one due to shorter carbon diffusion paths through ferrite matrix in the latter.

As mentioned above, cold deformation increases the number of austenite nuclei and therefore accelerates the initial stages of austenitization. However, the overall effect of cold working on the austenitization kinetics depends not only on the microstructure prior to cold deformation but also on the heating rate, holding time, and carbon content in steel. In particular, increase in the number of austenite nuclei due to cold working can interfere with subsequent growth of these nuclei and possible microstructure evolution. For example, cold deformation facilitates the spheroidization of pearlite lamellae; therefore in the slow heating, it can slow down the austenitization.

As it would be discussed later, nucleation and growth of austenite proceed differently depending on the completion of recrystallization in ferrite before $\alpha \rightarrow \gamma$ transformation. Therefore, the *amount* of cold deformation and its specific effect on recrystallization of ferrite are of particular importance.

2.2.2.4   Effect of Heating Rate

The growth of austenite fraction, V, with time, t, can be described by the Avrami equation (Avrami 1940):

$$V = 1 - \exp(-Kt^n) \tag{2.5}$$

In the study of ferrite–pearlite steel with 0.15C–1.42Mn–0.37Si–0.031Nb, it was found that the parameter K that represents the integral formation rate of austenite (nucleation and growth) becomes two orders of magnitude higher when the heating rate is increased only from 0.1 to 1 °C/s that resulted, in particular, in austenite fraction increase from 0.5 to 0.85 after heating at 800 °C (Oliveira et al. 2007). The acceleration of austenite growth with increasing heating rates was confirmed by several researchers.

Significant acceleration of austenite growth during isothermal holding of both low-carbon C–Mn–Mo DP and C–Mn–Si TRIP steels at given temperatures reached at higher heating rates was observed by Huang and coworkers (2004). In the later study (Azizi-Alizamini et al. 2011) using 0.17C–0.74Mn steel and heating rates ranging from 1 to 900 °C/s, it was found that higher heating rate affected austenite transformations during continuous heating of only initially hot-rolled (HR) microstructure while in case of cold-rolled (CR) steels the austenite fractions were relatively independent of heating rates. The authors explained these results (which can be applied to other controversial data as well) by the differences in initial microstructure prior to austenite formation. In HR microstructure, the increase in heating rate required a superheating until a network of austenite grains formed at ferrite grain boundaries. In CR steel, the increase in the heating rate also tends to increase the necessary superheat, but the initial (instantaneous) microstructure prior to austenite formation was also a function of the heating rate itself. The increase in the heating rate delayed recrystallization changing the microstructure of ferrite from fully recrystallized to partially recrystallized, while in pearlite, it delayed the spheroidization of cementite particles. Accordingly, both nucleation and growth patterns of austenite formation changed that resulted in a kind of mutual offset. Chemical composition of steels can substantially affect both recrystallization and A_{c1} temperatures, as well as the propensity to spheroidization of $(FeMn)_3C$. This could explain significant inconsistencies in the results of studies of heating rate effects using different steels.

A systematic study of heating rate effects was carried out with cold-rolled 0.08C–1.9Mn–0.01Nb steel also containing additions of Mo and Cr (Mohanty et al. 2011). The employed heating rates (10 and 50 °C/s) were related to those in a CGL equipped with radiant tubes (RTF) and direct fire furnaces (DFF). The effect of annealing temperature and time on austenite formation was considered separately for the isothermal growth of austenite fraction A_{is} and for the amount of austenite formed during heating by the beginning of isothermal soaking.

Higher heating rate of 50 °C/s produces a larger relative amount of A_{is} in the temperature range of 780–800 °C. At 820 °C, the portions of isothermally formed

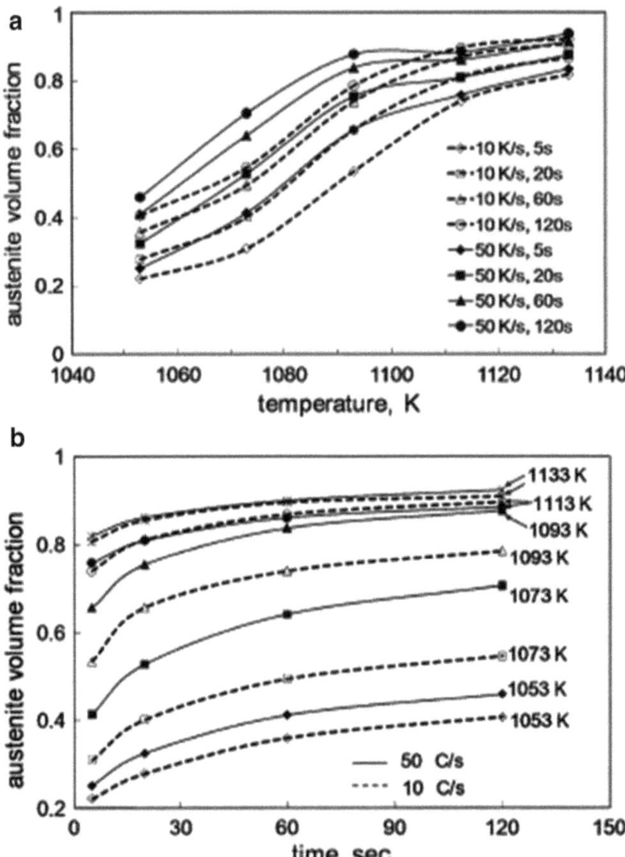

Fig. 2.14 Effect of heating rate on total fraction of austenite formed during continuous heating followed by isothermal holding as a function of (**a**) intercritical temperature and (**b**) isothermal holding time (Mohanty et al. 2011)

austenite become almost equal for both heating rates. At temperatures above 820 °C, the fraction of A_{is} decreases for both heating rates.

Figure 2.14a, b show the total amount of austenite (A_{tot}) formed during continuous heating followed by isothermal holding as a function of annealing temperature and holding time, respectively. The relative amounts of austenite continuously increase with temperature and time for both heating rates. It can be seen that the heating rate has a pronounced effect on A_{tot} for the temperature range of 780–820 °C, where higher heating rate results in larger total amount of austenite. At higher temperatures (840–860 °C), the difference in the total austenite fraction obtained at the two heating rates becomes negligible.

Schematic diagram presented in Fig. 2.15 explains the underlying mechanisms of the heating rate's effect on austenite formation that eventually result in the differences in morphology of ferrite–austenite mixture and in microstructures

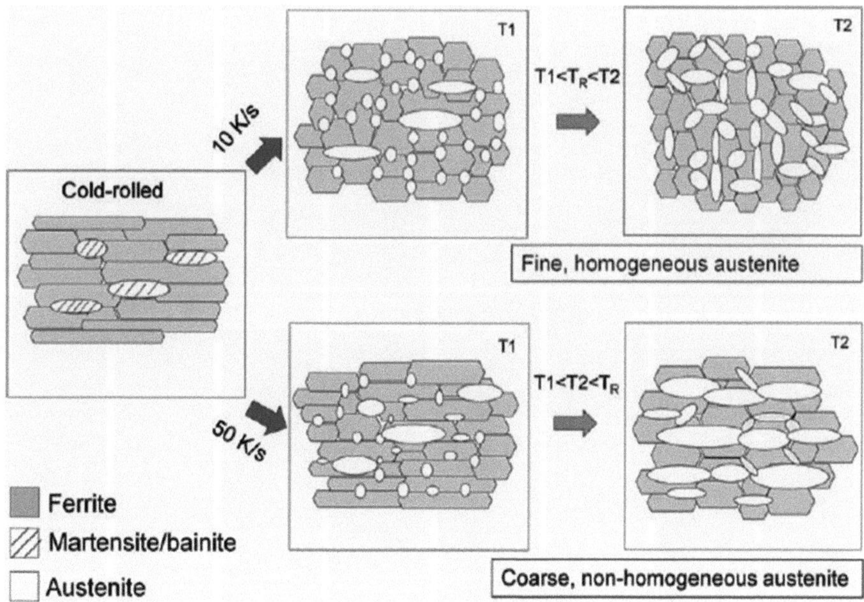

Fig. 2.15 Schematic representation of prevailing mechanisms of austenite formation for the heating rates of 10 and 50 °C/s (Mohanty et al. 2011)

observed after fast cooling. As was emphasized earlier, the key role is related to the overlapping of ferrite recrystallization and austenite formation.

Recrystallization of ferrite is delayed when steel is heated at a higher rate. The carbon-rich regions (martensite/bainite/pearlite) of the initially cold-rolled microstructure transform into austenite first regardless of the heating rate. Hereafter, for the sake of convenience, the austenite nucleated in carbon-rich areas is referred to as "pearlite-nucleated" austenite. The difference is in the number of austenite nuclei at ferrite grain boundaries. At higher heating rate, the time available for carbon diffusion is shorter, and the number of recrystallized ferrite grains is smaller. Hence, at a higher heating rate, relatively less austenite nuclei are formed at ferrite grain boundaries compared with lower heating rate, and the nucleation of austenite takes place mostly in carbon-rich areas. In other words, "pearlite-nucleated austenite" grows without significant competition from "grain boundary-nucleated austenite." Shorter time available at higher heating rates limits carbon diffusion to short distances, mostly between the blocks of "pearlite-nucleated austenite." Consequently, the growth of austenite occurs predominantly along the rolling direction, where the distance between the existing nuclei of austenite is smaller. As a result, higher heating rate leads to the formation of coarse and elongated islands of austenite, as schematically described in Fig. 2.15.

At lower heating rates, the recrystallization process starts earlier. Therefore, at the same annealing temperature, there is a larger fraction of recrystallized ferrite grains in microstructure. This fact and longer times available for carbon diffusion

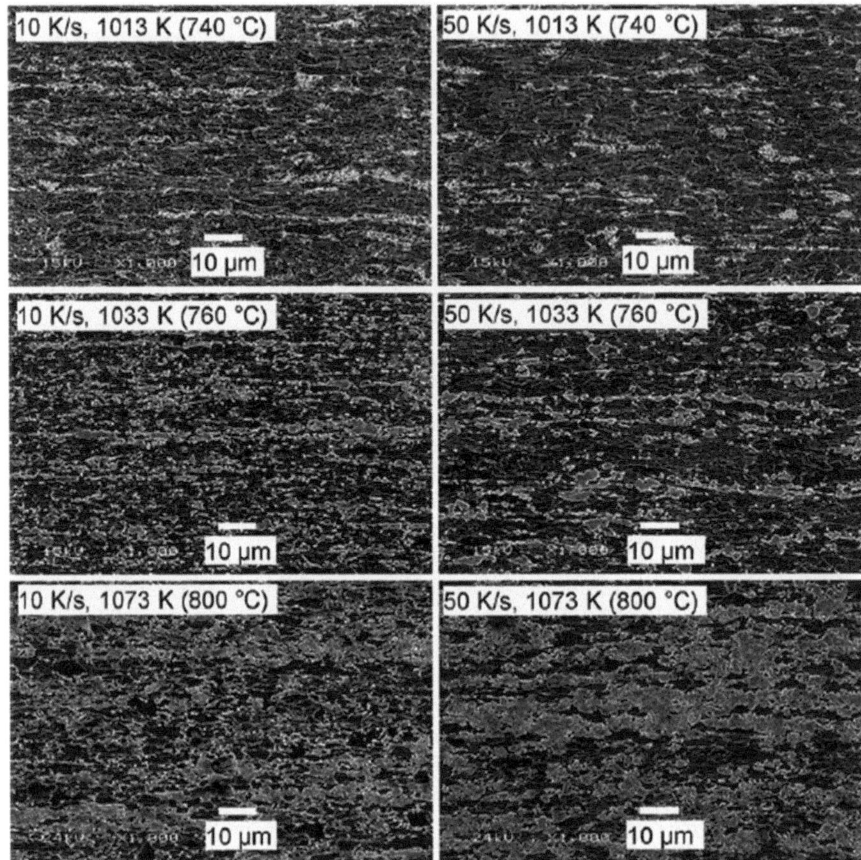

Fig. 2.16 SEM micrograph of quenched samples showing the effect of heating rate on recrystallization and austenite formation during continuous heating at different heating rates (Mohanty et al. 2011)

facilitate the nucleation of austenite at ferrite grain boundaries. As a result, austenite grains nucleated in prior carbon-rich areas and at ferrite grain boundaries grow simultaneously, forming a network of fine austenite islands along the ferrite boundaries. These considerations are consistent with the observations of microstructure of quenched samples illustrated in Fig. 2.16, where martensite (former austenite) islands after heating at a 10 °C/s are finer and more uniformly distributed than those after heating at 50 °C/s, with the latter being coarser and aligned along the rolling direction.

At temperatures close to full austenitization, the effect of heating rate on the volume fraction of formed austenite becomes insignificant as a result of very high diffusivity of carbon and shorter diffusion distance due to presence of large amount of austenite. Therefore, the differences in the process of austenitization due to changes in heating rate reported in different publications reflect the differences in the balance between nucleation and growth, which are, in turn, controlled by the interrelation between the kinetics of recrystallization and austenite formation at various heating rates and annealing temperatures that can be different for different steels.

2.3 Microstructural Changes in the Ferrite Constituent

Morphology and intrinsic properties of ferrite matrix exert critical influence on the final properties and primarily on the ductility of intercritically annealed steels. Apart from the effects of alloying and microalloying that will be discussed in Chap. 4, the characteristics of ferrite matrix are strongly affected by heating, holding in two-phase region, and subsequent cooling. These characteristics include ferrite grain size, extents of its recovery and recrystallization, concentrations of dissolved interstitials (nitrogen and carbon), partitioning of alloying elements, and precipitation hardening.

2.3.1 Recrystallization of Ferrite

During heating of a cold-rolled low-carbon unalloyed steel, ferrite has enough time to recrystallize well before the intercritical temperature is reached. The presence of Mn, Cr, Mo, Nb, as well as higher heating rates or insufficient soaking time, raise the temperature, at which recrystallization is complete, to 730–800 °C. Heating to lower temperatures results only in recovery of ferrite (Garcia and DeArdo 1981; Pradhan 1984; Fonstein 1985).

Studies of cold-rolled 0.08C–1.45Mn–0.21Si steel heated into the intercritical region showed that, regardless of the initial microstructure, the recrystallization of ferrite completed only after 30 s at 760 °C. This was accompanied by the change in aspect ratio of ferrite grains from about 10:1 to (1.5–2):1. No sensible growth of ferrite grains was observed during holding in the intercritical range varied from 30 to 600 s. It was suggested that the growth of recrystallized ferrite grains at that temperature was constrained by "particles" of austenite or initial cementite (Yang et al. 1985). According to Ogawa et al., who studied 0.10C–2.02Mn steel after cold rolling with 67 % reduction and then heated at 30 °C/s to 750 °C, the recrystallization of ferrite can be progressively retarded by austenite phase with annealing time increasing from 100 to 1000s as more austenite was formed (Ogawa et al. 2010).

The difference in the initial ferrite grain sizes prior to cold rolling remained after recrystallization: the finer the ferrite grains in the hot-rolled steel, the finer the ferrite grains in the dual-phase or TRIP steel after cooling of cold-rolled steel from the intercritical region.

As the holding time or heating temperature in the intercritical range is increased, the effective ferrite grain size seems to shrink as a result of ferrite "consumption" by the growing γ-phase.

Heating rate can significantly change the volume fraction of recrystallized ferrite by the moment of austenite nucleation and growth. Therefore, as shown above, the variations in heating rate affect not only the amount of austenite obtained at given annealing temperature but can also alter the morphology of ferrite–austenite mixture.

The preference of recrystallization completion before austenitization was shown above. Austenite first forms at cementite particles along the boundaries of non-recrystallized elongated (pancaked) ferrite grains so that the emerging austenite is aligned in the rolling direction. Thus, although cold working favors a large number of austenite nuclei, the austenitization in non-recrystallized volumes can lead to anisotropic microstructure of ferrite matrix.

As shown, when recrystallization and austenitization occur simultaneously during heating of cold-rolled steels, austenite can also emerge at isolated carbide particles inside the ferrite grains. This is in contrast with hot-rolled or normalized steels with initial ferrite–pearlite microstructure, where the austenite nucleation is always associated with carbides at ferrite grain boundaries (Caballero et al. 2006; Shirasawa and Thomson 1987).

2.3.2 Precipitation of Dispersed Particles

In steels containing carbide or nitride forming elements, such as niobium, titanium, zirconium, aluminium, molybdenum, and vanadium, fine particles of special carbides and carbonitrides can precipitate during heating or holding in the intercritical temperature range, as well as during subsequent cooling. These particles tend to precipitate primarily in ferrite due to substantially lower solubility of the carbides and nitrides in α-phase compared to their solubility in austenite.

Heating in the intercritical range can also lead to the growth of preexisting particles precipitated during hot rolling or preliminary heat treatment and located in the untransformed ("old") ferrite. As the heating temperature or holding time are increased, some dispersed carbides or carbonitrides can dissolve in the growing γ-phase according to their solubility at the appropriate temperatures (Meyer et al. 1975). In particular, the maximum solubility of nitrides is almost two orders of magnitude lower than that of carbides. If the amount of particles is greater than the maximum solubility product in γ-phase, the austenite, growing into ferrite, and hence the final martensite islands can contain undissolved carbides or carbonitries.

Carbonitrides (nitrides) are well known to coarsen about 50 times slower than carbides (Ballinger and Honeycombe 1980). For this reason, the "old" ferrite of intercritically annealed steels containing higher amount of nitrogen and hence

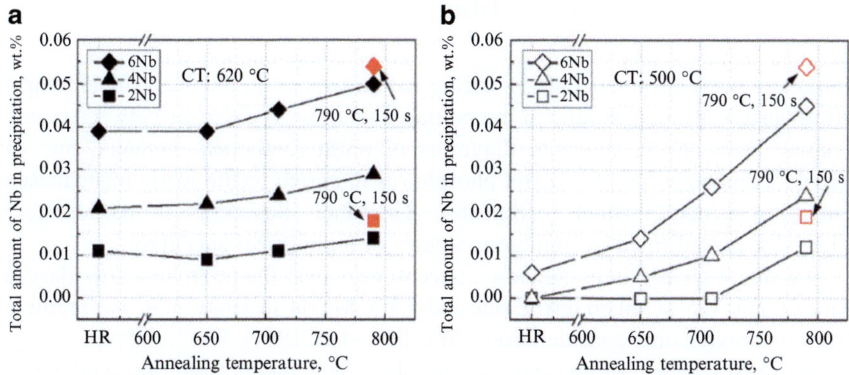

Fig. 2.17 Total amount of Nb in precipitation during heating and isothermal holding at 790 °C (**a**) CT: 620 °C; (**b**) CT: 500 °C (Song et al. 2014)

larger fraction of carbonitrides should exhibit a slightly stronger precipitation hardening due to finer particle size. Cold deformation increases the rate of carbonitride growth in nearly 30 times and can thus to some extent offset the effect of higher nitrogen content in steel.

The presence of carbide- and/or carbonitride-forming elements can significantly enhance the effects of processing parameters, such as coiling temperature, on the final properties of steels. For example, the low coiling temperature for V-bearing steels resulted in hardening ferrite due to precipitation hardening and grain refinement (Marder and Bramfitt 1981).

Figure 2.17 presents the results of the analysis of precipitates in cold-rolled steels with various content of Nb, coiled at different temperatures. Quantitative analysis of Nb precipitates was performed using inductively coupled plasma (ICP) (Jun et al. 2003) with measurement accuracy of ≤0.005 %.

Figure 2.17 shows the total amount of Nb in precipitates after hot rolling and after annealing at various temperatures and two holding times at 790 °C (0 and 150 s). It can be seen that about 50–65 % of the total amount of Nb precipitated after coiling at 620 °C (Fig. 2.17a), while it was less than 10 % after coiling at 500 °C (Fig. 2.17b).

Upon heating the cold-rolled steels to 790 °C, precipitation of Nb particles took place irrespectively of coiling temperature, but it was more pronounced after coiling at 500 °C with less preexisting Nb precipitates (Fig. 2.17b). After isothermal holding of steel with 0.02 % Nb at 790 °C for 150 s (as shown by the red square symbols), almost all Nb precipitated in ferrite regardless of coiling temperature (Song et al. 2014).

2.4 Partitioning of Interstitial and Alloying Elements

Important phenomenon occurring during austenitization in the intercritical temperature range is the partitioning of interstitials and alloying elements between α and γ phases. From the equilibrium diagrams of multi-component systems, one can expect a tendency for silicon and phosphorus to enrich ferrite while manganese is enriching austenite (final martensite). This fact illustrated, in particular, in Fig. 2.18, is confirmed by numerous studies of intercritical annealing.

An abrupt change in manganese concentration near the interphase boundary, as for example, 0.5 % Mn on ferrite side and 1.5 % Mn on austenite side, was predicted from thermodynamic considerations (Wycliffe et al. 1981). Similar abrupt changes in Mn or Si concentrations near the interphase boundaries were observed experimentally (Souza et al. 1982; Guo et al. 2009; Toji et al. 2011) as exemplified in Fig. 2.19. In the last years, direct proofs of similar behavior were obtained using atom-probe tomography that allowed measuring not only the local concentration distributions of elements but also the path lengths of several nanometers of highly

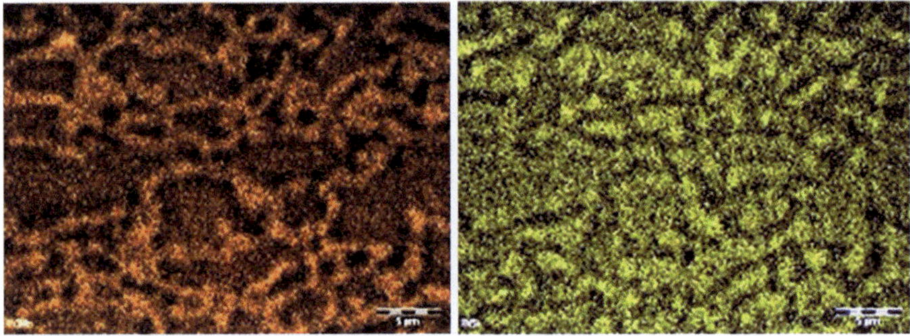

Fig. 2.18 Distribution of Mn (*left*) and Si (*right*) in microstructure of water-quenched DP steel—original

Fig. 2.19 Manganese concentration of ferrite and austenite phases in 0.06C–1.5Mn steel after 1 h at 740 °C; STEM (Speich et al. 1981)—modified

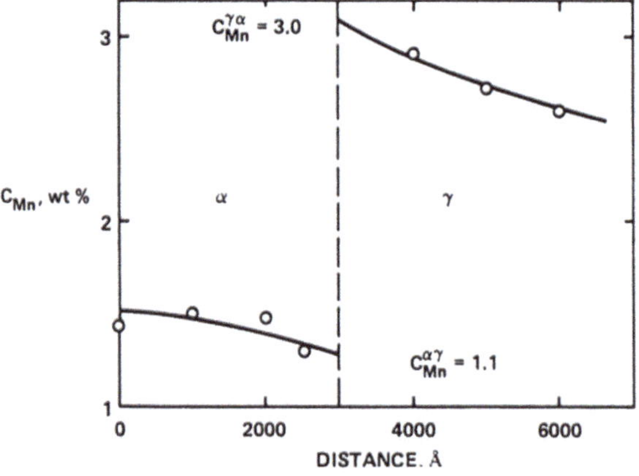

localized diffusion (Dmitrieva et al. 2011). It was suggested that both the interfacial segregations and diffusion induced concentration pileups contribute to the enrichment in alloying elements with the vicinity of moving boundaries.

Since the diffusion rates of substitutional elements are three or more orders of magnitude slower than that of carbon, the actual partitioning patterns are expected to be far from those in equilibrium and depend on kinetic factors such as heating temperature and (really short) holding time, initial microstructure (diffusion paths), etc. (Wycliffe et al. 1981; Kogan et al. 1984).

Numerous experimental data are available in the literature confirming the existence of concentration gradient of alloying elements during holding in the intercritical region. The presence of 0.2 μm wide manganese-rich zones along the perimeter of γ-phase and the associated depletion of surrounding ferrite with manganese were experimentally revealed when 0.08C–1.0Mn steel with initial ferrite–pearlite microstructure was heated at temperatures close to A_{c1}.

In some works, the diffusivity of alloying elements in austenite of two-phase microstructure was found to be one or two orders of magnitude higher than expected, and significant differences in alloying content between α- and γ-phases were observed (Enomoto and Aaronson 1987). In particular, it was also noted that near an austenite nucleus close to a carbide particle, the concentration of silicon in ferrite at the interphase boundary with γ-phase is twice its average concentration in steel (Miller and Smith 1977).

The approach to the equilibrium partitioning of elements is more pronounced during heating of pre-quenched microstructure since high density of interphase boundaries and short diffusion paths through acicular mixture of phases enhance diffusion. In fact, heating in the intercritical region (at 820 °C) of as-quenched 1.56Mn–1.12Si steel led to 1.6 times higher Mn concentration in austenite and two times higher silicon concentration in ferrite compared to the average concentrations of these elements in steel (Speich et al. 1981).

Some researchers intentionally use manganese partitioning between α- and γ-*phases* during prolonged heating into intercritical temperature range to radically change the stability of the austenite, to modify the final microstructure, and thereby to substantially improve the properties of medium-alloy steels containing 3–8 % Mn (Mould and Skena 1977; Crawley et al. 1981; Arlazarov et al. 2012).

Distribution of carbon is not homogeneous either because of different sources of austenite formation or growth as well as due to effects of alloying elements on local carbon activity. During cooling from the intercritical temperature range at medium cooling rates, the higher carbon portions of austenite transform to martensite, while the lower carbon portions transform by diffusional mechanism giving birth to freshly formed ("new") ferrite. The formation of the "new" ferrite at the periphery of an austenite grain is favorable because the resultant martensite islands become smaller. The situation, when carbon concentration decreases from the center towards the periphery of the γ-phase, is facilitated by raising the heating temperature closer to A_{c3} and by employing short holding times so that the γ-phase boundaries are scarcely enriched in manganese.

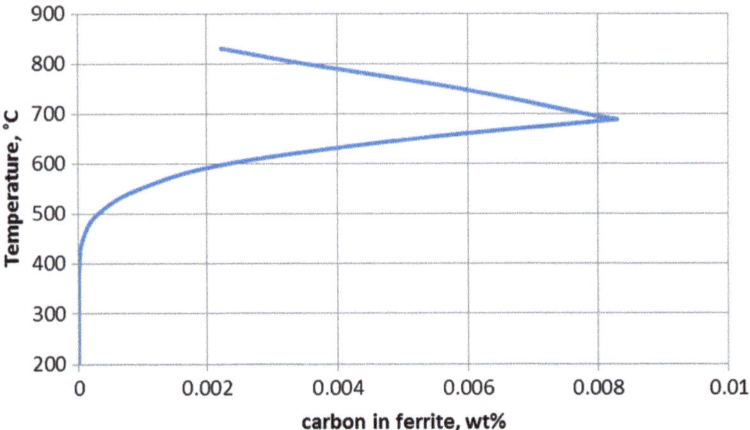

Fig. 2.20 Effect of annealing temperature on carbon concentrations in ferrite of 0.06C–1.5Mn steel (calculated using Thermocalc—original

Essential resource for enhancing the mechanical properties of dual-phase steels is the increase in strength by low-temperature strain aging (170–200 °C) in stamped parts (more details can be found in Chap. 3). Therefore, controlling the amount of carbon dissolved in ferrite is of great importance. Low carbon concentration in ferrite (C_f) of DP steels is considered to be a result of carbon partitioning to austenite during soaking of coexisting α- and γ-phases at the intercritical temperatures (Davies and Magee 1981). Exact value of C_f, however, is a function of the annealing temperature in α + γ temperature range, the content of alloying elements, and the cooling rate.

At higher temperatures within the two-phase region, the carbon content in ferrite, C_f, should decrease following the PG line of Fe(Me)–C diagram (Fig. 2.1). Figure 2.20 depicts the changes in C concentration in ferrite calculated using Thermocalc. Therefore, an increase in the volume fraction of austenite (and, hence, of martensite in dual-phase steel) with higher annealing temperature induces a simultaneous decrease in C_f and C_m, i.e., both phases should become softer. The decrease in C_f should be assisted by adding the elements that increase the thermodynamic activity of carbon in the α-lattice (such as silicon, aluminum, and nickel, which eject carbon from the ferrite) or the elements that bind carbon into insoluble carbides (such as vanadium, niobium, titanium, and zirconium).

2.5 Transformations of Austenite During Cooling from the Intercritical Region

The γ → α transformation during cooling from the intercritical temperature range controls all major characteristics of a dual-phase steel such as volume fractions of martensite and new ferrite, carbon content in martensite, and therefore the expansion of martensite during transformation and hence the residual stresses in ferrite. The cooling rate also determines the content of interstitials in ferrite, the presence of retained austenite, and some amount of bainite.

2.5.1   Features of Transformations of Austenite During Cooling from the Intercritical Temperature Range

The enrichment of austenite with carbon during heating into two-phase region substantially increases the austenite stability compared with austenite of the same steel cooled from a temperature above A_{c3}. This is an extremely important aspect of the intercritical heat treatment, which allows for austenite of low-carbon steel to undergo transformation that are mostly typical for medium- and high-carbon structural steel. The important peculiarity of transformations during cooling of ferrite–austenite mixture is also brought about by facilitated formation of ferrite or by its epitaxial growth at preexisting ferrite interface without the nucleation stage (Geib et al. 1980). This promotes further enrichment of austenite with carbon. As a result, cooling of austenite phase from the $\alpha + \gamma$ temperature range can substantially delay the pearlite transformation compared to cooling of the fully austenitized steel of the same composition.

Figure 2.21 presents superposition of two CCT diagrams for 0.14C–1.2Mn–0.5Si–0.5Cr–0.003B steel, obtained at cooling from γ region and from the intercritical temperature range.

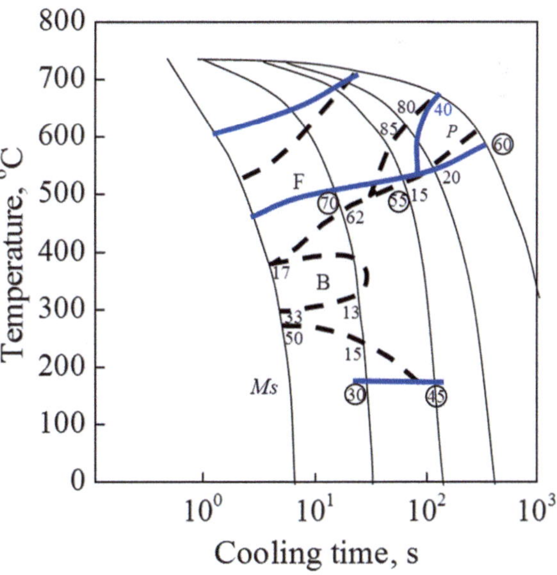

Fig. 2.21 CCT diagrams of steel with 0.14C–1.0Mn–0.5Cr–0.003B cooled from $\alpha + \gamma$ region (*solid lines*) and after full austenitization (*dashed lines*)—original

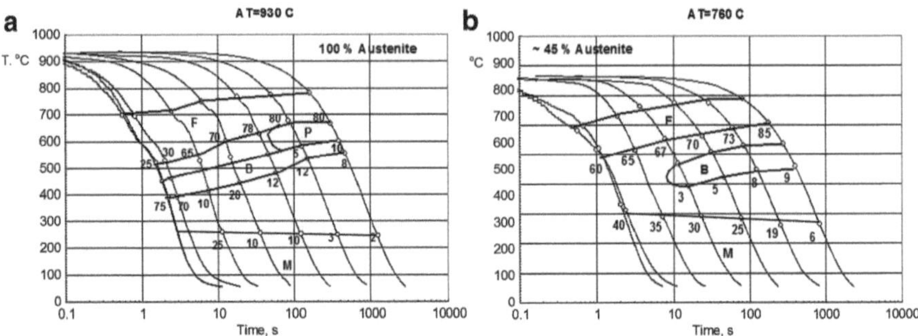

Fig. 2.22 CCT diagrams of 0.08–1.8Mn steel cooled from fully austenitic area (**a**) and from intercritical region (**b**)—(Girina et al. 2003)

Evidently, the enrichment of austenite with carbon during intercritical heating produces a significant shift of pearlite transformation toward lower temperatures and slower cooling rates, suppression of bainite region, and consequently the decrease in M_s temperature. In contrast, the starting temperature of ferrite formation increases.

Another comparison of CCT diagrams for 0.06C–1.8Mn steel after full and incomplete austenitization is depicted in Fig. 2.22 (Girina et al. 2003). As can be seen, steel composition can change the rate of new ferrite formation, but enrichment of austenite with carbon shifts bainite transformation to lower temperatures and slower cooling rates, whereas the increased volume of ferrite formed at slow cooling rates results in slight decrease in M_s temperature.

Better understanding of the features of austenite transformations during cooling from the intercritical region was gained by directly comparing the stability of austenite in ferrite–austenite mixture and the austenite of *the same* composition in a fully austenitized steel. The experiments were performed using 0.06C–1.2Mn–0.5Si–0.5Cr steel. During heating at 780 and 800 °C for 5 min, the amount of austenite of 20 and 33 %, respectively, was obtained. The corresponding concentrations of carbon in austenite were respectively 0.30 and 0.18 % C. The CCT diagrams of austenite transformation after annealing at 780 and 800 °C and cooling were compared with CCT diagrams of fully austenitized lab-melted steels containing 0.18 and 0.30 % C, i.e., with the same carbon concentrations in the *uniform* γ-phase.

As shown in Fig. 2.23, at cooling from the α + γ region, the stability of *γ-phase in the two-phase mixture* is greater (as shown above) than in cooling *100 % austenite of the same composition*. At the same time, it appeared to be substantially lower in terms of resistance to ferrite formation than that of *austenite with the same carbon content* in fully austenitized steel. The possible contributions to this behavior could be brought about by insufficient homogeneity of austenite after holding for a given time and by smaller austenite grain size. However, the dominant role should be attributed to preexisting austenite–ferrite interfaces that facilitate ferrite formation dramatically.

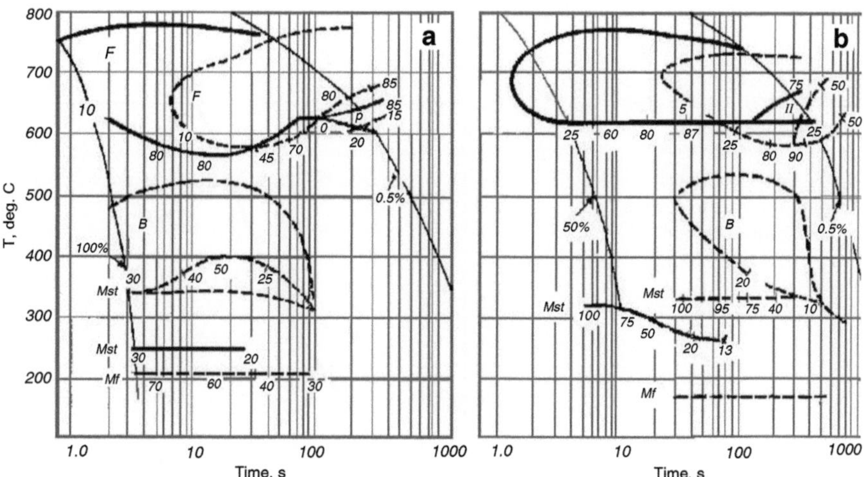

Fig. 2.23 Comparison of CCT diagrams of 0.06C–Cr–Mn–Si steel cooled from 800 °C (**a**) and 780 °C (**b**) (*solid lines*) and of fully austenitized Mn–Cr–Si steels with 0.18 % C and 0.30 % C (*dashed lines*, (**a**) and (**b**) accordingly) (original)

The role of interphase boundaries facilitating the formation of "new" ferrite was confirmed by observations of changes in martensite fraction depending on the refinement of microstructure. It was found that the microstructure of Mn–Cr–Si steel with 0.06 % C, cooled at 30 °C/s rate from a fixed annealing temperature in the $\alpha + \gamma$ region, was progressively refined with increasing number of heating–cooling cycles being accompanied by reduced amount of martensite in DP steel.

An increase in the relative amount of "new" ferrite formed during cooling from intercritical region causes significant enrichment of the remaining portion of austenite with carbon. That, in turn, alters the character of subsequent γ-phase transformations, e.g., tends to eliminate the bainite reaction and to lower the M_s temperature along with the slowing down of cooling rate, as shown in Figs. 2.21 and 2.23b. The combined effect of higher annealing temperature and moderate cooling rate on stability of the austenite phase (characterized by the M_s temperature) is given in Fig. 2.24.

At initial increase in annealing temperature, M_s value is increasing due to decrease in carbon content in larger volume of austenite. However, at further increase in austenite fraction and decrease in its carbon content, the hardenability of γ-phase becomes insufficient for being fully transformed to martensite at given cooling rate. Its decomposition with formation of the "new" ferrite results in a gradual increase in carbon content in remaining portion of austenite and consequent decrease in M_s temperature.

In a few studies, important role of austenite homogeneity is emphasized. In particular, it was supported by experiment, in which the stability of austenite in 0.06C–1.1Mn–0.45Cr–0.003B steel was found to increase with the time of holding in the intercritical region.

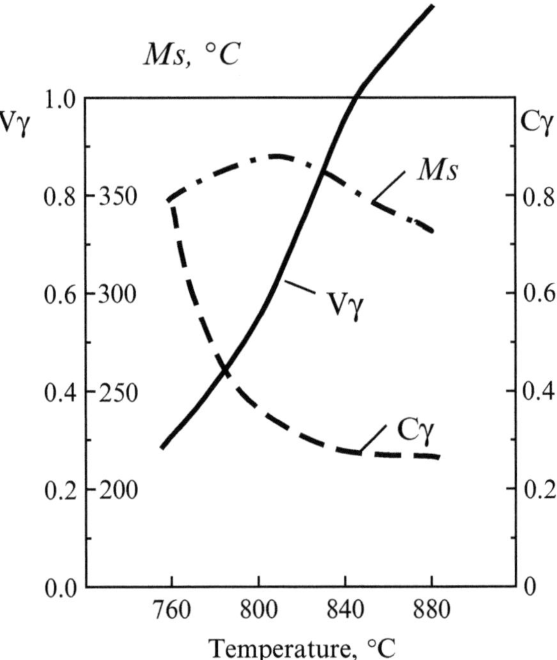

Fig. 2.24 Effect of annealing temperature in two-phase region on the austenite fraction, its carbon content and M_s temperature at cooling rate of 30 °C/s, 0.06C–Mn–Cr–B steel, holding for 10 min at each temperature—original

Since the total concentration of carbon in steel C_{st} is fixed, the isothermal growth of the volume fraction of austenite (V_γ) should be accompanied by a gradual decrease in the average carbon content in austenite (C_γ). Thus, the longer the isothermal holding is, the less should be the stability of γ-phase formed. Nevertheless, as shown in Fig. 2.25, longer holding time increases hardenability of austenite that manifested itself through a shifting of ferrite region to lower temperatures and cooling rates as decrease in the fraction of the formed ferrite results in higher M_s values.

Higher stability of austenite after longer holding in α + γ region was noted in other studies and various explanations for that were higher homogeneity of austenite, increased effective carbon content due to continuing dissolution of carbides, growth of dimensions of austenite islands, etc. Increasing phase stability with holding time can be deduced, for example, from the data published by Messien et al. (1981), who found higher content of martensite after longer holding of 0.08C–1.59Mn–0.39Si steel in a wide range of moderate cooling rates. The authors suggested that the increasing homogeneity of austenite during its growth compensated for anticipated decrease in average carbon content that should reduce the stability of austenite.

Important role of γ-phase homogeneity was also supported by studies of the effect of carbon content (C_{st}) in steels with base composition of 0.5Si–1.3Mn–0.55Cr–0.003 % B. For investigated steels during heating at 2 °C/s, the critical temperatures were found as $A_{c1} = 740$ °C and $A_{c3} = 895, 860,$ and 825 °C for carbon concentrations of $C_{st} = 0.06, 0.14,$ and 0.21 % C, respectively.

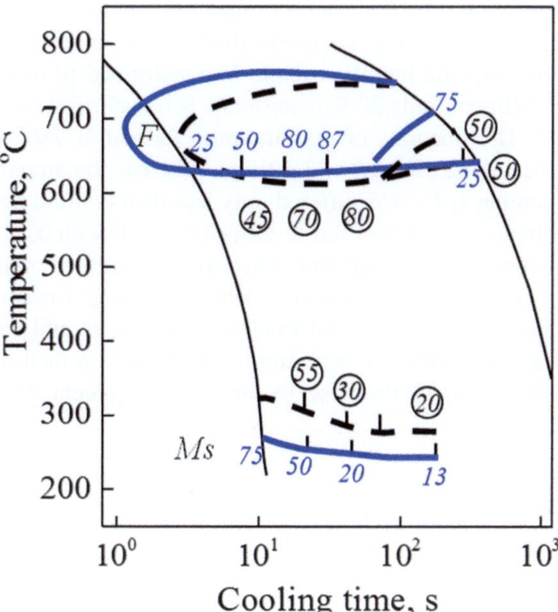

Fig. 2.25 Effect of holding time in two-phase region on stability of austenite in 0.06C–1.1.Mn–0.45Cr–0.003B: superposed CCT diagrams obtained after 1 min (*blue lines*) and 10 min (*dashed black lines*) of holding—original

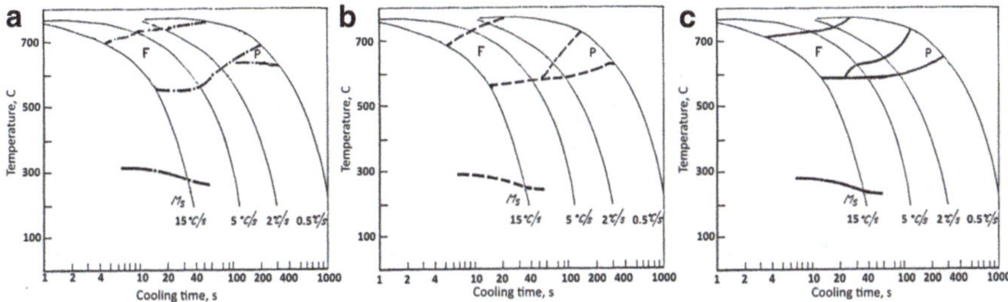

Fig. 2.26 Effect of carbon content on the CCT diagrams of Cr–Mn–Si–B steels cooled from 780 °C after 10 min holding: *solid, dashed,* and *dash-dotted lines* correspond to steels with 0.21 %, 0.14 %, and 0.06 % C, respectively—original

Specimens were cold rolled with 75 % reduction, heated at 2 °C/s to 780 °C, held for 10 min, and then cooled at 0.5–15 °C/s; reference specimens were quenched to measure the volume fraction of martensite that was assumed to be equal to the volume fraction of the initial austenite at the elevated temperature.

As was determined, holding for 10 min at 780 °C produced 18, 45, and 60 % austenite in steels with 0.06, 0.14, and 0.21 % C, respectively. As expected, the theoretical carbon content in the γ-phase, C_γ, was approximately the same, and the values calculated using Eq. (2.3) ranged from 0.29 to 0.33 %, i.e., were very close. However, as seen in Fig. 2.26 presenting the CCT diagrams of the above steels, the pearlite area widens and shifts towards higher cooling rates as the carbon content in steel and hence the volume of initial austenite increases. In fact, in steel with

0.06 % C, pearlite can be formed only in cooling from 780 °C at slower rate than 2 °C/s but at 10 °C/s in steel with 0.21 % C. In steel with 0.06 % C cooled at 15 °C/s, 40 % of the initial austenite is transformed to martensite, while under identical conditions, only 20 % of austenite is transformed to martensite in steel with 0.21 % C. The fractions of the austenite, formed at 780 °C in steels with 0.06 % and 0.21 % C, differed in 3.3 times, whereas the amounts of martensite produced in cooling at 15 °C/s differed only less than twofold being 7 and 12 %, respectively. Greater amounts of ferrite formed in steels with 0.21 and 0.14 % C lowered the M_s temperatures during final transformation of the remaining portions of austenite.

Lower γ-phase stability with increasing carbon content in steel was noted elsewhere (Hansen and Pradhan 1981). It should be assumed that an increase in carbon content in steel aiming to produce a dual-phase steel of higher strength should require the offset by higher alloy content when slow cooling rates are used.

2.5.2 The Effect of Annealing Temperature in the Two-Phase Region

The annealing temperature in two-phase region determines the amount of austenite and its carbon content. The latter is the key factor of austenite hardenability affecting both the fractions and types of transformation products.

In fast cooling, the decrease in annealing temperatures and therefore an increase in carbon content in austenite can change the substructure of produced martensite from a lath-type of low-carbon martensite to twinned martensite at carbon content higher than 0.35–0.40 % accompanied with retaining of some amount of untransformed γ-phase (retained austenite) (Thomas and Koo 1981).

As was shown in Fig. 2.24, in relatively slow cooling, the increase in heating temperature and corresponding decrease in C_γ can increase the amount of the formed new ferrite and lower the M_s temperature due to enrichment of the remaining portion of austenite with carbon.

In general, the stability of austenite decreases with increasing annealing temperature. However, interaction of various factors can shadow the role of calculated carbon concentration in initial austenite. In particular, when special carbides are present in the initial microstructure of steel, their dissolution at higher temperatures and the corresponding increase in effective carbon content in austenite can contribute to some spike of austenite hardenability as well (Garcia et al. 2011).

2.5.3 Effect of Cooling Rate

The rate of cooling from annealing temperature determines the type of austenite transformations, the fraction of austenite transformed to martensite (and hence the strength of steel), as well as the strength and ductility of ferrite, which depends on the content of interstitials in solid solution and therefore on cooling rate.

The effects of heating temperature in the intercritical range and those of the critical cooling rate are interrelated in processing of dual-phase steels. For a given steel composition, the lower the heating temperature in the $(\alpha + \gamma)$ region, the higher the C_γ and the lower the cooling rate required for martensitic transformation of austenite phase.

With high cooling rate, dual-phase steels can be produced at lower carbon content and/or at substantially reduced alloy content (Nakaoka et al. 1981). However, in rapid quenching, the amount of the formed austenite and that of the corresponding martensite, as well as the properties of dual-phase steels, are very sensitive to heating temperature. In contrast, with relatively slow cooling rates (2–30 °C/s), the formation of ferrite can take place with a near-equilibrium carbon partitioning from the new ferrite to the remaining austenite. The latter, prior to its transformation to martensite, can be therefore independent or less dependent on heating temperature. The new ferrite plays a critical "adjusting" role: its amount grows with increasing volume of austenite, which therefore has lower carbon content and hardenability, and vice versa (Lagneborg 1978). This self-control of the final DP microstructure was observed by several researchers (Davies 1978; Eldis 1981; Matlock et al. 1981). Consequently, at a given slow cooling rate, DP steels can be produced with roughly constant amount of martensite (and hence with the same strength) over a reasonably wide range of heating temperatures of 40–80 °C (Fig. 2.27). However, the "permanent" amount of martensite obtained this way somewhat depends on the applied cooling rate. In 0.06C–1.5Mn steel heated in the range of 740–820 °C and then cooled at 1.4 and at 10 °C/s rate, the volume fraction of martensite was ~10 % and ~18 %, respectively (Fonstein et al. 1984). Besides, as shown in Fig. 2.27, adjusting effect of "new ferrite" depends on chemical composition of steel and applied cooling rate.

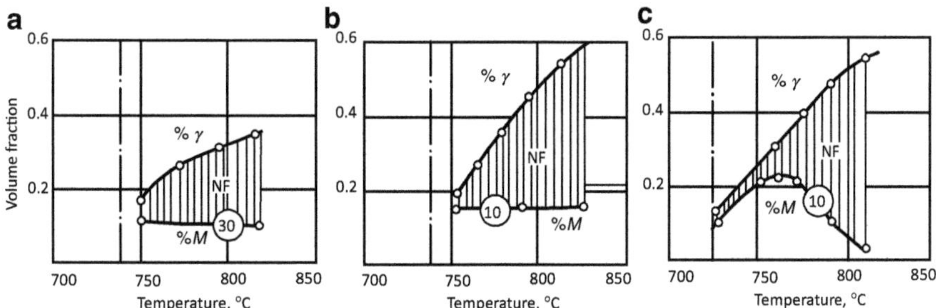

Fig. 2.27 Effect of annealing temperature on amount of formed γ-phase and the martensite fraction depending on cooling rate: (**a**) 0.08C–1.35Mn–0.35Si–0.08V steel, cooling 30 °C/s rate; (**b**) 0.065C–1.21Mn–0.6Si–0.003B, cooling rate 10 °C/s (**c**) 0.065C–1.2Mn–0.2Cr–0.1Mo steel, cooling rate 10 °C/s—original

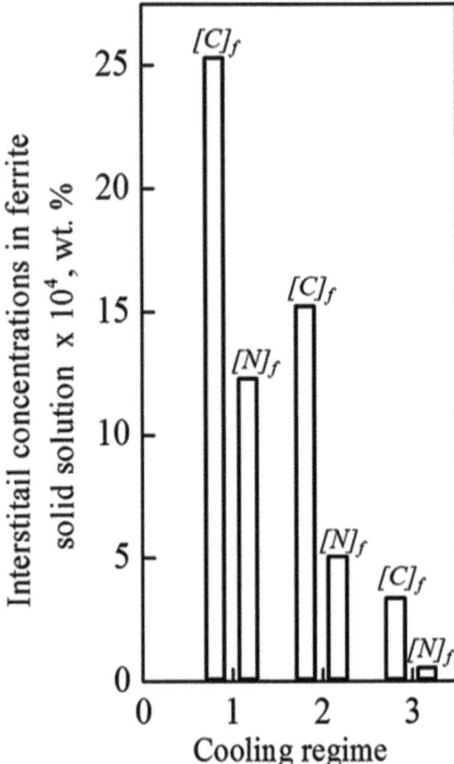

Fig. 2.28 Effect of cooling rate on concentration of interstitials in ferrite of 0.09C–1.9Mn–0.5Si steel: 1-water quenching from 800 °C, 2-cooling from 800 to 550 °C at 10 °C/s followed by water quenching, 3-air cooling from 800 °C—original

A decrease in cooling rate reduces concentrations of dissolved carbon C_f and nitrogen N_f in ferrite due to partitioning of interstitials from ferrite to the remaining austenite and due to possible precipitation of fine cementite and nitride particles. This is confirmed by measurements of nitrogen and carbon Snoek peaks of temperature-dependent internal friction (Fig. 2.28). Lower concentration of interstitials in ferrite enhances the ductility of ferrite and prevents the danger of aging during steel storage.

When slow cooling is employed, the final portion of remaining austenite is enriched with carbon due to its partitioning from ferrite to the γ-phase. This induces a decrease in the martensite transformation temperature (M_s) and can cause a transition from lath to twinned martensite. Some amount of retained austenite can also stem from the increased stability of the final portion of γ-phase at slower cooling (Bangaru-Rao and Sachdev 1982; Rigsbee et al. 1981).

An increase in the cooling rate from the $\alpha + \gamma$ region, especially in the portion near the M_s temperature, induces high residual stresses that accompany martensitic transformation and increases the density of unpinned dislocations in ferrite areas, adjacent to martensite (Hansen and Pradhan 1981). As will be shown in Chap. 3, these factors have important impact on properties of dual-phase steels.

Moreover, an excessive amount of carbon dissolved and remained in ferrite as a result of rapid quenching facilitates precipitation of fine carbides during short tempering thus preventing premature overaging during heating of stamped components (Nishimoto et al. 1981) and hence promoting higher BH effect, discussed below.

2.6   Obtaining of As-Rolled Dual-Phase Microstructure by Cooling of Deformed Austenite

Dual-phase microstructure can be obtained by thermomechanical processing that involves deformation in the two-phase region ("intercritical rolling") and subsequent rapid cooling (Ahmad et al. 2009). This, however, results in strain-hardened ferrite that impairs ductility of ferrite matrix of the dual-phase microstructure. In addition, fast cooling or quenching of deformed ferrite–austenite microstructures produces a highly anisotropic banded or fibrous microstructure. Therefore, a better way to produce dual-phase steels is deformation in the γ region under the conditions that allow obtaining a desired amount of ferrite during subsequent cooling to transform the remaining austenite into martensite.

For example, in manufacturing a hot-rolled dual-phase steel with tensile strength of ~600 MPa (so far, the most used commercial DP steel grade), the key is to ensure large amount of ferrite (80–85 %) to emerge within a short period of time while the steel strip is still moving down the run-out table.

It is well known that a faster generation of ferrite during cooling is promoted by low carbon content, heavy deformation of the austenite close to A_{r3} temperature, and by refinement of austenite grain size, i.e., by all factors reducing the hardenability of austenite and/or facilitate the nucleation and growth of ferrite. For example, a 40 % deformation of austenite at temperatures close to A_{r3} increases the rate of the $\gamma \rightarrow \alpha$ transformation during continuous cooling by roughly an order of magnitude (Coldren and Eldis 1980).

As shown in Fig. 2.29a, deformation shifts all transformations so that the selected steel composition should provide a range of cooling rates and temperatures of coiling ("processing window") that allow avoiding both pearlite and bainite transformation, as shown by dash-and-dot line (Kato et al. 1981). This approach implies that stability of the remaining 15–20 % of austenite should be sufficiently high due to enrichment with carbon and appropriate alloying to ensure the transformation to martensite during relatively slow cooling of coils at 20 °C/h. After 85 % volume fraction of ferrite has been formed, the CCT characteristics of the remaining portion of austenite enriched with carbon are similar to those of high-carbon steels with the same alloying.

Modern hot strip mills equipped with powerful cooling systems utilize the capability for interrupted cooling at temperatures close to "ferrite nose," to ensure the fastest formation of the necessary portion of ferrite. This is followed by fast

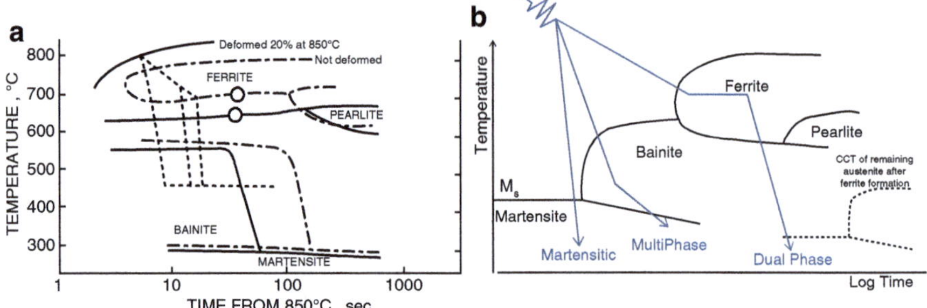

Fig. 2.29 Two approaches to produce hot-rolled dual-phase steels (see text for comments)—original

cooling and coiling in bainite region or below M_s temperature aiming at final ferrite–bainite or ferrite–martensite microstructures, respectively (Fig. 2.29b). This approach allows using lean steel compositions.

2.7 Austempering Annealing Cycle

Transformations of austenite under the conditions of austempering using thermal cycle presented in

Figure 2.2b, attract growing attention with regard to cooling from both super-critical and intercritical temperature ranges. On coating galvanizing lines, similar conditions are reproduced using equalizing sections.

When cooled from intercritical temperatures, the initial austenite is enriched with carbon so that, depending on the initial cooling rate(s), a substantial portion of austenite can stay untransformed down to the temperature of isothermal holding. The temperature of isothermal holding then dictates whether either upper or lower bainite or martensite can be generated in the end.

If alloying elements that prevent precipitation of carbides in bainite are added, the carbon content in the remaining portion of austenite can become so high that M_s temperature can drop below the room temperature. Under these circumstances, steel would contain certain amount of retained austenite. Thermal cycles of this type serve as a basis for obtaining steels with Transformation-Induced Plasticity (TRIP) effect described in Chap. 5.

While this annealing cycle is definitely favorable for manufacturing TRIP steels, some annealing lines are designed so that isothermal or pseudo isothermal holding is unavoidable in production of dual-phase steels as well. This applies to annealing line equipped with direct transfer of strip cooled by mist or cold rolls to the so-called overaging furnace (Kakuta and Takahashi 1982), as well as to some hot-dip galvanizing lines with long overaging before Zn bath.

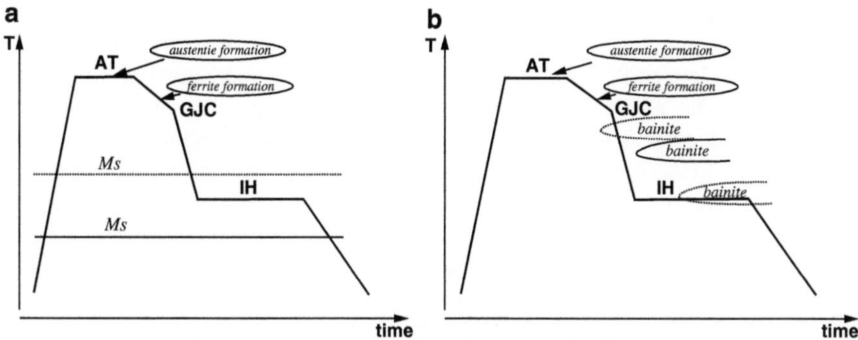

Fig. 2.30 Phase transformation during annealing using austempering cycle: (**a**) formation of ferrite–martensite microstructure; (**b**) formation of ferrite–bainite–martensite microstructure, *dashed line* designates conditions for steels with unsufficient hardenability (Girina et al. 2003)

In continuous annealing lines, the temperature of isothermal holding is typically below 400 °C and can be controlled in the desired range. Final microstructures can vary from ferrite–bainite to ferrite–martensite or to the ferrite plus mixture of bainite and martensite, which can be fully controlled by the relationship between processing parameters and austenite stability determined by its composition.

Various combinations of annealing parameters with respect to critical regions of phase transformation are illustrated in Fig. 2.30. Initial cooling rate can control the generation of "new ferrite" and therefore the additional enrichment of austenite with carbon. The latter, in turn, lowers the M_s temperature and slows down the bainite reaction similarly to effects of certain additional alloying. For example, at cooling of a low C–Mn mixture of 45–50 % of austenite plus ferrite, the M_s temperature was detected to be 270–310 °C (Girina et al. 2003).

To obtain ferrite–martensite structure, using this type of thermal cycle, the temperature of IH should be above M_s, and the austenite to martensite transformation should take place after IH to avoid the presence of tempered martensite.

If temperature of IH below M_s is used (Fig. 2.30a), austenite to martensite transformation during rapid cooling from GJC temperature or isothermal holding is leading to tempered martensite in the final structure.

To avoid the formation of bainite, the time of bainite reaction start in comparison to IH duration should be delayed by appropriate alloying, and cooling rate to IH temperature should be higher than critical position of bainite region for given steel composition (Fig. 2.30b).

Austempering type cycle can sometimes be applied to cooling from austenite temperature range. Depending on steel composition and isothermal holding temperature, the final microstructure can contain a mixture of bainite, martensite, and retained austenite. Different authors and steel makers use different terms to describe steels with this type of microstructure: TRIP steels with bainitic ferrite (TBF) (Sugimoto et al. 2000), ferrite-free TRIP (Baumer et al. 2011), or carbide-free bainitic (CFB) (Hell et al. 2011) steels discussed in Chap. 8.

Continuous galvanizing lines (CGL) have less freedom in selecting the isothermal holding temperatures. A few of the most modern lines have the capability for the so-called "low end" of initial cooling below Zn pot temperature, as was shown in Fig. 2.4. The availability of "low end" of initial cooling allows complete avoiding of undesired bainite and, depending on the temperature of this "end" with respect to M_s temperature, the final microstructure can contain tempered martensite, bainite, and retained austenite (Mohrbacher et al. 2011). The corresponding process is used as a basis for so-called Q&P (quenching & processing) technology and will be discussed in Chap. 10.

The majority of CGL have preliminary overaging sections of various lengths to keep the steel strip at melted Zn temperature. As was found for typical DP steel with C–Mn–Cr(Mo) composition at given annealing temperature, the nose of the bainite C-curve can be near 450 °C, i.e., very close to a typical CGL isothermal holding temperature (Pichler et al. 2000) facilitating bainite reaction.

It should be some combination of slowing down bainite reaction by in particular, using lower soak temperature (higher carbon content in the initial austenite) (Hoydick and Haezenbrouck 2003) or higher alloying or applying higher overaging temperature (above the nose of bainite transformation) to decrease the fraction of bainite in the final microstructure of DP steel.

The elevated strip temperature higher than Zn bath is also possible, promoting "quenching" austenite fraction, but requires special means of Zn bath cooling, e.g., the ZinQuench system (Radtke et al. 1986; Patil et al. 2004).

2.8 Tempering of Ferrite–Martensite Mixture

The most important features of dual-phase steels are defined by significant differences in tensile strength of martensite and ferrite, both of which decrease with reheating of steel or steel parts. The reheating, or tempering/auto-tempering, is often a mandatory step either during annealing or slow cooling, when dual-phase steels are produced, or in final processing of stamped parts during baking of protective paint coating.

The changes in microstructure and properties of dual-phase steels during tempering substantially differ from those occurring in tempering of fully quenched steels. Normally, the latter are rather uniform with regard to dislocation annihilation, precipitation of carbides, and transformation of retained austenite. The major distinctions of tempering DP steels stem from super-saturation of ferrite with interstitial atoms, residual stresses in ferrite, higher localized dislocation density in ferrite grains near the boundaries with martensite, fine sizes of retained austenite grains, and from inhomogeneous volume distribution of carbide precipitates.

The external parameters that affect the changes in DP microstructure and properties after tempering are the same: temperature and duration of tempering. The difference is determined by internal features of DP microstructure that cannot be described by a single parameter or be controlled only by chemical composition as opposed to Q&T microstructures.

In particular, the important processes occurring in dual-phase steels during low-temperature tempering are diffusion of carbon atoms in ferrite towards dislocations and relaxation of residual stresses. At the same time, the relative portion of ferrite grains with high dislocation density depends on the volume fraction of martensite, the M_s temperature, the corresponding strength of martensite, and the morphology of DP microstructure, i.e., whether it is the martensite islands dispersed in ferrite matrix or martensite skeleton surrounding ferrite grains. As a result, the absolute changes in mechanical properties of ferrite–martensite microstructure after tempering can vary significantly despite the predicted trends of changes in the individual phases (see more in Chap. 3).

The most important "internal" parameters that influence the tempering effects in dual-phase steels are the annealing temperature and the rate of cooling from the intercritical region, i.e., the factors that control the concentrations of interstitials in ferrite, the volume fraction of martensite and its carbon content, as well as the amount of retained austenite. Steel alloying can affect the solubility of carbon in ferrite and the kinetics of carbide precipitation in martensite during tempering (see more in Chap. 4).

2.8.1 Tempering of Martensite

The excess carbon in martensite lattice is the major contributor to the stored energy of martensite and is the major driving force for processes at tempering of martensite. In fully quenched steels, there are four generally accepted tempering stages: I—precipitation of carbides and partial loss of martensite tetragonality (up to 250 °C), II—decomposition of retained austenite (between 200 and 300 °C); III—replacement of carbides by cementite, complete elimination of martensite tetragonality (between 250 and 350 °C); IV—coarsening and sheroidizing (up to 700 °C) of cementite and recrystallization of ferrite, which replaces martensite (above 350 °C).

When DP steel contains low-carbon martensite, the latter can be auto-tempered during slow cooling because of high M_s temperature (usually above 300 °C).

As found in early TEM studies, tempering at 200 °C for 1 h has a negligible effect on microstructure and mechanical properties of martensite of both dual-phase and fully austenitized steels after water quenching (Koo and Thomas 1979).

Later, however, Speich et al. (1983) observed precipitation of carbides along the martensite lath or packet boundaries during tempering DP steel at 200 °C. According to Rashid, who studied steel with 0.1C–1.5Mn–0.5Si–0.1V, tempering of higher carbon-twinned martensite after annealing at low intercritical temperatures lead to precipitation of numerous cementite particles in the form of fine platelets or films (Rashid and Rao 1982). Hardness maxima were observed during tempering of dual-phase steels at 160–200 °C as a result of precipitation of fine carbides.

Volume contraction of martensite during tempering due to precipitation of carbides reduces distortions in martensite induced by local $\gamma \to \alpha'$ transformation, and thereby decreases the residual stresses. The height of the internal friction Koster peak, however, does not decrease indicating the retention of high dislocation density in martensite.

Tempering at 300–400 °C leads to more intensive precipitation and coarsening of the carbides. Twinned martensite substructure (for the fractions of martensite with carbon content above 0.4–0.45 %) can be no longer observed.

Waterschoot and coworkers studied 0.08C–1.4Mn–0.13Si–0.36Cr–0.21Mo steel using dilatometry, TEM, and X-ray diffraction (Waterschoot et al. 2006). In the temperature range of 60–120 °C, the authors revealed segregation of carbon towards such lattice defects as lath boundaries and dislocations that caused a minor decrease in volume and clustering of carbon atoms. Next structure evolution occurring within temperature range of 120–195 °C was characterized by the formation of transitional orthorhombic η-carbide Fe_2C and/or hexagonal ε-carbide $Fe_{24}C$. Following changes both occurred within the same temperature range of 250–300 °C and included precipitation of carbides and transformation of retained austenite. Starting from 290 °C and up to 390 °C, cementite precipitates were observed that led to cubic martensite lattice. This was considered by authors as the final tempering stage of martensite.

2.8.2 Tempering of Ferrite

Because of strains accompanying volume changes during $\gamma \to \alpha'$ transformation, as well as due to the corresponding increase in dislocations density, the ferrite phase in dual-phase steel should have higher hardness than the ferrite phase after softening annealing. Besides, during rapid cooling from intercritical temperatures, the ferrite phase becomes supersaturated with interstitial atoms, which also increases strength and lower its ductility.

Combination of excess interstitials in the α-lattice and high dislocation density can lead to quench aging during tempering of dual-phase steels promoting the precipitation of ultrafine carbides in ferrite. Additional dislocations generated during cold deformation of DP steels enhance the driving force of the interstitial atoms to diffuse towards dislocations in ferrite during heating, in particular at 170–200 °C for baking paint, thus producing a marked strengthening by strain aging termed "bake hardening" (BH) (Nakaoka et al. 1979) (for more detailed discussion see Chap. 3).

Accordance to the Fe(Me)–C diagram (Fig. 2.1), the content of soluble interstitials increases with decreasing temperature of annealing in intercritical region, but the diffusion rate of carbon atoms decreases and so does the probability of homogeneous precipitation. Heterogeneous precipitates at dislocations and ferrite grain and subgrain boundaries were observed after heating at 200 °C for 10 min. As the tempering temperature was raised to 300 °C, ε-carbide and cementite precipitates grew in size.

Precipitation of fine carbides in ferrite of dual-phase steels during tempering is accompanied by a considerable decrease in the Snook carbon peak attributable to solute atoms in ferrite. Measurements of the temperature dependence of internal friction revealed that the carbon concentrations in ferrite (C_f) were equal to 0.0025 wt% after heating to 800 °C and water quenching from 550 °C and 0.0008 and 0.0003 wt% after tempering at 200 and 400 °C, respectively.

Rashid and Rao suggested that tempering at 200 °C does not produce any significant changes in ferrite other than rearrangements of the dislocations into lower energy configurations. However, these authors also showed that the hardness of ferrite decreases after tempering at 200–300 °C due to reduction of lattice distortion induced by interstitial atoms. After heating at 300 °C, the nucleation of extremely fine (2.5 nm) precipitates at the dislocations was observed. The biggest changes in fine microstructure of ferrite were observed after tempering at 400 °C or higher, when the amounts of fine carbides (5 nm) and cementite particles nucleated at ferrite grain and subgrain boundaries, substantially increased. Tempering at 500 °C induced coarsening of special (V and Nb based) carbonitrides (Rashid and Rao 1982).

The dislocation density in ferrite in the vicinity of ferrite–martensite boundaries substantially decreases during tempering. However, carbides that precipitate at dislocations delay the annihilation of dislocations, and therefore higher strength of ferrite in a dual-phase steel is preserved during reheating up to 600 °C.

As will be shown in Chap. 3, the conditions which make the strength/hardness of martensite and ferrite to converge can dramatically affect the properties of DP steels.

2.8.3 Decomposition of Retained Austenite

Water-quenched DP steels usually do not contain the retained austenite. The presence of retained austenite is favorable for producing the so-called TRIP-assisted dual-phase steels (discussed in Chap. 5) in which the TRIP effect is realized. Retained austenite can be generated during isothermal holding (austempering) immediately after cooling of ferrite–austenite mixture. Small amounts of retained austenite can be observed after any annealing with relatively slow cooling that can induce an extreme enrichment of a portion of austenite with carbon causing the M_s temperature to drop below room temperature. During posttreatment reheating, the retained austenite can decompose depending on its stability and the attained temperature.

The stability of retained austenite after cooling from the intercritical region largely depends also on the austenite grain size (this is discussed in detail in Chaps. 5 and 9).

According to Rashid and Rao, tempering of 0.12C–1.44Mn–0.5Si–0.13V steel at 200 °C did not produce any changes in retained γ-phase, which was present in the form of isolated particles of 2–6 μm in diameter (Rashid and Rao 1982). 90 % of initial retained austenite remained in microstructure after tempering at 300 °C as well, while only 10 % transformed to bainite. At 400 °C, 90 % of retained austenite was transformed to bainite. At 500 °C, the entire retained austenite was transformed.

The type of decomposition reactions of retained austenite depends on the relation between the M_s and tempering temperatures. When 0.18C–1.2Mn–1.0Cr–0.2Si steel was tempered at 400 °C after cooling from the γ-region, the retained austenite transformed into upper bainite, while similar tempering of the same steel cooled from two-phase region (with higher carbon content, C_γ, in the austenite fraction) led to the formation of lower bainite.

2.9 Summary

Enrichment of austenite with carbon and some alloying elements increases the austenite stability and therefore significantly changes the nature and characteristics of phase transformations during cooling from the intercritical temperature range, as compared to those occurring in the same steel during cooling from the γ-region. However, in contrast to homogeneous austenite of the same composition, the presence of preexisting α–γ interfaces substantially accelerates the formation of ferrite during cooling of austenite in the two-phase austenite–ferrite mixture.

Generation of new ferrite is the key to the existence of close relationship between annealing temperature and cooling rate and can be used to control the amount of martensite produced over relatively wide ranges of annealing temperatures and holding times.

The effect of annealing time is determined by the balance between the increase in the amount of γ-phase and hence the decrease in C_γ (that reduces the austenite stability), on one hand, and the increase in chemical homogeneity of γ-phase with time that raises its stability on the other hand.

The type of initial microstructure impacts the austenitization kinetics and the morphology of ferrite–austenite mixture and ultimately the morphology of ferrite–martensite microstructure.

The effect of heating rate on austenitization during annealing of cold-rolled steels in the intercritical region depends on the heating temperature that determines the completion of recrystallization in ferrite prior to austenitization. Austenitization of non-recrystallized ferrite can result in coarsening and enhanced banding of DP microstructure.

The rate of cooling from annealing temperature determines the type of austenite transformations, the fraction of austenite transformed to martensite and hence the strength of martensite, as well as the strength and ductility of ferrite, which depends on the content of interstitials in solid solution and therefore on cooling rate.

Volume contraction of martensite during tempering due to precipitation of carbides reduces distortions in martensite induced by local $\gamma \rightarrow \alpha'$ transformation, and thereby decreases the residual stresses.

To improve the ductility of ferrite, a dual-phase steel produced using cooling rates higher than 30 °C/s requires post (or auto)-tempering to reduce the concentration of interstitials in ferrite, as well as to relieve residual stresses and to decrease dislocation density.

The temperatures and holding time of annealing in the intercritical region, as well as subsequent cooling and tempering determine the fractions of ferrite and martensite in the microstructure of DP steels and their individual properties, whereas morphology of microstructure constituents depends on the initial (prior heat treatment) microstructure and the heating rate employed.

References

Ahmad, E., T. Manzoor, and N. Hussein. 2009. "Thermomechanical Processing in the Intercritical Region and Tensile Properties of Dual-phase Steel." *Material Science and Engineering, A* 508 (1-2): 259–265.

Andrews, K W. 1965. "Empirical Formulae for Calculation of Some Transformation Temperatures." *Iron and Steel Institute Journal* 203 (Part 7): 721–27.

Arlazarov, A., M. Gouné, O. Bouaziz, A. Hazotte, G. Petitgand, and P. Barges. 2012. "Evolution of Microstructure and Mechanical Properties of Medium Mn Steels during Double Annealing." *Materials Science and Engineering: A* 542 (0): 31–39.

Avrami, M. 1940. "Kinetics of Phase Change. II Transformation-Time Relations for Random Distribution of Nuclei." *The Journal of Chemical Physics* 8 (2): 212–24.

Azizi-Alizamini, H., M. Militzer, and W.J. Poole. 2011. "Austenite Formation in Plain Low-Carbon Steels." *Metallurgical and Materials Transactions A: Physical Metallurgy and Materials Science* 42 (6): 1544–57.

Ballinger, N. K., and R. W. K. Honeycombe. 1980. "Coarsening of Vanadiem Carbide, Carbonitride and Nitride in Low-Alloy Steel." *Met. Sci.* 14 (4): 120–33.

Bangaru-Rao, N., and A. K. Sachdev. 1982. "Influence of Cooling Rate on the Microstructure and Retained Austenite in an Intercritically Annealed Vanadium Containing HSLA Steel." *Metallurgical and Materials Transactions A* 13 (11): 1899–1906.

Baumer, A., E. Bocharova, T. Hiller, D. Krizan, and A. Pichler. 2011. "New Development in High Strength TRIP Steels." In *Auto Circle*.

Caballero, F.G.a, A. García-Junceda, C. Capdevila, and C.G. De Andrés. 2006. "Evolution of Microstructural Banding during the Manufacturing Process of Dual Phase Steels." *Materials Transactions* 47 (9): 2269–76.

Cho, Y.-B. 2000. "Ph.D. Thesis. The Kinetics of Austenite Formation during Continuous Heating of a Multi-Phase Steel." University of British Columbia.

Coldren, A.Phillip, and George T. Eldis. 1980. "Using CCT Diagrams to Optimize the Composition of an as-Rolled Dual-Phase Steel." *Journal of Metals* 32 (3): 41–48.

Crawley, A.F., M.T. Shehata, N. Pussegoda, C.M. Mitchell, and W.R. Tyson. 1981. "Processing, Properties and Modelling of Experimental Batch-Annealed Dual-Phase Steels." In *Conference of Fundamentals of Dual-Phase Steels, Symposium at the 110th AIME Annual Meeting*, 181–97. Chicago, IL, USA: Metallurgical Society of AIME, Warrendale, Pa, USA.

Davies, R.G. 1978. "The Deformation Behavior of a Vanadium-Strengthened Dual Phase Steel." *Metallurgical Transactions A* 9 (1): 41–52.

Davies, R.G., and C.L. Magee. 1981. "Physical Metallurgy of Automotive High-Strength Steels." In *Conference of Structure and Properties of Dual-Phase Steels, Symposium at the AIME Annual Meeting*, 1–19. New Orleans, LA, USA: Metallurgical Society of AIME, Warrendale, Pa, USA.

Dmitrieva, O., D. Ponge, G. Inden, J. Millán, P. Choi, J. Sietsma, and D. Raabe. 2011. "Chemical Gradients across Phase Boundaries between Martensite and Austenite in Steel Studied by Atom Probe Tomography and Simulation." *Acta Materialia* 59 (1): 364–74.

Dyachenko, S. S. 1982. "The Austenite Formation in Fe-C Alloys." *M. Metallurgia*, 128.

Eldis, George T. 1981. "Influence of Microstructure and Testing Procedure on the Measured Mechanical Properties of Heat Treated Dual-Phase Steels." In *Conference of Structure and Properties of Dual-Phase Steels, Symposium at the AIME Annual Meeting*, 202–20. New Orleans, LA, USA: Metallurgical Society of AIME, Warrendale, Pa.

El-Sesy, I. A., H-J Klaar, and A.-H. Hussein. 1990. "Effect of Intercritical Temperature and Cold Deformation on the Kinetics of Austenite Formation during the Intercritical Annealing of Dual-Phase Steels." *Steel Research* 61 (3): 131–35.

Enomoto, M., and H.I. Aaronson. 1987. "Partition of Mn during the Growth of Proeutectoid Ferrite Allotriomorphs in an Fe-1.6 At. Pct C-2.8 At. Pct Mn Alloy." *Metallurgical Transactions A* 18 (9).

Fonstein, N. M. 1985. "Heat Treatment for Obtaining a Controlled Ferritic-Martensitic Structure in Steel." *Metal Science and Heat Treatment (English Translation of Metallovedenie I Termicheskaya Obrabotka* 27 (7-8): 610–16.

Fonstein, N. M., L. M. Storojeva, and T. M. Efimova. 1984. "Development of Ferritic-Martensitic Structure in Low-Alloy Steels during Heat Treatment in Continuous Furnaces." *Steel in USSR*, 9: 75–77.

Garcia, C.I.a, K.a Cho, K.a Redkin, A.J.a c Deardo, S.b Tan, M.c Somani, and L.P.c Karjalainen. 2011. "Influence of Critical Carbide Dissolution Temperature during Intercritical Annealing on Hardenability of Austenite and Mechanical Properties of DP-980 Steels." *ISIJ International* 51 (6): 969–74.

Garcia, C.I., and A.J. DeArdo. 1979. "Formation of Austenite in Low-Alloy Steels." In *Conference of Structure and Properties of Dual-Phase Steels, Symposium at the AIME Annual Meeting*, 40–61. New Orleans, LA, USA: Metallurgical Society of AIME, Warrendale, Pa.

Garcia, C.I., and A.J. Deardo. 1981. "Formation of Austenite in 1.5 Pct Mn Steels." *Metallurgical Transactions. A, Physical Metallurgy and Materials Science* 12 A (3): 521–30.

Geib, M.D., D.K. Matlock, and G. Krauss. 1980. "The Effect of Intercritical Annealing Temperature on the Structure of Niobium Microalloyed Dualphase Steel." *Metallurgical Transactions A* 11 (10): 1683–89.

Girina, O.A., N.M. Fonstein, and D. Bhattacharya. 2008. "Effect of Nb on the Phase Transformation and Mechanical Properties of Advanced High Strength Dual-Phase Steels." In *International Conference of New Developments on Metallurgy and Applications of High Strength Steels*, 1 Plenary Lectures Automotive Applications High Temperature Applications Oils and Gas Applications: 29–35. Buenos Aires.

Girina, O., and D. Bhattacharya. 2003. "Effect of Mn on the Stability of Austenite in a Continuously Annealed Dual-Phase Steel." In *45th MWSP Conf., Proc. ISS*, XLI: 415–20.

Girina, O, D. Bhattacharya, and N. M. Fonstein. 2003. "Effect of Annealing Parameters on Austenite Decomposition in a Continuously Annealed Dual-Phase Steel." In *Proc. of 41st MWSP*, 403–14.

Golovanenko, S. A., and N. M. Fonstein. 1986. "Dual-Phase Low Alloyed Steels." *M, Metallurgia*, 206.

Granbom, Y. 2008. "Effects of Process Parameters prior to Annealing on the Formability of Two Cold Rolled Dual Phase Steels." *Steel Research International* 79 (4): 297–305.

Guo, H., S. W. Yang, and C. J. Shang. 2009. "A Quantitative Analysis of Mn Segregation at Partitioned Ferrite/Austenite Interface in a Fe-C-Mn-Si Alloy." *Journal of Materials Sciences and Technology* 25 (03): 383.

Hansen, S.S., and R.R. Pradhan. 1981. "Structure/property Relationships and Continuous Yielding Behavior in Dual-Phase Steels." In *Conference of Fundamentals of Dual-Phase Steels, Symposium at the 110th AIME Annual Meeting*, 113–44. Chicago, IL, USA: Metallurgical Society of AIME, Warrendale, Pa, USA.

Hornbogen, E., and J. Becker. 1981. "Microscopic Analysis of the Formation of Dual-Phase Steels." In *Conference of Structure and Properties of Dual-Phase Steels, Symposium at the AIME Annual Meeting*, 20–39. New Orleans.

Hell, J.C., M Dehmas, S. Allain, and et. al. 2011. "Microstructure-Properties Relationship in Carbide-Free Bainitic Steels." *ISIJ International* 51 (10): 1724-1732.

Hoydick, D. P., and D. M. Haezenbrouck. 2003. "Effect of Annealing Parameters on Structure and Properties of Intercritically Annealed Steels Intended for Hot-Dip Dual-Phase Application." In *45th MWSP Conf.*, XLI:421–35.

Huang, J., W. Poole, and M. Militzer. 2004. "Austenite Formation during Intercritical Annealing." *Metallurgical and Materials Transactions A* 35 (11): 3363–75..

Jun, H.J., K.B. Kang, and C.G. Park. 2003. "Effect of Cooling Rate and Isothermal Holding on the Precipitation Behavior during Continuous Casting of the Nb-Ti Bearing HSLA Steels." *Scripta Materialia* 49: 1081–86.

Kakuta, K., and N. Takahashi. 1982. "Technology of Nippon Steel's Continuous Annealing and Processing Line (C.A.P.L.)." In *Symp. of The Metal Soc. of AIME*, 373–77. Metal/Society of AIME.

Kato, T., K. Hashiguchi, I. Takahashi, T. Irie, and N. Ohashi. 1981. "Development of as-Hot Rolled Dual-Phase Steel Sheet." In *Fundamentals of Dual-Phase Steels*, 199–220.

Kogan, L. I., G. M. Murashko, and R.I.. Entin. 1984. "Redistribution of Elements during Austenitization in the Intercritical Temperature Range." *Phys. Met. Metallography (USSR)* 58 (1): 130–35.

Koo, J.Y., R. V. N. RAO, and G. Thomas. 1979. "Designing High Performance Steels With Dual Phase Structures." *Metal Progress* 116 (4): 66–70.

Koo, J.Y., and G. Thomas. 1977. "Design of Duplex Fe/X/0.1C Steels for Improved Mechanical Properties." *Metallurgical Transactions A* 8 (3): 525–28.

———. 1979. "Design of Duplex Low Carbon Steels for Improved Strength: Weight Application." *Proc. of Symp of Met. Society of AIME " Formable HSLA Dual-Phase Steels*, 40–55.

Lagneborg, R. 1978. "Structure-Property Relationship in Dual-Phase Steels." In *Dual-Phase and Cold Pressing Vanadium Steels, Seminar of Vanitec*, 43–52.

Law, N.C.a, and D.V.b Edmonds. 1980. "The Formation of Austenite in a Low-Alloy Steel." *Metallurgical and Materials Transactions A* 11 (1): 33–46.

Lenel, U. R., and R. W. K. Honeycombe. 1984. "Morphology and Crystallography of Austenite Formed during Intercritical Annealing." *Metal Science* 18 (11): 503–10.

Maalekian, M.a, R.b c Radis, M.a Militzer, A.d Moreau, and W.J.a Poole. 2012. "In Situ Measurement and Modelling of Austenite Grain Growth in a Ti/Nb Microalloyed Steel." *Acta Materialia* 60 (3): 1015–26.

Marder, A.R., and B.L. Bramfitt. 1981. "Processing of a Molybdenum-Bearing Dual-Phase Steel." In *Conference of Structure and Properties of Dual-Phase Steels, Symposium at the AIME Annual Meeting*, 242–59. New Orleans, LA, USA: Metallurgical Society of AIME, Warrendale, Pa.

Matlock, David K., G. Krauss, L. F. Romas, and G. S. Huppi. 1981. "A Correlation of Processing Variables with Deformation Behavior of Dual-Phase Steels." In *Conference of Structure and Properties of Dual-Phase Steels, Symposium at the AIME Annual Meeting*, 62–90. New Orleans, LA, USA: Metallurgical Society of AIME, Warrendale, Pa.

Messien, Pierre, Jean Claude Herman, and Tony Greday. 1981. "Phase Transformation and Microstructures of Intercritically Annealed Dual-Phase Steels." In *Conference of Fundamentals of Dual-Phase Steels, Symposium at the 110th AIME Annual Meeting*, 161–80. Chicago, IL, USA: Metallurgical Society of AIME, Warrendale, Pa, USA.

Meyer, L., F. Heisterkamp, and W. Mueshenborn. 1975. "Columbium, Titanium and Vanadium in Normalized, Thermomechanically Treated and Cold-Rolled Steels." In *Microalloying-1975*, 153–67. Pittsburgh.

Miller, M.K., and G.D.W. Smith. 1977. "Atom Probe Microanalysis of a Pearlitic Steel." *Materials Science* 11 (7): 249–53.

Mohanty, R.R., and O.A. Girina. 2012. "Effect of Coiling Temperature on Kinetics of Austenite Formation in Cold Rolled Advanced High Strength Steels." *Materials Science Forum* 706-709: 2112–17.

Mohanty, R.R., O.A. Girina, and N.M. Fonstein. 2011. "Effect of Heating Rate on the Austenite Formation in Low-Carbon High-Strength Steels Annealed in the Intercritical Region." *Metallurgical and Materials Transactions A* 42 (12): 3680–90. doi:10.1007/s11661-011-0753-5.

Mohrbacher, H, T. Schulz, V. Flaxa, and M. Pohl. 2011. "Low Carbon Microalloyed Cold Rolled Hot Dip Galvanized Dual Phase Steel for Larger Cross-Sectional Areas with Improved Properties." In *Galvatech'11 6th International Conference on Zinc and Zinc Alloy Coated Steel Sheet - Conference Proceedings*. Genova.

Mould, R. R., and C. C. Skena. 1977. "Structure and Properties of Cold Rolled Ferrite -Martensite *dual Phase) Sheet Steels." In *Formable HSLA and Dual-Phase Steels, Metallurgical Society of AIME*, 181–204. New York: Metallurgical Society of AIME.

Nakaoka, K., K. Araki, and K. Kurihara. 1979. "Strength, Ductility and Aging Properties of Continuously-Annealed Dual-Phase High-Strength Sheet Steels." In *Formable HSLA and Dual-Phase Steels, Metallurgical Society of AIME*, 126–41. Metallurgical Society of AIME.

Nakaoka, K., Y. Hosoya, M Ohmura, and A. Nishimoto. 1981. "Reassessment of the Water-Quenching Process as a Means of Producing Formable Steel Sheets." In *Conference of Structure and Properties of Dual-Phase Steels, Symposium at the AIME Annual Meeting*, 330–45.

Nishimoto, A., Y. Hosoya, and K. Nakaoka. 1981. "A New Type of Dual Phase Steel Sheet for Automobile." *Transactions of ISIJ* 21 (11): 778–82.

Ogawa, T.a, N.b Maruyama, N.c Sugiura, and N.c Yoshinaga. 2010. "Incomplete Recrystallization and Subsequent Microstructural Evolution during Intercritical Annealing in Cold-Rolled Low Carbon Steels." *ISIJ International* 50 (3): 469–75.

Oliveira, F.L.G., M.S. Andrade, and A.B. Cota. 2007. "Kinetics of Austenite Formation during Continuous Heating in a Low Carbon Steel." *Materials Characterization* 58 (3): 256–61.

Patil, R., O. Girina, and D. Bhattacharya. 2004. "Development of a Dual-Phase High Strength Galvanized Steel Using Zinquench Technolog®." In *Galvatech '04: 6th International Conference on Zinc and Zinc Alloy Coated Steel Sheet -* 439–47. Chicago, IL.

Pawlovski, B. 2011. "Critical Points in Hypoeutectoid Steel - Prediction of the Pearlite Dissolution Finish Temperature Ac1F." *Journal of Achievement In materials and Manufacturing Engineering* 40 (2): 331–37.

Pichler, A, S. Traint, and G Arnoldner. 2000. "Phase Transformation during Annealing of a Cold-Rolled Dual-Phase Steel Grade." In *Proceedings of 42nd MWSP Conf*, XXXVIII: 573–93.

Pradhan, R. 1984. "Continuously Annealed Cold-Rolled Microalloyed Steels with Different Microstructures." In *International Conference on Technology and Applications of HSLA Steels*, 193–201. Philadelphia, PA, USA: ASM, Metals Park, OH, USA.

Radtke, S, P. Sippola, and S Makimattila. 1986. "ZINQUECH a New Process for Coated Steels." *Iron and Steel Engineer* 63 (6): 32–35.

Rashid, M.S., and V.N. Rao. 1982. "Tempering Characteristics of a Vanadium Containing Dual Phase Steel." *Metallurgical Transactions* 13A (10): 1679–86.

Rigsbee, J.M., J.K. Abraham, A.T. Davenport, J.E. Franklin, and J.W. Pickens. 1981. "Structure-Processing and Structure-Property Relationships in Commercially Processed Dual-Phase Steels." In *Conference of Structure and Properties of Dual-Phase Steels, 304-329,* New Orleans, LA, USA.

Sadovsky, V. D. 1973. "Structure Inheritance in Steels." *M. Metallurgia*, 205.

Savran, V.I., Y. Leeuwen, D.N. Hanlon, C. Kwakernaak, W.G. Sloof, and J. Sietsma. 2007. "Microstructural Features of Austenite Formation in C35 and C45 Alloys." *Metallurgical and Materials Transactions A* 38 (5): 946–55.

Shirasawa, Hidenori, and J.G. Thomson. 1987. "Effect of Hot Band Microstructure on Strength and Ductility of Cold Rolled Dual Phase Steel." *Transactions of the Iron and Steel Institute of Japan* 27 (5): 360–65.

Shtansky, D.V.a, K.b Nakai, and Y.b Ohmori. 1999. "Pearlite to Austenite Transformation in an Fe-2.6Cr-1C Alloy." *Acta Materialia* 47 (9): 2619–32.

Singh, J., and C. M. Wayman. 1987. "Reverse Martensite Transformation in Iron-Nickel Manganese Alloy." *Metallography* 20 (4): 485–90.

Song, R, N. Fonstein, H. J. Jun, D.B. Bhattacharya, S. Jansto, and Pottore N. 2014. "Effects of Nb on Microstructural Evolution and Mechanical Properties of Low-Carbon Cold-Rolled Dual-Phase Steels." *Metallography, Microstructure and Analysis* 3: 174–84.

Souza, M., J. Guimarães, and K. Chawla. 1982. "Intercritical Austenitization of Two Fe-Mn-C Steels." *Metallurgical and Materials Transactions A* 13 (4): 575–79.

Speich, G., V. Demarest, and R. Miller. 1981. "Formation of Austenite During Intercritical Annealing of Dual-Phase Steels." *Metallurgical and Materials Transactions A* 12 (8): 1419–28.

Speich, G.R. 1981. "Physical Metallurgy of Dual-Phase Steels." In *Fundamentals of Dual-Phase Steels*, 3–46. Chicago, IL, USA.

Speich, G.R., and R.L. Miller. 1981. "Mechanical Properties of Ferrite-Martensite Steels." In *Conference of Structure and Properties of Dual-Phase Steels, Symposium at the AIME Annual Meeting*, 145–82. New Orleans, LA, USA: Metallurgical Society of AIME, Warrendale, Pa.

Speich G.R., A.J. Schwoeble, and G.P. Hoffman 1983. "Tempering of Manganese and Mn-Si-V Steels." *Metallurgical Transactions* 14 (6): 1079–87.

Stevenson, R., D.J. Bailey, and G. Thomas. 1979. "High Strength Low Carbon Sheet Steel by Thermomechanical treatment: II. Microstructure." *Metallurgical Transactions A* 10 (1): 57–62.

Sugimoto, K.-I., J. Sakaguchi, T. Ida, and T. Kashima. 2000. "Stretch-Flangeability of a High-Strength TRIP Type Bainitic Sheet Steel." *ISIJ International* 40 (9): 920–026.

Surovtsev, A. P., V. V. Yarovoy, and V.E.. Sukhanov. 1986. "Kinetics of Polymorphous Transformation during Heating of Low Carbon Steels." *Metal Science and Heat Treatment (USSR)* 28 (1–2).

Thomas, G., and J.Y. Koo. 1981. "Developments in Strong, Ductile Duplex Ferritic - Martensitic Steels." In *Conference of Structure and Properties of Dual-Phase Steels, Symposium at the AIME Annual Meeting*, 183–201. New Orleans, LA, USA.

Toji, Y.a, T.b Yamashita, K.c Nakajima, K.b Okuda, H.b Matsuda, K.a Hasegawa, and K.b Seto. 2011. "Effect of Mn Partitioning during Intercritical Annealing on Following $\gamma \rightarrow \alpha$ Transformation and Resultant Mechanical Properties of Cold-Rolled Dual Phase Steels." *ISIJ International* 51 (5): 818–25.

Waterschoot, T., K. Verbeken, and B.C. De Cooman. 2006. "Tempering Kinetics of the Martensitic Phase in DP Steel." *ISIJ International* 46 (1): 138–46.

Wycliffe, P., G.R. Purdy, and J.D. Embury. 1981. "Austenite Growth in the Intercritical Annealing of Ternary and Quaternary Dual-Phase Steels." In *Conference of Fundamentals of Dual-Phase Steels, Symposium at the 110th AIME Annual Meeting*, 59–83. Chicago, IL, USA: Metallurgical Society of AIME, Warrendale, Pa, USA.

Yang, D.Z.a, E.L.b Brown, D.K.b Matlock, and G.b Krauss. 1985. "Ferrite Recrystallization and Austenite Formation in Cold-Rolled Intercritically Annealed Steel." *Metallurgical Transactions A* 16 (8): 1385–92.

Yi, J.J.a, I.S.b Kim, and H.S.c Choi. 1985. "Austenitization during Intercritical Annealing of an Fe-C-Si-Mn Dual-Phase Steel." *Metallurgical Transactions A* 16 (7): 1237–45.

Chapter 3
Effect of Structure on Mechanical Properties of Dual-Phase Steels

Contents

© Springer International Publishing Switzerland 2015

N. Fonstein, *Advanced High Strength Sheet Steels*,
DOI 10.1007/978-3-319-19165-2_3

Chapter 2 outlined the general principles of heat treatment of low-carbon low-alloyed steels from intercritical temperature region that allows producing a microstructure containing specified amount of martensite. Local $\gamma \rightarrow \alpha'$ transformation of austenite islands within ferritic matrix determines the main microstructural features of "dual-phase" ferrite–martensite steels. Practical interest in these steels is related to the potential for achieving combinations of high strength and ductility making these steels extremely suitable for cold forming processes.

Improved ductility at the same or even higher strength compared to those of normalized or hot-rolled steels is not an inherent feature of every steel heat treated in the $(\alpha + \gamma)$ region. To optimize mechanical properties, i.e., to maximize uniform and total elongation or to control the Yield Strength/Tensile Strength ratio, the composition of steel and heat treatment processing should ensure appropriate proportions of microstructure constituents with particular chemistry and morphology.

Whereas the role of chemical composition is discussed in Chap. 4, the present chapter summarizes the experimental and theoretical data regarding the relationships between the parameters of microstructure and mechanical properties of dual-phase steels.

It should be emphasized that numerous studies of dual-phase steels employ different processing parameters (hot rolling finishing and coiling temperatures, cold reduction, temperature and holding time of annealing, heating/cooling rates, etc.) that makes it difficult to compare and summarize the existing data. On top of that, a broad variety of geometries and dimensions of specimens for mechanical testing reported in the literature further exacerbate the analysis of the data.

Therefore, the test results and conclusions made in different studies should be regarded primarily as general guidelines describing the established trends, which may be quantitatively meaningful only if all test conditions are known.

3.1   Existing Models of the Tensile Behavior of Heterogeneous Materials

Heterogeneous alloys containing phases with substantially different mechanical properties can be regarded as natural composites, whose properties are determined by the properties of individual phases and their stereological parameters. Search for a dependable model describing the mechanical behavior of dual-phase steels is aimed at predicting the properties of two-phase microstructure based on the relevant data for ferrite and martensite. These predictions can be used to specify the

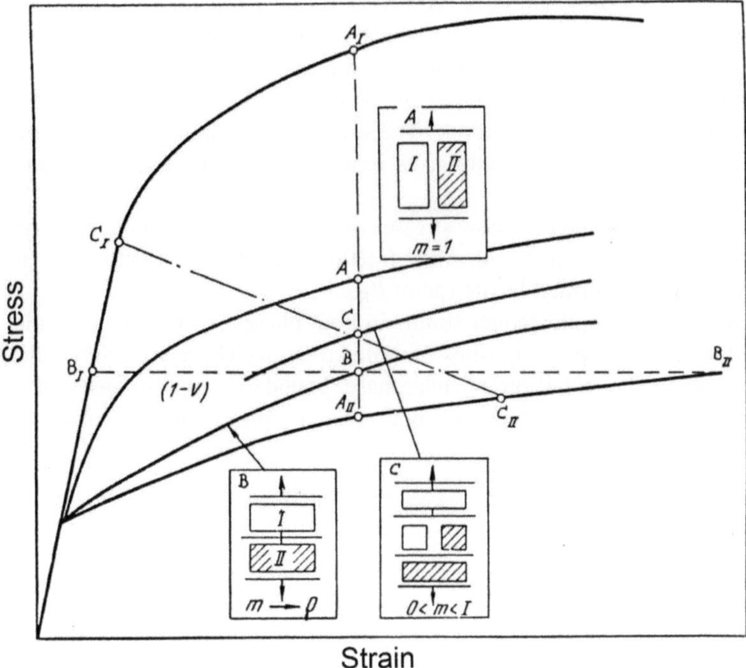

Fig. 3.1 Models for two-phase microstructure and the corresponding stress–strain curves (Tomota and Tamura 1982)

required proportions of phases and hence to prescribe the target steel composition and optimal heat treatment.

The rule of mixtures used for two-phase materials assumes

$$\sigma_c = \sigma_1 V_1 + \sigma_2 V_2 \tag{3.1}$$

where σ_c, σ_1, and σ_2 are the strengths of the composite and of each of the two phases, respectively, with the phase volume fractions $V_1 + V_2 = 1$. This equation is only valid if the flow stress of the composite at given strain ε, σ_ε^c, is the sum of flow stresses of individual phases.

If the strain–stress curves $\sigma(\varepsilon)$ for each phase are known, then, assuming that in two-phase mixture the same relationship of stress and strain remains, it can be possible to predict properties of a DP steel based on the parameter $m = \varepsilon_2/\varepsilon_1$ and using either model A or B or their combination C described in Fig. 3.1. Two extreme cases were considered (Tomota and Tamura 1982). In case A, the *strains* in phases I and II are equal, as in fibrous materials under loading, parallel to the fibers. This case pertains to the $A_I A_{II}$ line and to the resultant curve A for composite in whole. In case B, the *stresses* are equal, as in lamellar composites when the applied stress is perpendicular to the composite layers: line $B_I B_{II}$ and the resultant stress–strain curve B describe the behavior of the composite of this type.

For a random distribution of hard particles in a soft matrix, as in the case of dual-phase steels, an intermediate alternative mechanical behavior "C" would take place, where neither strains nor stresses are identical in the two phases but are distributed between the two phases in some way. In this case, the softer phase is more deformed than the harder one. The stresses in the harder phase are lower than the average stress of the mixture, as represented by the $C_I C_{II}$ line and by the resultant curve C.

The conditions when the flow stresses of individual phases in a dual-phase steel are identical can only met with small fraction ($B_{II}B/B_{II}B_I$) of extremely (unrealistically) heavily deformed ferrite (point B_{II}).

Model A that implies equal strains in both phases, so that $m = 1$, satisfies the conditions for fibrous composites of Mileiko theory (Mileiko 1969). This model also assumes the cohesion of interphase boundaries up to onset of necking and the stress–strain curves for both phases follow the Hollomon equation (Hollomon 1945)

$$\sigma = k\varepsilon^n \tag{3.2}$$

where σ and ε are the true stress and true strain, respectively, n is the strain-hardening exponent, and k is the constant. The Mileiko model gives an analytical description of plastic instability, i.e., the attainment of ultimate uniform strain in the composite as a function of volume fraction of hardening fibers and of the ratio of the maximum uniform strain to the tensile strength of the two phases.

Numerous attempts to utilize this model to describe the properties of dual-phase steels are grounded on an attractive possibility to predict maximum uniform strain UE or numerically equal to UE a strain-hardening exponent n.

In fact, Davies (1978a) used experimental values of UE $\approx n$ for ferrite and martensite under the assumption of their independence of phase strength. This way he calculated the UE of dual-phase steels as a function of their tensile strengths for various combinations of individual tensile strengths of ferrite and martensite. The results revealed that:

- The uniform elongation of dual-phase steel increases with increasing tensile strength of ferrite;
- The uniform elongation of steel increases with increasing tensile strength of martensite;
- At the same tensile strength of dual-phase steels, an increase in ferrite strength has a greater effect on the increase in ultimate uniform strain than does an increase in the martensite strength.

The latter conclusion might also be used as a criterion for optimizing the composition or heat treatment of dual-phase steels. Koo et al. (1980) refined Mileiko's model by including the strengthening effects of ferrite grain size L_F and carbon content in martensite C_M.

The Hall–Petch equation was used for ferrite

$$\sigma = \sigma_0 + K_y L_F^{-1/2} \tag{3.3}$$

Regarding the strength of ferrite, it was proposed also that it is controlled rather not by grain size but by the free path in ferrite matrix that is decreased at an increase in martensite fraction.

The equation for martensite strength was described as

$$\sigma = \sigma_0 + K' C_M \tag{3.4}$$

For example, for yield strength in accordance with Leslie and Sober (1967)

$$YS^M = 620 + 2585 C_M;$$

whereas for tensile strength data of Koo et al. (1980) indicates

$$TS^M = 770 + 3290 C_M.$$

Re-arranging the basic Eq. (3.1) gives

$$\begin{aligned} TS_{DP} &= TS_F V_F + TS_M V_M = TS_F(1 - V_M) + TS_M V_m \\ &= TS_F + (TS_M - TS_F) V_M \end{aligned} \tag{3.5}$$

Eventually, in spite of numerous modifications, it appeared that Mileiko's model could not be used for an adequate description of the properties of dual-phase steel because of the unequal strains in the ferrite and martensite, which have been revealed experimentally (Balliger and Gladman 1981; Korzakwa et al. 1980; Gurland 1982). Because of strain distribution between the phases, the stresses in the ferrite and the martensite in Eq. (3.1) now refer to different strains, e.g., in accordance with line $C_I C_{II}$ in Fig. 3.1 and then

$$\varepsilon_c = \varepsilon_1 \times V_1 + \varepsilon_2 \times V_2 \tag{3.6}$$

Martensite deformation is very small, but martensite experiences relatively high stresses and so provides the major portion of the load carrying capacity of the dual-phase steel.

The model of Bhadeshia and Edmonds (1980) takes into account the strain distribution between the phases on the basis of an ellipsoidal or spherical hard particle shape, which represents a better approximation to dual-phase steel structures than do fibers in the above model. Practical use of that model for predicting the mechanical behavior of dual-phase steels, however, is prevented by some uncertainty slope of the line C_{II}–C_I, in Fig. 3.1, as well as ratio of phase strains $\varepsilon_1/\varepsilon$. The latter appeared to be a function of the relative phase fraction and properties and to vary during deformation as a result of different work hardening response.

Tamura and Tomota (1973) as well as Araki (Araki et al. 1977b) have found that the strain ratio $m = \varepsilon_M/\varepsilon_F$, which is varying from $m = 0$ for equal stresses to 1 for equal strains, depends on ratio of hardness of phases and rapidly decreases when this ratio reaches 2.5.

In investigation provided by Speich and Miller (1981), who studied the group of dual-phase steels with various content of C and Mn, a satisfactory agreement with the Tomota–Tamura model was found using $m \approx 0.01$. In this case, at TS of DP steel (with UE $= 21$ %), the strain of martensite was approximately 0.2 %.

However, on the whole, it is worth to note that in addition to the difficulty of experimental determining m-value, the problem lies in the fact that the stress–strain curve of a single-phase may be different from that of the same phase in a two-phase structure (Fisher and Gurland 1981). In particular, strain hardening of soft phase in two-phase mixture is higher than in single-phase condition (Tomota and Tamura 1982). Another shortcoming of this approach is that it does not take into account residual stresses as a result of local $\gamma \to \alpha'$ transformation and during the process of deformation.

As was mentioned, both models of Mileiko and Araki (Araki et al. 1977b) assume perfect cohesion between the phases up to the onset of plastic instability. The deviations from this assumption are of paramount importance for fracture behavior (see Sect. 3.8). The observed origination of voids at the ferrite–martensite interface in dual-phase steels before martensite fracture suggests that in that moment, the ultimate tensile strength level had not been achieved for the hard phase in these steels.

In Uggowitzer–Stuwe study (Uggowitzer and Stuwe 1982), the authors have used, besides traditional parameters as V_M, such stereological factors as continuity of martensite units C^β and a free length of paths across ferrite $L_{\{\alpha\}}$. They measured YS for 0.5Mn–0.25Si–0.003/0.5 % C dual-phase steels as functions of V_M and k— the martensite/ferrite microhardness ratio, which were controlled by tempering temperature. In accordance with Lee and Gurland's findings (Lee and Gurland 1978), three regions of $\sigma_{0.2} = f(V_M)$ have been observed, which depended on the continuity of the phases, as presented in Fig. 3.2. Curve I was corresponding to discreet hard phase in continuous soft matrix, curve 2 was related to the soft matrix with continuous strengthening fibers, and curve 3 was considered for discreet soft phase within continuous hard phase.

The values of $m = \varepsilon_M/\varepsilon_F$ were evaluated by comparing the curves for dual-phase steels with theoretical curves, calculated by different arbitrary m-values from Tomota and Tamura's model, using stress–strain curves for model alloys, reproducing ferrite and martensite. It was found that m increases with increasing strain and generally is the higher, the greater the volume fraction of martensite, the lower the martensite/ferrite strength ratio, and the greater the continuity of the martensite regions. It was found that m varied from 5×10^{-4} to 1 for the V_M and k ranges 30–75 % and 2.5–4.2, respectively.

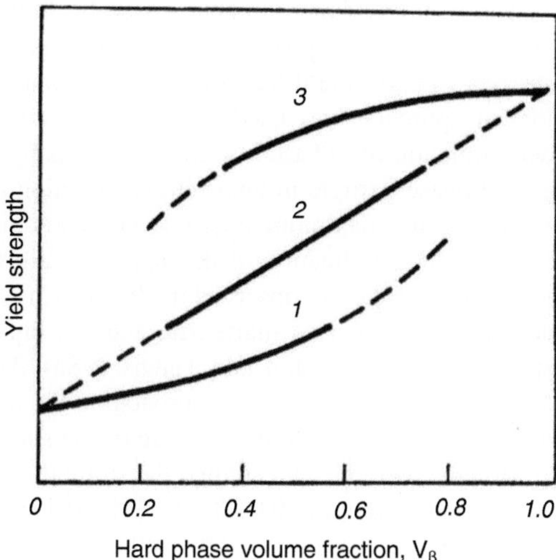

Fig. 3.2 Yield strength vs. hard-phase volume fraction in a composite depending on the tree types of structure (Lee and Gurland 1978)

In accordance with the simple law of mixtures, in a two-phase structure, the yield or the tensile strength of the ferrite–martensite mixture should vary from values typical for ferrite ($V_M = 0$) to those typical for martensite ($V_M = 1$), if they (properties of individual phases) are independent of V_M. It should be noted that irrespective of assuming equal strain or strain redistribution, the more elaborate models (Tamura and Tomota 1973; Araki et al. 1977b) also predicted a linear relation between the properties of composite and the amount of the hard phase. As shown by Tamura, deviations from the linear dependence YS (V_M) were observable at $YS^M/YS^F \geq 3$. In this context, the linear or nonlinear dependences of the tensile strength of dual-phase steels on the martensite content cannot be used either to support or reject any particular model.

The lack of direct strain distribution measurements in the ferrite and martensite of dual-phase steels in the process of deformation is another factor, which limits the judgments in favor of specific models. Drawing dual-phase steels from 5.6 to 2.8 mm diameter, which is equal to the strain of 400 %, has revealed that the ferrite matrix flows around the martensite particles (Korzakwa et al. 1980). This was interpreted as qualitative proof of substantial strain differences between the individual phases at a strain ratio, close to the one predicted by Speich and Miller (1981), i.e., 1:100 or $m \approx 0.01$.

Balliger and Gladman (1981) made an attempt to evaluate quantitatively the strains in the individual phases using metallography. They used the ratio of mean chords for a given phase both parallel and perpendicular to the tensile axis as a function of the specimen deformation. Measurements were taken on longitudinal sections through a broken tensile specimen to ensure a wide strain range around the necking area. The major results of this study had shown that there was a considerable delay in the initial deformation of martensite relative to the specimen as a whole and that m varied during deformation.

The method of measurements used by Balliger and Gladman could not distinguish the true hard-phase stretching and its rotation in the direction of external stresses and so should lead to an overestimated martensite strain. Another stereological approach was used by Szewczyk and Gurland (1982) and involved the determination of m based on l/w variations by measuring the maximum length of a hard-phase particle in an arbitrary direction l (or an equivalent ellipse diameter D_{2B}) and the maximum chord w perpendicular to the maximum length, when $l/w = D_{2B}/w$. Application of this approach allowed to separate the effects of martensite grain rotation, observed in alloy of Cu–9.6 % Al after 5 % deformation, and actual deformation of martensite starting only after 15 % total deformation of specimen (Yegneswaran and Tangri 1983). The strain of martensite determined as $\ln((D_{2B}/w)_0/(D_{2B}/w)_f)$ was approaching the average strain of the alloy only near ultimate strain, close to the fracture of the specimen.

Bortsov and Fonstein studied the strain distribution between the phases of dual-phase steel containing 0.072C–1.38Mn–0.49Si–0.71Cr–0.003B. Dual-phase structure of samples was obtained by isothermal holding at 700 °C after cooling from the fully austenitized condition (Bortsov and Fonstein 1986). To investigate the deformation behavior of the phases at heavy deformations, heat-treated samples were cold rolled from 2.5 mm to thicknesses of 1.5, 0.9, 0.6, and 0.35 mm, which corresponded (with corrections for widening) to the true strains in the rolling direction of 0.42, 0.97, 1.31, and 1.81. The effect of martensite hardness on the strain distribution in the phases was evaluated by a comparative study of quenched samples with those tempered at 400 °C for 1 h prior to cold rolling. The martensite hardness values were 7680 MPa for the quenched and 5950 MPa for the quenched and tempered specimens.

Figure 3.3 shows microstructures of dual-phase steels with different martensite hardness after rolling up to a true strain of 1.81. The relation between the strain of the hard phase and its hardness is quite obvious.

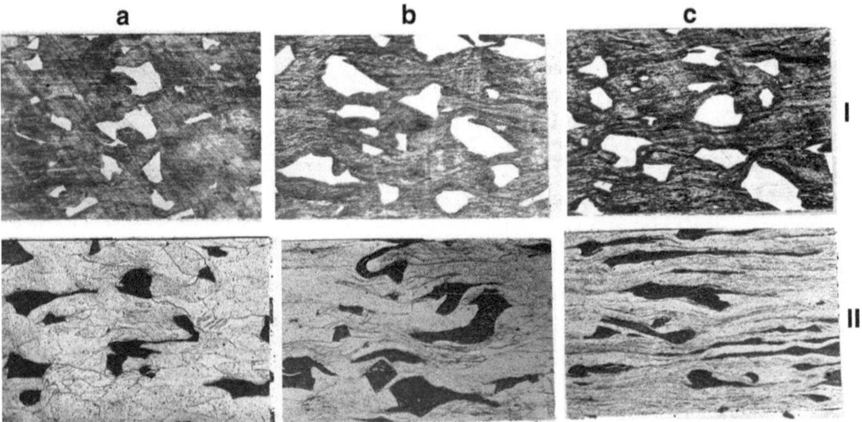

Fig. 3.3 Microstructure of dual-phase steels after true deformation $\varepsilon = 0.4$ (**a**), 1.0 (**b**), and 1.81 (**c**): I—non-tempered martensite (using La Pera etching); II—martensite tempered at 400 °C (etched by Nital) OM, ×200—original

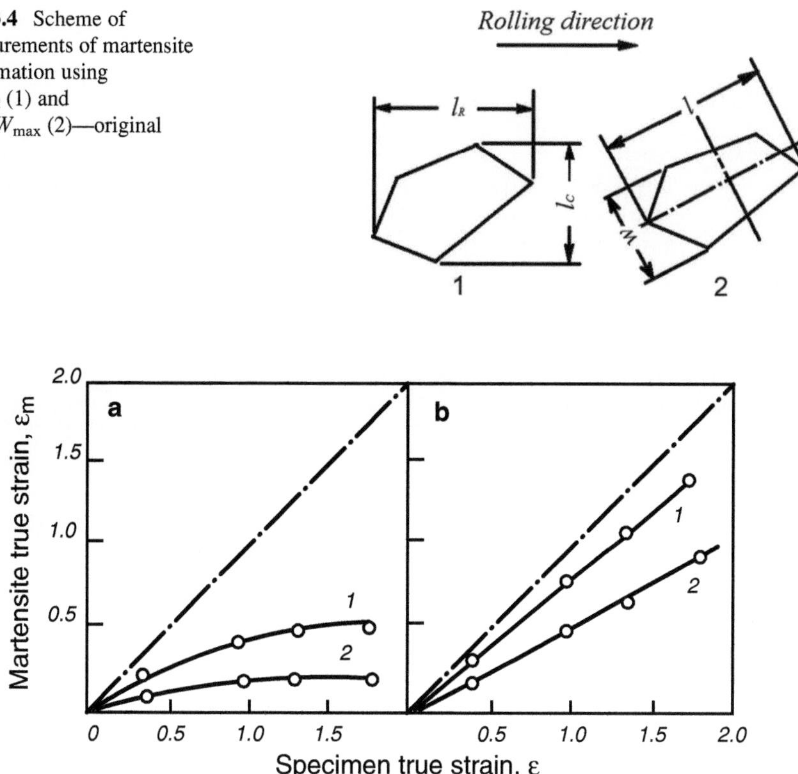

Fig. 3.4 Scheme of measurements of martensite deformation using L_R/L_C (1) and L_{max}/W_{max} (2)—original

Fig. 3.5 Relationship of deformation of non-tempered (**a**) and tempered (**b**) martensite and the total deformation; 1—calculations without deduction of rotation; 2—calculations taking into account the contribution of martensite grain rotation—(Bortsov and Fonstein 1986)

Quantitative measurements of the strains in the martensite were made using image analyzer and software developed based on the method proposed by Gurland (Fig. 3.4).

Mean values of l/w ratio were measured at different strain, assuming the shape of martensite particles close to ellipse so the ratio of the big axis of ellipse, l, to its small axis, w, can be calculated based on formulas for ellipse area, A, and its perimeter, P.

Figure 3.5 shows strain measurements in the martensite as a function of specimen deformation. The dashed line is corresponding to the condition of equal strains in the ferrite and the martensite when $m = \varepsilon_M/\varepsilon_F = 1$.

As shown, for the dual-phase steel, tempered at 400 °C, the martensite strain ε_M increases progressively with the total strain of specimens ε, when ε_M being approximately 0.5–0.7 of ε. For untempered martensite at $\varepsilon < 1$, the martensite strain is maximum 0.2 of the total strain. Accurate measurements could only be made at $\varepsilon \geq 0.4$, with the initial m being of 0.11 for untempered martensite.

Table 3.1 Hardening of the individual structure constituents in steel of 07CrMnSiB during rolling—original

Ferrite–pearlite structure				Ferrite–martensite structure			
Axial strain (%)	H_{50}^{F} N/mm^2	H_{50}^{P}	K	Axial strain (%)	H_{50}^{F} N/mm^2	H_{50}^{M}	K
0	1890 ± 50	3210 ± 100	1.7	0	1970 ± 60	7800 ± 200	3.96
19	2570 ± 60	3300 ± 120	1.28	15	2850 ± 80	7890 ± 200	2.77
283	3050 ± 150	3300 ± 150	1.08	75	3180 ± 150	8000 ± 230	2.52

Note: $K = H^2/H^F$, where H^2 is the hardness of second phase (pearlite or martensite)

Fig. 3.6 Fracture of martensite grain (indicated by *arrow*) at heavy deformation OM ×200—original

The results suggest that the strain distribution in dual-phase steels depends on the relation between the properties of the phases, which vary, in particular, during the deformation, as had estimated in Table 3.1 by ratio of hardness of "second" phase and ferrite during strain. Ferrite–pearlite and ferrite–martensite steels are compared.

As presented in Fig. 3.5, after strain $\varepsilon > 1$, ε_M of untempered martensite no longer changes with the total strain. This can be interpreted in terms of the achievement of the maximum possible strain in the martensite resulting in its fracture, as was observed in microstructure (Fig. 3.6).

In conclusion, it is necessary to note that the discrepancy between the predicted stress–strain curves according to different models, in particular, between determinations of the maximum uniform strain ε_u, is not significant but becomes more conspicuous with increasing ratio of phase strengths or hardnesses.

An important limitation of most models and numerous experimental studies, aiming to establish the relationship between mechanical properties and structure in dual-phase steels, is the neglect of such stereological parameters of dual-phase microstructures as martensite particle size, shape, and continuity. Stereological approach to structure of DP steel was generalized, in particular, by Fonstein et al. (1984).

As emphasized by many researchers, models for the mechanical behavior of dual-phase steels should take into account not only the behavior of a mechanical continuum but also an increase in density of dislocations and their movement affected by such parameters as martensite volume fraction, martensite size, and dislocation free paths in ferrite. Thus, theoretical models of mechanical behavior of two-phase microstructures need to be further refined, and structure optimization so far is based mostly on empirical relations and experimental data presented, in particular, below.

3.2 Strength Characteristics of Dual-Phase Steels

Various mechanisms can contribute individually or jointly to the strength of dual-phase steels: strengthening by hard martensite particles, solid solution strengthening of martensite by carbon, solid solution strengthening of ferrite, strengthening of ferrite by grain boundaries, precipitation hardening of ferrite, as well as its quenching and strain aging.

These hardening mechanisms are not additive in dual-phase steel. Moreover, they evolve differently during deformation because:

1. It was shown that the strength of dual-phase steels is essentially controlled by distribution of strains and stresses between ferrite and martensite. Therefore, the actual hardening role of martensite should depend on all factors that can change its distribution: the ratio of martensite and ferrite strengths, phase continuity, and the total deformation of steel.
2. The intrinsic properties of ferrite and martensite in a two-phase mixture can differ from those in the single-phase condition. Some researchers note, for example, that the strength of martensite in dual-phase steels is lower than that in through-hardened steel with the same carbon content (Speich and Miller 1981; Davies 1977; Koo et al. 1980). On the other hand, depending on the volume fraction of martensite, the strength of ferrite in dual-phase steels can be higher than that of ferritic steels of the same composition (Koo et al. 1980; Ma et al. 1983) being dictated in particular by the spacing between martensite islands.
3. Properties of dual-phase steels, especially at low stains, can be significantly affected by residual tensile stresses in ferrite generated by expansion of adjacent volumes during transformation of austenite to martensite. The magnitude of residual stresses seems to depend on the volume fraction of martensite and its carbon content that controls the M_s temperature, as well as on the cooling rate from M_s to room temperature, which determines the possible extent of relaxation of the generated stresses (Hansen and Pradhan 1981).

It is worth to note that the attempts to analyze and summarize the experimental data are further complicated because changes in microstructure or conditions of the intercritical heat treatment usually induce *simultaneous* changes in *several* microstructural parameters and phase properties. For example, an increase in the volume fraction of martensite, V_M, by raising the intercritical temperature reduces both the strength of martensite by lowering C_M and the strength of ferrite due to lower concentrations of interstitials, as prescribed by the PG and BC lines, respectively in Fe–C and Fe–N diagrams. Precipitation hardening of ferrite by carbides is accompanied by a decrease in the volume fraction of martensite and/or by lowering its carbon content as carbon becomes partially tied up in carbides. Consequently, it is often difficult or impossible to separate contributions from changes in individual phase to strength or ductility of DP steels. For example, the refinements of ferrite grains and martensite particles typically occur simultaneously. In this context, the experimental results presented below mostly outline the general trends in mechanical properties of dual-phase steels depending on the parameters of their microstructure.

3.2.1   Resistance of Dual-Phase Microstructure to Microplastic Deformation

Because of the difference in the specific volumes of γ-phase V_γ and martensite V_M, the martensitic transformation of austenite in $\alpha + \gamma$ mixture induces local volume expansion of prior γ-phase islands by $\varepsilon = (V_M - V_\gamma)/V_\gamma$. The value of ε is proportional to C_M, and at $C_M = 0.6\ \%$, the local volume deformation can reach 2.5–4 %. As a result, the ferrite grains adjacent to the martensite islands deform plastically generating significant number of dislocations.

These dislocations are mobile because relatively low temperature of martensite transformation and high cooling rate provide little possibility for their pinning by interstitials. This is confirmed, in particular, by strain dependence of internal friction of DP steels at low strain amplitude, Fig. 3.7.

As a result, the onset of microplastic deformation in dual-phase steels is controlled by the significantly lower lattice resistance to movement of free dislocations than the stress needed to unpin the dislocation. Another important factor that facilitates unpinning of dislocations or reduces the stress required for their glide is high level of residual tensile stresses which are additive to the applied stress (Tanaka et al. 1979; Stevenson et al. 1979; Hashiguchi and Nashida 1980; Araki et al. 1977a). The effect of residual stresses in dual-phase steels is generally interpreted in terms of the local Bauschinger effect as

$$\sigma_\varepsilon = \sigma_\varepsilon' - V \times \sigma_R \qquad (3.7)$$

where σ_R is the residual stress, and V is the volume fraction of material where this stress exists. The residual stresses and their evolution during deformation cause the "rounding" of σ–ε curve of DP steels and reduce the magnitude of external stresses

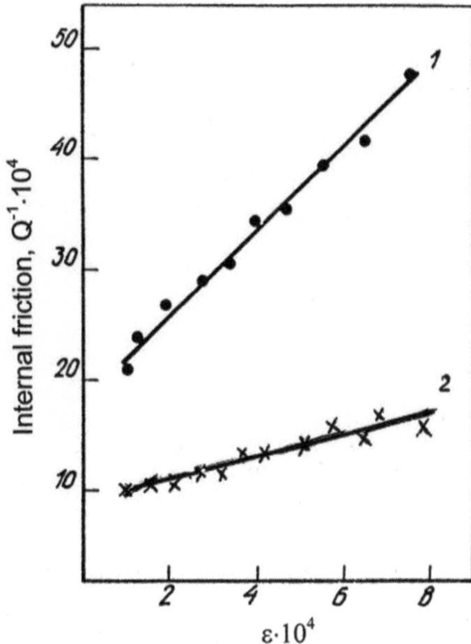

Fig. 3.7 Strain dependence of internal friction (measured at 3 Hz) for water-quenched DP 0.05C–1Mn–0.3Cr steels: 1—after water quenching, 2—after tempering at 400 °C—original

required for small plastic strains of up to 2–5 % (Romaniv et al. 1977; Shieh 1974; Tseng and Vitovec 1981; Gerbase et al. 1979)

The Eshelby model (1957) further developed by Brown and Clarke (1975) gives the value of hydrostatic residual stress at lattice deformation of ε as

$$\sigma = G \, \frac{4(1+\nu)}{5(1+\nu)} \, \varepsilon V_{M} \tag{3.8}$$

where ν is the Poisson ratio (≈ 0.29). Thus, with $V_{M} \approx 0.2$ and linear strain of $\varepsilon \approx 0.008$, the residual stress becomes

$$\sigma = 0.023G \approx 190 \ \text{MPa} \tag{3.9}$$

Flow stresses induced in DP steels by microplastic deformations are very low. According to Davies (1979a; Bailey and Stevenson 1979), $\sigma_{0.001} \approx 100$ MPa. Experiments show (Fig. 3.8) that $\sigma_{0.001} = 145$ MPa, which is rather close to Davis' estimate and indicates that dislocations in DP steels begin to move at stresses equal to the lattice friction stresses. It is important to point out that $\sigma_{0.001}$ is independent of volume fraction of martensite V_{M}, which also agrees with Davies' data. As the strain increases, so does the slope of flow stress dependence on V_{M}, since the higher the value of plastic deformation the higher the contribution of strain hardening.

The level of internal stresses is a function of V_{M}, hardness, and the distribution of martensite, as well as of stress relaxation on cooling. If cooling is slow, the internal

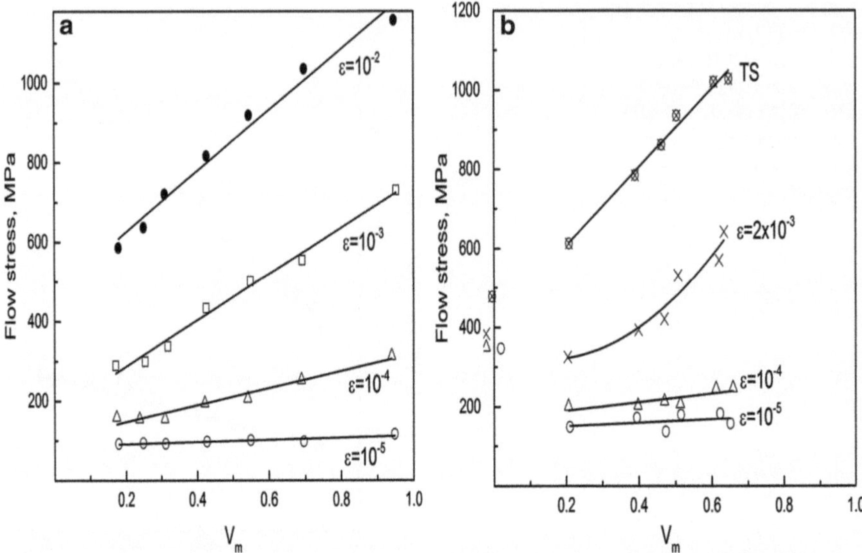

Fig. 3.8 Effect of volume fraction of martensite in DP steels on flow stresses at different strains: (a) 0.11C–1.34Mn–0.55Si–0.085V steel (Davies 1979a); (b) of 0.083C–1.61Mn–0.73Si steel (author)

stresses may be expected to decrease and so that the flow stress at given strain increases.

Thus, at strains of up to $\varepsilon \approx 0.2\,\%$, the flow stresses of DP steels with $V_M < 0.3$–0.4 are substantially lower not only with respect to ferrite–pearlite steels but also (by about 80 MPa) with respect to pure ferrite (Sarosiek and Owen 1983; Bortsov and Fonstein 1984a).

3.2.2 Yield Strength of Dual-Phase Steels

The absence of a physical discontinuous yield strength, i.e., the absence of yield point elongation (YPE), and low YS/TS ratio, which can be as low as 0.4–0.5, are believed to be intrinsic features of dual-phase steels. Low yield strength, similarly to low stress in microplastic deformation of DP steels, was attributed to higher density of mobile dislocations. This can be interpreted in terms of the well-known Johnston–Gilman model (1959). From the following equation:

$$\dot{\varepsilon} = \rho b v \tag{3.10}$$

where b is the Burger vector (constant), an increase in the density of mobile dislocations ρ at a constant strain rate $\acute{e}$ reduces their average velocity v. On the other hand,

$$v = \left(\frac{\tau}{\tau^*}\right) m \qquad (3.11)$$

where τ^* is the thermal component of shear stress, τ is the applied shear stress (reflecting the velocity of a single dislocation), and m is the dislocation mobility. Parameters τ^* and m can be regarded constant. Therefore, with lower v (velocity of a single dislocation) due to high dislocation density, lower stress τ (and YS) is required for dislocation glide at a given strain rate. Contributions of internal tensile residual stresses due to ferrite contraction during $\gamma - \alpha'$ transformation should act in the same direction (Stevenson et al. 1979; Nishimoto et al. 1981; Bailey and Stevenson 1979; Araki et al. 1977a; Tanaka et al. 1979)

It was shown experimentally that the volume fraction of martensite of some 3 % is sufficient to achieve zero yield point elongation in low-carbon DP steels (Hansen and Pradhan 1981). Depending on steel composition, this minimum amount of martensite necessary to suppress YPE should be a function of the M_s temperature and thus of the level of residual stresses. Also important is the amount of ferrite with higher dislocation density adjacent to martensite islands that can be correlated with the specific area of interphase boundaries. Marder and Bramfitt (1979) believe that the critical amount of martensite is 10–12 %, while Greday et al. (1979) state that it ranges from 10 to 30 %.

Sakaki et al. (1983) derived an analytical expression for minimum V_M to suppress YPE in dual-phase steels:

$$V_{\min} = k \frac{(1 - v)Y_0}{E\varepsilon} \qquad (3.12)$$

where ν is Poisson ratio, Y_0 is the yield strength of ferrite at the M_s temperature, which is controlled by its chemical composition, ferrite grain size, and concentration of solute carbon; ε is the strain in ferrite, determined by M_s (the $\gamma \rightarrow \alpha'$ transformation temperature).

In this respect, it is of interest to recall the results of study by Sudo and Tsukatani (1984), who showed that YPE is suppressed at 2 % of martensite in coarse-grain microstructure and at 8 % of martensite in fine-grain DP steels. This was attributed primarily to higher concentration of interstitials dissolved in ferrite of fine-grain dual-phase steel due to more intense dissolution of cementite and shorter austenite–ferrite diffusion paths and hence to a higher strength Y_0 of fine-grain ferrite.

It can be expected that YPE can return after tempering of DP steels due to relaxation of residual stresses and decrease in density of mobile dislocations. Takada et al. (1982) showed that low YS/TS ratio and the absence of YPE require the hardness of martensite H_M to be much higher than that of ferrite, H_F, $H_M/H_F \geq 2.2$.

The yield strength, YS, of DP steels reflects not only the resistance to dislocation motion but also some certain strain-hardening component in the range of small plastic strains beyond the elasticity limit. Measurements of flow stress in dual-phase steels as a function of volume fraction of martensite (Fig. 3.8) reveal increasing YS with V_M indicating that mobile dislocations near martensite–ferrite interface are the sources of new dislocations even at small strains (see Sect. 3.3).

3.2.2.1   Effect of Volume Fraction of Martensite

Systematic measurements of YS as a function of microstructural parameters in dual-phase steels were published by Uggowitzer and Stuwe (1982). The properties of martensite were varied by changing the temperature of tempering after quenching from constant temperature in the intercritical region, whereas the volume fraction of martensite V_M of up to 100 % was produced by changing the carbon content in steel. It should be noted, however, that this study dealt with neither dual-phase steels with V_M between 0 and 30 %, where YS usually exhibits minimum, nor non-tempered DP steels that should have higher density of mobile unpinned dislocations.

The experimental data obtained by Uggowitzer and Stuwe, as well as by other workers, who varied properties of martensite by varying the tempering temperature (Speich and Miller 1981; Tamura and Tomota 1973) demonstrate that yield strength of DP steels increases with V_M, and this increase is more pronounced when the microhardness (strength) of martensite is higher. These observations generally align with the simple law of mixtures.

After water quenching of 0.09C–0.45Si–1.17Mn–0.6Cr–0.37Mo–0.12 V steel from 790 °C, the volume fraction of 17 % of martensite was detected in microstructure. Approximate calculation assuming some loss of carbon to vanadium carbides gave the concentration of ~0.4 % C in martensite. In addition to this steel, two lab heats with 0.02 and 0.38 % C and similar alloy composition were studied after quenching from the same temperature to simulate properties of ferrite and martensite in the original two-phase mixture (Bortsov and Fonstein 1984a).

After substitution of the appropriate values into the law of mixture equation, the calculated yield strength value of 480 MPa appeared to be significantly higher than the experimental value of 305 MPa. Thus, YS of DP steels cannot be considered a linear (additive) combination of the properties of the two constituents. Due to pinning dislocation in ferrite and relaxation of residual stresses, the YS of DP steels increases after tempering regardless of the effects of tempering on YS of medium-carbon martensite (Table 3.2). As a result of tempering, the YS/TS ratio of dual-phase steels can increase from 0.45–0.5 to 0.65–0.7.

Figure 3.9 describes the effect of V_M on YS of cylindrical samples of Cr–Mn–Si–B steels with ~0.03 to ~0.20 % C quenched from different temperatures in the $\alpha + \gamma$ region. It was found (Bortsov and Fonstein 1984a), as in the earlier study (Davies 1978a), that the effect of V_M on YS can be described by a single curve independent on strength of martensite (its carbon content).

Table 3.2 Tensile properties of Mn–Cr–Mo–V steel grade with different microstructure and its individual constituents—original

Steel grade	Structure	YS (MPa)	UTS (MPa)	YS/TS ratio	Total elongation (%)	Uniform elongation (%)	Reduction of area (%)
10Cr1MoV	$F+P$	275	455	0.60	34	18	76
id	$F+M^a$	305	635	0.48	29	15	60
id	$(F+M)+LT$	390	610	0.64	33	13	70
02Cr1MoV	F	330	440	0.75	30	17	80
id	$F+LT$	300	410	0.73	33	19	84
40Cr1MoV	M''	1640	1900	0.89	9	4	54
id	$M''+LT$	1580	1850	0.85	12	4	61

LT—low tempered (250 °C, 1 h); F—simulated ferrite, M''—simulated martensite of $F+M$ structure

[a]As-quenched from 790 °C

Fig. 3.9 Effect of volume fraction of martensite on yield strength of Cr–Mn–Si–B steel with various carbon content quenched from 760 to 820 °C: *open circle*—0.07 % C; *open triangle*—0.10 % C; *open square*—0.18 % C—original

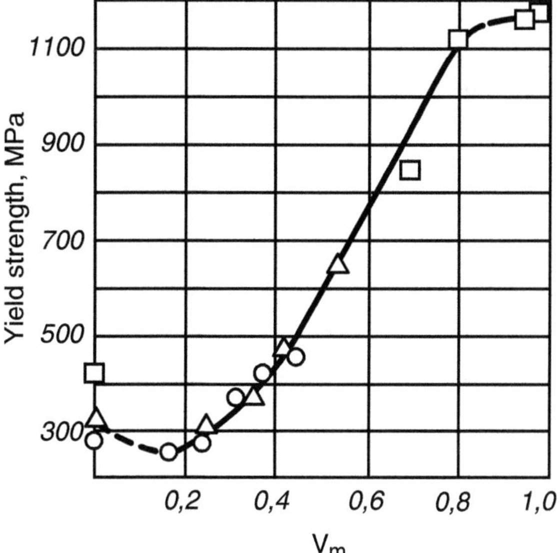

One of the possible reasons for insensitivity of YS of *untempered* dual-phase steels to strength of martensite is that at small offset strains (0.2 %) the martensite islands deform elastically so that the actual stresses in martensite are essentially independent on its strength (C_M).

As will be shown later, some conclusions related to the role of martensite strength and its carbon content can be changed depending on the total martensite volume and its connectivity. It should be noted that studies employed steels with DP steels of TS of up to 1200 MPa with volume fractions of martensite of up to 75 % cannot be directly compared with published data mostly related to DP steels with

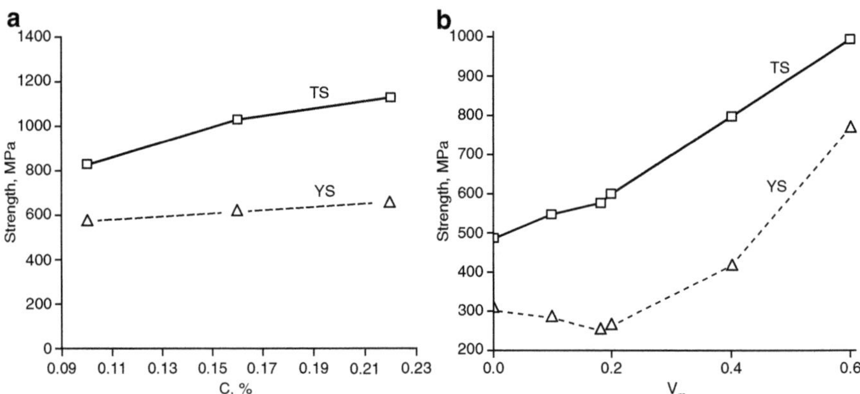

Fig. 3.10 Effect of carbon content on properties of quenched steels with 1.5Mn–1.35Al containing the same ~50 % martensite (**a**) and effect of martensite fraction on tensile properties of 0.1C–2.1Mn–0.7Si steel quenched from different temperature of intercritical region (**b**); all steels were tempered at 200 °C for 3 min

TS below 600–700 MPa and V_M lower than 30–35 %. At martensite volume above 0.45, the properties of martensite began to affect the dependence of tensile strength of Dp steels on martensite fraction. This difference in mechanical behavior is most likely because with bigger portions of martensite, the actual ability of steel to bear the loads is closer to that predicted from the law of mixture. Besides, with high volume fraction of martensite, the ferrite islands are frequently located inside the martensite matrix and therefore undergo a substantially more uniform transformation-induced hardening.

For example, as presented in Fig. 3.10a, the steels of 1.5Mn–1.35Al with different carbon content, annealed at different temperatures to get the same ~50 % austenite (martensite after water quenching) demonstrate the effect of carbon content in martensite and therefore its tensile strength (Fonstein 2005). As another example depicted in Fig. 3.10b, the effect of martensite fraction on YS of DP steels demonstrates the usual initial decrease in YS followed by its continuous increase in the range of $V_m > 0.2$ (Fonstein et al. 2007).

3.2.2.2   Effect of Ferrite Grain Size

Numerous publications concerning the effects of *ferrite grain size* on flow stress of DP steels can be found. Hall–Petch type equations were derived based on inverse proportionality of YS to dislocation free path in ferrite of dual-phase steels $L_{<\alpha>}$ that is related to V_M. Increase in k_y in Hall–Petch equation from 25 to 69 N/mm$^{3/2}$ for ferrite–martensite boundary with microhardness of martensite increasing from 4000 to 7000 MPa was attributed to strengthening the barrier for dislocation motion by phase interfaces (k_y is 19 N/mm$^{3/2}$ for ferrite/ferrite and 80 N/mm$^{3/2}$ for ferrite/ cementite).

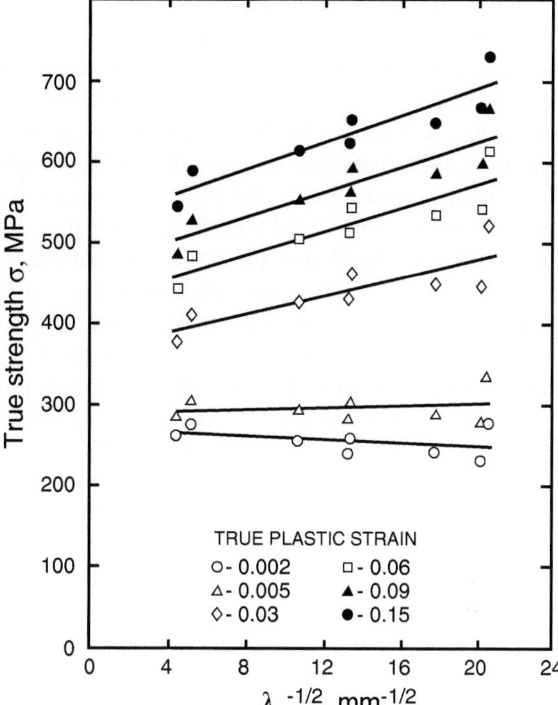

Fig. 3.11 Hall–Petch analysis of flow stress data, steel 0.08C–1.45Mn–0.2Si oil quenched after holding at 760 °C for 345 s (Budford et al. 1985)

The physical meaning of the Hall–Petch coefficient k_y for tempered DP steels differs from that for quenched steels because the variation in martensite strength during tempering is accompanied by pinning of dislocations in ferrite and by relaxation of residual stresses. These conditions define the usual meaning k_y as the intensity of stress necessary to generate new dislocations. In fact, the decrease in the strength of martensite by tempering (and therefore expected lowering $K_y^{\alpha\alpha}$) increases YS in dual-phase steels compared to the untempered condition (Rashid and Rao 1981; Davies 1981a).

As was found at investigation of untempered dual-phase 0.08C–1.45Mn–0.2Si steel (Budford et al. 1985), the Hall–Petch coefficient k_y strongly depends on true strain, increasing from about 0 close to yield strength with fast growth at strain increase (Fig. 3.11). The yield strength (at 0.2–0.5 % offset strain) is considered equal to the lattice friction stress, σ_0, before straining and is associated with rapid movement of unpinned dislocations, as well as with the contribution of residual tensile stresses generated in ferrite by martensitic transformation.

Qualitatively similar results were obtained later when studied 0.1C–0.15Si–1.5Mn–0.8 (Cr + Mo) steel with about 50–55 % of martensite (Tsipouridis et al. 2006). The authors also analyzed the strength difference in the DP steel with different grain size of ferrite at different levels of plastic strain. Although some increase in YS with grain refinement was found, $k_y = 9.88$ N/mm$^{3/2}$ was found to be remarkably smaller than the value of 17.4 N/mm$^{3/2}$ reported for ferritic steel. The authors explained their data in comparison with previous publication by lower stresses required to move dislocations due to existence of unlocked dislocations sources in quenched DP steels. Similar conclusions were made by Bleck et al., who found $k_y = 4$ N/mm$^{3/2}$ for YS of DP steels compared to 12 N/mm$^{3/2}$ for YS of Al-killed mild steels after temper rolling (Bleck and Phiu-On 2005).

At higher plastic strains, the grain size dependence of flow stresses in DP steels becomes more pronounced with k_y growing from 9.88 to 13.4 N/mm$^{3/2}$ that was attributed to the rapid increase in the work hardening of a finer-grained material.

In the study of DP grades from DP590 to DP980 with volume fractions of martensite ranging from 13 to 48 %, it was found that at 8 % strain, the Hall–Petch coefficient k_y for DP steels was seven times higher than that for single-phase ferrite (Bergstrom and Granbom 2008). This also points out the increasing contribution from higher strain hardening due to microstructure refinement in DP steels as discusses below. In contrast, strain hardening of *single-phase ferrite* should decrease with grain refinement (Morrison 1966).

It should be noted that it is quite difficult to separately control grain or particle sizes of ferrite and martensite. In fact, Jiang et al. (1995) found that when ferrite grain size decreased from 17–20 to 4–8 μm, the martensite particle size also decreased from 7–9 to 1.5–4.5 μm. And vice versa evidently, obtaining coarse ferrite grains in DP steels should be inevitably accompanied by coarsening of martensite grains, whereas the impact of the martensitic transformation on adjacent ferrite volumes should be different for coarse and fined microstructures.

The study of the effect of the ferrite grain size on the yield strength of dual-phase steels were carried out using two heats of 0.07C–1.2Mn–0.5Cr–0.003B steel with two levels of silicon content. The size of microstructural constituents in DP steel was varied by preparation of initial ferrite–pearlite structure using austenitization at different temperatures followed by isothermal holding at 650 °C and air cooling (Fig. 3.12,I).

Dual-phase microstructures after quenching of ferrite–pearlite-steels with different ferrite grain sizes from 760 °C are presented in Fig. 3.12,II. As shown in Fig. 3.13, the ratio of specific surface area of ferrite–ferrite interface and ferrite–pearlite interface for the initial ferrite–pearlite microstructure S_{ff}/S_{fp} is around 1.5–2.0, while the S_{ff}/S_{fm} ratio in the dual phase is significantly smaller: from 0.05 to 0.20. As noted in Chap. 2, this should be explained by preferential austenite growth along the ferrite boundaries and by the tendency of dual-phase steel to form a network type microstructure. As could be expected, the ferrite grain size L_F in DP steels was essentially the same as that for the original ferrite–pearlite microstructure. In spite of some appeared correlation of big martensite islands with sizes of pearlite colonies, the *average* martensite particle size $L_{<M>}$ showed little or no variation due to numerous tiny martensite particles near ferrite grains.

Figure 3.14 shows the effect of ferrite grain size on YS of steels with F–P and water-quenched F–M structure. Yield strength calculations as a function of ferrite grain size using the Hall–Petch equation had shown the k_y value in DP steels 1.6 N/mm$^{3/2}$ in contrast to ferrite–pearlite microstructure with $k_y = 17.7$ N/mm$^{3/2}$.

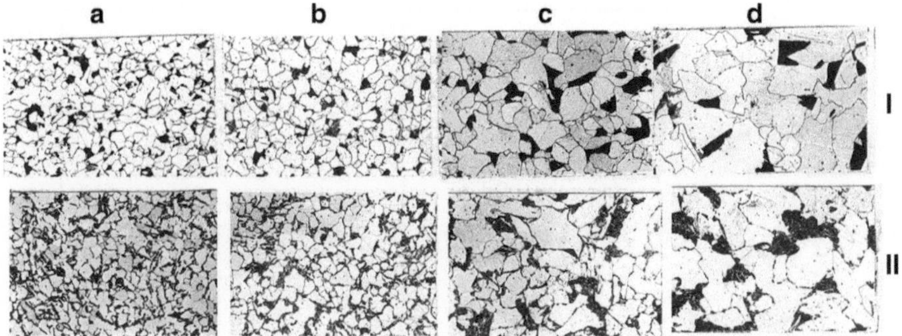

Fig. 3.12 Microstructure of 07CrMnSiB steel austenitized at 900 °C (**a**), 1000 °C (**b**), 1100 °C (**c**) and 1200 °C (**d**) followed by isothermal holding at 650 °C for 1 h to get ferrite–pearlite structure (I) and the corresponding F–M structure (II) obtained by annealing of initial ferrite–pearlite samples at 760 °C and water quenching; OM, ×200—original

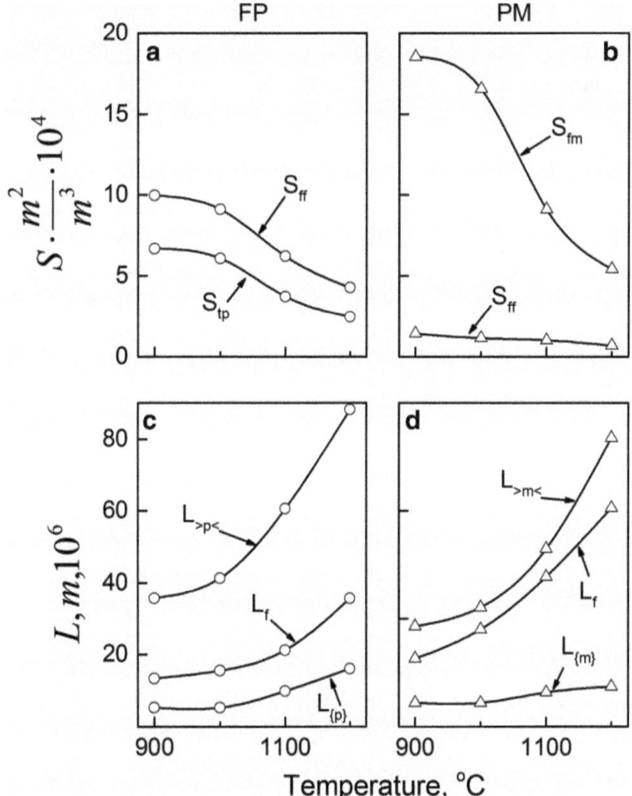

Fig. 3.13 Effect of temperature of preliminary austenitization on the specific interface areas and sizes of microstructure constituents: 07CrMnSiB steel with ferrite–pearlite (**a, c**) and ferrite–martensite microstructures (**b, d**)—original

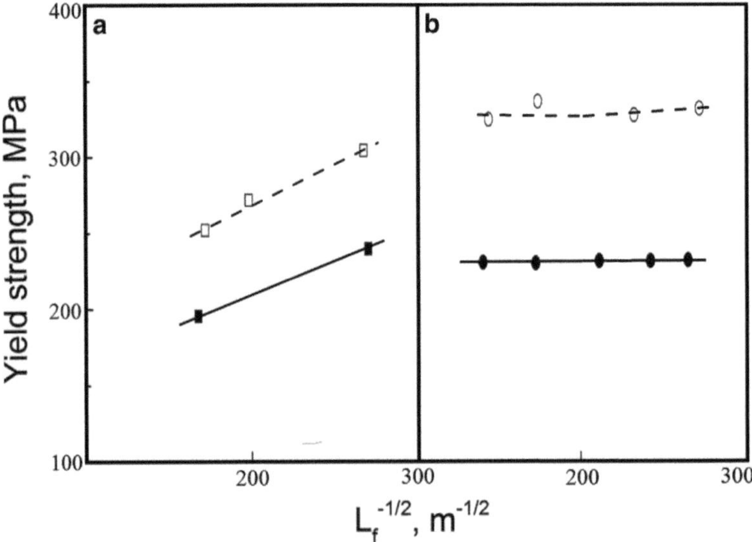

Fig. 3.14 Effect of ferrite grain size on yield strength of 07CrMnB (*solid lines*) and 07CrMnSiB (*dashed lines*) steels with ferrite–pearlite (**a**) and ferrite–martensite (**b**) microstructures—original

Thus *at constant* V_M, the yield strength of dual-phase steels was almost constant within the studied range of ferrite grain sizes L_F. This again confirms that ferrite grain boundaries in *untempered* DP steels do not serve as efficient barriers for grain-to-grain transfer of dislocation slip. This can be also attributed to the abundance of free dislocations in ferrite of quenched dual-phase steels near to interphase boundaries and to the existence of numerous new dislocations sources. The differences in the absolute levels of yield strength of steels with different silicon content are due to solid solution strengthening ferrite.

3.2.2.3   Effect of Conditions of Martensite Transformation

The efficiency of the existence of transformation-induced mobile dislocations and residual stresses accompanying the austenite–martensite transformation depend on multiple factors. First of all, these are the amount of the emerging martensite and the magnitude of volume changes during formation of martensite (concentration of carbon in austenite). Besides, the martensite transformation temperature M_s and the cooling rate near this affect the extent of martensite self-tempering after transformation. The design of experiments aiming to distinguish these factors is presented schematically in Fig. 3.15.

When a DP steel is produced by water quenching per thermal cycle in Fig. 3.15a, the major factor affecting the residual stresses is the temperature of annealing,

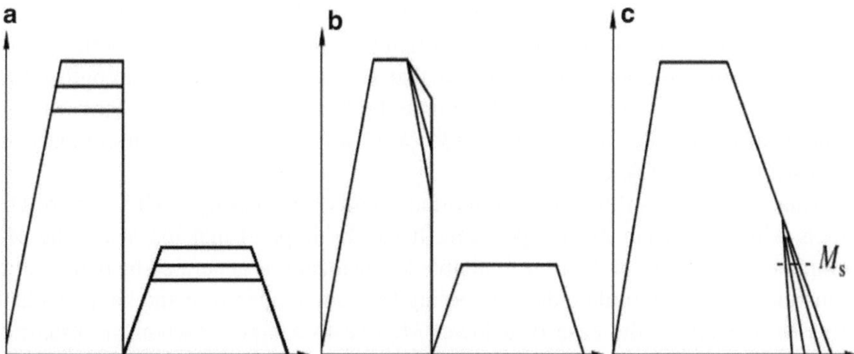

Fig. 3.15 Thermal cycles to obtain DP with (**a**) various volume of martensite, (**b**) various thermal stresses due to different water-quenching temperatures, (**c**) various cooling rate near the M_s temperature—original

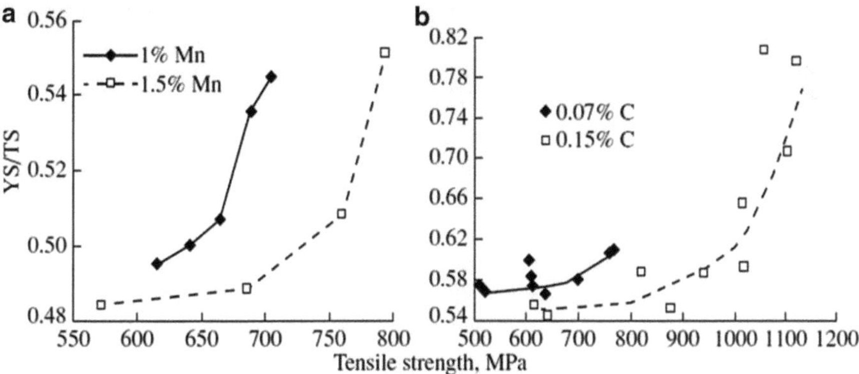

Fig. 3.16 Effect of (**a**) manganese in steel with ~0.08 % C and (**b**) carbon in steels with ~1.5 % Mn on the YS/TS ratio depending on strength of steel (different volume fractions of martensite obtained by water quenching from different temperatures) (Fonstein et al. 2007)

which determines the amount and properties of martensite, and the final tempering (see Sect. 3.5).

A large series of experiments was performed using steels with manganese content varying from 1 to 3 % and with two levels of carbon concentrations (0.07–0.08 and 0.15 %). These experiments allowed comparing the behavior of DP steels that are substantially different in the volume fraction of martensite and in its carbon content i.e., in the M_s temperatures (Fonstein et al. 2007).

As shown, the YS/TS ratio for compared steels grows with increasing volume fraction of martensite. It can, nonetheless, differ significantly depending, in particular, on the martensite start temperature, M_S. Figure 3.16a, displays the plots of variations in YS/TS ratio against tensile strength for two levels of manganese in steel with ~0.08 % C; Fig. 3.16b shows the plots of variations in YS/TS ratio for two levels of the carbon in steel with 1.5 % Mn. It is well known that both carbon and Mn lower the M_S temperature. In particular, for the conditions pertinent to Fig. 3.16b the reasonable assumption is that at equal level of steel strength up to 800 MPa (and presumably equal volume fractions of martensite), the carbon content in martensite of steel with 0.15 % C was twice as high in comparison with steel with 0.07 % C.

Thus, both graphs illustrate an important trend of decreasing YS/TS ratio of DP steels with decreasing M_S temperature. It can be implied that the lower the M_S temperature, the shorter the time available for dislocations generated by martensitic transformation to annihilate during cooling to room temperature and have residual stresses relaxed. In the case of a lower M_S due to a higher carbon in austenite, additional contribution can come from bigger volume changes during the $\gamma \rightarrow \alpha'$ transformation that generates more dislocations.

Significant role of unrelaxed residual stresses and mobile dislocations generated by martensitic transformation of austenite is confirmed by the data in Table 3.3 obtained using thermal cycle presented in Fig. 3.15c. Steel with 0.10 % C and 1.9 % Mn was air cooled from intercritical region (790 °C) to 400 °C with subsequent cooling at different controlled rates. Based on 18–20 % martensite and its calculated carbon content of 0.5–0.55 % and using the equations proposed by Steven and Haynes (1956), the computed M_s temperature was then within the range of 250–270 °C. As follows from the presented data, an increase in cooling rate around the M_S temperature has an insignificant effect on the strength of steel since the amount of the remained austenite and generated martensite is already determined by the rate of initial cooling. However, a monotonous decrease in yield strength with increasing rate of the final cooling is clearly seen. Simultaneously, an increase in the strain-hardening rate is observed, here estimated as the difference of flow stresses at 5 and 0.2 % strain that was predominantly controlled by the increased density of mobile dislocations as well as by increased residual stresses.

Table 3.3 Effect of the cooling rate near the M_s temperature on the properties of the DP steels (steel grade 09G2; 2.2 mm thick sheet; heating to 790 °C)—original

Mechanical characteristics	Initial hot-rolled steel	Untempered DP steel after accelerated cooling from 400 °C			
		Fan-cooled	Oil-cooled	Cooling in hot water	Cooling in cold water
Yield strength (MPa)	400	310	310	280	255
Tensile strength (MPa)	525	575	580	560	565
YS/TS	0.76	0.54	0.53	0.50	0.45
Difference in flow stress at 5 % and 0.2 % strain, MPa	–	95	95	125	150

Based on the presence of 18–20 % martensite, the calculated carbon concentration was taken to be 0.5–0.55 %

3.2.2.4 Effect of Martensite Dispersion

Martensite dispersion, MD, described by the ratio of the number of martensite particles to the number of ferrite grains, was used by Takada et al. (1982) as an important criterion for continuous yielding and a reduced YS/TS ratio in DP steels. It implies that the finer the martensite at given volume fraction, the higher the reheating (tempering) temperature, when the characteristic absence of YPE and low (less than 0.6) YS/TS ratio can still be observed, including the higher resistance to aging. An increase in MD and its beneficial effect on the YS/TS ratio is favored, with all factors being kept constant (including sizes of martensite particles), by ferrite grain coarsening, when longer carbon diffusion paths to regions of higher dislocation density near phase interfaces are required.

3.2.3 Tensile Strength

Many authors noted a linear relationship between TS of DP steel and the volume fraction of martensite, but in most cases, the researchers did not obtain 100 % of martensite or even skeleton type of martensite microstructure. Serious debates continue as to whether this relationship should depend on the properties of martensite or ferrite matrix.

3.2.3.1 The Role of Ferrite Strength

Strength of DP steels is determined by contributions from both phases.

For example, TS of dual-phase steels tends to increase with ferrite grain refinement because of bigger contribution of stronger ferrite matrix, $TS_F \cdot V_F$.

Some authors suggest that the strength of ferrite is controlled by the dislocation mean free path in ferrite phase rather than by ferrite grain size as the former decreases with increasing martensite volume fraction. In fact, the microhardness of ferrite, H_μ^F, is sometimes observed increased with growing V_M due to expansion of the interfacial area, and increase in the portion of ferrite deformed during the $\gamma \rightarrow \alpha'$ transformation.

The strength of ferrite also affects strengthening of dual-phase steels by martensite. As follows from Eq. (3.5), the rate of hardening DP steel per 1 % martensite should depend on the difference $(\sigma_M - \sigma_F)$.

Thus, an increase in ferrite strength as a result, for example, of solid solution hardening by silicon or by grain refinement should reduce the rate of martensite hardening, measured as MPa/%. Based on known regression equations for dual-phase steels of different compositions (Marder and Bramfitt 1979; Gribb and Rigsbee 1979; Bucher et al. 1979; Mould and Skena 1977; Davies 1978a; Eldis 1981; Lawson et al. 1981; Marder 1982), Fig. 3.17 (solid line) shows the

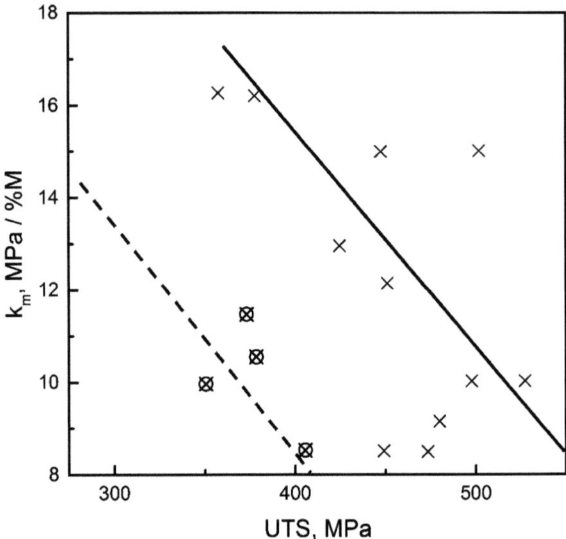

Fig. 3.17 Relationship between extrapolated tensile strength of ferrite in DP steel and martensite hardening rate, K_M—original

relationship between the martensite hardening rate, k_M, and the ferrite strength obtained by listed researchers using the extrapolation of the tensile strength of dual-phase steel to zero martensite content. The results obtained by author with coworkers on steels with different silicon and phosphorus contents, or with different ferrite grain sizes (Bortsov and Fonstein 1984a; Fonstein et al. 1985) are identified separately in Fig. 3.17 by dashed line. As shown, there is some trend for k_M to decrease with increasing σ_F, but there is no direct correlation between the two.

3.2.3.2   Effects of Volume Fraction of Martensite and Its Carbon Content

Figure 3.18 presents experimental data of measured TS (solid line) of quenched CrMnSiB steels with different carbon content vs. volume fraction of martensite (Bortsov and Fonstein 1984a). Dashed lines correspond to TS calculated using modified equation for the rule of mixtures [see Eq. (3.5)]. Substitution of σ_M with Eq. (3.4) gives:

$$TS_{DP} = TS_F + K'C_{st} + (\sigma_o - TS_F)V_M \tag{3.13}$$

The coefficients in the last equation were obtained experimentally from tensile tests of fully quenched CrMnSiB steels with various carbon content ($V_M = 1$). This led to the relationship:

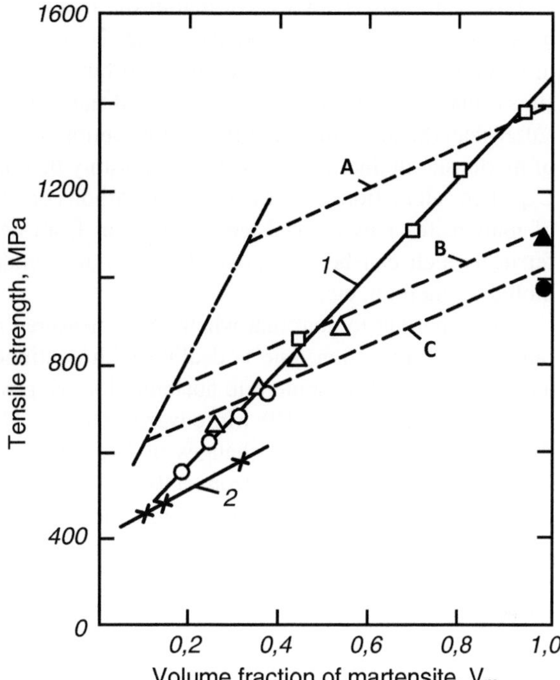

Fig. 3.18 Effect of volume fraction of martensite on tensile strength of DP CrMnSiB steels with various concentrations of carbon: *open circle*—steel with 0.07 % C; *open triangle*—0.10 % C; *open square*—0.18 % C; *solid symbols* indicate fully martensitic microstructure; *dashed lines*—calculations using rule of mixtures (Bortsov and Fonstein 1984a)

$$TS_{DP} = 870 + 2700C_{st} \tag{3.14}$$

where C_{st} is the carbon content in steel. As shown, the *rate* of strengthening by martensite experimentally observed was found to be three times higher than the value calculated from the rule of mixture (11.5/% and 3.5/%, respectively), but *absolute experimental values* of TS_{DP} were lower than calculated.

In accordance with Fig. 3.18, the tensile strength of dual-phase steels is independent of the intrinsic strength of martensite. This was also observed by others (Hansen and Pradhan 1981; Davies 1978a; Marder and Bramfitt 1979; Eldis 1981). In fact, Eldis found that the TS *versus* V_M plots for dual-phase steels followed the same curve for a steel containing 0.12 % C quenched from different temperatures in the intercritical range (with C_M ranging from 0.15 to 0.45 %) and for various air-cooled steels with different carbon contents exhibiting roughly constant $C_M \approx 0.45$ %. No effect of C_M at constant L_F on the rate of strengthening of DP steel by increased amount of martensite was noted by Davies (1978a) in the study of steels containing 0.06–0.47%C quenched from different temperatures of the intercritical range. According to Davis, the extrapolation of strength to 100 % martensite gave a value typical for martensite containing 0.4 % C, whereas Bortsov and Fonstein found that it was corresponding to the carbon content in martensite of about 0.25 % (Bortsov and Fonstein 1984a).

Lower strength (hardness) of martensite in dual-phase steels compared to hardness of martensite in fully quenched steels with the same carbon content was noted in several publications. This fact was attributed to weaker strain constraint and to stress relaxation during transformation of austenite into ferritic matrix, which is softer than the austenite. A certain contribution to the apparently reduced strength of martensite in dual-phase steels can possibly be related also to overestimation of C_M when calculations are based on total carbon content in steel and volume fraction of martensite. This inconsistency can stem from neglecting carbon solubility in ferrite (which can be as high as 0.01 %) and binding of carbon in carbides of niobium, titanium, etc.

It is important to note that when TS is measured as a function of V_M using the only steel composition, one is dealing with continuous changes in strength martensite due to carbon content in austenite lowering with increasing temperature of annealing (Pouranvari 2010). Moreover, for the used low hardenability steel such as 0.11C–0.534Mn–0.7Si, the changes in C_M with quenching temperature alter the calculated M_S temperatures from 330 to 467 °C. Therefore, in these cases, the comparison is really made between high-carbon low-auto-tempered martensite and low-carbon highly auto-tempered martensite (at high M_s temperatures, the morphology and strength of the transformation products should be quite different from those of conventional martensite). For this reason, at $V_M > 45$ %, the author did not observe further strengthening of the above steel by martensite. Combined with the formation of martensite skeleton, this perhaps resulted in similar extremal values of strength at around 45–55 % volume fraction of martensite observed by (Bag et al. 1999).

Deviation from "single curve dependence" is seen in Fig. 3.18 as well. For example, steel with 0.06%C demonstrates TS of 880 MPa at fraction of martensite of ~0.4 and 980 MPa in fully martensitic condition. The same for steel with 0.10 % C, which has ~895 MPA at $V_m \approx 0.5$ and 1180 MPa for 100 % martensite. It is difficult to assume that the tensile strengths of steels with 100 % martensite and different carbon content converge to the same value, as shown by Davies (1978a), so the demonstrated independence of properties of DP steels on properties of martensite should really have definite limited ranges.

Deviations from rule of mixtures, assuming that the variation of martensite strength TS_M should change the contribution of $TS_M \cdot V_M$ to the strength of dual-phase steel, can be explained by ignoring the stereological parameters of two-phase microstructure responsible for redistribution of loads between the matrix and the reinforcing particles. One of these parameters is the martensite continuity. In particular, the formation of a continuous martensite network reduces the rate of increase of strength with volume fraction of martensite.

According to the continuum mechanics, the hardening capacity of plate or needle-like martensite particles is greater than that of spherical particles. For example, Mileiko (1969) extended his model to composites with directional continuous fibers that can be assumed to correspond to hard-phase particles of 1 μm in diameter and the length of 8–10 μm. The model then implies that when the dimensions of actual martensite particles deviate from these values, the plastic

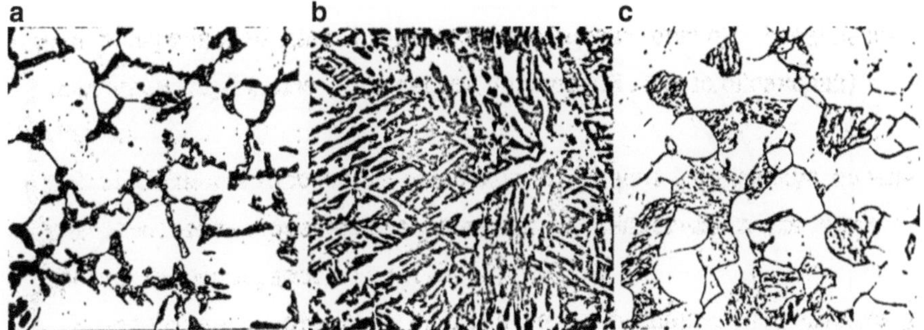

Fig. 3.19 Microstructure of DP steels obtained after different cycles of 06C–Mn–Cr DP steel original

instability can prematurely develop because instead of fracture of the hardening phase, the voids can form along the phase interfaces affecting the attainment of TS.

Performed experiments revealed that TS of DP steel is practically insensitive to the geometry of martensite islands in DP microstructures presented in Fig. 3.19, although it has to be noted that the corresponding tests were done in a narrow range of martensite volume. "Dispersed" martensite islands (Fig. 3.19a) were obtained by direct annealing of cold-rolled samples; "acicular" martensite (Fig. 3.19b) was produced by repeated annealing of preliminary quenched samples after full austenitization, and the "globular" morphology (Fig. 3.19c) of martensite was achieved by isothermal holding in the intercritical range during cooling from the austenite region.

Kim and Thomas also did not observe a big effect of martensite morphology, but in one of the papers (Kim and Thomas 1981), they reported that geometry and continuity of the hardening phase did in fact have some effect on the relationship between TS and V_M in dual-phase steels. They found that acicular and dispersed microstructures demonstrated higher hardening rates than the granular martensite microstructure. It should be noted, however, that the entire range of martensite hardening rates (from 6 to 2.5 MPa per 1 % martensite) was much lower in the work by Kim and Thomas than in most of other studies.

As will be shown below, the geometrical parameters of the second phase have an important effect on the strain-hardening rate, $d\sigma/d\varepsilon$, of dual-phase steels and therefore on their ultimate uniform strain.

It was already mentioned above, the values of the martensite strength, calculated using strength of fully quenched samples of the same composition and carbon content, were substantially greater than the contribution of martensite fraction to the strength of DP steels, calculated from test data.

Actual tensile strength values for martensite, TS_M, in dual-phase steels can be calculated from the expression

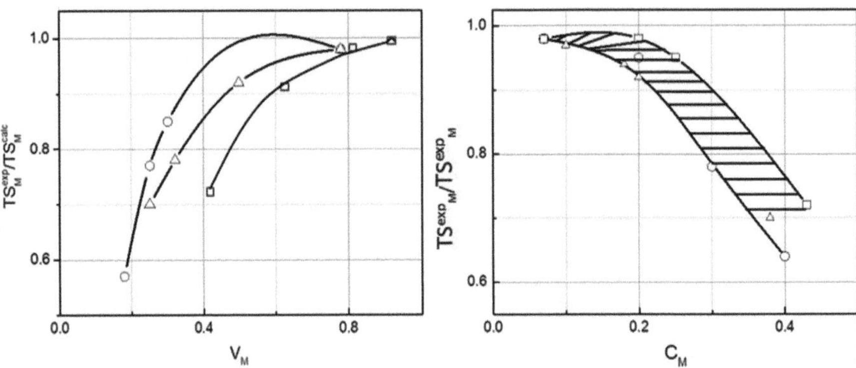

Fig. 3.20 Ratio of actual and calculated values of TS_M in DP steels depending on the volume fraction of martensite (**a**) and its carbon content (**b**)—original

$$TS_M = \frac{TS^{DP} - TS^F(1 - V_M)}{V_M} \tag{3.15}$$

Similarly, the actual YS values can be calculated.

The experimentally determined YS and TS of ferrite in the studied steels were, respectively, 280 and 405 MPa. The real fraction of the calculated ("theoretical") martensite strength observed in two-phase microstructure was found to depend on the amount of martensite and its carbon content, and, as presented in Fig. 3.20, approaches 100 % with increasing V_M or decreasing C_M.

As mentioned above, the most vivid explanations of the apparent independence of $TS(V_M)$ curves on the properties of martensite in dual-phase steels were grounded on the assumption of strain redistribution between the phases controlled by their relative volume fractions, geometries, and strength (Balliger and Gladman 1981; Tomota and Tamura 1981; Ostrom 1981).

The results of metallographic analysis presented above confirmed the redistribution of strain and stresses between the phases: the stronger the martensite is, the smaller its contribution to bearing the loads, and the lower the evident fraction of its potential strength and its contribution to the total deformation of steel (Fig. 3.5).

Interesting method of quantitative separation of strains in ferrite and martensite employing in situ high-energy X-ray diffraction technique (HEXRD) was developed by Cong et al. (2009). Using a DP980 grade 0.14C–2.2Mn steel, the authors revealed significantly higher strain of ferrite than in martensite, whereas the calculated stress in martensite was almost three times as high as the stresses in ferrite. The different properties of martensite, due to, for example, lower carbon, should change the distribution of strains and stresses leading to an apparent independence of TS of DP steels of strength of martensite per se.

3.2.3.3   Effect of Bainite

Very few data are available regarding *the effect of bainite* when ferrite–bainite mixture is formed or martensite in DP steels is partially replaced by bainite. Bhadeshia and Edmonds stated that the effects of martensite and bainite on the strength of dual-phase steels are described by a common curve (Bhadeshia and Edmonds 1980). Some data suggest that hardening by bainite is substantially weaker than that by martensite (Rigsbee et al. 1981). Similar results were obtained when a part of martensite was replaced by bainite: a change in martensite volume fraction by 10 % caused a decrease in TS by 40 MPa. In other words, strengthening by bainite was 4 MPa/% lower than by martensite (Fonstein et al. 2011).

3.3   Strain Hardening of Dual-Phase Steels

Extremely important advantage of low-carbon ferrite–martensite steels is their high work hardening rate compared to other low-alloy steels. The combination of low yield strength in the initial state that facilitates yielding and high strain-hardening rate, which determine high uniform elongation and the possibility of a significant increase in steel strength during deformation. Understanding the factors controlling strain-hardening behavior is the key to optimization of the microstructure of DP steels.

3.3.1   Characteristics of Strain Hardening

Finding a relationship between strain-hardening rate and parameters of microstructure requires appropriate selection of quantitative characteristics. There are more than twenty different models to approximate a stress–strain curve including proposed by Ludwig (1909), Swift (1952), Voce (1948), as well as modified Jaoul–Crussard model (Jaoul 1957).The most common parameter of strain hardening is the exponent n in the parabolic approximation of stress–strain σ–ε curve introduced by Hollomon (1945)

$$\sigma = k\varepsilon^{n} \tag{3.16}$$

where σ and ε are the true stress and true strain, respectively. The coefficients k and n can be obtained by plotting the true stress–true strain curve on a logarithmic scale (Fig. 3.21):

$$\log\sigma = \log k + n\log\varepsilon \tag{3.17}$$

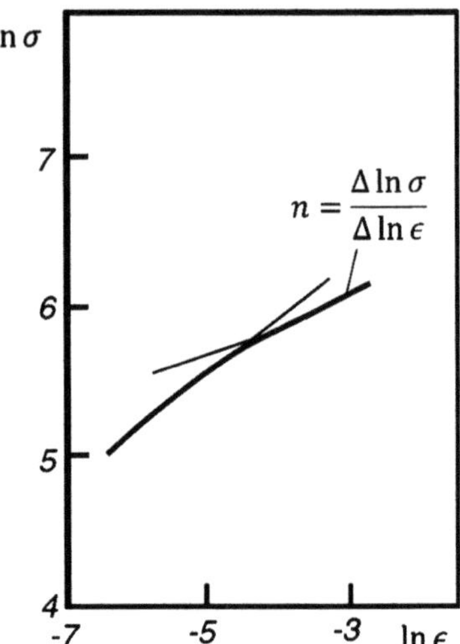

Fig. 3.21 Determination of strain hardening parameter n derived from stress–strain curve using logarithmic coordinates—original

then

$$n=\Delta\log\sigma/\Delta\log\varepsilon \tag{3.18}$$

When a strain-hardening curve is plotted on a logarithmic scale, several hardening stages can often be identified (Matsuda and Shimomura 1980). This makes the n values depend on the number of data points and their sampling rate used for numerical approximation. Sometimes the strain ranges for evaluation of n are prescribed to be fixed, e.g., as from 6 to 12 % elongation, from 10 % up to UE (ultimate uniform elongation), etc. The complete plot of instantaneous n_i-values can be obtained using

$$n_i = \frac{d\sigma'}{d\varepsilon_{\varepsilon-\varepsilon_i}} \times \frac{\varepsilon_i}{\sigma'_i} \tag{3.19}$$

In fact, the physical significance of the point of ultimate tensile strength TS is the attainment of the condition when the true tensile stress σ' (with account of decreasing cross-section of the specimen, i.e., "geometrical softening") becomes equal to strain-hardening rate $d\sigma'/d\varepsilon$, i.e.,

$$\sigma_{\varepsilon=e_u} = d\sigma'/d_{\varepsilon-e_u} \tag{3.20}$$

Thus, at this point,

$$\frac{1}{\sigma'}\frac{d\sigma'}{d\varepsilon} = 1 \tag{3.21}$$

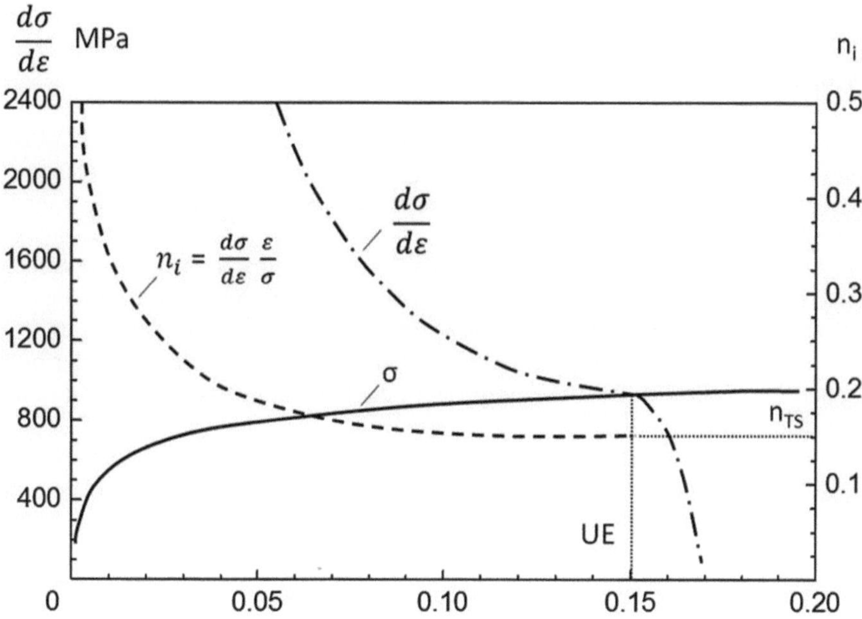

Fig. 3.22 True stress–true strain curve and various parameters of strain hardening of DP steels—original

After reaching ε_u, the strain hardening can no longer compensate for geometrical softening, which initiates necking. At this point, the combination of Eqs. (3.20) and (3.22) gives $n_i = \varepsilon_u$. The corresponding points are shown on experimental plots in Fig. 3.22. Equality of n to maximum true uniform elongation was confirmed experimentally by Davis (1978b).

Strain-hardening rate $d\sigma/d\varepsilon$ has the dimensionality of true stress which allows plotting them together with stress–strain curve in one graph, Fig. 3.22. The point of intersection of $d\sigma/d\varepsilon$ curve with true stress curve thus describes the attainment of ultimate uniform deformation, i.e., the onset of plastic instability.

With Ludwig approximation of *stress–strain* curve (Ludwig 1909),

$$\sigma = \sigma_0 + k'\varepsilon \tag{3.22}$$

and assuming

$$\sigma_0 = \alpha b G \sqrt{\rho} \tag{3.23}$$

where ρ is the dislocation density, G the shear modulus, b the Burgers vector, and α the constant in the range of 0.2–1.0 (typically taken as 0.5 for polycrystalline iron), the value of k' in Eq. (3.23) reflects the rate of increase in dislocation density ρ per unit plastic strain ε_{pl}.

Examples of various approximations of flow curves were presented by Matlock (Matlock et al. 1979) and Gribb and Rigsbee (1979). The analysis based on Crussard–Jaoul expression

$$\sigma = \sigma_0 + K\varepsilon^n \tag{3.24}$$

allowed to reveal a few different strain-hardening stages with different slopes in ln $(d\sigma/d\varepsilon)$ versus ln ε plot.

In particular, as was shown by Korzekwa et al., stage I ends at strains somewhat lower than ~0.5 %, while the transition from stage II to stage III spans to approximately 2 %. Detailed TEM investigation showed that stages I and II are featured by nonhomogeneous dislocation distribution with high dislocation density near martensite (Korzekwa et al. 1984). Two stages of strain hardening were found in the study of Jing et al. (1995).

In one of the recent studies, fitting of stress–strain curves into four existing models were compared using six DP steels, all with very low volume of martensite and with TS ranging from 450 to 600 MPa (Colla et al. 2009).

Apart from the Hollomon (Eq. 3.16) and Crussard–Jaoul (Eq. 3.24) equations, the authors also used the Pickering model

$$\sigma = A + B\varepsilon + C\ln\varepsilon \tag{3.25}$$

and the Bergstrom model

$$\sigma = \sigma_0 + k\sqrt{1 - \exp(-0.5\varepsilon)} \tag{3.26}$$

Among all of the tested equations, the Pickering model rendered the best fit and predictive ability. Although fitting to the Hollomon equation was not the best, the predicted and the fitted curves were very close because of the simplicity of the model. Since no evident transition of plastic behavior at various strain-hardening stages was found, the authors preferred the parabolic Hollomon equation over the Crussard–Jaoul analysis.

The above brief outline suggests that comparing the effects of various parameters on strain-hardening behavior of dual-phase steels should be performed for the identical conditions, e.g., strain range, and the same strain-hardening characteristics and so on.

3.3.2 Models of Strain Hardening of Dual-Phase Steels

The following stages are generally identified during deformation of a two-phase material: I-both phases undergo elastic straining; II-the softer phase is plastically deformed; III-both phases are plastically deformed; IV-either fracture of the harder phase or decohesion along phase interfaces takes place.

Of biggest interest, of course, are strain-hardening parameters describing the transition from stage I to stage II and then to stage III of deformation. As mentioned above, determining these parameters is likely to depend on the extent of discontinuity of those transitions and the method used to analyze stress–strain curve.

Typically, strain-hardening rate of DP steels is the highest in strain range of $\varepsilon = 0.01$–0.05 (Fig. 3.22). This is interpreted in terms of limited deformation of ferrite due to presence of hard martensite islands and, hence, due to higher rates of dislocation accumulation and multiplication in this strain range. The retained austenite, if presents, also transforms to martensite at this stage generating additional dislocations.

From the data similar to those presented above in Figs. 3.3 and 3.5, it follows that the martensite constituent of dual-phase steels remains essentially undeformed until the maximum uniform strain has been attained. Deformation is localized in ferrite around the hard-phase particles, which is in agreement with earlier findings (Korzakwa et al. 1980; Gurland 1982).

The decrease in strain-hardening rate with increasing strain was attributed to the formation of more ordered dislocation substructure or to the beginning of deformation of the second phase. It can also be surmised that during the strain of dual-phase steels at stages II and III, different types of dislocation sources operate simultaneously, e.g., in ferrite and at the ferrite–martensite interfaces.

The interpretation of high initial work hardening rate of dual-phase steels based on additive contributions from ferrite and martensite (Davies 1978b) contradicts the experimental data. In fact, a comparison of dual-phase ferrite–martensite steels and ferrite–pearlite steels with the same amount of ferrite shows that the strain-hardening rate in the latter is substantially lower, although the difference in intrinsic strain-hardening rate of pearlite and martensite, $n = 0.25$ and 0.06, respectively, does not account for this difference. Moreover, the hardening rate of dual-phase steels containing 75–80 % of ferrite is higher than that of 100 % ferrite of the same composition.

The best interpretation for high strain-hardening rate in DP steels is based on the concept of nonhomogeneous strain distribution in the phases and on the analysis of factors influencing this distribution.

Studies of slip lines (Tomota and Tamura 1981) suggested that localization of the initial deformation in ferrite and hence the rate of subsequent dislocation accumulation depend on the ratio of phase properties, e.g., of their yield strengths YS_M/YS_F, and on the volume fraction of the second phase, V_M. It was shown that ferrite can exhibit higher strain hardening in a two-phase microstructure with harder martensite than in the presence of the same amount of softer martensite, Fig. 3.23 (Bortsov and Fonstein 1986).

During loading of ferrite–martensite steel, the deformation of martensite begins earlier if its strength is lower and its volume fraction is higher. Phase continuity also impacts the deformation of the harder phase. In fact, in microstructures with continuous harder phase, deformation ferrite is less pronounced and terminates earlier even with higher hardness of martensite and lower volume fraction V_M.

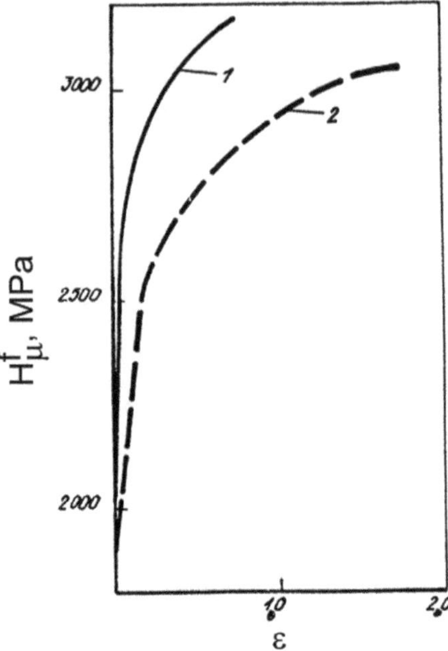

Fig. 3.23 Effect of strength of martensite on microhardness of ferrite during deformation of DP steel: 1—untempered martensite, 2—martensite, tempered at 400 °C

The nature of high strain hardening in DP steels is the same as that of continuous yielding of DP steels or of low stresses at small plastic strains. The major role is played by increased density of dislocations and nonhomogeneity of their distribution in ferrite, as well as by residual stresses in both phases generated by local $\gamma \rightarrow \alpha'$ transformation.

Some researchers interpreted the strain-hardening behavior of dual-phase steels in terms of hardening by undeformable dispersed particles that cannot be cut by dislocations. As shown in the theory due to Ashby (Ashby 1966), which is often employed in this approach (Balliger and Gladman 1981; Brown and Stobbs 1971; Rizk and Bourell 1982; Sarosiek and Owen 1983; Lanzilotto and Pickering 1982), displacement of grain boundaries and phase interfaces under external loading involves the formation of dislocation pile-ups necessary to sustain the continuity (accommodation) between undeformable particles and matrix.

According to Ashby, the precipitation hardening is proportional to the ratio of the volume fraction to the size of undeformable particles, or to the dislocation free path in the matrix, i.e., to the particle spacing (martensite particle spacing in DP steels)

$$\sigma - \sigma_0 = kG\sqrt{\frac{bV_\beta \varepsilon_{pl}}{0.4\bar{L}_\beta}} \tag{3.27}$$

where σ_0 is a constant, G is the shear modulus of the matrix ($\approx$82,400 MPa for ferrite), b is the Burgers vector of the matrix dislocations (about 0.247 nm), V_β is the volume fraction of hard phase, $\overline{L_\beta}$ is the mean particle size in mm, ε_{pl} is the plastic strain, and k is a constant close to 1.

Differentiation of Eq. (3.27) gives

$$\frac{d\sigma}{d\varepsilon}=0.78k\frac{Gb^{1/2}}{\varepsilon^{1/2}}\sqrt{\frac{V_\beta}{\overline{L_\beta}}} \tag{3.28}$$

Substitution of constants gives, e.g., for $\varepsilon=0.2$ (about 20 % engineering strain),

$$\frac{d\sigma}{d\varepsilon_{0.2}}=2200\sqrt{\frac{V_\beta}{\overline{L_\beta}}} \tag{3.29}$$

The Ashby model assumes the relaxation of stresses by formation of dislocation loops around the obstacles. By contrast, Brown and Stobbs (1971) pointed out that an important contribution to precipitation hardening is due to unrelaxed stresses of inverse sign. Accounting for these stresses substantially increases the initial slope of the calculated stress–strain curve. Nevertheless, the predicted level of $d\sigma/d\varepsilon$ at the initial strain-hardening stage is much lower than the actually measured strain-hardening rate of DP steels (Sarosiek and Owen 1983). Thus the precipitation hardening model is incapable to satisfactorily explain high strain-hardening rate in ferrite–martensite steels.

In this context, Sarosiek and Owen considered local changes in the volume of austenite during the $\gamma \to \alpha'$ transformation viewing these changes as the sum of dilatational and shear strains. Grounding on evaluation of the number dislocation loops in ferrite produced to accommodate the transformation, they concluded that the strain hardening of DP steels due to geometrically necessary dislocations (GND) caused by the different deformability of the two phases is immeasurably lower than that induced by the accommodation dislocations generated by local martensitic shear transformation stresses between phases admitting, in particular, the possibility of martensite deformation.

3.3.3   *Experimental Data on the Effects of Microstructure Parameters on Strain-Hardening Behavior of DP Steels*

3.3.3.1   Effect of Volume Fraction, Sizes, and Morphology of Martensite Phase

As mentioned earlier, stress–strain curves of DP steels are better approximated as having at least two strain-hardening stages described with the Hollomon equation with n_1 (initial stage) and n_2 (final stage up to the limit of uniform elongation).

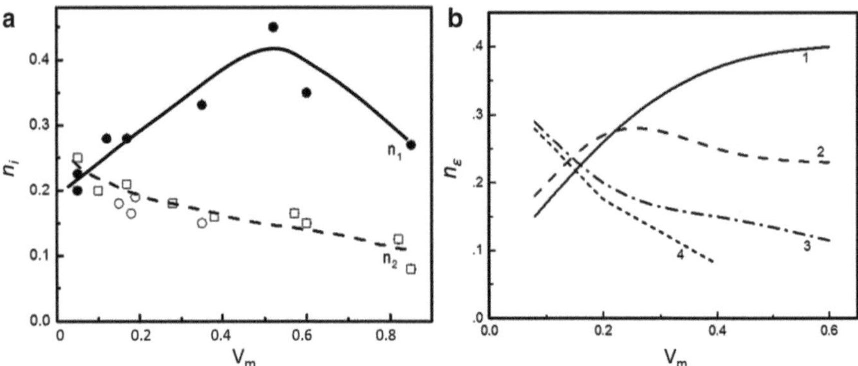

Fig. 3.24 Effect of volume fraction martensite in DP steels on strain hardening exponent: (**a**) 10Mn2V steels, n_1 and n_2 are the average strain hardening exponents for strain ranges 0.01–0.05 and 0.1–0.15, respectively, (**b**) 0.06C–CrMnSiB steel, n_i values for true strain of 1–0.01, 2–0.02, 3–0.05, 4–0.10—original

As shown in Fig. 3.24a, n_i-values measured in the range of $\varepsilon = 0.01$–0.05 increase with volume fraction of martensite V_M until the formation of martensite skeleton (at about 50–55 % of martensite). In contrast, the n_2 values at final stage of uniform deformation (at strains above 0.05) decrease with V_M within the entire range of martensite volume fractions. As shown in Fig. 3.24b, instantaneous n_i values demonstrate different variations with volume fraction of martensite in 0.06C–CrMnSiB steel depending on strain value.

As noted, the Ashby theory explains high strain hardening of DP steels at the initial deformation stage by significant difference in hardness of microstructure constituents. This is experimentally confirmed by decrease in both n_1 and n_2 values after tempering of DP steels when the difference in hardness between phases becomes smaller (Fig. 3.25).

From the Ashby theory [Eq. (3.27)] it follows that strain-hardening rate of DP steels should increase with higher V_M and lower $\overline{L}_M$. Although this model, which is based on the assumption of non-deformable equiaxed particles, cannot quantitatively describe strain hardening of DP steels, qualitatively it aligns well with the experimental data for DP steels with various volume fractions and sizes of martensite phase. An important point is that refinement of martensite particles should be beneficial for strain hardening and therefore for attainment of higher uniform elongation, ε_u, of dual-phase steels.

Lanzilotto and Pickering (1982), similarly to Sarosiek and Owen (1983), discussed the effects of microstructure parameters on work hardening of DP steels using the GND (Geometrically Necessary Dislocations) concept using the Brown–Stobbs equations (Brown and Stobbs 1971). Neglecting some special features of dislocation substructures typical for local $\gamma \rightarrow \alpha'$ transformation and for strains ε less than 0.2, neglecting the statistically accumulated dislocations, Lanzilotto and Pickering arrived at the following analytical expression:

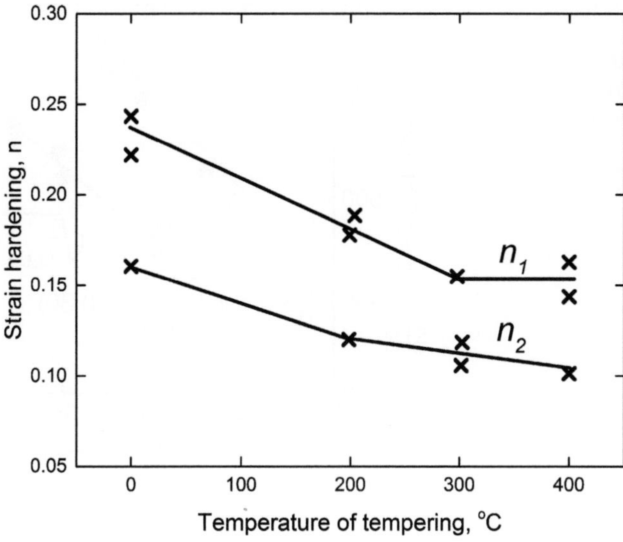

Fig. 3.25 Effect of tempering temperature on strain-hardening coefficient of water-quenched DP steel (10Mn2V), annealed at 765 °C: n_1 was measured in 0.01–0.05 range; n_2—in 0.06–0.12 strain range—original

$$\sigma - \sigma_0 = A\sqrt{\frac{\varepsilon V_M}{\overline{L}_M}} + B V_M \sqrt{\frac{\varepsilon}{V_M}} \tag{3.30}$$

where $A = 63.9$ MPa and $B = 36.8$ MPa. This allows for calculation of $d\sigma/d\varepsilon$ at given strain ε:

$$\frac{\sigma}{d\varepsilon_{\varepsilon=\varepsilon_i}} = \left(k_1\sqrt{V_M} + k_2 M\right)\sqrt{\frac{1}{\overline{L}_M}} \tag{3.31}$$

where k_1 and k_2 are constants for the given ε (e.g., for $\varepsilon = 0.2$, $k_1 = 71.4$, and $k_2 = 41.1$ MPa) and $\overline{L}_M$ is the average martensite particle size, mm. From Eq. (3.31), $d\sigma/d\varepsilon = 40.1\sqrt{\frac{1}{\overline{L}_M}}$ for $V_M = 0.20$ and $51.4\sqrt{\frac{1}{\overline{L}_M}}$ for for $V_M = 0.3$. Some of the results of calculations obtained in that work are presented in Fig. 3.26.

Theoretical relationships for strain hardening agree well with the experimental data presented above (Fig. 3.24) showing that initial strain-hardening rate increases with V_M (in this case, up to the formation of martensite skeleton). Maxima in $d\sigma/d\varepsilon - V_M$ plots were reported in numerous publications such as, for example (Adamczyk and Graijcar 2007) and (De Cooman et al. 2013).

It was found experimentally that the initial strain-hardening rate is higher in case of small, well-dispersed martensite particles than in case of coarse particles. Further

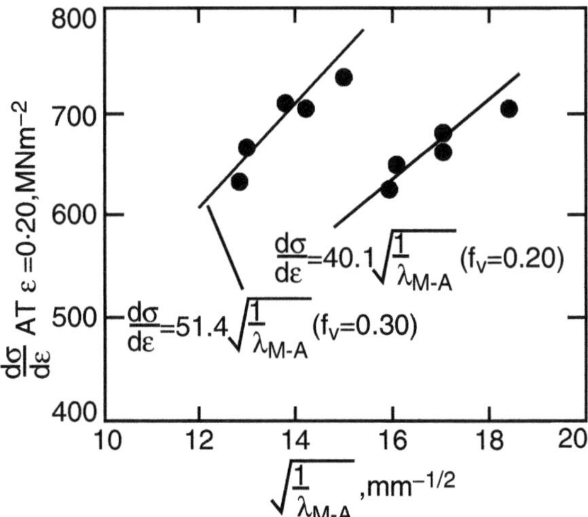

Fig. 3.26 Relationship between work hardening rate at $\varepsilon = 0.20$ and size of martensite particles (Lanzilotto and Pickering 1982)

hardening at the second stage of deformation (at strains above 4–5 %) seems to only slightly depend on martensite particle size.

Lanzilotto and Pickering suggested that the discrepancy between the theoretical and experimental value of $d\sigma/d\varepsilon$ begin to deviate from each other at true strain higher than 0.05, when martensite begins to deform or decohesion of phase interface begins to develop violating material continuity during deformation.

In some approaches, only two microstructural parameters, i.e., volume fraction V_M of martensite and its particle size $\bar{L}_M$ are considered. A common drawback of these theories is that they neglect the intrinsic phase properties such as strengthening of ferrite by interstitial or substitutional solutes or by grain size refinement, and the changes in strength of martensite controlled by C_M.

3.3.3.2 Effect of Properties and Size of Ferrite Phase

It is known that strengthening of ferrite by additions of silicon, phosphorus, or their combination increases the strain-hardening rate and hence the uniform elongation of dual-phase steels (Davies 1977) probably due to localization of the initial deformation in ferrite and the contribution of intrinsic properties of ferrite alloyed by those elements.

As described below, an increase in the amount of newly transformed ferrite has similar effect. No rigorous explanation of the effect of new ferrite on strain hardening of dual-phase steels exists. Probably, the major role is played by pronounced enhanced strain nonhomogeneity at deformation of DP steels, containing new ferrite (Lawson et al. 1981).

As to the effects of the structure parameters of ferrite, it is difficult to attribute the possible variations in strain hardening of dual-phase steels only to, for example, ferrite grain refinement. Experiments indicate that a decrease in ferrite grain size is almost always accompanied by refinement of martensite particles that should enhance strain hardening.

Burford et al. (1985) used repeated austenitization to refine ferrite grain size from 80 to 6 µm. An increase in the Hall–Petch coefficient k_y with plastic strain pointed to higher strain hardening in case of finer ferrite grains. The use of Jaoul–Crussard analysis of strain-hardening curves revealed that the onset of deformation at stages II and III is shifted toward smaller strains with decreasing ferrite grain size due to higher rate of dislocation accumulation.

Systematic studies of the effects of ferrite grain size were carried out by Jiang et al. (1995) who also used the Jaoul–Crussard analysis. They concluded that ferrite grain refinement effect on strain-hardening behavior of dual-phase steels diminishes at high strains where martensite deformation can take place.

Increase in the initial strain-hardening rate was noted in studying of 0.17C–1.49Mn–0.22Si steel, where grain refinement of ferrite from 12.4 to 1.2 µm at constant martensite content was produced by severe warm deformation at different temperatures with subsequent intercritical annealing (Calcagnotto et al. 2011). The authors attributed the effect of ferrite grain refinement to early onset of dislocation interactions, high number of dislocation sources, and back stresses from martensite islands.

As reasonable to point out that the findings about positive effects of fine and ultrafine ferrite grains (Song et al. 2005) on strain hardening and consequently plasticity of DP steels are quite different in comparison with known negative influence of grain refinement on single-phase ferrite (Morrison 1966)

3.3.3.3   Effect of Martensite Hardness

It was noted above that an increase in the strength of martensite (C_M) should enhance the work hardening of dual-phase steels by increasing plastic incompatibility between the phases. However, the real effect of the martensite strength on the average work hardening of DP steels induced by the changes in C_M in *untempered* martensite is not significant. A more substantial decrease in martensite strength as *a result of tempering* reduces work hardening.

Martensite strength determined by its carbon content, C_M, should have the most pronounced impact on work hardening of dual-phase steels at the initial deformation stage. Martensite strength directly affects distribution of strain between the phases, the rate of dislocation accumulation in the ferrite, and contributes to initial evolution of substructure brought about by the $\gamma \rightarrow \alpha'$ transformation (Ramos et al. 1979).

Changes in C_M involve a change in the shear component of the local strain during formation of martensite and influence the number of the generated accommodating dislocation loops. Higher C_M (higher initial $C\gamma$) also reduces the M_S temperature thus interfering with relaxation of internal stresses and increasing the number of dislocation loops that remain unpinned.

Figure 3.25 illustrates variations in strain-hardening rate of dual-phase steels with hardness of martensite, which in that case is controlled by the tempering temperature.

3.3.3.4 Effect of Bainite

A direct comparison of the effects of bainite and martensite on strain-hardening behavior is usually difficult because of variable geometry of the hard phase. It is reasonable to assume that substitution of martensite by the same volume fraction of less hard bainite would reduce strain-hardening rate of steel. This assumption was confirmed, in particular, by higher YS/TS ratio of DP steels with bainite (Gunduz et al. 2008, Girina et al. 2008).

In the study of Brnfin et al. of 0.057C–1.38Mn–1.51Si–1.12Cr steel, authors transferred intercritically annealed (800 °C, 30 min) samples, containing 16 % of austenite to salt pot with the temperature of 550 °C. The samples were held there for various periods of time to obtain from 0 to 15 % of bainite and then they were rapidly cooled at 28 °C/s. The triple-phase steels showed increasing yield strength with discontinuous yielding and slightly lowering tensile strength along with the increase in bainite fraction. The authors explained the observed lower initial strain-hardening rate due to replacement of martensite by bainite by reduced density of mobile dislocations and lower internal stresses (Bronfin et al. 1986).

3.3.3.5 Effect of Retained Austenite

The influence of retained austenite on strain-hardening behavior of DP steels is typically attributed to the TRIP-effect during deformation, especially in the early stages of straining (Marder 1977; Rigsbee et al. 1981; Gribb and Rigsbee 1979). It is believed that local transformation of retained austenite increases the dislocation density, creates additional nonuniformity of deformation and hence the rate of strain hardening.

The good performance of dual-phase steel produced by slow cooling from the $\alpha + \gamma$ region was also interpreted in terms of beneficial effects of higher strain hardening due to retained austenite (Matlock et al. 1979), the amount of which increases with a slower cooling rate from the intercritical range as a result o gradual enrichment of remaining austenite by carbon due to significant portion of formed new ferrite (see Chap. 2).

Bronfin et al. stated that 5–7 % of retained austenite in the microstructure of air-cooled 0.055C–1.91–0.29Mo–0.17Si–0.091V dual-phase steel undergoes decomposition during deformation with strain of 2–3 %. Therefore, in their opinion, the transformation of retained austenite fraction could control strain hardening of dual-phase steels only at stage I of deformation.

In conclusion, it should be noted that in spite of the imperfect microstructure models for the strain hardening of dual-phase steels, a few facts seem to be beyond doubt. Namely, the strain-hardening rate of the two-phase structures during initial deformation (at strains before 4–5 %) is increased by refinement of martensite particle, by an increase in the volume fraction of martensite provided, the martensite particles do not coarsen and to a lesser extent, by an increase in martensite strength. The same effect is induced by the factors favoring increase in the strain-hardening rate of ferrite, as well as increase in the amount of retained austenite.

3.4   The Ductile Properties of Dual-Phase Steels

High ductility at given strength level is one of the greatest advantages of dual-phase steels.

Total elongation in tension is the sum of the uniform elongation (strain before necking) and the elongation at the stage of localized deformation during necking that leads to failure.

Analysis of the literature is complicated because presented results were obtained on quite different geometry of tested samples, whereas the value of total elongation, TE, depends on the gauge length of a tensile specimen. Tensile test results for different gauge lengths L can, with a certain degree of approximation, be adjusted to a single gauge length, L_0, using the expression

$$TE = UE + (TE - UE)\sqrt{\frac{L}{L_0}} \tag{3.32}$$

However, adjustments for different specimen gauge lengths are only possible when both values of TE and UE are available. It is worth to mention that values of UE are more important characteristics of DP steels. On the other hand, the influence of gauge length is greater with higher localized (necking strain) (TE–UE) and therefore, for DP steels exhibiting high UE/TE ratios, the effect of gauge length is smaller than for conventional steels. Since TE and UE of dual-phase steels are controlled by different microstructure parameters, in what follows they are considered separately.

3.4.1   Uniform Elongation

As already noted, uniform elongation, UE, is confined by the onset of plastic instability and is therefore governed by strain-hardening rate during uniform elongation. However, strain-hardening rate changes with strain so that the reasonable average parameter should be selected to find the ways to control ductility of DP steels.

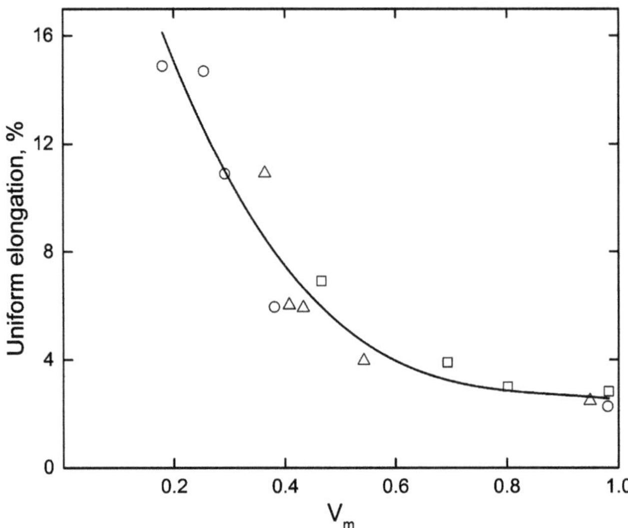

Fig. 3.27 Effect of volume fraction of martensite on uniform elongation of Cr–Mn–Si–B steel with various amount of carbon (designation of carbon content is the same as in Fig. 3.16)

From the Ludwig equation (Ludwig 1909), the UE value can be defined by ratio

$$\mathrm{UE} = \frac{n'(\mathrm{TS} - \sigma_0)}{(1+m)\mathrm{TS}} \tag{3.33}$$

where m and σ_0 are the constants.

Thus, higher uniform elongation UE is facilitated by all factors that increase n'-value, or $d\sigma/d\varepsilon$, or simply the difference TS–YS in comparison with value of increased TS.

However, as shown in Fig. 3.27, an increase in V_M progressively lowers UE. It agrees first with data presented in Fig. 3.24 pointing out that increase in V_M at strain above 0.1 (approaching uniform elongation) decreases strain hardening and secondly with the experimental facts indicating that larger martensite fraction facilitates the decohesion on ferrite–martensite interface.

There are some discrepancies concerning the effect of the *martensite strength*. In Fig. 3.27, all the UE(V_M) data fit, within experimental scatter, a common curve irrespectively of the C_M values (Bortsov and Fonstein 1984a). Similar results were obtained by Davies (1979b) and Marder (1977). However, according to Mileiko (1969), at constant TS, an increase in martensite strength (C_M) should raise UE of dual-phase steels due to higher strain-hardening rate. Some increase in UE as a result of higher carbon concentration in martensite and corresponding increase in strain-hardening rate was experimentally observed by Bronfin et al. (1983).

By contrast, in spite of higher strain-hardening rate especially in the initial deformation stage, a decreasing ductility of DP steels with increasing V_M at higher relative strength of martensite (at higher $\frac{\mathrm{YS_M}}{\mathrm{YS_F}}$ ratio) was reported in other studies

(Speich and Miller 1979). The most likely explanation for this discrepancy is the premature formation of voids in steels with significantly harder martensite due to incompatibility of deformation of phases.

Speich and Miller derived an empirical equation correlating true uniform strain of DP steel, ε_u^{DP}, and true uniform strain of ferrite, ε_u^F, on the one hand, with the volume fraction of martensite, V_M, and its carbon content, C_M, on the other hand:

$$\frac{\varepsilon_u^{DP}}{\varepsilon_u^F} = 1 - 2.2 C_M \sqrt{V_M} \qquad (3.34)$$

This dependence suggests, in particular, that uniform elongation can be increased by reducing C_M without increasing V_M, e.g., by decreasing carbon content in the steel. Further analyzing Eq. (3.34) and acknowledging the dependence of tensile strength on C_M and V_M, the authors concluded that increasing ε_u^F is the most efficient way to achieve higher UE in dual-phase steels. This can, in particular, be accomplished by adding silicon or phosphorus (Davies 1979b), which are believed to primarily increase $d\sigma/d\varepsilon$ of DP steels rather than TS. The negative role of interstitials in ferrite is confirmed by observing higher ductility of DP steel with the same amount of martensite after slow cooling compared to that after water quenching.

Predictions based on the modified Mileiko model and assuming strain redistribution between the phases show that ε_u^F can be also improved by increasing the *strength of ferrite*, e.g., by ferrite grain refinement.

Very scarce information can be found concerning the influence of martensite particles geometry on ductility of dual-phase steels besides the above effects of *martensite size* on strain-hardening rate that should correlate with uniform elongation. This suggests that the uniform elongation as well as strain hardening of DP steels can be improved by any means that reduce the effective grains size of the γ phase at the onset of austenite–martensite transformation and hence reduce the martensite particle size.

3.4.2   Total Elongation

Total elongation, TE, decreases with increasing strength TS (or V_M) of dual-phase steels. This can be interpreted in terms of both decrease in UE and smaller post-necking strain (TE–UE).

The duration of latter stage of localized deformation is determined by the rate of nucleation and amount of microvoids, whereas the moment of their coalescence to produce a main crack is defined by the distance between martensite islands and properties of ductile ferrite matrix.

Irrespectively of the void nucleation mechanism, say, by decohesion along grain boundaries or due to fracture of martensite particles, an *increase in V_M* reduces the void spacing and hence the plastic strain necessary for their coalescence. It was shown that the difference TE–UE, i.e., the span of localized deformation, decreases with increasing C_M as a result of easier decohesion or martensite fracture.

At a constant volume of martensite ductility of a dual-phase steel, a composite material is seemly governed by the properties of *ferrite matrix*. The total elongation of dual-phase steels generally decreases with increasing strength of ferrite (its lower ductility) if it is not compensated by general refinement of microstructure.

So-called new ferrite (NF) is formed during cooling from the $(\alpha + \gamma)$ region in case of insufficient austenite stability.

An increase in V_{NF} usually improves ductility, which can be attributed to enhanced ductility of the newly formed ferrite itself due to its depletion with ferrite-forming elements compared with the "old ferrite" (see Chap. 2). However, NF may not necessarily always be more ductile than the initial ferrite in any steels. In fact, in the presence of carbonitride-forming elements, new ferrite can be strengthened by interphase precipitates. Moreover, since the newly formed ferrite is adjacent to martensite generated from the γ-phase, it undergoes local plastic deformation as a result of the volume changes and exhibits an increased dislocation density. Because of the relatively high cooling rate usually used (10–30 °C/s), the diffusion flux of interstitial atoms from the newly formed α- to the γ-phase is not likely to approach equilibrium, i.e., the concentrations of these atoms in the "new" α-phase are in fact higher than in the original, "old" ferrite.

This view was confirmed in the study, in which different amounts of new ferrite was obtained by raising the annealing temperature followed by cooling at ~30 °C/s rate (Fonstein and Efimova 2006). Microhardness measurements (Table 3.3) indicated that in compared samples, newly transformed ferrite was harder and apparently less ductile than the original ferrite.

Data presented in Fig. 3.28 confirm in particular a stabilizing effect of the newly transformed ferrite on final martensite fraction that practically does not change in a wide range of annealing temperatures. The beneficial effect of new ferrite formation on the ductile properties of dual-phase steels is evident but in the limited temperature interval.

It was suggested that effect of "new" ferrite on ductility and work hardening was not just due to intrinsic properties of the "new" ferrite" but was rather due to

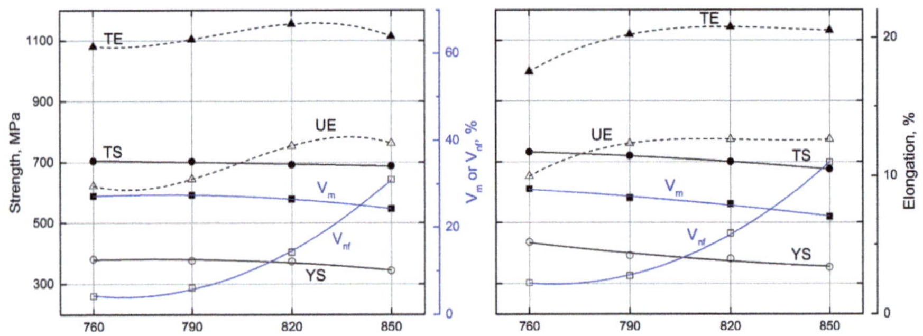

Fig. 3.28 Effect of new ferrite on properties of 06CrMn2Si1 (**a**) and 06CrMn2Si1V(**b**) obtained using cooling under fan from different temperatures of intercritical region

phenomena that accompany the generation of initial significant amount of *NF* (up to 15–20 %). These include (a) effective refinement of the remaining γ-phase (final martensite islands), which becomes divided into smaller fragments by the newly formed ferrite and favor higher strain hardening and therefore uniform elongation, (b) actual decrease in the interstitial content in old ferrite as a result of lower cooling rate and higher temperatures of heating in the $(\alpha+\gamma)$ range applied that should result in its higher ductility , and (c) higher stability of the remaining carbon-enriched γ-phase because less stable fraction has transformed. The remaining particles have the lower M_s temperature and hence the increased level of unrelaxed stresses and the density of mobile dislocations in ferrite facilitating higher strain hardening (uniform part of elongation).

After some amount of NF, varied in this study by increasing annealing temperature in the intercritical region, the TE has tend to decrease due to coarsening of martensite islands.

The differences in TE observed in both DP steels at roughly constant TS (and the same V_M) point at the importance of *martensite particle sizes*, geometry, distribution and continuity on the initial size of microvoids, and the length of the stage of localized deformation before coalescence of microvoids (post-necking part of elongation). Dual-phase steels with finely dispersed martensite particles exhibit the highest total ductility because of both higher uniform elongation and smaller voids and their delayed coalescence to generate a main crack after reaching ultimate strength.

No systematic data are available on the effects of *non-martensitic products* in DP steel microstructure on ductility. Since bainite is less hard than martensite, the steels with bainite should possess lower strain hardening and lower uniform elongation. It is, however, unclear how long the post-necking stage due to delayed phase decohesion could be.

It was suggested that higher amount of *retained austenite*, although it improves the uniform elongation, does not change the total elongation since the increase in UE involves a proportional decrease in the localized post-necking strain (TE–UE) (Rigsbee et al. 1981).

3.4.3   Reduction of Area

The data on the reduction of area (RA) for dual-phase steels are sparse, primarily because these steels are mostly intended for sheet applications. At the same time, since the correlation was found between *RA* and such an important parameter as hole expansion (Sugimoto et al. 2000), the interest in this characteristic of localized deformation of AHSS is a currently growing.

Relatively small reduction of area of dual-phase steels is a disadvantageous consequence of their high work hardening. Plastic incompatibility of ferrite and martensite and the correspondingly high rate of increase in dislocation density with strain lead to earlier void nucleation compared to ferrite–pearlite steels and hence to easier void coalescence during necking.

As shown in Fig. 3.29, high strain-hardening rate combined with low YS/TS ratio is accompanied by low reduction of area designated here as ψ.

The effect of martensite strength (i.e., carbon content in steels) is stronger than that of the volume fraction of martensite, Fig. 3.30. The increase in RA with

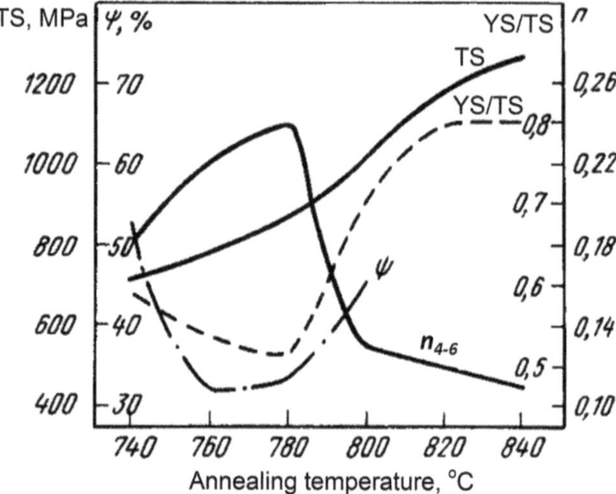

Fig. 3.29 Effect of annealing temperature on tensile properties of 0.09C–1.6Mn–0.05V–0.03Nb steel (diameter of wire 5 mm, water quenching, tempering at 200 °C for 30 min—original

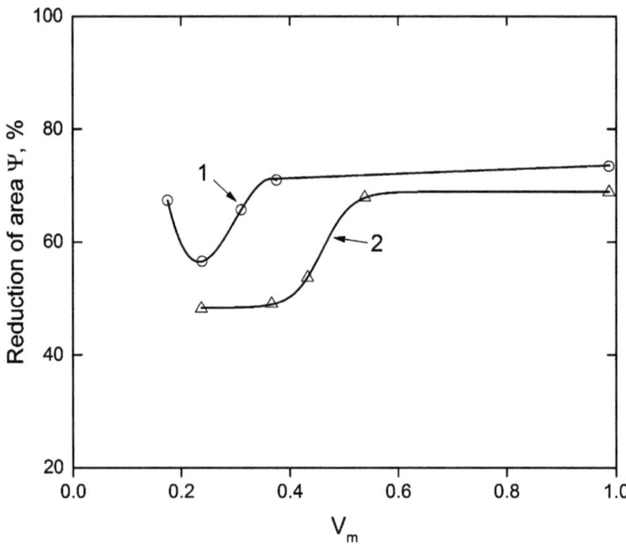

Fig. 3.30 Effect of volume fraction of martensite and carbon content in steel on its reduction of area; 06CrMnSiB DP steels: 1–0.07 % C, 2–0.11 % C, diameter of samples 5 mm, water quenching and tempering at 200 °C for 30 min—original

decreasing strength of martensite was noted also by Speich and Miller (1979). With this respect, higher tempering temperature markedly improves RA.

According to Sugimoto (Sugimoto et al. 1985), ferrite grain refinement is favorable for higher reduction of area along with increasing YS and TS due to relative suppression of size-related critical stresses and therefore later void formation around martensite islands.

3.5 Quench and Strain Aging

Dual-phase ferrite–martensite steels, as materials for cold forming applications, should not possess any propensity for static aging during long storage at ambient temperatures. On the other hand, the service ("in-use") characteristics of the final parts manufactured from dual-phase steels, such as dent resistance, can be improved by hardening during deliberate strain aging at slightly elevated temperatures that makes such hardening highly desirable. Therefore, studying of the dependence of hardening on the aging temperature and time, as well as on the pre-strain and interstitial concentrations in the ferrite is of both scientific and practical value and has been the subject of numerous publications.

In dual-phase steels, ferrite that undergoes rapid cooling from the $\alpha + \gamma$ region is supersaturated with interstitials (see Chap. 2). Nevertheless, as shown by many researchers, dual-phase steels do not exhibit any aging effects such as the development of YPE or a yield strength increase during holding at room temperature for as long as more than one year.

Aging effects of DP steels after holding at increased temperatures are usually characterized by an increase in yield strength, ΔYS, and then by return and an increase in yield-point elongation, *YPE*. As shown in variety of works, the *YPE* return in DP steels is retarded to a much greater extent than in ferrite or ferrite–pearlite steels and is observed only after heating at 180–300 °C.

In reheating of nondeformed DP steels, several stages must be considered in ferrite (aging process) and in martensite (tempering process): (1) formation of Cottrell atmospheres; (2) carbon clustering; (3) precipitation, and (4) tempering of martensite—volume contraction of martensite, changes in martensite strength and additional carbon clustering or precipitations near ferrite/martensite interfaces where the dislocation density is high.

Prestrain accelerates all of the above processes but does not change the nature of the phenomena.

The data obtained in Fig. 3.31 after various temperatures of re-heating are presented vs. the extent of pre-strain and overaging time.

The changes in the strength of nondeformed dual-phase steels upon heating (Fig. 3.31a) can be an important tool for strengthening the sections of parts that do not undergo stretching.

After aging at low temperatures of about 100 °C, the increase in YS is relatively small and the yield point is absent, which is typical for dual-phase steel.

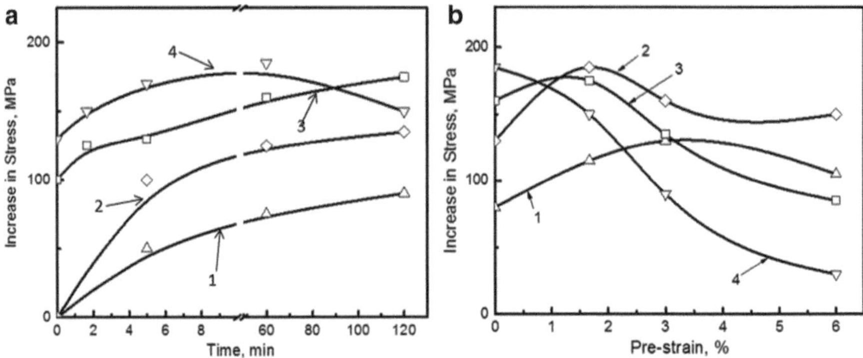

Fig. 3.31 Effect of aging time (**a**) and prestrain (**b**) on YS increase of dual-phase 0.08C–1.5Mn–0.5Si steel heated at: 1–100, 2–200, 3–300, 4–400 °C—original

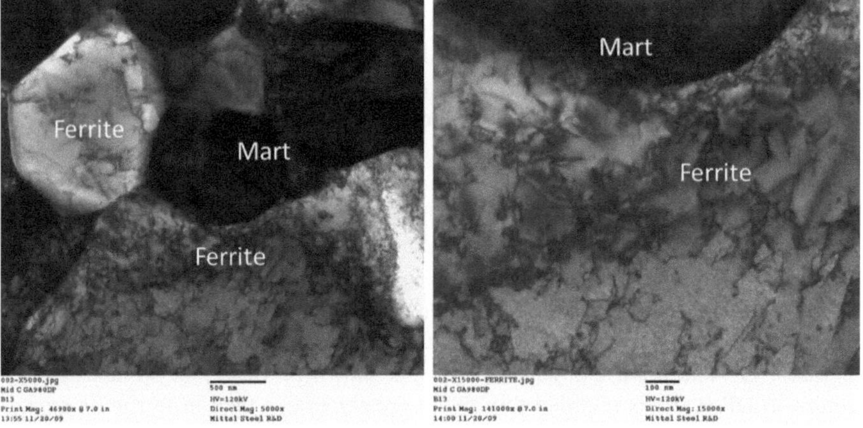

Fig. 3.32 Increased dislocation density near martensite islands in DP steels—original

As noted above and presented in Fig. 3.32, the dislocation density in ferrite of dual-phase steel, primarily near dispersed martensite particles, is high. It is generally believed that high stability of dual-phase steels against quench aging is caused by nonuniform distribution of dislocations. Therefore a long-range diffusion of interstitials from internal areas of ferrite grains to their periphery is required to produce sensible dislocation pinning. As a result, at relatively low temperatures of reheating, many dislocations remain unpinned. Moreover, ferrite in dual-phase steel is under residual tensile stresses induced by localized $\gamma \rightarrow \alpha'$ transformation, and these stresses also contribute to dislocation unpinning upon repeated loading.

Tanaka et al. (1979) employed the concept of nonhomogeneous dislocation substructures inherent in dual-phase steels, which implies that the dislocation density in ferrite, while being high in the vicinity of martensite/ferrite interface, decreases sharply towards the interior of ferrite grains. Interstitials should segregate to dislocations during aging; however, their concentrations in the regions of high dislocation density would be too small to pin all the dislocations and to induce the yield point elongation or substantial strengthening.

It is generally accepted that nonhomogeneous dislocation substructure is preserved during the temper rolling of dual-phase steels, as opposed to tensile deformation that leads to a more uniform dislocation distribution. In the latter case, the phase interfaces no longer act as preferential sources of new dislocations and many active sources emerge within the grains, with the short-range diffusion being sufficient for dislocation pinning.

As a result, dual-phase steels typically exhibit no yield-point elongation after aging at temperatures of up to 260 °C if the annealed steel has been skin-passed with reductions of up to 10 %. In 0.1C–1.4Mn steel deformed in tension, strengthening achieved maximum (80 MPa) at this temperature at strain of 5 %, and the yield point elongation developed at 260 °C after 10–15 s of holding (Krupitzer 1981).

As shown in Fig. 3.31, increase in the aging temperature to a level required for baking corrosion-resistant paint coatings, i.e., to 170–220 °C, results in substantial strengthening of dual-phase steels (Davies 1978c; Bailey and Stevenson 1979; Nakaoka et al. 1977). The increase in yield strength and yield-point elongation with aging temperature or time is indicative of formation of interstitial atmospheres at dislocations (aging stage 1). No increase in TS or decrease in elongation as a result of aging of non-deformed DP steels was observed during this stage.

Substantial strengthening of nondeformed dual-phase steels during aging at 200–400 °C (ΔYS = 100–180 MPa), illustrated in Fig. 3.31a, is generally attributed to high density of formerly mobile dislocations in ferrite that appear to be pinned by sufficient amount of interstitials that form clusters and precipitates (Stages II and III). The necessity to raise the temperatures above 100 °C to reach the required extent of aging indicates the key role of carbon atoms, since the diffusion rates of nitrogen atoms are fairly high even at room temperature and are believed to be responsible for aging at temperatures below 100 °C (Baird 1963).

At temperatures beyond the second stage of aging, internal friction measurements did not reveal any presence of interstitial carbon in ferrite phase (Waterschoot et al. 2003). Contraction of martensite lattice and therefore of the volume of martensite islands due to formation of carbon clusters or transitional carbides reduces internal stresses in ferrite, which usually facilitate YS decrease in DP steels. In combination with the presence of pinned dislocations in ferrite, a significant increase in YS can be observed at this stage, this increase being more pronounced with higher volume fraction of martensite.

Aging is a thermally activated process. As shown in Fig. 3.31a, the quantitatively same aging effects can be achieved after holding for shorter times at higher aging temperatures and after holding for longer times at lower temperatures. For example, in 0.05C–1.2Mn steel, the long range diffusion from grain interior requires 128 h at 100 °C or 40 min at 170 °C (Tanaka et al. 1979).

With longer *aging time*, ΔYS generally increases to a maximum and then decreases manifesting the overaging.

The value of ΔYS is proportional to the initial height of carbon Snoek peak (Nakaoka et al. 1977) and therefore to the *cooling rate*. An increase in carbon concentration in ferrite resulting from rapid quenching leads to substantially higher aging strengthening than in the case of slow cooling.

It is reasonable to expect that a decrease in carbon concentration in ferrite due to higher temperature of intercritical annealing or longer tempering can reduce the effect of strain aging. On the contrary, *short-term tempering* of quenched steels increases strengthening of dual-phase steels during aging at 170 °C (Bailey and Stevenson 1979; Araki et al. 1977a; Nakaoka et al. 1977). This effect can be related to carbon precipitation at dislocations at low oversaturation facilitated by short-term tempering. With increasing oversaturation, carbon precipitates as carbides resulting in weaker further pinning and strengthening. Thus, low temperature tempering of dual-phase steels during manufacturing produces good strengthening during baking of corrosion-resistant coatings/paint on steel parts ("the baking hardenability"), which is an important advantage of dual-phase steels.

The dislocation density increases with cold working thus accelerating aging. It can be assumed that the final strain aging effect should reflect the difference between ferrite strengthening and martensite *softening* for martensite is not deformed in DP steel strained at 2–5 % and hence does not undergo strain aging.

As was shown in Fig. 3.31, the level of pre-strain corresponding to the maximum increase in YS depends on the aging temperature, but at 170–200 °C is observed typically at strains of about 1.5–2 %, i.e., very close to strain level used in standard BH test (Fig. 3.33).

The increase in tensile strength accompanied by some decrease in ductility after heating of pre-strained specimens is indicative of aging stages II and III, i.e., of carbon clustering and precipitation of dispersed carbides.

The shape of aging curves is rather complex reflecting simultaneous contributions from several mechanisms. Table 3.4 shows the calculated activation energies for aging in dual-phase steels. With this regard it should be mentioned that the

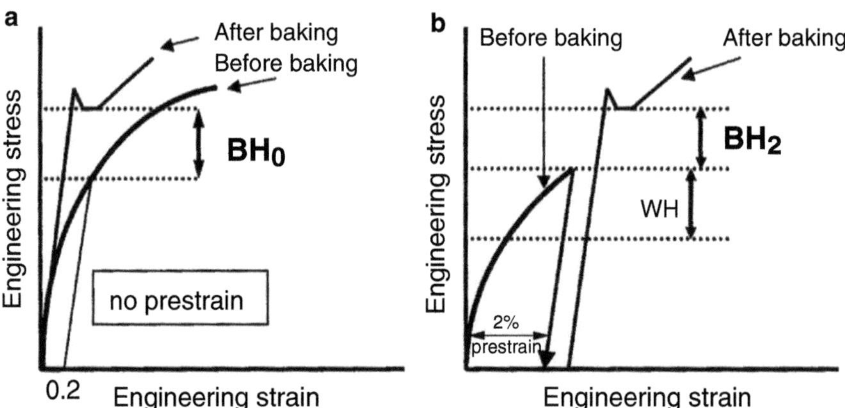

Fig. 3.33 Scheme of baking hardenability test

Table 3.4 Microhardness of the initial "old" ferrite (OF) and newly transformed ferrite (NF)

Steel	Heating temperature (°C)	Microhardness		Content of elements, mass %				
				Mn			Si	
		OF	NF	OF	NF	M	OF	NF
06KhGS1	820	156 ± 11	179 ± 14	1.56	1.82	2.02	1.09	0.97
06KhGS1F	850	–	–	1.65	1.79	2.00	1.06	0.77
	820	172 ± 13	219 ± 17	1.58	1.88	2.10	1.21	–
	850	149 ± 10	165 ± 18	1.60	1.79	1.95	1.05	0.94

Notes:
1. The metal was held at the heating temperature for 25 min
2. The value of microhardness was the average of 17 measurements made under a load 5 g
3. The concentrations of the elements were determined as the average of five measurements at each point

Table 3.5 Activation energies for strain aging of Dual-Phase steels (Himmel et al. 1981; Davies 1979a; Krupitzer 1981)

Aging criterion	Aging time (h)	Strain		Q (kJ/mole)	References
		%	Type		
Return of yield-point elongation	50–100	2	Tensile	80	Himmel et al. (1981)
Yield-point elongation of 0.2 %	50–100	2	Tensile	75	
Yield-point elongation of 2.5 %	150–200	2	Tensile	125	
Return of yield-point elongation	–	2	Rolling	138	Davies (1979a)
Yield-point elongation of 0.2 %	150	1	Tensile	90.8	Krupitzer (1981)

activation energies for carbon and nitrogen diffusion in iron are 75.6 and 84 kJ/mol, respectively, and the energies of dislocation interaction with carbon and nitrogen atoms are 126 and 147 kJ/mol, respectively. Therefore, complex effects such as precipitation at dislocations, formation of matrix precipitates of various types and morphologies, diffusion of carbon and nitrogen atoms, clustering at dislocations, relaxation of residual stresses, etc., can be considered as controlling factors.

In steels with higher Mn content (1.5 %), the increase in activation energy for aging to 107 kJ/mol was attributed to the interaction of interstitial and substitutional atoms that slows down the diffusion of carbon atoms to the dislocations (Yang et al. 1985).

Strengthening by strain aging in dual-phase steels is not proportional to *strain*. As shown in Fig. 3.31b, an increase in strain beyond 1.5–2 % generally leads to a decrease in strengthening as the result of overaging. This is an important factor that reduces the nonuniformity and variability of strength of parts made from DP steels exposed to different strains during forming.

Decrease in strengthening as a result of aging after a preliminary tensile deformation at more than 1.5 % strain or after rolling at more than 5 % reduction was interpreted in terms of the increased dislocation density and longer holding times required for pinning of the same portions of dislocations.

Another interpretation of decrease in strengthening with increasing pre-strain is based on the stronger tendency to form dislocation cell structure at small strains of about 5 % that reduces the effective dislocation density. It may be also assumed that the maximum in "ΔYS—pre-strain" curves, as in Fig. 3.31b, similarly to those in "YS—aging temperature" and "ΔYS—aging time" plots, is a result of conventional overaging due to transitions to precipitation coarsening or changes in precipitation sites. This is supported by the fact that the amount of pre-strain corresponding to the maximum strengthening decreases with increasing aging temperature (Storozeva et al. 1985).

In fact, aging evolves from the stage of individual *interstitial–dislocation* interactions to clustering or atmosphere formation and subsequent coalescence of dispersed precipitates. As particles coarsen, the length of free dislocation segment increases and the strengthening effect decreases. The strongest strengthening is achieved with the smallest precipitates. Simultaneous increase in prestrain and aging temperature or time results in earlier particle coarsening. As the aging temperature increases, the strengthening maximum moves from 3 % of strain during aging at 100 °C to 1.5 % at 200 °C and finally to zero strain at 400 °C.

The effect of *microstructure* on aging kinetics and strengthening has not been studied in detail.

It was shown that higher *amount of interstitial carbon* due to higher cooling rate applied shortens the incubation times for the first and second stage of aging and enlarges the yield strength increase at the second stage (Waterschoot et al. 2003).

Considering the contribution from residual stresses and dislocation density in ferrite, the strengthening produced by aging should depend on the *volume fraction of martensite* V_M. As V_M increased from 5 to 25 %, the strengthening effect increased from 35 to 100 MPa in a dual-phase steel quenched from 785 °C and deformed in tension at 2 %, with subsequent tempering at 260 °C. At prestrain of 4 % in steel of 0.17C–1.4Mn–0.4Si with 14 and 22 % martensite after annealing at 735 and 780 °C, the maximum strengthening effect after aging at 200 °C was 65 and 85 MPa, respectively (Gunduz et al. 2008).

The found changes in bake hardening of steels of 1.3Mn–0.1Si–0.65Cr with 0.05 and 0.10 % C, quenched from the same temperature in $\alpha + \gamma$ range (in particular BH_2 of 30 vs. 60 MPa) (Krieger et al. 2006), should be also explained as being due to difference in martensite volume fraction (10 and 30 %, respectively).

The effect of martensite volume fraction is illustrated in Fig. 3.34. An increase in V_M by varying carbon content in a 1.7 % Mn steel increases aging strengthening ΔYS due to the increased fraction of deformed ferrite grains and the higher dislocation density. In 0.05C–1.2Mn–0.4Cr steel with V_M varied by quenching from different intercritical temperatures, the effect of higher V_M was offset by lower concentration of interstitials in ferrite. The resultant strengthening after aging thus remained approximately constant.

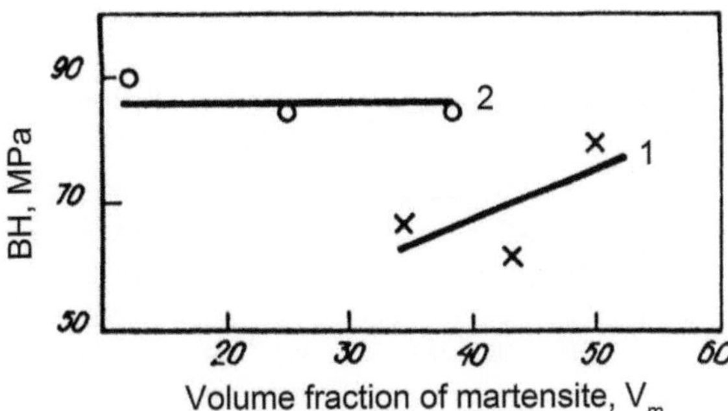

Fig. 3.34 Effect of volume fraction of martensite, V_M, on strain aging strengthening: 1—steels with 1.7 % Mn, V_M was varied by increase in steel carbon content; 2—steel 0.05C–1.2Mn–0.4Cr, V_M varied by variation of annealing temperature—original

Grain refinement reduces the effect of nonhomogeneity of dislocation distribution responsible for delaying natural aging. Together with shorter diffusion paths, grain refinements should accelerate aging and increase the resultant strengthening (Yegneswaran and Tangri 1983).

3.6 Tempering of Dual-Phase Steels

The influence of tempering on the properties of DP steels is becoming more important because of the current trends to utilize higher tempering temperatures (up to 450–500 °C) to improve the behavior of AHSS during localized deformation (cut edge, hole expansion).

Tensile strength of DP steels decreases upon tempering being mostly controlled by changes in the strength of martensite phase.

Conventional shot-term tempering of DP steels in continuous annealing lines at temperatures of ~200 °C mostly results in slight yield strength increase without changes in TS. Further heating leads to softening of DP steels mainly due to softening of martensite. However, as shown in Chap. 4, the actual level and the rate of softening depend on alloying composition that affect the behavior of martensite during tempering (precipitation and coarsening of carbides).

Based on the variations of calculated activation energy during tempering of martensite in DP steel, five stages of tempering were distinguished in the temperature range of 20–500 °C (Waterschoot et al. 2006).

Stage I. At tempering temperatures below 120 °C, the redistribution of carbon atoms takes place. Carbon segregation induces slight volume changes of less than 0.05 vol.% and significant reduction of tetragonality of martensite lattice.

Stage II. In the temperature range of 120–200 °C, η- or ε-carbides precipitate, which is accompanied by further volume reduction of about 0.30 % and complete loss of martensite tetragonality.

Stage III. At temperatures of 200–300 °C, Haäg-carbides precipitate following the precipitation of η-carbide.

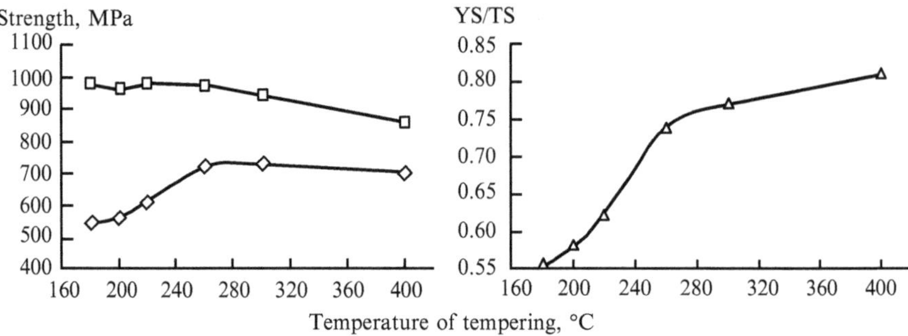

Fig. 3.35 Effect of tempering temperature on strength characteristics (YS, TS, and YS/TS ratio) of 0.08C–2.1Mn–0.7Si–Mo steel heated to 800 °C and water quenched from 760 °C (Fonstein et al. 2007)

Stage IV. At 250–350 °C, the volume increases due to decomposition of retained austenite.

Stage V. At final stage (290–390 °C), all transitional carbides formed during decomposition of martensite transform into cementite.

Typical changes in mechanical properties upon tempering are presented in Fig. 3.35.

Along with this general trend, changes in properties of the particular DP steel with tempering depend on the initial properties of martensite and its volume fraction that determine the total extent of distortions and hence the level of initial stresses in ferrite and the density of mobile dislocations. The performed study (Fonstein et al. 2007) aimed to compare the response to tempering of steels with simultaneously varied volume fraction of martensite and its carbon content. This was achieved by changing the temperature of quenching and the carbon concentration in the steel. After quenching from 735 °C, the 0.14 % C steel had strength of 785 MPa; the steel with 0.10 % C quenched from 760 and 780 °C demonstrated strength of 885 and 955 MPa, respectively. The volume fractions of martensite V_m in these steels were, respectively, 0.38, 0.47, and 0.52 corresponding to the calculated carbon content of 0.37, 0.21, and 0.19 %.

As shown in Fig. 3.36a, steel with the highest carbon content in martensite (~0.38 %) had the lowest YS/TS ratio. This again indicates the dominating role of the transformation-induced hardening, which is higher, the lower the M_s temperature and the higher the tetragonality of martensite, when abundance of free dislocation and high residual stresses lower the yield strength.

At the same time, as seen in Fig. 3.36b, the return of YPE (although quite small within a studied range of tempering temperatures) is observed at lower temperatures of tempering and is most intense in the same steel quenched from the lowest temperature. This, apparently, should be explained by increasing carbon content in ferrite when the temperature approaches A_{c1} and the related acceleration of free dislocations pinning. At constant carbon content in steel (0.10 %), quenching from higher temperature shifts the return of YPE to higher tempering temperatures due to

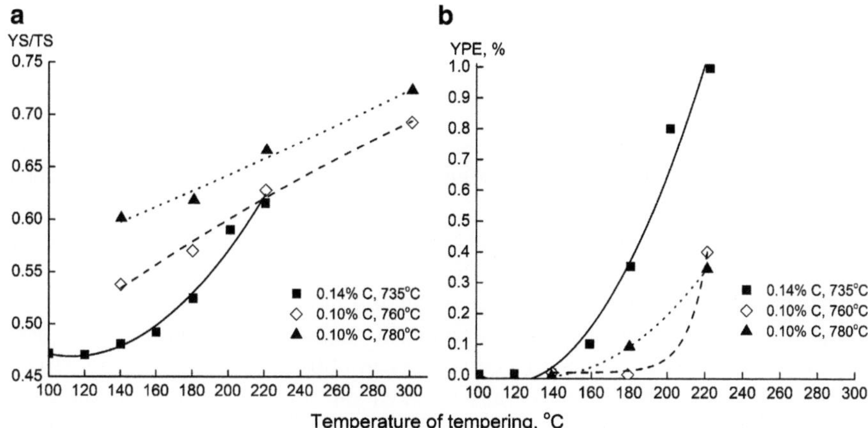

Fig. 3.36 Effect of the temperature of tempering on (**a**) the YS/TS ratio and (**b**) the return of the YPE in steels with 1.4–1.8 % Mn differing in the carbon content and in the temperature of quenching from the two-phase region ((Fonstein et al. 2007)

reduction of carbon concentration in ferrite and simultaneously due to favorable increase in density of mobile dislocations as a result of higher volume fraction of martensite.

Sometimes, an improvement of ductility of dual-phase steels is observed after tempering at temperatures above 300 °C. This is believed to be primarily by larger post-uniform deformation (Pradhan 1985) as a result of better strain compatibility between the phases and the suppression of microvoid nucleation and growth.

The increased ductility along with lower strain hardening and smaller differences in hardness of ferrite and martensite with increase in temperature of tempering result in substantial improvement of stretch flangeability evaluated by hole expansion.

3.7 Fracture Behavior of Dual-Phase Steels

Applications of high strength materials are often restricted by their fracture behavior. Higher strength implies higher resistance to deformation and hence a risk of fracture which is controlled by the fracture toughness expressed in particular through the critical stress intensity factor K_{Ic}. The value of K_{IC} describes the relationship between the applied load and the critical defect size required for catastrophic, i.e., macroscopically brittle fracture.

Many parts stamped from dual-phase steels are used in service under alternating loads. This, in addition, calls for the importance of the fatigue fracture resistance characterized, e.g., by the fatigue limit and low-cycle fatigue life. The more sophisticated parameters of fatigue are based on the concepts of fracture mechanics.

Among those are the fatigue crack growth rate, dl/dN (mm/cycle), and the constants c and m in the Paris equation:

$$dl/dN = c(\Delta K)^{m} \tag{3.35}$$

as well as ΔK_{th}, that is the stress-intensity threshold below which the crack propagation is virtually impossible.

From a practical point of view, it is also often important to distinguish the stages of crack initiation and crack propagation.

There exist the models capable of predicting the crack propagation resistance (K_{1c}) from the parameters of microstructure. Using these models, the crack propagation characteristics can be to a certain extent utilized in microstructure design.

At insufficient resistance to crack propagation, the crack initiation stage can become critical. Stress concentrators, small notches, or coalescence of internal microvoids can facilitate crack initiation.

At present, no analytical tools are available to predict the crack initiation characteristics.

3.7.1   Theoretical Concepts of Fracture of Heterogeneous Materials

General approach to predicting the effects of microstructure parameters on the fracture behavior of heterogeneous materials has not been fully formulated and needs experimental verification.

Relevant discussions usually focus on four possible aspects of the coexistence of phases with vastly different properties (Hornbogen 1982):

- How strong is the influence of the second phase on the homogeneity of deformation?
- How strong is the influence of the second phase including internal cracks and nonmetallic inclusions on the driving forces for crack propagation and on the crack paths?
- How well can the fracture characteristics be described by the properties of individual phases?
- What is the relation between fracture parameters, such as the size of the plastic zone, r_y, or the critical crack opening displacement, δ_c, and the size of microstructure constituents or individual phases or particle spacing?

Nonhomogeneity of plastic deformation has the strongest impact on crack initiation as it accelerates void nucleation (see Sect. 3.7.2). At the same time, however, plastic nonhomogeneity often increases the resistance to crack propagation. Under certain conditions, the nonhomogeneous microstructure can either accelerate or delay crack growth, e.g., through crack branching. Another possibility is the local strain-induced martensitic transformation of retained austenite at the crack front in DP steels containing some amount of austenite. The resulting volume change generates internal compressive stresses in martensite, thus reducing the level of the applied tensile stresses.

To predict fracture characteristics of nonhomogeneous materials, it is critical to evaluate whether or not the contributions of the individual phases are additive and also what relation exists between their geometrical parameters and the size of the plastic zone.

Using the rule of mixtures, the property p of two-phase microstructure can be expressed with the properties of individual phases and their volume fractions:

$$p = p_\alpha V_\alpha + p_\beta V_\beta \tag{3.36}$$

This approach can be applied to fracture characteristics if crack propagation is not selective with respect to individual phases, i.e., if the contributions from both phases to the fracture path are proportional to their volume fractions. It follows then that the crack propagation energy is

$$G_{1c} = G_{1c\alpha} V_\alpha + G_{1c\beta} V_\beta \tag{3.37}$$

Thus, the presence of a tougher phase such as fine layers of retained austenite between laths of martensite (Romaniv et al. 1977) or regions of tough martensite in Fe–Mn or Fe–Ni alloys (Stratmann and Hornbogen 1979) significantly increases the fracture toughness (G_{1c} or K_{1c}). By contrast, the toughness of steels can decrease if the crack propagates along the path through low-energy regions.

For certain network-type two-phase microstructures, it is possible that the crack essentially avoids one of the phases. In this case,

$$p = p_\alpha V_\alpha \tag{3.38}$$

i.e., the properties of an alloy depend on the properties and the amount of only one phase.

In practice, however, no such an extreme case exists, and the crack propagates primarily in one of the phases only if this phase has a special orientation with respect to external stress. As a result, there is some increase in the fraction of fracture *surface* compared to the *volume fraction* of the phase indicating relatively preferable fracture in one of the phases (Chernyavskii and Fonstein 1984).

The contribution of a phase to crack propagation and hence to fracture surface can be predicted by comparing the sizes of the plastic deformation zone, r_y, and the element of microstructure, D. The latter can be related to the grain size of soft phase or the spacing between particles of harder phase.

According to fracture mechanics, the plastic zone size

$$r_y = \frac{1}{2\pi} \left(\frac{K_{1c}}{\mathrm{YS}} \right)^2 \tag{3.39}$$

is determined by the level or in case of cyclic loading, by the amplitude of external stresses. In the latter case, ΔK should be used instead of K_{1c}.

The rule of mixture is adequate with respect to fracture characteristics if $r_y > D$. In the case when $r_y < D$, the plastic zone is localized in one of the phases and the crack propagates selectively through the less resistant phase. The condition of homogeneous deformation is $r_y \geq D$, when many dislocation sources are activated both at the boundaries and in the interiors of both phases.

Good illustrations of different dependences of fracture characteristics on the relative properties of phase and on the type of the applied stress were given by comparing nickel-bearing (Stratmann and Hornbogen 1979) and medium-carbon (Romaniv et al. 1977) structural steels quenched from the $\alpha + \gamma$ region. In both cases, the steels were finally tempered after heat treatment. The ferrite/martensite ratios were 50:50 and 20:80, respectively. Compared to a homogeneous structure, K_{1c} was found to increase in dual-phase nickel-bearing steels and to decrease dramatically in medium-carbon martensite–ferrite steel. In cyclic loading, however, the dual-phase microstructure of nickel-bearing steels exhibited poorer performance, while medium-carbon steel rendered higher fatigue resistance.

Ferrite cleavage can be observed at a small volume fraction of martensite when ferrite is present as a continuous phase. This is due to highly constrained plastic deformation of the ferrite. Internal tensile stresses in ferrite induced by local $\gamma \rightarrow \alpha'$ transformation are similar to those near the notch tip, and the stress state in α'-phase under external loading is tri-axial. The result of this constraint of plastic relaxation in ferrite is that the softer phase contributes to crack initiation by cleavage mechanism rather than delays crack initiation and growth.

If ferrite phase is discontinuous at higher volume fraction of the hard phase, then martensite acts as the tougher phase that can inhibit cleavage in ferrite (Bortsov and Fonstein 1984b).

3.7.2   Microstructural Features Controlling Fracture Initiation in Dual-Phase Steels

The harder phase primarily affects *crack initiation* occurs as a result of unequal phase strains, fracture of hard particles, or decohesion of phase interfaces. The amount of strain that leads to nucleation of microvoids in dual-phase steels is determined by the strength of interfaces, by the strength and shape of the second-phase particles, and by the properties of ferrite that control the unequal distribution of plastic strains between phases during deformation. The extent of interfacial void nucleation can also be effectively changed by solutes that form interface segregations.

Detailed data on fracture initiation in dual-phase steels are rather scarce and inconsistent.

In particular, Araki et al. (1977b) found that the product TS–TE decreased with volume fraction of martensite increasing above 0.2. They attributed this behavior to early void nucleation due to fracture of martensite particles that accelerated the onset of plastic instability. However, direct metallographic observations did not reveal any fracture of martensite.

Stevenson also reported (Stevenson 1977) that voids in as-quenched dual-phase steels were formed primarily as a result of fracture of martensite, but in tempered dual-phase steels, the dominating mechanism was phase decohesion because of higher martensite ductility. On the other hand, Tomota et al. (1981) demonstrated that lower hardness of martensite after tempering also promotes void nucleation due to fracture of martensite, although higher pre-strain is required. This behavior is more likely when the difference in ductility between the phases is small and when strains are equally distributed in two phases, which decreases the internal stresses at the phase interfaces.

Gurland (1982) affirmed that fracture was initiated along the martensite–ferrite interfaces although his micrographs showing the voids do not contradict the assumption of preliminary fracture of fine martensite particles.

Void formation along martensite–ferrite interfaces is favored by strain concentration in the ferrite. Because ferrite has lower strength and higher ductility than martensite, the voids are found to be initiated at the ferrite side of the interface. The preferential void nucleation in ferrite of quenched and low temperature tempered DP steels was also revealed by Hong and Lei (Hong et al. 1986) who used successive sectioning of samples along the planes parallel to the fracture surface. They determined the void density, ρ, and the void size as functions of the distance, z, from fracture surface and noted that in tempered samples, ρ decreases more rapidly with increasing z, apparently because of a higher strain required for void nucleation due to softening of martensite and decrease in internal stresses after tempering.

Figure 3.37 shows sections of broken tensile specimens of ferrite–martensite and ferrite–pearlite steels. It is seen that in ferrite–martensite steel, the voids had been initiated fairly far from the fracture surface slightly before the ultimate uniform strain was attained. This is different from void localization close to fracture surface in ferrite–pearlite steel that evidently requires higher plastic strain for initiation of voids. Whether the voids were generated by interfacial decohesion or particle cleavage remains uncertain. It may very well be that the maximum attainable strain in martensite at ultimate uniform strain of DP steel of given strength is insufficient to fracture martensite in steel of that strength of ~600 MPa.

Fatigue crack nucleation is a result of nonhomogeneous plastic deformation in ferrite that gives rise to accumulation of slip bands at the interphase boundaries. During the tensile stage of a fatigue cycle, the highest stresses are generated at phase interfaces and thus the initiation of a fatigue crack, as well as its early growth stage, are related to interface surface.

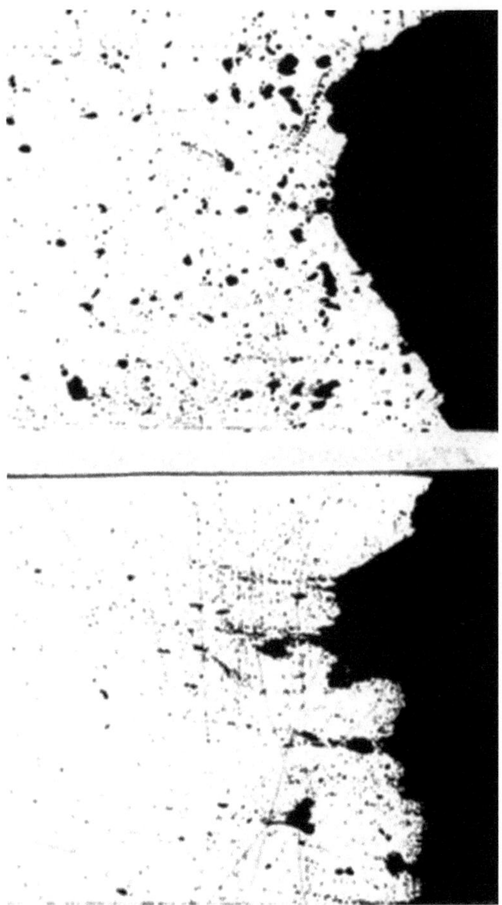

Fig. 3.37 Sections of broken tensile specimens of ferrite–martensite (**a**) and ferrite–pearlite (**b**) steels, ×200 (Bortsov and Fonstein 1986)

3.7.3   *Effect of Microstructure of DP Steel on Crack Propagation*

In general, three characteristics of martensite, i.e., its volume fraction, morphology (shape, size, and distribution), and toughness, which are determined by carbon content in martensite, along with the properties of ferrite affect the toughness of dual-phase steels. The parameters of martensite are more or less interrelated. For example, when steel composition remains constant, an increase in volume fraction of martensite, V_M, results in its higher intrinsic toughness of martensite due to its lower carbon content, C_M. Decrease in V_M in given steel usually also induces an additional favorable effect of reduced martensite continuity. This should retard crack propagation because of the necessity for a crack to bypass or cleave the dispersed martensite particles that increases toughness. The effect of the martensite particle dispersion on toughness of dual-phase steels was discussed by Tomota and

Tamura (1981) who showed that coarse dual-phase microstructures exhibit lower toughness level evaluated from K_{1c}.

Increase in strength at improved ductility due to microstructure refinement is attributable to the combined effect of strengthened ferrite and enhanced toughness of martensite. This leads to less severe stress/strain partitioning and better interface cohesion (Calcagnotto et al. 2011) so that crack propagation become more sluggish.

Dual-phase steels are featured with high strain-hardening rate. One can expect them to exhibit enhanced toughness according to the equation (Kraft 1964):

$$K_{1c} = En\sqrt{d} \tag{3.40}$$

where E—Young modulus, n—strain-hardening coefficient, and d—critical structure unit size.

To compare the dual-phase properties with those in fully martensitic condition, Tkach et al. (1984) studied the resistance to static cracking of laboratory-melted 0.06C–1.8Cr–1.6Ni–0.6Mo steel ($A_{c1} = 715$ °C and $A_{c3} = 850$ °C). The relatively high alloy content was used to ensure sufficient hardenability in heavy sections over a wide range of C_γ after annealing at different temperatures in the intercritical region. Preliminary heat treatment included normalization from 1050 °C and annealing at 950 °C with subsequent furnace cooling. Final heat treatment included heating to 725, 750, and 870 °C, holding for 30 min and water quench.

K_{1c} was determined based on the J-integral concept using three point bending tests of pre-cracked specimens with dimensions of 12×18 mm. It was found that an increase in strength with volume fraction of martensite led to gradual increase in K_{1c} (Fig. 3.38), which contrasted to typically inverse relation between YS and K_{1c}.

Increased resistance of ferrite–martensite microstructures to static cracking with higher volume fraction of martensite was attributed to ability of martensite to

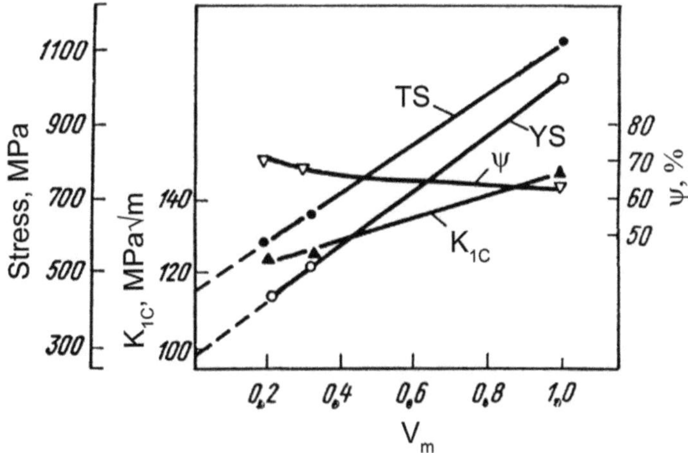

Fig. 3.38 Effect of martensite volume fraction, V_M, on mechanical properties of 06Cr2Ni2Mo steel (Tkach et al. 1984): here ψ indicates RA, reduction of area

efficiently delay crack growth. During fracture toughness testing of ferrite-based microstructure, the cracks grew predominantly by the cleavage mechanism, resulting in relatively low values of K_{1c}. In contrast, low-carbon martensite exhibited micro-ductile fracture with relatively high energy. Therefore, an increase in V_M in steel of a given composition by increasing the annealing temperature involved not only contribution from the phase of higher toughness but also an improvement in toughness of martensite as a result of decreasing carbon content.

3.7.4   Fatigue Resistance

It was established that commercial hot-rolled 0.1C–1.9Mn steel with dual-phase microstructure after intercritical heat treatment, 5 % cold reduction, and tempering at 200 °C for 1 h, exhibited a significant increase in the fatigue limit and a 3- to 3.5-fold increase of the number of cycles to failure in the low-cycle fatigue regime in comparison with hot-rolled ferrite–pearlite steel of the same composition after cold deformation and tempering.

The yield strengths of compared cold-rolled dual-phase steel with 0.05C–1.1Mn–0.3Cr and HSLA steel produced using batch annealing steels of 0.08C–0.9Mn–0.4Si–Ti were 285 and 315 MPa and the tensile strengths 545 and 515 MPa, respectively. The maximum tensile stress amplitudes after 10^6 cycle fatigue test at 20–60 Hz with $R = 0.1$ differed only by 10 MPa in as-heat-treated condition but was 40 MPa after 5 % deformation and tempering at 200 °C for 1 h. At maximum stress amplitude of 420 MPa the endurance limit of these steels after work hardening and aging differed by more than two orders of magnitude.

Hardness of martensite can significantly impact the growth of fatigue crack varying its path through the microstructure. For example, in a dual-phase steel with $V_M = 0.44$ and $YS_M/YS_F = 4.5$, the fatigue crack tended to zigzag as it moved through ferrite (Tamura and Tomota 1973).

Other authors note that σ_{-1} of dual-phase steels strongly depends on spacing between martensite particles, $L_{>M<}$ and with identical $L_{>M<}$-values σ_{-1} can be independent of C_M and V_M. It was suggested that a fatigue crack initiated in ferrite phase is often arrested by the phase interface and hence its propagation can be primarily controlled by microscopic parameters rather than by phase properties, i.e., by the amount, shape, and continuity of martensite particles.

3.7.5   Resistance of Dual-Phase Steels to Hydrogen Embrittlement

High strength materials are often susceptible to hydrogen embrittlement. In particular, Davies (1981b) studied dual-phase steels with TS = 690 MPa and found that hydrogen pickup resulted in a loss of ductility and in a reduction in long-term strength, σ_τ, during testing for 50–100 h, with the fracture mechanism changing from dimple-type to transcrystalline cleavage. It was suggested that hydrogen embrittlement is brought about by the presence of 15–20 % of high-carbon martensite (with $C_M \approx 0.6$ %). Davies supported this statement with observations of increasing σ_τ with increasing tempering temperature. These observations did not contradict the key role of internal stresses in dual-phase steels that are additionally weakening phase interfaces, where hydrogen can accumulate.

In fact, the extent of hydrogen embrittlement in dual-phase steels was not very significant. After 50–100 h of holding in hydrogen, delayed fracture was observed at stresses higher than the yield strength.

Interesting results were obtained when studying of the effects of cold working usually employed for low-carbon steels with ferrite–martensite structures. Additional dislocations introduced by plastic deformation increased σ_τ (the load carrying capacity of steel in a hydrogen environment) by 10–20 %.

Morphology of microstructure of DP steels is also important. The best resistance against hydrogen embrittlement was demonstrated by steel with ferrite–martensite microstructure subjected to pre-quenching, repeated heating, and subsequent controlled cooling at 30–40 °C/s from the γ-region. The resultant microstructure had fine ferrite grain size and maximum amount of low-carbon uniformly distributed needle-like martensite (up to 45 %). Coarser ferrite grain size, lower amount of martensite, and higher carbon content in martensite impaired the resistance to hydrogen embrittlement of dual-phase steels.

Similarly to earlier study (Davies 1981b), low-temperature tempering at 180–300 °C substantially improved the resistance to hydrogen embrittlement. The beneficial effect of low temperature tempering is generally attributed to deeper relaxation of internal stresses that usually facilitate hydrogen embrittlement. Figure 3.39 compares hydrogen embrittlement of steels with and without cold working to reach TS > 800 MPa. The materials were a dual-

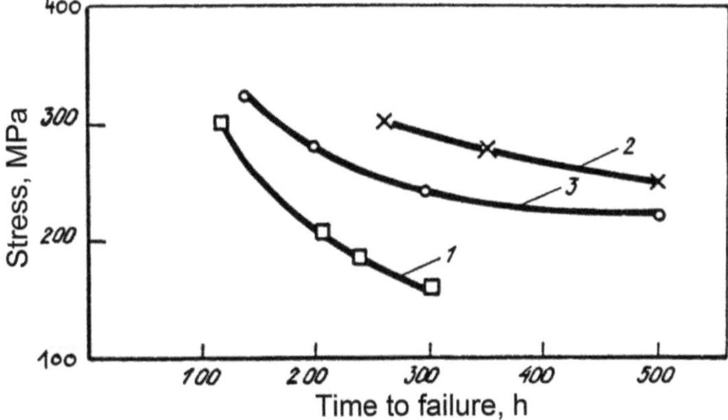

Fig. 3.39 Rupture strength of steels during continuous electrolytic hydrogen charging: 1—spheroidized steel 38Cr1 after cold drawing to TS ≈ 800 MPa; 2—as-annealed DP 06CrMnB steel; 3—same DP steel cold drawn to TS ≈ 800 MPa—original

phase 0.06C–1.3Mn–0.40Si–0.5Cr–0.0035B steel and medium-carbon 0.36C–0.59Mn–1.08Cr steel, both of which are used for fastener produced by cold heading. It is evident that the susceptibility to hydrogen embrittlement of low-carbon low-alloy DP steel is substantially weaker albeit equal or even higher strength.

Usually the susceptibility to hydrogen embrittlement increases with tensile strength of steel. Results of the study of DP steel with TS > 1180 MPa (Takagi et al. 2012) showed that hydrogen embrittlement was facilitated by higher stress, higher strain, and higher hydrogen content during testing, as well as by using U-bending mode of strain as opposed to cup drawing. On the other hand, it was shown how fast the content of initially charged diffusible hydrogen (about 1 ppm) decreased at elevated temperatures becoming equal to 0.1 ppm after 3 min of holding at 170 °C, which is typically applied after painting car parts.

In general, for accurate evaluation of delayed fracture of automotive parts, it is important to clarify the type of deformation of particular parts, real sequence of heat treatment, and potential hydrogen pick-up, as well as to consider the relevance of hydrogen content obtained by forced charging to the possible hydrogen accumulation during service.

The concentration of hydrogen absorbed, for example, during pretreatment in painting cycle is much lower than the critical concentration under all of the analyzed conditions (Lovicu et al. 2012). Moreover, no hydrogen-induced delayed cracks were detected during service of parts manufactured from DP steels.

3.8 Summary

Different characteristics of mechanical behavior are related to different parameters of microstructure of dual-phase steels.

Strength of DP steels is primarily controlled by the volume fraction of martensite, whereas the ductile properties are determined by ductility of ferrite matrix and additionally by the geometry of martensite particles.

Strain-hardening rate, yield strength, and strengthening during aging of DP steels depend on volume fraction of martensite and its morphology ($\sqrt{V_M}/L_M$ ratio), on the carbon content in martensite, the initial martensite transformation temperature, the volume expansion during the $\gamma \rightarrow \alpha'$ transformation, and, hence, on the level of residual stresses and the density of unpinned mobile dislocations.

References

Adamczyk, J., and A. Graijcar. 2007. "Heat Treatment and Mechanical Properties of Low-Carbon Steel with Dual-Phase Microstructure." *Journal of Achievements in Materials and Manufacturing Engineering* 22 (5): 13–20.

Araki, K., S.H. Fukunaka, and K. Uchida. 1977a. "Development of Continuously Annealed High Strength Cold Rolled Sheet Steels." *Trans ISIJ* 17 (12): 701–9.

Araki, K., Y. Takada, and K. Nakaoka. 1977b. "Work Hardening of Continuously Annealed Dual-Phase Steels." *Trans ISIJ* 17 (12): 710–17.

Ashby, M.F. 1966. "Work Hardening of Dispersion-Hardened Crystals." *Philosophical Magazine* 14 (132): 1157–78.

Bag, A., K.K. Ray, and E.S. Dwarakadasa. 1999. "Influence of Martensite Content and Morphology on Tensile and Impact Properties of High-Martensite Dual-Phase Steels." *Metallurgical and Materials Transactions A* 30A (5): 1193–202.

Bailey, D.I., and R. Stevenson. 1979. "High Strength Low Carbon Sheet Steel by Thermomechanical Treatment: I. Strengthening Mechanisms." *Metallurgical Transactions* 10A (1): 47–55.

Baird, J.D. 1963. "Strain Aging in Steel - Critical Review." *Iron and Steel* 36 (7): 326–34.

Balliger, N.K., and T. Gladman. 1981. "Work Hardening of Dual-Phase Steel." *Metal Science* 15 (3): 95–108.

Bergstrom, Y., and Y. Granbom. 2008. "Model for the Stress-Strain Behavior of Dual-Phase Steels." In *IDDRG*, 173–84. Olofstrom, Sweden.

Bhadeshia, H.K.D.H., and D.V. Edmonds. 1980. "Analysis of Mechanical Properties and Microstructure of High-Silicon Dual-Phase Steel." *Metal Science* 14 (2): 41–49.

Bleck, W., and K. Phiu-On. 2005. "Grain Refinement and Mechanical Properties in Advanced High Strength Sheet Steels." In *HSLA' 2005*, 50–57. Sanya, China.

Bortsov, A.N., and N.M. Fonstein. 1984a. "Influence of Carbon Concentration on the Mechanical Properties of the Low-Carbon Ferrite-Martensite Steels." *Physical Metallurgy and Metallography (USSR)* 57 (4): 142–48.

———. 1984b. "Influence of Cold Deformation and Low-Temperature Tempering on Mechanical Properties of Dual-Phase Ferritic-Martensitic Steels." *Soviet Materials Science* 20 (2): 142–47.

———. 1986. "The Distribution of Strain between the Phases of Ferrite-Martensite Steel." *Physical Metallurgy and Metallography (USSR)* 61 (2): 74–81.

Bronfin, B.M., M.I. Goldshtein, and V.P. Shveikin. 1986. "Effect of the Cooling Rate from the Intercritical Temperature Range on Strain Hardening and Ageing of Steel 05G2S2." *The Physics of Metals and Metallography (USSR)*, no. 11.

Bronfin, B.M., M.I. Goldstein, A.A. Emelyanov, and et al. 1983. "The Strength and Ductility of Dual-Phase Ferrite-martensite Steel." *The Physics of Metals and Metallography (USSR)* 56 (1): 167–73.

Brown, L.M., and D.R. Clarke. 1975. "Work Hardening due to Internal Stresses in Composite Materials." *Acta Metallurgica* 23 (7): 821–30.

Brown, L.M., and W.M. Stobbs. 1971. "Work Hardening of Cupper Silica. II The Role of Plastic Relaxation." *Philosophical Magazine* 23 (185): 1201–33.

Bucher, J.H., E.G. Hamburg, and J.F. Butler. 1979. "Property Characterization of VAN-QN Dual-Phase Steels." In *Structure and Properties of Dual-Phase Steels*, 346–59. New Orleans.

Budford, D.A., D.K. Matlock, and G. Krauss. 1985. "Effect of Microstructural Refinement on the Deformation Behavior of Dual-Phase Steels." In *Strength of Metals and Alloys*, 1:189–94. Montreal, Canada.

Calcagnotto, M., Y. Adachi, D. Ponge, and D. Raabe. 2011. "Deformation and Fracture Mechanism in Fine- and Ultrafined-Grained Ferrite/martensite Dual-Phase Steels and Effect of Aging." *Acta Materiala* 59: 658–70.

Chernyavskii, K.S., N.M. Fonstein. 1984. "Stereological Study of Fracture." *Industrial Laboratory* 51 (3): 45–49.

Colla, V., and et al. 2009. "Strain Hardening Behavior of Dual-phase Steels." *Metallurgical and Materials Transactions A* 40A (11): 2557–67.

Cong, Z.H., and et al. 2009. "Stress and Strain Partitioning of Ferrite and Martensite during Deformation." *Metallurgical and Materials Transactions A* 40A (6): 1383–87.

Davies, R.G. 1977. "On the Ductility of Dual-Phase Steels." In *Formable HSLA and Dual-Phase Steels*, 25–39.

———. 1978a. "Influence of Martensite Composition and Content on the Properties of Dual-Phase Steels." *Metallurgical Transactions* 9A (5): 671–79.

———. 1978b. "The Mechanical Properties of Zero Carbon Ferrite plus Martensite Structure." Metallurgical Transactions 9A (3): 451–66.

———. 1978c. "The Deformation Behavior of a Vanadium-Strengthened Dual Phase Steel." *Metallurgical Transactions A* 9 (1): 41–52. doi:10.1007/BF02647169.

Davies, R.G. 1979a. "Early Stages of Yielding and Strain Ageing of a Vanadium-Containing Dual-Phase Steels." Metallurgical Transactions 10A (10): 1549–55.

———. 1979b. "Influence of Silicon and Phosphorus on the Mechanical Properties of Both HSLA and Dual-Phase Steels." Metallurgical Transactions 10a (1): 113–18.

———. 1981a. "Tempering of Dual-Phase Steels." In *Fundamentals of Dual-Phase Steels*, 265–78.

———. 1981b. "Hydrogen Embrittlement of Dual-Phase Steel." Metallurgical Transactions 12A (9): 1667–72.

De Cooman, B.C., S. Lee, and Y. Estrin. 2013. "Strain Hardening Control in AHSS for Automotive Applications." In Veil, CO, USA.

Eldis, George T. 1981. "Influence of Microstructure and Testing Procedure on the Measured Mechanical Properties of Heat Treated Dual-Phase Steels." In *Conference of Structure and Properties of Dual-Phase Steels,* 202–20. New Orleans, LA, USA

Eshelby, J.D. 1957. "The Determination of the Elastic Field of an Ellipsoidal Inclusion, and Related Problems." *Proceedings of the Royal Society (London)* A241: 376–96.

Fisher, I., and Gurland, G. 1981. "The Effect of Alloy Deformation on the Average Spacing Parameters of Nondeforming Particles." *Metallurgical Transactions* 12 A (2): 167–171.

Fonstein, N. 2005. "Effect of Al, C, and Mn on Structure and Mechanical Properties of Dual-Phase and TRIP Steels." In *HSLA '05*. Sanya, China.

Fonstein, N., H.J. Jun, G. Huang, and et al. 2011. "Effect of Bainite on Mechanical Properties of Multi-Phase Ferrite-Bainite-Martensite Steels." In *MST'11*.

Fonstein, N.M., A.N. Bortsov, and K.S. Chernyavskii. 1984. "Stereological Description of Two-Phase Structure." *Industrial Laboratory* 51 (12): 34–37.

Fonstein, N.M., and T.M. Efimova. 2006. "Study of the Effects of 'New' Ferrite on the Properties of Dual-Phase Steels." *Metallurgist* 50 (9–10): 481–89.

Fonstein, N.M., M. Kapustin, N. Pottore, I. Gupta, and O. Yakubovsky. 2007. "Factors That Determine the Level of the Yield Strength and the Return of the Yield-Point Elongation in Low-Alloy Ferrite-Martensite Steels." *The Physics of Metals and Metallography* 104 (3): 323–336.

Fonstein, N.M., L.M. Storozeva, and B.A. Bukreev. 1985. "The Effects of Vanadium on the Properties of Two-Phase Ferrito-Martensitic Steels." *Russian Metallurgy*, no. 2: 111–16.

Gerbase, I., I.D. Embury, and R.M. Hobbs. 1979. "The Mechanical Behavior of Some Dual-Phase Steels - With Emphasis on the Initial Work Hardening Rate." In *Structure and Properties of Dual-Phase Steels*, 118–44.

Girina, O.A., N.M. Fonstein, and D. Bhattacharya. 2008. "Effect of Nb on the Phase Transformation and Mechanical Properties of Advanced High Strength Dual-Phase Steels." In *International Conference of New Developments on Metallurgy and Applications of High Strength Steels*, 29–35. Buenos Aires.

Greday, T., H. Mathy, and P. Messien. 1979. "About Different Ways to Obtain Multi-Phase Steels." In *Structure and Properties of Dual-Phase Steels*, 260–80.

Gribb, W.R., and I.M. Rigsbee. 1979. "Work-Hardening Behavior and Its Relationship to the Microstructure and Mechanical Properties of Dual-Phase Steels." In *Structure and Properties of Dual-Phase Steels*, 91–117.

Gunduz, S., B. Demir, and R. Kacar. 2008. "Effect of Aging Temperature and Martensite by Volume on Strain Aging Behavior of Dual-Phase Steel." *Ironmaking and Steelmaking* 35 (1): 63–68.

Gurland, G. 1982. *Metallurgical Transactions* 13A (10): 1821–26.

Hansen, S.S., and R.R. Pradhan. 1981. "Structure/property Relationships and Continuous Yielding Behavior in Dual-Phase Steels." In *Conference of Fundamentals of Dual-Phase Steels,* 113–44. Chicago, IL, USA

Hashiguchi, K., and M. Nashida. 1980. "Effect of Alloying Element and Cooling Rate after Annealing on Mechanical Properties of Dual Phase Steel." Kawasaki Steel Techn. Report, no. 1: 70–78.

Himmel, L., K. Goodman, and W.L. Haworth. 1981. "Strain Ageing in a Dual-Phase Steel Containing Vanadium." In *Fundamentals of Dual-Phase Steels*, 309–14. Chicago, IL, USA.

Hollomon, J.H. 1945. "Tensile Deformation." *Transactions of ASM*, no. 162: 268–90.

Hong, B.D., T.C. Lei, J.B. Yang, and D.M. Jiang. 1986. "Deformation-Aging Strengthening of High Strength Steel 30CrMnSi." In *7th International Conference*, 2:1013–18. Montreal.

Hornbogen, E. 1982. "Microstructure and Mechanism of Fracture." In *Strength of Metals and Alloys*, 3:1059–73. Melbourne, Australia.

Jaoul B. 1957. *Journal of Mechanics and Physics of Solids* 5 (1): 95–114.

Jiang, Z., Z. Guan, and J. Lian. 1995. "Effect of Microstructure Variables on the Deformation Behavior of Dual-phase Steels." *Materials Science and Engineering A*, no. A190: 55–64.

Johnston, W.G., and J.J. Gilman. 1959. "Dislocation Velocities, Dislocation Densities and Plastic Flow in Lithium Fluoride Crystals." *Journal of Applied Physics* 30 (2): 129–44.

Kim, N., and G. Thomas. 1981. "Effect of Morphology on the Mechanical Behavior of Fe/2Si/0.1C Steel." *Metallurgical Transactions* 12A (3): 483–89.

Koo, I.Y., M.I. Young, and G. Thomas. 1980. "On the Law of Mixtures and Dual-Phase Steels." *Metallurgical Transactions* 11 A (5): 852–54.

Korzakwa, D.A., R.D. Lawson, D.K. Matlock, and G. Krauss. 1980. "A Consideration of Models Describing the Strength and Ductility of Dual-Phase Steels." *Scripta Materialia* 14 (9): 1023–28.

Korzekwa, D.A., D.K. Matlock, and G. Krauss. 1984. "Dislocation Substructure as a Function of Strain in a Dual-Phase Steel." *Metallurgical and Materials Transactions A* 15A (6): 1221.

Kraft, I.M. 1964. "Correlation of Plane Strain Crack Toughness with Strain Hardening Characteristics of a Low, Medium and High Strength Steel." *Applied Materials Research* 3 (2): 83–96.

Krieger, M., and et al. 2006. "Mechanical Properties and Bake Hardening Behavior of Two Cold Rolled Multiphase Sheet Steels Subjected to CGL Heat Treatment Simulation." *Steel Research International* 77 (9–10): 668–74.

Krupitzer, R.P. 1981. "Strain Ageing Behavior in Continuously-Annealed Dual-Phase Steel." In *Fundamentals of Dual-Phase Steels*, 315–30.

Lanzilotto, C.A., and F.B. Pickering. 1982. "Structure-Property Relationship in Dual-Phase Steels." *Material Science* 16 (8): 371–82.

Lawson, R.D., D.K. Matlock, and G. Krauss. 1981. "The Effect of Microstructure on the Deformation Behavior and Mechanical Properties of Dual-Phase Steel." In *Fundamentals of Dual-Phase Steels*, 347–81.

Lee, H.C., and G. Gurland. 1978. "Hardness and Deformation of Cemented Tungsten Carbide." *Materials, Science and Engineering* 33 (1): 125–33.

Leslie, W., and R. Sober. 1967. "The Strength of Ferrite and of Martensite as Functions of Compositions, Temperature and Strain Rate." *Transactions of ASM* 60: 459–84.

Lovicu, G., M. Bottazi, F. D'Aiuto, and M. DeSanctis. 2012. "Hydrogen Embrittlement of Automotive Advanced High-Strength Steel." *Metallurgical and Materials Transactions A* 43 (11): 4075–4087

Ludwig, P. 1909. *Element Der Technologichen Mechanick*. Berlin: Springer.

Ma, M.T., D.Z. Wang, and B.R. Wu. 1983. "On the Law of Mixture in Dual-Phase Steels." In *Mechanical Behavior of Materials*, 2:1067–73. Stockholm, Sweden.

Marder, A.R. 1977. "Factors Affecting the Ductility of 'Dual-Phase' Alloys"." In *Formable HSLA and Dual-Phase Steels, Metallurgical Society of AIME*, 87–98.

———. 1982. "Deformation Characteristics of Dual-Phase Steels." *Metallurgical Transactions* 13A (1): 85–92.

Marder, A.R., and B.L. Bramfitt. 1979. "Processing of a Molybdenum-Bearing Dual-Phase Steel." In *Conference of Structure and Properties of Dual-Phase Steels,* 242–59. New Orleans, LA, USA.

Matlock, David K., G. Krauss, L.F. Romas, and G.S. Huppi. 1979. "A Correlation of Processing Variables with Deformation Behavior of Dual-Phase Steels." In *Conference of Structure and Properties of Dual-Phase Steels,* 62–90. New Orleans, LA, USA.

Matsuda, R., and T. Shimomura. 1980. "Production of High Strength Cold-Rolled Steel Sheets by the NKK Continuous Annealing Line Process (NKK-CAL Process)." *Nippon Kokan Technical Report*, no. 20: 1–9.

Mileiko, C.T. 1969. "The Tensile Strength and Ductility of Continuous Fibre Composites." *Journal of Materials Science* 4 (11): 974–77.

Morrison, W.B. 1966. "The Effect of Grain Size on Stress-Strain Relationship in Low Carbon Steels." *Transactions of ASM* 59: 224–46.

Mould, R.R., and C.C. Skena. 1977. "Structure and Properties of Cold-Rolled Ferrite-Martensite (Dual-Phase) Steel Sheets". In *Formable HSLA and Dual-Phase Steels* 181–204, Chicago, IL, USA.

Nakaoka, K., K. Araki, and K. Kurihara. 1977. "Strength, Ductility and Ageing Properties of Continuously-Annealed Dual-Phase High Strength Sheet Steels." In *Formable HSLA and Dual-Phase Steels,* 126–41.

Nishimoto, A., Y. Hosoya, and K. Nakaoka. 1981. "A New Type of Dual Phase Steel Sheet for Automobile." *Transactions of ISIJ* 21 (11): 778–82.

Ostrom, P. 1981. "Deformation Models for Two-Phase Materials." *Metallurgical Transactions* 12A (12): 355–57.

Pouranvari, M. 2010. "Tensile Strength and Ductility of ferrite-Martensite Dual-Phase Steels." *MjoM* 16 (3): 187–94.

Pradhan, R. 1985. "Metallurgical Aspects of Quenched and Tempered Dual-Phase Steels Produced via Continuous Annealing." In *Technology of Continuously Annealed Cold -Rolled Sheet Steel*, 297–318. Detroit, MI: Metal/Society of AIME.

Ramos, L.F., D.K. Matlock, and G. Krauss 1979. "On the Deformation Behavior of Dual-Phase Steels." *Metallurgical Transactions* 10A (2): 259–61.

Rashid, M.S., and B.V.N. Rao. 1981. "Tempering Characteristic of a Vanadium Containing Dual-Phase Steels." In *Fundamentals of Dual-Phase Steels*, 249–64.

Rigsbee, J.M., J.K. Abraham, A.T. Davenport, J.E. Franklin, and J.W. Pickens. 1981. "Structure-Processing and Structure-Property Relationships in Commercially Processed Dual-Phase Steels." In *Conference of Structure and Properties of Dual-Phase Steels,* 304–29, New Orleans, LA, USA.

Rizk A., and Bourell D.L. 1982. "Dislocation Density Contribution to Strength of Dual Phase Steels." *Scripta Materialia* 16 (12): 1321–24.

Romaniv, O.N., et al. 1977. *Soviet Materials Science*, no. 3: 31–36.

Sakaki, T., K. Sugimoto, and T. Fuluzato. 1983. "The Role of Internal Stresses for Continuous Yielding of Dual-Phase Steels." *Acta Metallurgica* 31 (10): 1737–46.

Sarosiek, A.M., and W.S. Owen. 1983. "On the Importance of Extrinsic Transformation Accommodation Hardening in Dual-Phase Steels." *Scripta Materialia* 17 (2): 227–31.

Shieh, W.T. 1974. "The Relation of Microstructure and Fracture Properties of Electron Beam Melted Modified SAE 4620 Steels." *Metallurgical Transactions* 5A (5): 1069–85.

Song, R., D. Ponge, and D. Raabe. 2005. "Improvement of the Work Hardening Rate of Ultrafine Grained Steels through Second Phase Particles." *Scripta Materialia* 52: 1075–80.

Speich, G.R., and R.L. Miller. 1979. "Mechanical Properties of Ferrite-Martensite Steels." In *Structure and Properties of Dual-Phase Steels*, 145–82. New Orleans, LA, USA.

Speich, G.R., and R.L. Miller. 1981. "Mechanical Properties of Ferrite-Martensite Steels." In *Conference of Structure and Properties of Dual-Phase Steels, Symposium at the AIME Annual Meeting*, 145–82. New Orleans, LA, USA.

Stevenson, R. 1977. "Crack Initiation and Propagation in Thermal Mechanically Treated Sheet Steels." In *Formable HSLA and Dual-Phase Steels, Metallurgical Society of AIME*, 101–10.

Stevenson, R., D.J. Bailey, and G. Thomas. 1979. "High Strength Low Carbon Sheet Steel by Thermomechanical treatment: II. Microstructure." *Metallurgical Transactions A* 10 (1): 57–62. doi:10.1007/BF02686406.

Steven, W., and A.G. Haynes. 1956. "The Temperature of Formation of Martensite and Bainite in Low-Alloy Steels." *Journal of the Iron and Steel Institute* 183 (8): 349–59.

Storozeva, L.M., N.M. Fonstein, and S.A. Golovanenko. 1985. "Examination of Quench and Strain Aging of Low-Carbon Ferritic-Martensitic Steels." *Russian Metallurgy*, no. 1: 89–93.

Stratmann, P., and E. Hornbogen. 1979. "Mechanische Eigenschaften Zweiphasigerduplex-Und Dispersiongefuge in Nickel Stahlen." *Stahls Und Eisen* 99 (12): 643–48.

Sudo, M., and I. Tsukatani. 1984. "Influence of Microstructure on Yielding Behavior in Continuous-Annealed Multi-Phase Sheet Steels." In *Technology of Continuously Annealed Cold-Rolled Sheet Steel*, 341–60. Detroit, MI.

Sugimoto, K-I., J. Sakaguchi, T. Ida, and T. Kashima. 2000. "Stretch-Flangeability of a High-Strength TRIP Type Bainitic Sheet Steel." *ISIJ International* 40 (9): 920–026.

Sugimoto, K., T. Sakaki, T. Fukusato, and O. Miyagawa. 1985. "Influence of Martensite Morphology on Initial Yielding and Strain Hardening in a 0.11C-1.36Mn Dual-Phase Steel." *Journal of ISIJ* 71 (8): 70–77.

Swift, H.W. 1952. *Journal of the Mechanics and Physics of Solids*, no. 1: 1–32.

Szewczyk, A.F., and I. Gurland. 1982. "A Study of the Deformation and Fracture of a Dual-Phase Steel." *Metallurgical Transactions* 13A (10): 1821–26.

Takada, Y., Y. Hosoya, and K. Nakaoka. 1982a. "Possibilities of Achieving Low Yield Ratio with Low Manganese Dual-Phase Steels." In *Metallurgy of Continuously Annealed Steels*, 251–69.

Takagi, S., Y. Toji, M. Yoshino, and K. Hasegawa. 2012. "Hydrogen Embrittlement Resistance Evaluation of Ultra-High Strength Steel Sheets for Automobiles." *ISIJ International* 52 (2): 316–22.

Tamura, I., and Y. Tomota. 1973. "On the Strength and Ductility of True Phase Iron Alloys." *Transactions of the Iron and Steel Institute of Japan* 13 (4): 283–92.

Tanaka, T., M. Nishida, K. Hachiguchi, and T. Kato. 1979. "Formation and Properties of Ferrite plus Martensite Dual-Phase Structures." In *Structure and Properties of Dual-Phase Steels*. Vol. 221–241.

Tkach, A.N., N.M. Fonstein, V.N. Simin'kovich, and et al. 1984. "Fatique Crack Growth in a Dual-Phase Ferritic-Martensitic Steel." *Soviet Materials Science*, no. 5: 448–54.

Tomota, Y., Y. Kawamura, and K. Kuroki. 1981. "On Ductile Fracture of Steels Containing the Coarse Second Phase." *Bulletin of JSME* 24 (188): 282–89.

Tomota, Y., and I. Tamura. 1981. "Mechanical Properties of Ductile Two-Phase Steels." *Journal Iron and Steel Institute of Japan* 67 (13): 439–55.

Tomota, Y., and I. Tamura. 1982. "Mechanical Properties of Dual-Phase Steels." *Transactions of ISIJ* 22 (5): 677–89.

Tseng, D., and F.H. Vitovec. 1981. "The Bauschinger Effect and Work-hardening of Dal-Phase Steels." In *Fundamentals of Dual-Phase Steels*, 399–411.

Tsipouridis, P., E. Werner, C. Krempaszky, and E. Tragl. 2006. "Formability of High-Strength Dual-Phase Steels." *Steel Research International* 77 (10): 654–67.

Uggowitzer, P., and H.P. Stuwe. 1982. "Plastizitat von Ferritisch-Martensitischen Zweiphasenstahlen." *Zeitschrift für Metallkunde* 73 (5): 277–85.

Voce, E. 1948. "The Relationship between Stress and Strain for Homogeneous Deformation." *Journal of the Institute of the Metals* 74: 537–49.

Waterschoot, T., De, A.K., Vanderputte, S., and De Coman, B.C. 2003. "Static Strain Aging Phenomena in Cold-rolled Dual-Phase Steels." *Metallurgical and Materials Transactions A* 34A (3): 781–91.

Waterschoot, T., K. Verbeken, and B.C. De Cooman. 2006. "Tempering Kinetics of the Martensitic Phase in DP Steel." *ISIJ International* 46 (1): 138–46.

Yang, D.Z., D.K. Matlock, and G. Krauss. 1985. "The Effect of Cold-Rolling on Aging of an Intercritically Annealed Mn-Si-C Steel." In *Technology O Continuously Annealed Cold-Rolled Sheet Steel*, 319–39. Detroit.

Yegneswaran, A.H., and K. Tangri. 1983. "Strain Distribution and Load Transfer Characteristics of a Cu-9.6% Al Dual-Phase Structure." *Zeitschrift für Metallkunde* 74 (8): 521–24.

Chapter 4
The Effect of Chemical Composition on Formation of Ferrite–Martensite Structures and Mechanical Properties of Dual-Phase Steels

Contents

© Springer International Publishing Switzerland 2015
N. Fonstein, *Advanced High Strength Sheet Steels*,
DOI 10.1007/978-3-319-19165-2_4

4.1   Introduction

In Chap. 3, the relationships between microstructure and properties of dual-phase steels were considered. These relationships serve as the foundation of "microstructure design" for ferrite–martensite steels with prescribed properties.

Basic concepts of the effects of the parameters of heating in the intercritical $\alpha + \gamma$ range and subsequent cooling on dual-phase microstructure in low-carbon low-alloyed steels have been reviewed in Chap. 2. This chapter concerns the role of steel composition in producing ferrite–martensite microstructure under specified heat treatment conditions, the effects of steel composition on the properties of individual phases, and mechanical behavior of two-phase microstructure as a whole.

So far, no systematic approach exists with regard to selection of the optimal alloy composition for obtaining dual-phase steels with specified properties. The wide ranges of capabilities offered by heat treatment process and various annealing equipment led to development of dual-phase steels with similar final properties using different chemical compositions. Some of these steels contain alloying elements such as chromium and molybdenum and are microalloyed with niobium. Other steels contain only manganese and silicon. Both groups can meet given specifications determining tensile strength and elongation. However, apart from cost difference, these steels can differ slightly or significantly in achieving such parameters as strain hardening, hole expansion, etc.

This chapter presents an attempt to analyze and summarize the available data on the influence of chemical composition on structure and properties of dual-phase steels. The author used various data, including own findings, aiming at determining the effects of composition not only on final microstructure and properties but also on individual stages of microstructure evolution that control the mechanical behavior of DP steels.

4.2   Effect of Steel Composition on Processes During Heating in the Intercritical Temperature Range

Selection of chemical composition for DP steels related to their influence on steel responses during heating in the $\alpha + \gamma$ temperature range is based on the effects of a particular alloying element on:

1. Temperature range of the $\alpha + \gamma$ region (A_{C1}–A_{C3}) and the corresponding sensitivity of the amount of γ-phase to the heating temperature;

2. Austenitization kinetics that governs the amount of austenite formed at specified temperatures and its homogeneity after soaking for specified duration in continuous annealing lines;
3. Austenite morphology that is to some extent inherited by final structure formed in cooling;
4. Temperature range and kinetics of recrystallization, which affect the austenitization and morphology of the final microstructure
5. Effective carbon content in γ-phase and the homogeneity of its distribution

The effects of alloying elements are discussed below in this sequence.

4.2.1   Effect of Chemical Composition on the Volume of the Formed Austenite

The effect of steel composition on the amount of austenite formed in the two-phase region at given heating parameters (t, T) is to a considerable extent controlled by the corresponding changes in the critical points of Fe(Me)–C diagram. For example, significant increase in volume fraction of austenite caused by increasing manganese content is induced by gradual lowering of the A_{c1} temperature, Fig. 4.1. This is opposite to the effects of Si and Al additions that raise A_{c1}.

Another factor affecting the amount of austenite is the reduction of carbon content in eutectoid. As shown in Fig. 4.2, practically all alloying elements reduce carbon content in eutectoid (*ASM Handbook; Heat Treating* 1991). Then, the increase in Mn content will result in additional increase in autenite volume fraction due to that factor.

Experimental data, illustrating the effect of steel composition on the amount of austenite, are exemplified in Fig. 4.3a. It can be seen that the additions of 0.5 % Mo or 0.5 % Si significantly reduce the amount of austenite at given temperatures in the intercritical range, which qualitatively agrees with tendencies presented in Fig. 4.1. The results of experiments and calculations both align with the changes in critical points and the width of the intercritical range.

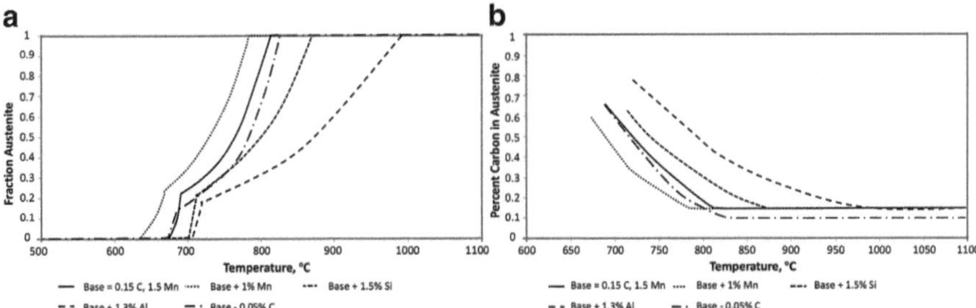

Fig. 4.1 Effect of changes in steel composition (C, Mn, Si, and Al) on (**a**) the amount of volume fraction of austenite formed during heating in the intercritical range; (**b**) carbon content in austenite (Thermocalc, orthoequilibrium calculations)—original

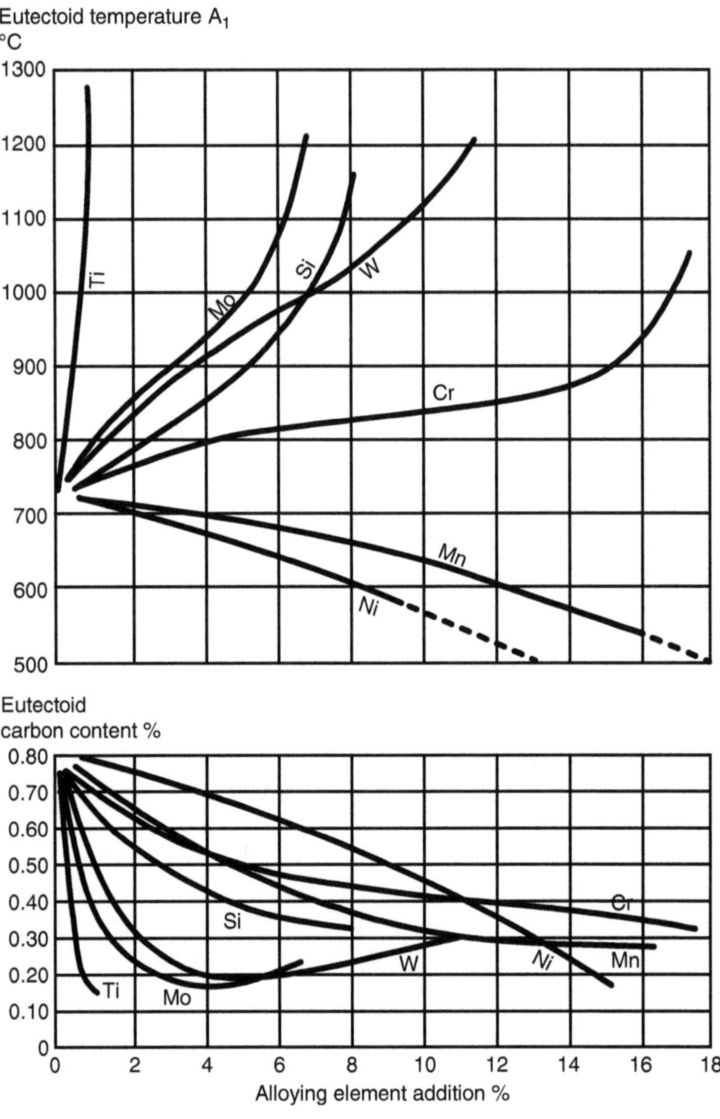

Fig. 4.2 Effects of alloying elements on eutectoid temperature and eutectoid carbon content (*ASM Handbook; Heat Treating* 1991)

Due to rather low diffusivity of Mn, its partitioning to austenite requires significant time and the conditions of paraequilibrium are attained later than the maximum of austenite fraction at given intercritical temperature can be achieved. As the general consequence, the increase in *holding time* of steels with relatively high manganese enhances both the growth of austenite fraction and Mn content in γ-phase, Fig. 4.3b (Toji et al. 2011).

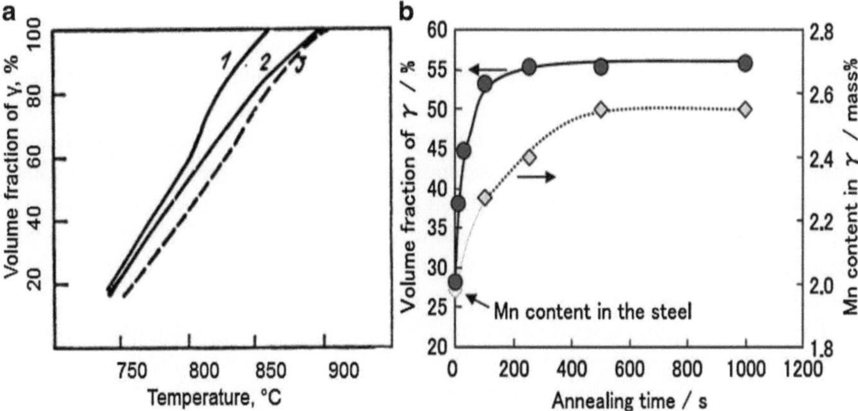

Fig. 4.3 Effect of steel composition on the amount of austenite formed (**a**) at different heating temperatures after holding for 15 min; 1–0.06C–1Mn–0.5Cr steel; 2–0.06C–1Mn–0.5Cr–0.5Mo; 3–0.06C–1Mn–0.5Cr–0.5Si original; (**b**) during isothermal annealing of 0.123C–1.98Mn–1.40Si steel at 800 °C (Toji et al. 2011)

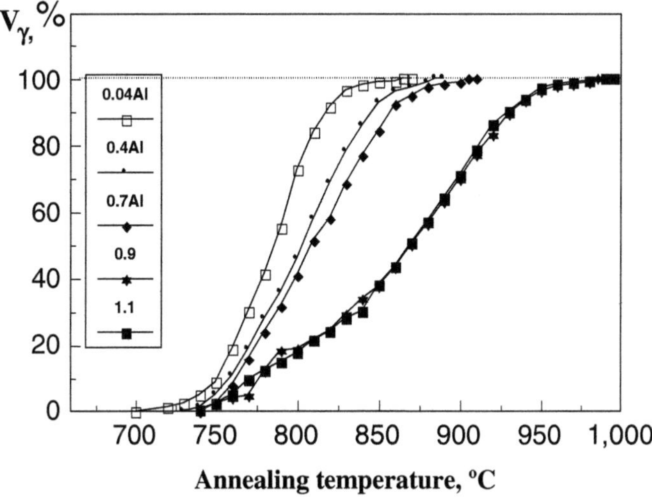

Fig. 4.4 Austenitization curves of 0.09C–2.0Mn–0.6(Cr + Mo) steels with different Al content during continuous heating at 10 °C/s (Girina and Fonstein 2005)

The effects of alloying elements on the amount of austenite can be predicted. This will only require the adjustments of the heating parameters or corrections to steel chemistry if the given alloying composition calls for the annealing conditions beyond the capability of the existing equipment. Particular case are steels with high aluminum content as Al significantly raises the A_{c3} temperature so that significantly higher annealing temperatures may be required to obtain the desired portion of austenite as presented in Fig. 4.4 (Girina and Fonstein 2005).

Robustness and minimum scatter of microstructure obtained after annealing that is translated into minimum scatter of properties underscore the importance of the ability to find the means to decrease the impacts of unavoidable variations of annealing parameters. In particular, as follows from the quasi-binary Fe (Me)–C diagram (Fig. 2.1), the rate of variation in the amount of austenite with heating temperature ($\%\gamma$-phase per 1 °C) depends on the width of the intercritical range. Simple calculation using the lever rule shows that the dependence of the amount of γ-phase on heating temperature is generally nonlinear (Fig. 4.1), but with wider intercritical area the slope of the $\%\gamma$ (T) curve is lower over the entire range of intercritical temperatures.

Carbon has the greatest effect on the A_{c1}–A_{c3} range. In fact, a decrease in carbon content in Cr–Mn–Si–B steel from 0.18 % to 0.10 % and then to 0.07 % reduced the rate of variation in the amount of γ-phase with temperature from 6 %/°C to 3.3 %/°C and then to 2.3 %/°C, respectively, within the temperature range of 760–850 °C.

Among the elements used in low-alloy steels, silicon and especially aluminum elevate the intercritical temperature range dramatically and therefore reduce the rate of variation of austenite volume fraction with temperature. The disadvantage, however, is that the additions of these two elements induce much higher increase of the A_{c3} temperature than that of the A_{c1} temperature.

As can be seen, an increase in manganese content slightly narrows the $\alpha+\gamma$ temperature range, thus increasing the rate of variation of the amount of austenite with heating temperature. Combined additions of manganese and silicon can widen the intercritical range and reduce the rate of variation of $\%\ \gamma$ with temperature.

In studying the effect of boron on phase transformation in 0.1C–1.35Mn–0.5Si–0.05Ti steels, Shen and Priestner (1990) found that with B content increasing from 5 to 29 ppm the volume fraction of formed austenite slightly decreased due to shifting of eutectoid point to higher temperatures.

The rate of variation of austenite fraction with heating temperature trends to decrease in the upper half of the intercritical range. Therefore, better robustness of dual-phase microstructure and properties can be achieved by heating to temperatures corresponding to more than 50–60 % austenite and then by obtaining the desired amount of martensite using controlled cooling rate, rather than by using rapid quenching of variable amount of austenite from lower temperatures in the two-phase region.

4.2.2   Austenitization Kinetics

General consideration regarding the factors that govern the kinetics of isothermal austenitization in the $\alpha+\gamma$ region is presented in Chap. 2.

To understand potential effects of chemical composition, it is worth recalling the equation that describes the growth rate of the γ-phase, i.e., the velocity of austenite interface:

$$v = D \frac{dC}{dx} \left(\frac{1}{\Delta C^{\gamma \to \alpha}} + \frac{1}{\Delta C^{C \to \gamma}} \right) \tag{4.1}$$

This velocity is the higher, the higher the carbon diffusion coefficient in austenite, D, through the austenite particle size, x, and the greater the concentration gradient dC/dx within an austenite area. The latter is determined by the ratio of the differences in carbon concentrations in the γ and α phases at a given temperature and, to a smaller extent, by the carbon concentration difference at the interfaces of γ/α, $\Delta C^{\gamma \leftrightarrow \alpha}$, or γ/Fe_3C, $\Delta C^{C \leftrightarrow \gamma}$.

This equation suggests that the effects of alloying elements on the rate of austenitization are determined by their effects on phase diagram, which control the differences in carbon concentration in the α- and γ-phases and in carbides. Another important role of alloying elements is their impact on carbon diffusion rate. Additions of ferrite-forming elements such as aluminum, chromium, silicon, and molybdenum reduce the difference in carbon concentration across phase interfaces and hence reduce the rate of carbon diffusion across γ-phase and decrease its growth rate. By contrast, the elements that expand the austenite area such as nickel, manganese, and cobalt increase the carbon concentration gradients across austenite–ferrite and austenite–carbides interfaces and hence raise the rate of austenitization.

The study of steels containing 0.06–0.20 % C and 1.57 % Mn suggests that the austenitization rate during dissolution of pearlite (initial stage of austenitization) at 740–760 °C is independent of carbon content in steel (Speich et al. 1981). On the other hand, an increase in carbon content in steel (and in volume fraction of carbon-containing phase in the initial microstructure) should increase the rate of diffusion-controlled processes due to a bigger contribution of grain boundary diffusion. At the same time, as noted in Chap. 2, the advancement of austenite boundary towards ferrite is substantially faster than the dissolution of carbides, especially in the presence of strong carbide-forming elements. Therefore, the enhancement of austenite growth rate in the steel with higher carbon content can be accompanied by increasing non-homogeneity of austenite. As a result, as was shown in Fig. 2.11, in short-time austenitization some of carbides can appear within austenite and therefore inside martensite after cooling.

Carbide-forming elements such as Cr, Mo, V, and Nb should not affect the initial austenitization rate when γ-phase is formed during transformation of pearlite, but seem to heavily retard the attainment of equilibrium.

In steels containing carbide-forming elements, a longer holding in the $\alpha + \gamma$ range can increase the effective carbon concentration in steel and hence in austenite as a result of gradual dissolution of the carbides (Garcia et al. 2011). It should be borne in mind that alloyed carbides dissolve at rates several orders of magnitude

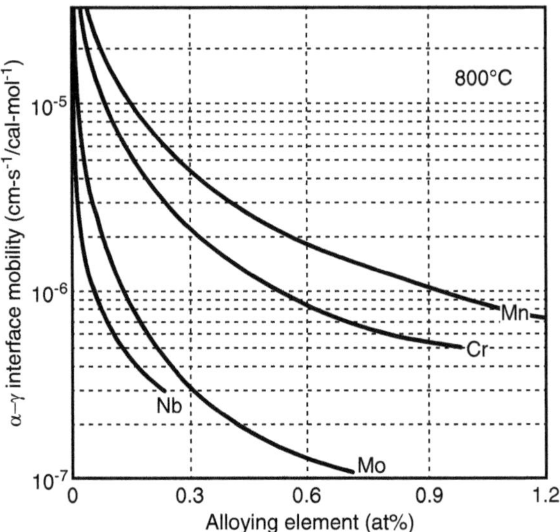

Fig. 4.5 Effect of alloying elements on mobility of ferrite/austenite interface

lower than that of cementite. Increased duration of heating in this case can also change the mobility of austenite boundary due to partitioning of carbide-forming alloying elements to the solid solution.

Solute drag effects of some of the alloying elements on mobility of α/γ interface are described in Fig. 4.5 (Togashi and Nishizawa 1976).

With regard to the intrinsic effect of alloying elements on carbon diffusion rate in γ-phase, it is generally believed that Ni, Cu, Co, and Si (up to 1050 °C) enhance carbon diffusion and Al and Mn have relatively small effect, but the carbide-forming elements such as Cr, Mo, V, and Nb markedly inhibit carbon diffusion (Krauss 2005).

These considerations agree with experimental data. It is known that additions of Mo or Cr require higher temperatures or longer time to obtain the given amount of austenite. It can be explained, in particular, by the data in Fig. 4.5 showing that molybdenum significantly decreases the mobility of ferrite–austenite interface thus retarding the growth of austenite. Besides, carbide-forming elements such as Mo and Cr reduce carbon diffusivity by forming clusters of carbon atoms (Lee et al. 2011).

On the contrary, silicon as a ferrite stabilizer should enhance the diffusivity of carbon in austenite. The observed decrease in austenitization rate with additions of Si is most likely related to shifting the A_{c3} temperature.

Aluminum should also accelerate the austenitization by enhancing carbon diffusivity in austenite, but increase in A_{c3} temperatures can prevail and reduce the thermodynamics potential of austenite formation and austenite volume fraction, when steels with different content of Al are compared at the same temperatures (Fig. 4.4).

Effect of steel composition on sensitivity of austenitization kinetics to *heating rate* is related to its influence on the critical temperatures, carbon diffusion rate, and recrystallization kinetics. For identical initial microstructures, the lowest sensitivity to heating rate should be expected in steels containing Si, Al, Cu, and Ni, while steels containing additions of carbide-forming elements should be most sensitive.

4.2.3   Austenite Morphology

As already shown in Chap. 2, the morphology of austenite in two-phase microstructure is governed by the initial microstructure and is inherited, especially after rapid quenching, by the geometry of martensite islands. In this case, the steel composition plays only an indirect role by affecting the rate of recrystallization and/or martensite tempering (when the initial hot-rolled microstructure contains martensite/bainite) which control austenitization and hence austenite morphology (Huang et al. 2004; Azizi-Alizamini et al. 2011; Mohanty et al. 2011).

For example, the presence of silicon and molybdenum that delay tempering of martensite makes it possible to retain the lath-like morphology of γ-phase during heating in a wide range of heating rates, from 1 to 1000 °C per minute. As shown by Thomas (Thomas and Koo 1979) in experiments with repeated quenching from intercritical temperatures, addition of 0.5 % Cr to 0.06 % C steel, in the absence of other alloying elements, promoted the formation of fibrous-like austenite (martensite) regions but resulted in a continuous martensite network at $V_M = 0.35$. An increase in silicon content from 0.5 to 2.0 % reduces the continuity of martensite facilitating fibrous-like *morphology* of martensite that is favorable for mechanical properties, particularly for fracture resistance. Eventually, a single alloying addition of 2 % Si, in the absence of other alloying elements, led to discontinuous fibrous martensite particles.

With higher carbon content in steel, the size of γ-phase grains (and hence if martensite islands) becomes smaller at *the same fraction of austenite* since the number of nucleation sites increases owing to lower temperature of austenitization needed to obtain the given volume fraction of austenite. On the contrary, a*t the given intercritical holding temperature* and time, an increase in the amount of the γ-phase in steel on account of higher carbon is associated with coarsening of austenite and consequently martensite (Bortsov and Fonstein 1984).

Higher nitrogen content refines austenite and hence martensite particles (Lanzilotto and Pickering 1982).

Because of the aforementioned strong effect of molybdenum on mobility of α/γ interface, the additions of Mo often also refine austenite/martensite grain size.

DP steels typically contain significant amount of Mn. Adding extra 1.2 % Mn is equivalent to the addition of 0.1 % Mo in terms of reducing the mobility of interphase boundaries.

At the temperatures of intercritical annealing, Nb is present mostly as precipitates while Ti precipitates completely. Therefore, these elements hamper austenite grain growth, especially when their carbonitrides are fine and are uniformly distributed.

4.2.4   Recrystallization of Ferrite and Ferrite Grain Size

Heating of cold-rolled steel in the intercritical temperature range leads to ferrite recrystallization, which controls the grain size and, to a significant extent, the properties of the ferrite phase. As described in Chap. 2, the austenitization kinetics and the morphology of austenite depend on the extent to which ferrite recrystallization is complete before austenitization.

Data, presented in Chap. 2, demonstrated the difference in austenitization process depending on recrystallization completion before austenitization or overlapping these two processes. Austenite first forms at cementite particles along the boundaries of ferrite grains; however, the transition from austenitization in recrystallizied ferrite matrix to the occurrence of parallel recrystallization and austenitization drastically changes the final structure. A random spatial distribution of austenite is observed if recrystallization was complete before austenitization. In non-recrystallized structure, austenite first forms at cementite particles along the boundaries of non-recrystallized elongated (pancaked) ferrite grains so that in can lead to a formation of austenite aligned in the rolling direction facilitating banded morphology (Azizi-Alizamini et al. 2011).

In conventional low-carbon steels, the recrystallization of ferrite starts at 500 °C and ends by 635 °C. Additions of manganese, silicon, vanadium, and especially niobium expand the recrystallization temperature range to higher temperatures. Pradhan reported that the effects of niobium and vanadium in 0.55Si–0.4Mn–0.06P steel can be described by a common curve with the recrystallization end temperature elevating from 650 to 750 °C with the addition of the first 0.08 % of either element. Increase in manganese from 0.5 to 1.5 % in 0.05C–0.06P–0.04Nb steel raised the recrystallization start and finish temperatures from 680° to 690° and from 760 to 815 °C, respectively. Additional alloying with 0.085 % V further raised these temperatures by 40 °C (Pradhan 1983).

Thus, in rephosphorized steels with 1.5 % Mn, microalloyed with niobium and/or vanadium, the recrystallization of deformed ferrite can complete only at temperatures of about 850 °C. Substantial portion of ferrite can undergo only recovery and retain elongated (pancaked) grain shape.

Effect of boron on recrystallization nucleation rate and hence on recrystallization kinetics is strongly related to boron segregation to grain boundaries, but the general tendency is that both ferrite recrystallization start and finish temperatures rise with additions of B.

Four heats with basic composition of 0.08C–2.2Mn and four levels of Nb (0, 0.02, 0.04, and 0.06 %) were used to study the effects of Nb on ferrite recrystallization. Heating to 730 °C induced evident ferrite recrystallization and grain growth in Nb-free steel. In 0.02Nb steel, a few small recrystallization nuclei were observed, while 0.06Nb steel showed no signs of ferrite recrystallization. With delayed recrystallization, the interplay between ferrite recrystallization and austenite formation during annealing takes place resulting in changes of morphology of austenite/martensite. Even after isothermal holding at 790 °C for

150 s, several elongated ferrite grains were still observed in the Nb-added steels becoming more pronounced with higher Nb content (Song et al. 2014).

Another aspect of the shown microalloying elements' effect on recrystallization is related to the amount of austenite formed in non-recrystallized structure (Girina et al. 2008). In the study, full hard samples of 0.085C–2.0Mn steel were continuously heated in dilatometer to 950 °C at the rate of 10 °C/s. Increase in Nb content to 0.03 % had insignificant effect on A_{c1} and A_{c3} temperatures (within the accuracy of experiments, for all studied steels the A_{c1} temperature was 715 °C and the A_{c3} temperature was 840 °C). However, as was found, the addition of Nb significantly retarded recrystallization. As a result, the plots in Fig. 4.6 show that the amount of austenite formed in steel with 0.03 % Nb is significantly larger than that in steel without Nb at the same temperatures. This means that additions of Nb accelerated the austenitization process. Evidently, microalloying with Nb refined the initial hot-rolled microstructure before cold rolling and created additional interfaces. Consequently, the non-recrystallized portion of ferrite contained significant amount of structure defects that could serve as nucleation sites in austenitization. The retardation of recrystallization by Nb preserved higher densities of defects and apparently accelerated the kinetics of isothermal austenitization as well.

Cr and Mo also retard recrystallization of deformed ferrite due to strong solute drag, but their effect is weaker than that of Nb.

Study of 0.086C–2.1Mn–0.2Cr–Nb–Ti–B steel showed that additions of 0.3 % Si accelerated recrystallization of ferrite. While the steel without Si was only partly recrystallized at 760 °C, in Si-added steel the recrystallization was complete (Drumond et al. 2012).

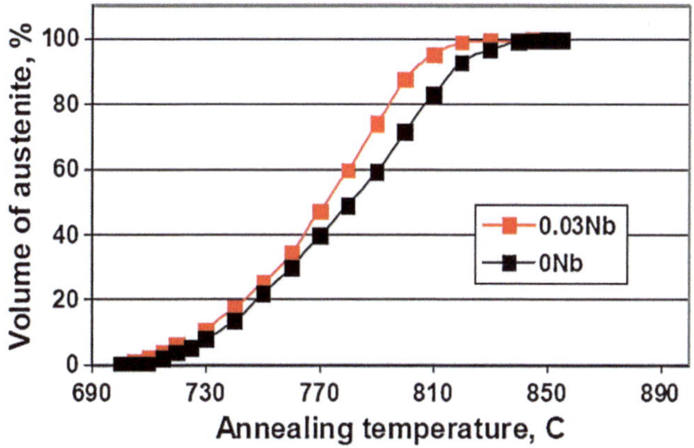

Fig. 4.6 Effect of niobium additions on austenite growth of 0.085C–2Mn steel during continuous heating—(Girina et al. 2008)

Depending on the heating rate and steel composition, soaking in two-phase region can induce annihilation or rearrangement of dislocations in ferrite and formation of stable subgrains or polygonized substructure, which can also contribute to delaying recrystallization.

4.2.5   Effective Carbon Content in Austenite and Its Homogeneity

The effects of alloying elements on the formation of the ferrite–austenite structure and the kinetics of austenitization also determine the effective carbon content in austenite and its homogeneity.

As was observed in numerous studies, local enrichment with manganese in the vicinity of austenite boundaries, which is enhancing when the annealing temperature approaches A_{c1} (Wycliffe et al. 1981; Speich et al. 1981), is accompanied by emerging of nonhomogeneous distribution of carbon in austenite due to effect of Mn on the activity of carbon.

Carbide-forming elements significantly influence the effective carbon content in austenite, C_γ. Calculations of carbon concentration in austenite, C_γ, suggest that all carbon, present in steel, is fully partitioned to austenite with volume fraction V_γ so that $C_\gamma = C_{st}/V_\gamma$. The presence of carbide-forming elements such as V, Nb, and Ti substantially changes the effective carbon content in steel and correspondingly in the γ-phase. In fact, the presence of vanadium in steel in the amount of, e.g., 0.1 % can bind in carbides 0.021 % and 0.016 % C, respectively, at annealing temperatures of 750 and 800 °C. Decrease in the available free carbon results, at a constant heating temperature, in smaller amount of austenite and in its lower carbon content (Fonstein et al. 1985; Priestner and Ajmal 1987). For example, addition of 0.18 % V to steel with 0.08 % C decreases V_M from 0.16 to 0.13 after quenching from 750 °C. Calculations using the stoichiometric ratio V:C = 4.5 and the solubility of vanadium at these temperature show that the effective C_{st} decreased by approximately 0.038 %, i.e., down to 0.042 %. With total carbon content of 0.08 % in steel, the calculated carbon content in martensite/austenite, $C_\gamma = 0.08/0.13$, should be equal to 0.61 %. After correcting the carbon content in steel for the amount of carbon bound by vanadium, the carbon concentration in austenite $C_\gamma = 0.042/0.13 = 0.32$ %. Since C_γ is the strongest hardenability factor, this decrease in C_γ should lower stability of the γ-phase.

The strong changes in hardenability of intercritically formed austenite and the mechanical properties of 980DP steels depending on the state of the dissolution reaction of carbides during intercritical annealing of 0.15C–Nb–Cr–Mo–V steels were found also in the study of Garcia et al. (2011)

Ferrite-forming elements such as Si and Al that reduce thermodynamic activity of carbon in austenite enhance its solubility and promote higher effective carbon content in the γ-phase, as well as they prevent the segregations of carbon-containing phases near interphase boundaries.

4.3 Effect of Steel Composition on Austenite Transformation and Properties of the Formed Phases in Cooling from $\alpha + \gamma$ Region

The previous section describes the effects of steel composition on the characteristics of the ferrite–austenite microstructure during *heating* in the $\alpha + \gamma$ region; however, the role of alloying is significantly more pronounced at cooling from intercritical temperature range.

In cooling of ferrite–austenite mixture, the main factors by steel composition are as follows:

1. Hardenability of the austenite in cooling from the $\alpha + \gamma$ region that controls the transformations of austenite, i.e., the type and the amount of product phases, in particular, the amount of "new ferrite" and/or bainite.
2. State of ferrite solid solution.
3. Precipitation hardening of ferrite.
4. Austenite non-homogeneity.
5. Martensite transformation temperature.

Analysis of the available data is presented below to characterize each of the above effects.

4.3.1 Hardenability of Austenite in Cooling from Intercritical Region

A great deal of well-organized data is available related to the characteristics of austenite transformation and to the effects of chemical composition at cooling after full austenitization. Important peculiarities of austenite behavior in cooling from intercritical region were discussed in Chap. 2.

Fortunately, the new generations of high precision dilatometers, development of EBSD techniques, and modified Jominy testing (Cho et al. 2012) have made it possible to collect the data of broadening scope regarding phase transformation of austenite in cooling of ferrite–austenite mixtures.

The typical criterion of austenite stability is the cooling rate from the $\alpha + \gamma$ region, which allows producing a dual-phase steel with strain–stress curve featured by continuous yielding and low YS/TS ratio, i.e., obtaining ferrite–martensite microstructure without pearlite constituent (and, as very often desired, without bainite as well).

Hashiguchi and Nashida (1980) calculated theoretical CCT diagrams for steels containing 0.5 % C (simulating ~20 % austenite in low-carbon steel) with grain size of 9 μm. They found that the strongest positive effect on the austenite stability, that is, the strongest decrease in the critical cooling rate, is produced by molybdenum and then weakens in the sequence chromium, manganese, copper, and nickel.

Hashiguchi and Nashida also showed that addition of up to 1 % silicon does not improve stability of austenite. The results of their study of the effects of Mo, Cr, Si, Cu, and Ni in the concentration range from 0.3 to 3 % at a constant manganese content of 1.2 % were summarized in regression equation:

$$\log V_{cr}(^\circ C/s) = -1.73 Mn_{eq}(\%) + 3.95 \qquad (4.2)$$

where V_{cr} is the critical cooling rate capable of producing a ferrite–martensite microstructure; the combined contribution of alloying elements was expressed in terms of Mn equivalent:

$$Mn_{eq} = (\%)Mn + 3.28(\%Mo) + 1.29(\%Cr) + 0.46(\%Cu) + 0.37(\%Ni) \\ + 0.07(\%Si) \qquad (4.3)$$

These equations should be used cautiously as they imply constant effective carbon content in austenite (0.5 %), which in reality can change with carbon content in steel and heating temperature and consequently change the relative contributions of individual elements. When modern steels with 0.08–0.12 % C are heated to obtain the volume fractions of austenite of 30–50 % and more, significant, almost double decrease in C_γ can alter the *quantitative* effect of a particular alloying element on hardenability of austenite. These equations also do not account for dramatic changes that can be induced by formation of new ferrite at preexisting ferrite interfaces, described in Chap. 2.

Nevertheless, the data by Hashiguchi and Nashida concerning the relative effects of alloying elements on austenite stability after heating in two-phase region *qualitatively* agree with the results due to Kunitake and Ohtani who determined the critical (minimal) cooling rate for obtaining fully martensitic structure without bainite after complete austenitization, based on different chemical composition and variable carbon content in austenite (Kunitake and Ohtani 1969). In their work, this critical cooling rate was described through the time τ of cooling, seconds from the heating temperature to 500 °C:

$$\log \tau = 3.274\, C_\gamma + 0.046(\%Si) + 0.626(\%Mn) - 1.818 \qquad (4.4)$$

However, regression analysis based on microstructure of dual-phase steels cooled from various intercritical temperatures targeting at least 5 % of martensite indicated a far greater contribution of silicon to austenite hardenability (Messien et al. 1981):

$$\log V_{5\%M}(C/s) = 4.93 - 1.70(\%Mn) - 1.34(\%Si) - 5.68 C_\gamma \qquad (4.5)$$

The data presented in Fig. 4.7 were obtained from experimental CCT diagrams for austenite transformation in cooling from the $\alpha + \gamma$ region and from experiments using modified end quenching method (Fonstein and Efimov 1985; Cho et al. 2012).

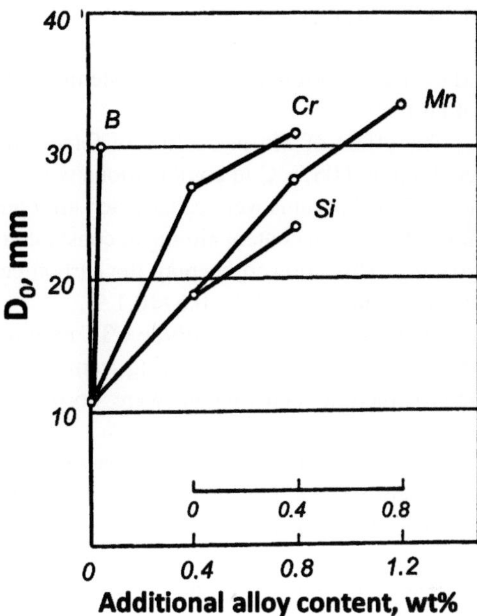

Fig. 4.7 Effect of alloying elements on the critical diameter of hardenabilty, Dc ($C_\gamma \approx 0.35$ %, $V_\gamma \approx 20$–30 %). Base composition: 0.2Si–0.2Cr–0.8Mn— original

It is seen that with $C_\gamma = 0.3$–0.4 %, the effects of alloying elements on the critical cooling rate of γ-phase or on the relevant critical diameter weaken in the order of boron, chromium, manganese, and silicon, with significant effect of silicon on hardenability of γ-phase.

The effects of individual element will be discussed below.

4.3.1.1   Carbon

Carbon content in austenite is the major factor governing phase stability during cooling from the $\alpha + \gamma$ region. The amount of carbon in austenite, in turn, is dictated by the carbon content in steel, as well as by the temperature and time of holding in the intercritical range.

The effective carbon content in the γ-phase, C_γ, can vary within wide range and can be considered in two different aspects: (1) carbon content in the initial austenite phase determined by the heating parameters, the initial microstructure, and the presence of carbide-forming elements (Garcia et al. 2011), and (2) carbon content in the final portion of γ-phase before the onset of martensitic transformation; this amount of carbon is determined by the amount of newly transformed ferrite and by the conditions for carbon partitioning from ferrite to austenite.

As will be shown later, influence of alloying elements on effective hardenability and phase transformation can be drastically changed by carbon content in austenite portion by the moment of transformations to bainite or martensite.

4.3.1.2   Manganese

Manganese is known as strong austenite stabilizer promoting higher hardenability of austenite.

The effect of Mn in cooling of ferrite–austenite mixture was studied in steels containing 0.06 % C in combinations with 1.0, 2.0, and 2.8 % Mn. Using dilatometry, CCT diagrams were constructed after austenitization at different temperatures for different Mn content aiming at constant volume 60 ± 2 % of γ-phase after 5 min soaking. This allowed keeping the carbon content in austenite of all steels practically constant, $C_\gamma \approx 0.06/0.6 \approx 0.1$ %.

Figure 4.8 shows parts of the CTT diagrams of steels with different manganese content for cooling rates ranging from 0.2 to 3 °C/s. The volume fractions of the microstructural constituents were normalized to the initial amount of γ-phase (~60 %) which was taken as 100 %.

In steel with 1 % Mn, 70 % of new ferrite and 30 % of martensite were detected after cooling at 3 °C/s, while 76 % of ferrite and 24 % of bainite were produced in cooling at 1 °C/s. As the manganese content was increased to 2 %, 58 % of new ferrite and 42 % of martensite were revealed after cooling at 2 °C/s and 72 % ferrite and 28 % bainite after cooling at 0.5 °C/s. In the 2.8 % Mn steel, all austenite was transformed to martensite already in cooling at the rate of about 0.5 °C/s, whereas 6 % of ferrite coexisting with 24 % of bainite and 70 % of martensite were detected only after cooling at 0.2 °C/s.

Increase in manganese moves the range of new ferrite formation to lower temperatures and slower cooling rates. Typically, with increase in Mn the bainite transformation regions after full austenitization also move to lower temperatures. In the 1 % Mn steel, however, the temperature range of bainite transformation is approximately 100 °C lower than in steels with 2 % and ~3 % Mn. This can be explained by the dominating role of higher volume fraction of new ferrite in the low Mn steel and the corresponding enrichment of the remaining portion of austenite with carbon that offset the Mn effect on the transformation temperatures.

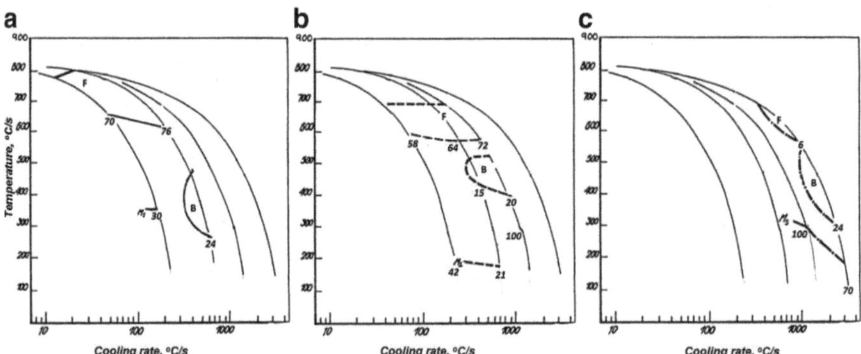

Fig. 4.8 CCT diagrams for steels with different Mn content cooled from different intercritical temperatures to obtain $C_\gamma \approx 0.1$ %: 1–06Mn1; 2–06Mn2 and 3–06Mn3—original

The M_s temperature is controlled by the concentrations of both Mn and carbon in the last remaining portions of austenite and possible changes in size of these remaining austenite grains. In the 1 % Mn steel, C_γ prior to $\gamma \rightarrow \alpha'$ transformation was found to be close to 0.3 % with $M_s = 340$ °C. In the 2 % Mn steel at cooling rate of 1–2 °C/s, $C_\gamma = 0.25$ %, but M_s was close to 200 °C due to stronger effect of Mn. In the 2.8 % Mn steel, where austenite had fully transformed to martensite at $V_{cr} \geq 0.5$ °C/s, $C_\gamma = 0.1$ and $M_s = 300$ °C. When ferrite and bainite transformations occur in cooling at the rate of 0.2 °C/s, C_γ in steel with 2.8 % Mn increased to 0.16 % and M_s decreased to 190 °C.

A very different effect of manganese was observed when the same steels were cooled from *identical* intercritical temperature so that the amount of initial austenite phase and therefore C_γ were different at different Mn content. After 5 min holding at 800 °C, the volume fractions of 25, 65, and 75 % of austenite were detected in steels with 1 %, 2 %, and 2.8 % Mn, and the carbon content in austenite, C_γ, was 0.24 %, 0.092 %, and 0.08 %, respectively.

As shown in Fig. 4.9, the temperature of ferrite start and amount of transformed ferrite are higher in the 1 % Mn steel at temperature higher than in steels with higher Mn content in spite of higher carbon content in austenite demonstrating the prevailing role of alloying of Mn on stability of austenite against ferrite transformation.

In the 1 % Mn steel, no bainite was detected and nearly half (47 %) of the remaining austenite transformed to martensite because of significant carbon enrichment due to significant fraction of new ferrite. In the 2 % Mn steel, the increase in manganese content did not offset the decrease in C_γ from 0.24 to 0.092 % and martensite was only formed in cooling at the rates above ≥ 5 °C/s and in smaller amounts. In the 2.8 % Mn, steel bainite could not be avoided even at the cooling rate as high as 5 °C/s. Due to variable quantity of new ferrite, the final amount of austenite before martensitic transformation was approximately the same with roughly the same carbon concentration C_γ of about 0.5 % for all studied steels. As a result, the manganese effect on the gradual decrease in M_s temperature became evident.

Some decrease in M_s, with decreasing cooling rate often observed in CCT for cooling from $\alpha + \gamma$ region, is also caused by increase in the amount of the new ferrite and by associated increase in C_γ immediately prior to martensitic transformation.

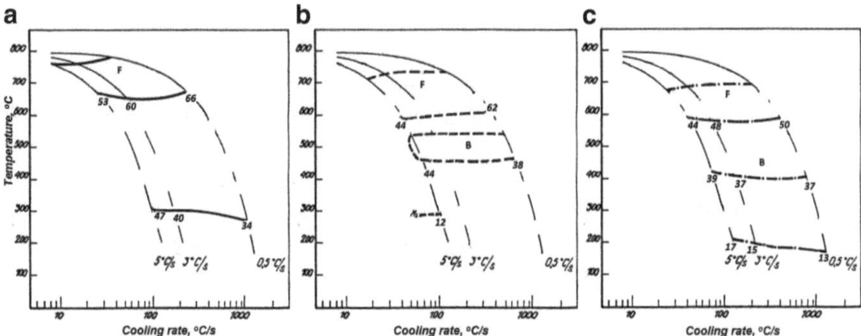

Fig. 4.9 Partial CCT diagrams for steel grades 06Mn1 (**a**), 06Mn2 (**b**), and 06Mn (**c**) cooled from 800 °C—original

It was observed that additions of manganese are accompanied with austenite grain refinement. This effect of Mn is induced by (1) decrease in A_{c1} and hence in the required annealing temperatures, (2) broadening of the $\alpha + \gamma +$ cementite triple-phase region where grain growth is inhibited, (3) refinement of cementite in the initial microstructure, which causes a stronger pinning effect, and (4) decreasing grain boundary mobility due to solute drag effect (Calcagnotto et al. 2012).

4.3.1.3 Silicon

Mn and Si are the most typical alloying elements of dual-phase steels and often their combined effects are observed and influence of silicon can depend on the Mn content. In particular, the effect of manganese was studied at two levels of silicon on the minimal cooling rate that allowed to get some portion of martensite in the final structure. At constant $C_\gamma \approx 0.1$ %, when 50–62 % of γ-phase was formed in these steels by annealing at different temperatures, the increase in Mn content from 1 to 3 % allowed transformation to martensite at lower cooling rate changed from 3 to 1 °C/s. Addition of 1 % Si does not change the critical cooling rate in steel with 1 % Mn (it was about 3 °C/s after formation of relative portion of 70 % new ferrite), but the same addition of 1 % Si to steel with 2 % Mn allows obtaining martensite at lower cooling rate (~1 °C/s). According to Nouri et al. (2010) in case of steels with 0.14C–1.6Mn and (0.34–2.26)Si, this could be explained by significant enhancement of Mn partitioning between ferrite and austenite at higher Si.

4.3.1.4 Aluminum

Formation of new ferrite during austenite decomposition in cooling at moderate rates is an important factor for stabilization of the remaining portion of γ-phase due to carbon partitioning and resultant increase in carbon content C_γ. This can prevent the bainitic transformation and extend the range of martensitic transformation towards lower cooling rates, as was shown in Fig. 2.21. Thus, the ferrite-forming elements such as Si and Al should indirectly enhance the stability of γ-phase suppressing the pearlitic reactions and allowing for martensitic transformation of the remaining austenite at lower cooling rates.

Additions of 0.85 % Al to 0.074C–1Mn–0.15Si–0.4Cr steel raise both A_{r3} and A_{r1} temperatures during cooling from full austenitization temperatures accelerating ferrite formation in a wide range of cooling rates (Mein et al. 2012).

Figure 4.10 presents three CCT diagrams for cooling of two-phase mixture of steels with 0.042, 0.7, and 1.1 % Al. At the given annealing conditions (780 and 820 °C) 85, 80, and 65 % of austenite was obtained, respectively, i.e., any influence of lowering carbon content in γ-phase with increasing Al was excluded. It is seen that the formation of new ferrite is accelerated by adding more Al (normalized per cent of new phase related to the initial austenite are shown under the curves of the end of ferritic transformation). As a result, the higher carbon content in

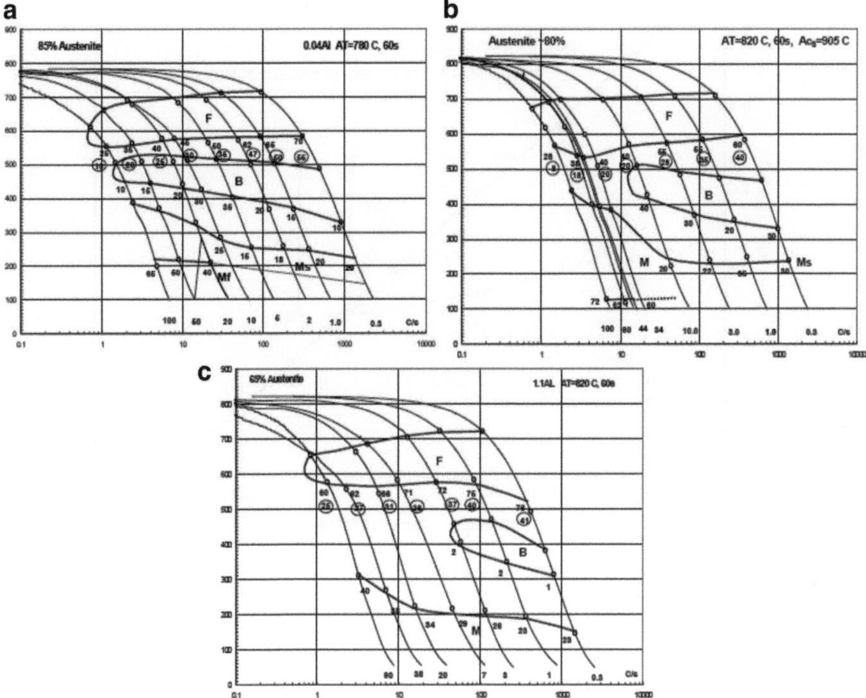

Fig. 4.10 CCT diagrams of 0.09C–2.0Mn–0.6(Cr + Mo) steels with different Al content: (**a**) 0.04 %, (**b**) 0.7 %, and (**c**) 1.1 % C (Girina and Fonstein 2005)

the remaining austenite suppresses bainite and with 1.1 % of Al the ferrite–martensite microstructure can be obtained at cooling rate as low as 5 °C/s (Girina and Fonstein 2005).

4.3.1.5   Molybdenum

Molybdenum is known to enhance the hardenability of austenite by suppressing the pearlitic reaction, lowering the bainite transformation temperature and expanding the bainite region.

In fact, Mo has the strongest effect on austenite stability facilitating austenite-to-martensite transformation at the lowest cooling rate, Fig. 4.11. Its effect is 2.6 and 1.3 times stronger than those of Mn and Cr, respectively (Irie et al. 1981).

Due to its effect on hardenability of austenite, Mo shifts the ferritic transformation to lower cooling rates. Adding more Mo substantially lowers the bainite start temperatures and/or extends the holding time for the onset of bainitic transformation under quasi-isothermal conditions of zinc coating or galvannealing, thus allowing for avoiding of bainite in the final microstructure of coated DP steels.

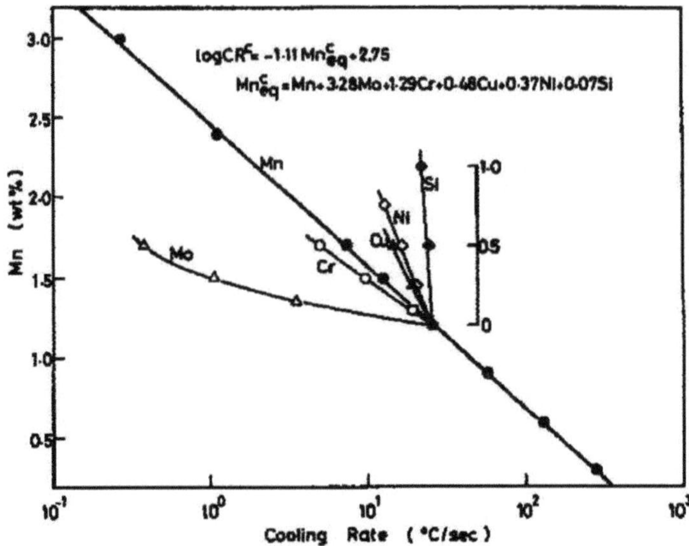

Fig. 4.11 Effects of alloying elements on the critical cooling rate for martensitic transformation in low-carbon steels (Hashiguchi and Nashida 1980)

Some data regarding Mo effect at cooling from full austenitization and from $\alpha + \gamma$ region are presented below in Fig. 4.12 (Girina et al. 2015).

At cooling from supercritical region, additions of Mo are enhancing the hardenability of austenite suppressing pearlite and ferrite formation and slowing down the kinetics of ferrite transformation, shifting the whole diagram to the lower temperatures. The retarding effect of Mo is stronger for cooling from intercritical region, which could be related to higher carbon content in the austenite fraction.

Important features of molybdenum presence in low-carbon steel are reduced carbon activity due to formation of Mo–C clusters and retardation of carbon diffusion in austenite (Hara et al. 2004). This effect of Mo makes carbon less available to form carbides not only with B but also with Nb that leaves the microalloying elements in solid solution making them more efficient to impact recrystallization, hardenability, and, as Nb, be available for precipitation hardening at later processing stages.

4.3.1.6   Chromium

Chromium effect presented in Fig. 4.13 is similar to that of Mo in terms of tendency but weaker. As seen, additions of Cr enhance the hardenability of austenite, slowing down the kinetics of ferrite transformation.

Similarly to Mo, retarding effect of Cr on austenite transformations is more significant after intercritical annealing due to higher carbon content in austenite.

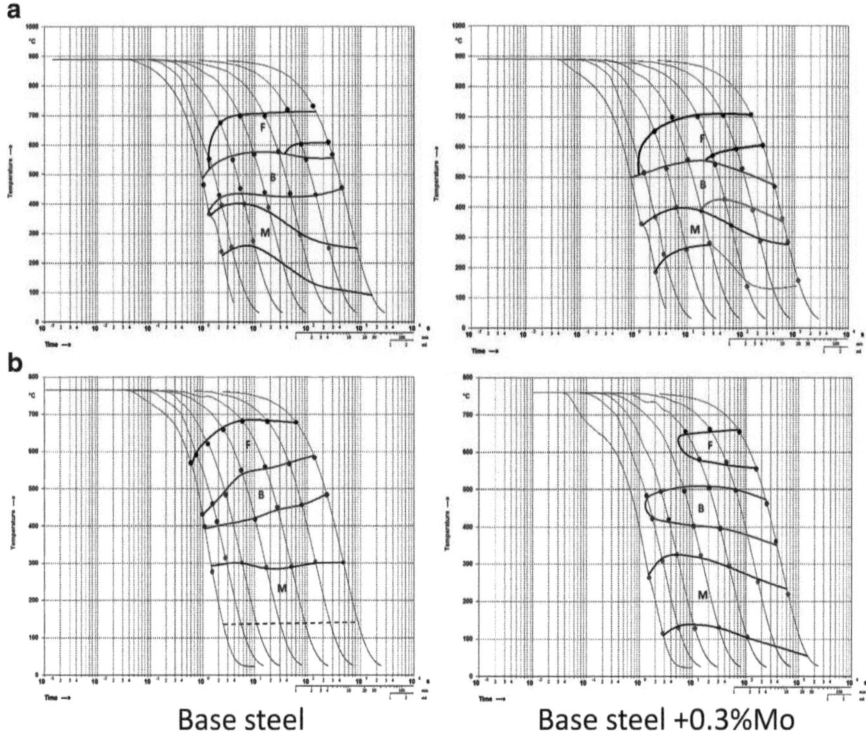

Base steel   Base steel +0.3%Mo

Fig. 4.12 Effect of 0.3 % Mo on CCT diagrams of 0.15C–2Mn–0.3Si–Nb steel built in cooling from supercritical (**a**) and intercritical (**b**) temperatures—original

4.3.1.7   Boron

As pointed out in numerous publications, additions of boron increase hardenability of austenite reducing the sensitivity of final microstructure and tensile properties to the cooling rate or to the temperature of annealing in the intercritical region. It was found that boron additions increase the fraction of austenite transformed to martensite and decrease the volume fraction of new ferrite. At the same time, as discussed in various studies, the growth of ferrite phase in cooling from intercritical range does not require nucleation, whereas the mechanism of boron effect is known to prevent the nucleation of ferrite at austenite grain boundaries.

Particle tracking autoradiography did not reveal resegregation of boron from prior austenite γ/γ grain boundaries to γ/α boundaries and therefore also ruled out the expectations that boron can prevent nucleation and growth of ferrite during cooling in the same way as it does after full austenitization (Shen and Priestner 1990). These authors mentioned that boron affects the critical points in Fe–Fe$_3$C diagram decreasing the fraction of formed austenite and thus raising its carbon content. However, they did not describe in detail the mechanism of boron effects during cooling from $\alpha + \gamma$ region.

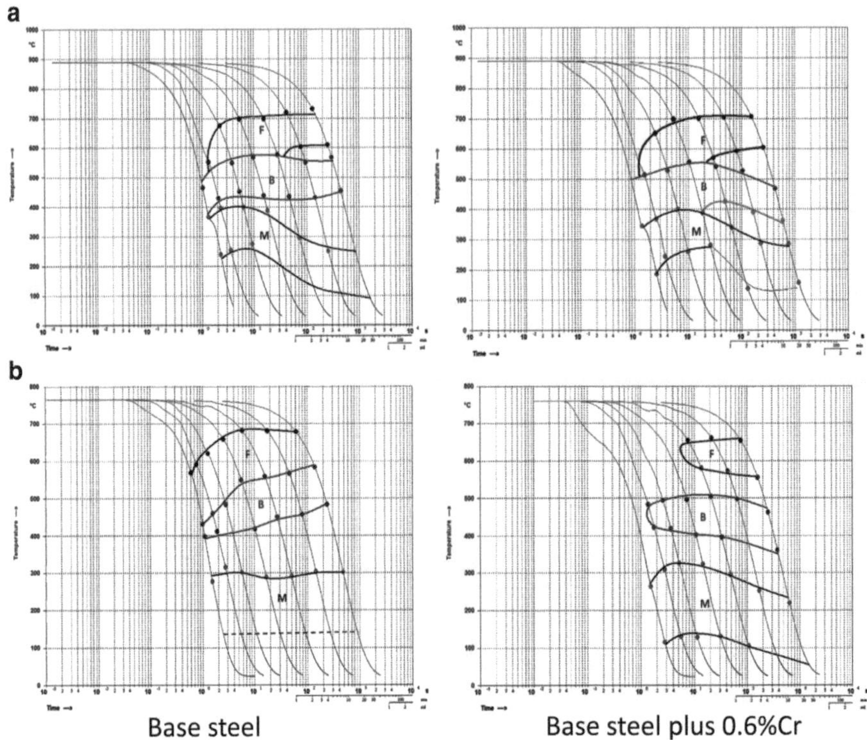

Fig. 4.13 Effect of 0.6 % Cr on CCT diagrams of 0.15C–2Mn–0.3Si–Nb steel built in cooling from supercritical (**a**) and intercritical (**b**) temperatures—original

According to the data due to Granbom et al., boron does not effectively prevent nucleation and growth of ferrite during cooling of austenite in the presence of preexisting ferrite and therefore the prepared ferrite interfaces (Granbom et al. 2010).

It is known that higher (full) austenitization temperature and therefore coarser austenite grain size deteriorate hardenability of boron steel due to precipitation of boro-carbide $M_{23}(CB)_6$ because higher temperatures significantly intensify boron segregation to austenite grain boundaries. Similar consideration of higher propensity to form boro-carbide can be used to explain the relatively weaker effect of boron in austenite fraction with higher carbon content. Grounding on this reasoning, it can be assumed that in case of cooling from two-phase region the relative effect of boron should increase with decreasing C_γ, i.e., with increasing temperature in the $\alpha + \gamma$ region that agrees with experimental facts.

Experimental CCT diagrams built at cooling after full austenitization and heating in the intercritical temperature range are presented in Fig. 4.14 for the base composition and the steel with boron additions. As shown, additions of boron are less effective in suppression of ferrite transformation in cooling of two-phase

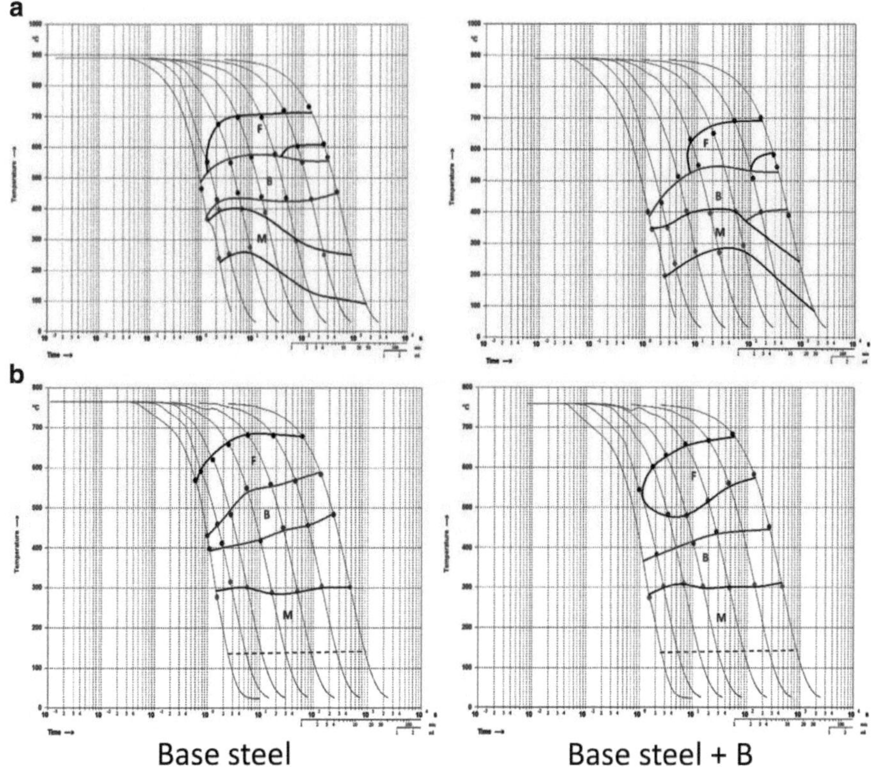

Base steel   **Base steel + B**

Fig. 4.14 Effect of boron additions (0.002 %) on CCT diagrams of 0.15C–2Mn–0.3Si–Nb steel built in cooling from supercritical (**a**) and intercritical (**b**) temperatures—original

mixture than of 100 % austenite. However retardation of bainite reaction and lowering Bs temperature is more significant.

Modern high- and ultrahigh-strength steels are alloyed with significant amounts of elements that retard the recrystallization of ferrite and therefore these steels have to be annealed at higher temperatures of two-phase region where higher fractions of austenite of lower carbon content are formed. The ability of boron additions to increase the stability of austenite phase at lower carbon content can be viewed as being very advantageous. At the same time in the cases of very low-carbon content in γ-phase insufficient formation of ferrite and hence enrichment of remaining austenite by carbon, boron containing steels facilitate bainite formation at high annealing temperatures.

It is worth to note that the additions of Mo to boron steel suppress the precipitations of boro-carbide by inhibiting carbon diffusion and thus they contribute to increasing the concentration of solute boron at austenite grain boundaries promoting a synergetic effect of B + Mo on hardenability of steels (Hara et al. 2004).

Synergism of boron and molybdenum additions is confirmed by CCT diagrams presented in Fig. 4.15, which demonstrate significant retardation of phase

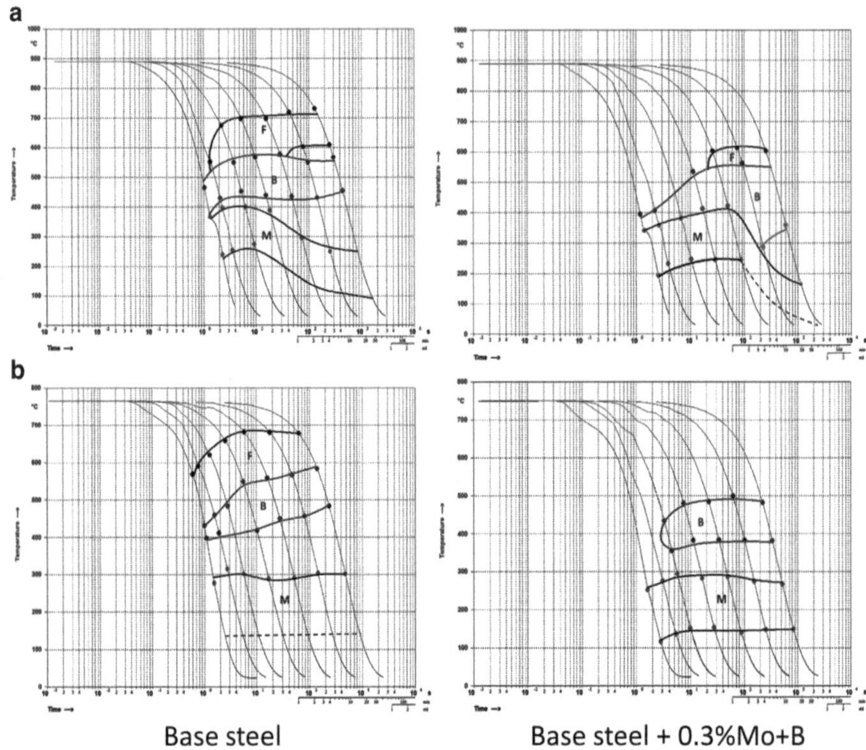

Fig. 4.15 Effect of B + 0.3 % Mo additions on CCT diagrams of 0.15C–2Mn–0.3S–Nb i steel built in cooling from supercritical (**a**) and intercritical (**b**) temperatures—original

transformations at cooling from both full austenitization and intercritical temperature region.

4.3.1.8   V, Nb, Ti

The data concerning the influence of carbonitride-forming elements on austenite stability during cooling from the intercritical range are very limited. In general, the effects of these elements can be viewed as being induced by a number of factors: possible arrest of the $\gamma \rightarrow \alpha$ transformation front that is controlled by the concentration of microalloying elements in austenite, the refinement of microstructure and hence increase in interfacial area, and the presence of disperse carbonitride particles.

The first factor should favor austenite stability and the last two should cause it to decrease. The relative contribution of each factor is governed by the solubility of particular element in austenite at the given intercritical temperature and by the type of potential precipitates such as carbides or nitrides.

It is believed that after annealing at the higher intercritical temperatures, the dissolved vanadium can enhance austenite stability. Speich et al. (1983), investigating steel of 0.12C–1.7Mn–0.58Si–0.04V, noted that vanadium addition increased the amount of retained austenite.

Priestner and Ajmal (1987) studied the 0.11C–1.46Mn–0.35Si steel with additions of the carbide-forming elements (0.102 % Nb or 0.046 % Nb + 0.092 % V) that reduce the effective carbon content in austenite and therefore decrease the amount of austenite formed at given temperature. With identical amount of austenite in the initial $\alpha + \gamma$ microstructure, lower C_γ in steel containing both V and Nb was more efficient in reducing the austenite stability compared to direct individual effects of V and Nb. This manifested itself as an increase in the cooling rate necessary to produce the same portion of martensite in microalloyed steels.

On the contrary, there are some indications that niobium can increase hardenability of steels. Thus, in the study of Garcia et al., the microstructure and properties of steels with 0.15C–1.76Mn–0.42Si–0.5Mn–0.5Cr–0.02Nb without and with 0.06–0.1V were compared, when authors came to conclusions that additions of Nb plus vanadium increase the hardenability of austenite (Garcia et al. 2012). Construction of CCT diagrams of 0.05C–1.56Mn (base composition) with additions of 0.03 and 0.06Nb demonstrated evident decrease in transformation temperatures after full austenitization for the steel with 0.03 % Nb but no effect for the addition of 0.06 % Nb (Isasti et al. 2013).

4.3.2 Ferrite Solid Solution

The characteristics of ferrite matrix are largely responsible for the ductile properties and aging-induced strengthening of dual-phase steels. This emphasizes the importance of the ferrite solid solution.

Substitutional alloying elements can have a direct effect on the properties of ferrite due to solid solution strengthening. Indirectly, they impact the behavior of interstitials in ferrite. This impact, in turn, has many different aspects, such as (a) direct decrease in interstitial solubility in ferrite, e.g., by manganese; (b) variation of the thermodynamic activity of carbon in iron, e.g., by silicon, which can affect carbon partitioning between α- and γ-phases; (c) chemical interaction of, e.g., carbonitride-forming elements with interstitial elements that leads to precipitation; and (d) indirect, alloying permits using slower cooling and thereby decreasing the content of interstitial atoms in ferrite.

Pickering noted that strengthening of ferrite by lattice distortions induced by interstitial solutes, viz., carbon and nitrogen, is 100 times stronger than by solute substitute elements (Pickering 1978). At the same time, the quenching from temperatures close to A_{C1} can fix in solution about 0.0025 at.% carbon, dissolved in the iron ferrite (Nakaoka et al. 1977).

As mentioned above, alloying influences the solid solution strengthening of ferrite by interstitials due to changing carbon solubility in ferrite. Figure 4.16

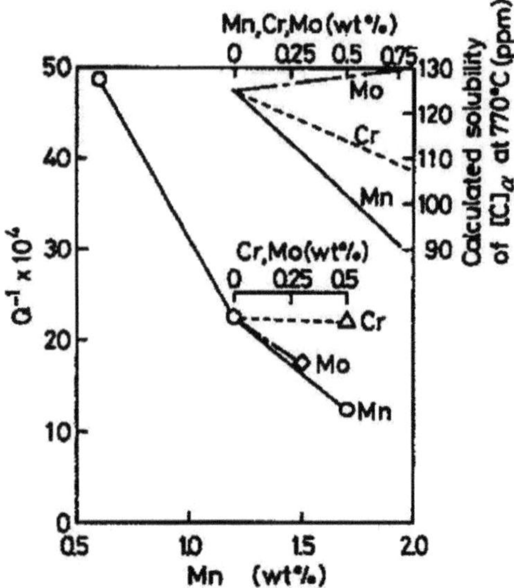

Fig. 4.16 Effects of substitutional elements on internal friction and calculated solute carbon in ferrite (Hashiguchi and Nashida 1980)

shows the results of internal friction measurements illustrating how substitutional elements such as Mn, Cr, and Mo alter the solubility of carbon in ferrite compared to that in the absence of alloying (Hashiguchi and Nashida 1980). As can be seen, Cr effect is weaker than that of Mn, whereas Mo slightly increases solubility of carbon in ferrite. Such elements as Al and Si increase the thermodynamic activity of carbon in α-phase and hence should significantly decrease the solubility of carbon in ferrite.

Important advantage of alloying elements to weaken strengthening of ferrite by interstitial atoms is the result of the beneficial opportunity of producing ferrite–martensite microstructures at lower cooling rates. The significance of the cooling rate in determining the relative carbon content in 0.08C–1.3Mn–0.35 steel is depicted in Fig. 4.17 that shows measurements of Snoek peak in temperature dependence of internal friction. As shown, additions of microalloying elements (in particular, vanadium) can decrease soluble interstitials by binding them in carbonitrides.

Some substitutional elements strengthen ferrite, especially silicon and phosphorus (Nakaoka et al. 1979; Davies 1979). According to Pickering (1978), an addition of 1 % of these elements increases the tensile strength of ferrite by 80–120 and 600 MPa, respectively.

Effects of Si and P were studied using dual-phase steels containing up to 2.0 % Si (Ramos et al. 1979; Davies 1979) and up to 0.4–1.0 % P (Davies 1979; Becker et al. 1981).

One of important benefits of solid solution strengthening of ferrite is its higher contribution to the overall strength of ferrite–martensite mixture and the possibility to achieve the desired strength at smaller fraction of martensite, which should

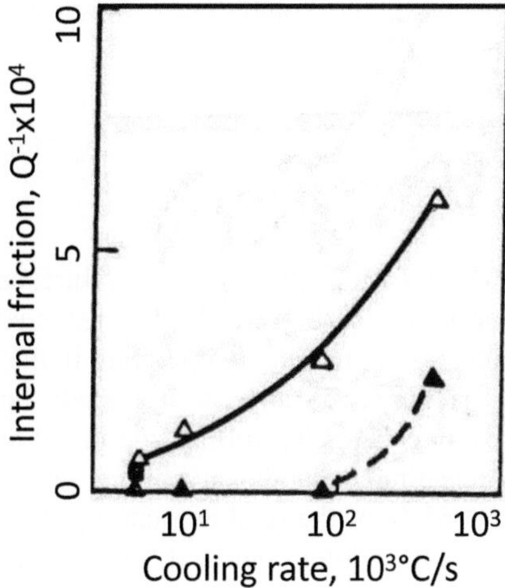

Fig. 4.17 Effect of cooling rate on the Snoek carbon peak height in 0.08C–1.35Mn–0.35Si (*solid line*) and the same steel with additions of vanadium (*dashed line*) steel—original

improve ductility (see Chap. 3). Additionally, the advantage of using silicon is that this element increases strain hardening of ferrite and thus contributes to overall strain hardening of dual-phase steels.

Additions of ferrite strengthening elements to dual-phase steels can sometimes deteriorate ductility and toughness. In this respect, the additions of chromium are especially beneficial as Cr positively affects hardenability of austenite while being the weakest ferrite strengthening element. According to Pickering, an addition of 1 % Cr to steel can even lower the yield strength of ferrite by approximately 35 MPa.

4.3.3   Precipitation Hardening of Ferrite

Everything that strengthens ferrite and reduces its ductility can impair the overall ductility of dual-phase steel if this reduction is not compensated by simultaneous grain refinement, or by decrease in the martensite fraction.

The current trends in demand for DP steels with higher strength requirements dramatically change the attitude towards utilization of microalloying elements in multiphase steels, in general, and in DP steels, in particular. While vanadium is not used much in modern DP steel, the majority of these steels with the strength above 600 MPa contain Nb and Ti. Titanium is added mostly to bind free nitrogen aiming to improve ductility of ferrite or to protect boron from binding with nitrogen.

Precipitation kinetics of Nb and Ti was studied using the 0.08C–2.2.Mn–0.02Ti steel with Nb content ranging from zero to 0.06 %.

The solubility of these two microalloying elements at low temperatures of intercritical region is known to be low. As the temperature in the $\alpha + \gamma$ range is increased, the solubility of carbide-forming elements grows according to the solubility product equation:

$$\log[\mathrm{Me}]x\,[\mathrm{C}] = A - \frac{B}{T} \tag{4.6}$$

where T is the absolute temperature, and [Me] and [C] are the concentrations of Nb or Ti and carbon in at.%. Increase solubility is aided by decreasing C_γ with the increase in heating temperature in the two-phase region.

Whereas Ti almost completely bound in carbonitrides regardless of thermal parameters of hot rolling, the initial amount of Nb in solution prior to intercritical annealing depends on coiling temperature as was shown in Fig. 2.17: about 50–65 % of the total amount of Nb precipitated after 620 °C, whereas it was less than 10 % after coiling at 500 °C.

However, the parameters of subsequent heat treatment used to produce cold-rolled dual-phase steels reduce potential strengthening by dispersed particles. Nb partially precipitates in situ during heating when prior existed precipitates are coarsening and fully precipitates at annealing temperatures.

If steels contain additions of Ti, accelerated precipitation of Nb carbides is observed since TI is serving as substrate for Nb-based particles facilitating the overall increase in sizes of precipitates.

Substantial supersaturation of ferrite with interstitial solutes during intercritical annealing facilitates binding of carbon and nitrogen with carbonitride-forming elements that leads to precipitation strengthening of ferrite (Rigsbee and VandlerArend 1977).

One can expect that newly transformed ferrite with extra carbon retained from original austenite should have higher propensity for precipitation strengthening. In fact, Rashid and Rao observed characteristic interfacial vanadium carbide precipitates formed at the γ/α transformation front during slow cooling rate from the $\alpha + \gamma$ region (Rashid and Rao 1982). The sizes of precipitates were decreasing with increasing distance from the original "old" ferrite boundary indicating the lowering precipitation temperature. No dispersed particles near martensite islands were detected because of unfavorable kinetic conditions for precipitation at low temperatures of martensitic transformation. Exact amount and size of precipitates are governed by the annealing temperature in $\alpha + \gamma$ region and the rate of subsequent cooling.

4.3.4   Austenite Non-homogeneity

The non-homogeneity of carbon distribution in austenite in DP steel was observed in numerous studies. It manifested itself, in particular, by coexistence of neighboring different transformation products. For example, as the result of such non-homogeneity one part of an austenite grain can transform to pearlite while

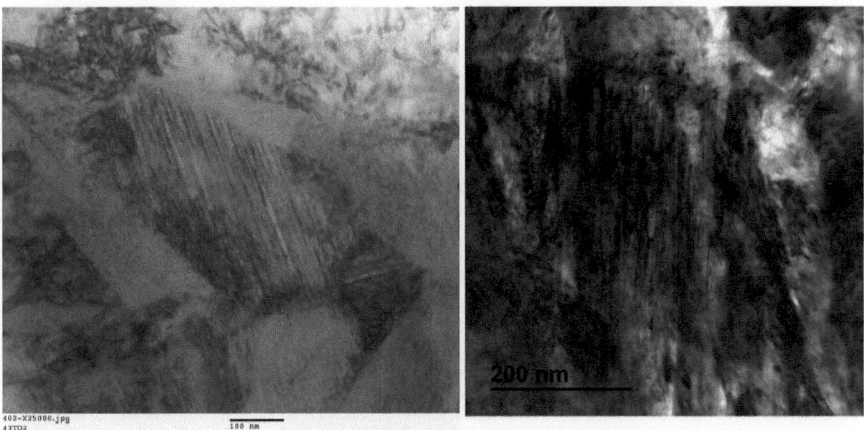

Fig. 4.18 Morphology of martensite in DP steel—original

the other can transform to martensite (Speich et al. 1981; Hansen and Pradhan 1981). As presented in Fig. 4.18, in spite of low-carbon average content in martensite in the majority of DP steels, TEM investigation finds a dominant twinned morphology of martensite.

The observed phenomenon of non-uniformity of carbon distribution is due to nonuniform distribution of substitute elements which affect carbon activity, insufficient soaking time, as well as effects of alloying elements which slow down the diffusion rate of carbon,

4.3.5   Martensite Transformation Temperature

The martensite transformation start temperature M_s controls the substructure of martensite, potential auto-tempering during cooling and the possibility to preserve some amounts of retained austenite in final microstructure. The importance of M_s cannot be overestimated because the most critical mechanical properties of dual-phase steels are governed by the number of mobile dislocations and the level of residual stresses generated in local $\gamma \rightarrow \alpha$ transformation (see Chap. 3). As was shown in numerous equations, carbon, manganese, chromium, and silicon, in that order, have the greatest effects on M_s (Andrews 1965; Steven and Haynes 1956).

Carbon content in austenite fraction by the moment of the onset of its transformation in cooling from $\alpha + \gamma$ region is the main factor determining the M_s temperature and hence the properties/strength of martensite, the level of residual stresses, dislocation density in ferrite, etc.

Depending on the carbon content in typical DP steels, as well as on the annealing temperature and the amount of newly transformed ferrite, the carbon content in the

γ-phase undergoing the martensite transformation, C_γ, can vary from 0.1 to 0.6 %. This leads to wide variations in M_s.

Some researchers point to the beneficial effects of manganese, chromium, or molybdenum on M_s. However, not less significant are indirect effects of alloying elements on M_s temperature as the alloying varies the permissible cooling rate, the amount of newly transformed ferrite, and the relevant carbon enrichment of γ-phase.

In particular, Al is known to raise the M_s temperature at about 30 °C per 1 % Al. This is true for quenching from supercritical region. However, with incomplete austenitization, an increase in Al content facilitates formation of new ferrite, thus increasing carbon content in remaining portion of austenite and actually decreasing the M_s temperature (Fonstein 2005; Girina and Fonstein 2005).

On the contrary, additions of alloying elements that substantially increase hardenability of austenite and prevent the formation of new ferrite in some cases, as was depicted in Fig. 4.8 for 3 % Mn, can raise the M_s temperature compared to the less alloyed steels due to lower carbon enrichment in remaining portion of austenite.

The fact of dominating role of enrichment of remaining portions of austenite by carbon is confirmed by observed decrease in the M_s temperature at lower cooling rate as shown on numerous presented CCT diagrams.

4.4 Effect of Chemical Composition on Mechanical Properties of Dual-Phase Steels

The effect of composition on mechanical properties of dual-phase steels is the result of the variation in their microstructural parameters such as the volume fraction of martensite, morphology of martensite and ferrite, the strength of martensite and ferrite, etc.

The influence of alloying elements on strain aging (hardening) behavior is related to the interstitial content in ferrite, the energy of interaction between interstitials and dislocations, the microstructure morphology and dispersion, as well as variations in the applied rates of cooling from the $\alpha + \gamma$ region.

Summarizing the potential role of solid solution strengthening of ferrite by substitutional elements, it is reasonable to repeat the important advantage of the opportunity to reach required strength of DP steels at lower volume fraction of martensite. In this case, the potential decrease in ferrite ductility can be substantially compensated by beneficial effects of lower V_m on ductility, whereas the accompanying decrease in carbon content in steels can improve weldability.

Regarding modern requirements for improved flangeability, it is worth to refer to data that reducing the strength difference between the two phases in DP steels (in particular, on account of higher strength of ferrite) improves their behavior during localized deformation, particularly in hole expansion.

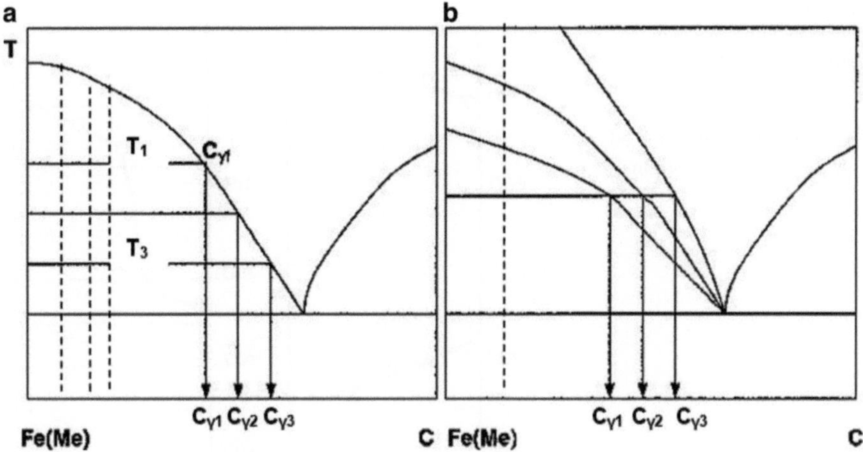

Fig. 4.19 Schematic Fe(Me)–C diagrams illustrating the possible difference in carbon content in austenite, C_γ, depending (**a**) on temperature of annealing and (**b**) different alloying at the same temperature of annealing—original

4.4.1   Tensile Properties

For proper design of DP steels, it is very important to know potential contributions of different alloying elements to strength of dual-phase mixture. If higher strength achieved by specific alloying/microalloying can allow decreasing the volume fraction of martensite or the carbon content in steel, it can render additional benefits of better ductility and/or weldability.

As illustrated by Fig. 4.19, the estimations of effects of chemical composition on properties of DP steels can be drastically changed depending on whether it is is conducted at the same temperature in the intercritical region for the steel of different composition (Fig. 4.19b) or at constant volume fraction of martensite achieved by changing carbon content in the steel (dashed lines).

4.4.1.1   Carbon

The effect of carbon on tensile properties of ferrite–martensite microstructure is clearly defined by the ability to form a certain volume fraction of martensite, V_M, containing certain amount (concentration) of carbon (C_M). This was discussed in detail in Chap. 3.

The majority of the published data show that the strength of DP steels is linearly dependent on the volume fraction of martensite.

With regard to the strength of martensite, it is worth recalling that the final properties are *independent of carbon content in martensite* when V_M is below

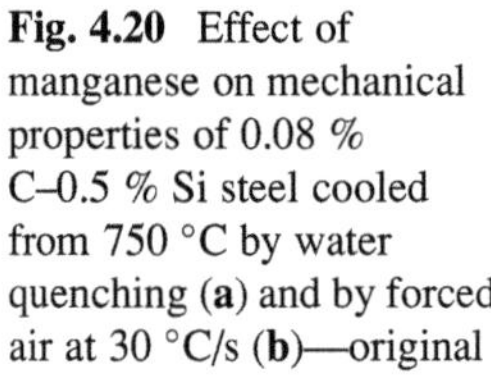

Fig. 4.20 Effect of manganese on mechanical properties of 0.08 % C–0.5 % Si steel cooled from 750 °C by water quenching (**a**) and by forced air at 30 °C/s (**b**)—original

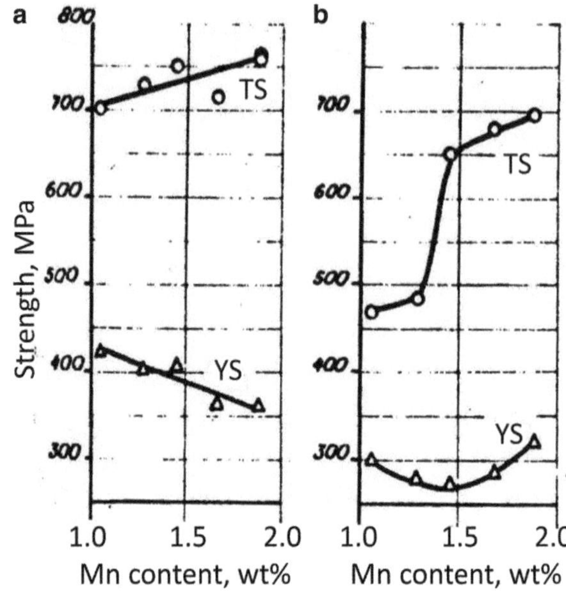

40–45 %. At higher V_M, the microstructure stereology changes so that instead of martensite islands embedded into soft ferrite matrix with possible partitioning of stresses, a skeleton-type martensite becomes prevailing.

4.4.1.2   Manganese

The effect of manganese itself on the strength of ferrite phase is not significant and is about 25–35 MPa per 1 % Mn (Pickering 1978). At the same time, at given temperature of heating in the $\alpha + \gamma$ region, the increase in manganese content results in higher volume fraction of austenite, V_γ, as Mn lowers the A_{c1} temperature and the carbon content in eutectoid. This typically produces larger volume fractions of martensite due to higher hardenability of austenite, more essential under slower cooling, Fig. 4.20. Thus, the overall appeared influence of Mn on the strength of DP steel can be far stronger than just merely ferrite solute solution strengthening.

Takada et al. noted that the YS/TS ratio decreases with increasing Mn content. They attributed this behavior to higher extent of dispersion of microstructure, emphasizing the important role of higher ratio of the number of martensite particles to that of ferrite particles, as described in Chap. 3 (Takada et al. 1982). However, probably the domination role of Mn is the enhancement of the austenite stability, as resulted in lower M_s temperature. In particular, with the same extent of microstructure dispersion, the transformation of austenite at lower temperature should increase the density of mobile dislocations near to the martensite/ferrite interface as well as residual stress in ferrite, which should contribute to lower YS/TS ratio as well.

4.4.1.3 Silicon

Alloying with silicon induces substantial increase in both yield strength and tensile strengths of dual-phase steels due to solid solution strengthening of ferrite, but simultaneously increase in Si content is accompanied by increase in balance of uniform/total elongation and tensile strength of DP steels (Koo and Thomas 1977; Davies 1979; Fonstein et al. 2013).

The retention of high ductility in silicon bearing DP steel has been often attributed to enhanced work hardening rate of silicon-alloyed ferrite within the entire strain range. It has been observed as a higher increase of tensile strength rather than in yield strength of DP steels with increasing Si content. To study the nature of that phenomenon, Hironaka et al. (2010) used 0.15C–2.2Mn DP steels with Si ranging from 0.01 to 1.6 %. At 8 % tensile strain in low Si steels, the dislocation cell structure was observed whereas in high Si steels tangles dislocation still dominated instead. Retardation of the formation of dislocation cell structure resulted in significantly higher rate of increase in dislocation density in ferrite during initial tensile deformation. The authors concluded that Si additions reduced the stacking fault energy of ferrite inhibiting dislocation cross slip, thus increasing the dislocation density. Higher strain hardening, in turn, promotes higher uniform elongation and better tensile strength–ductility balance of DP steels at higher Si content.

Higher cleanness of ferrite in the presence of Si due to enhanced partitioning carbon to austenite should promote higher ductility of ferrite phase, also contributing to ductility of DP steel.

The general trend of better balance between tensile strength and elongation due to increased Si content is illustrated in Fig. 4.21 (Fonstein et al. 2013). Davies, who investigated dual-phase steels with TS ranging from 600 to 800 MPa, also pointed out the better balance of strength and uniform elongation at comparison DP steels with 1 and 2 % Si (Davies 1979).

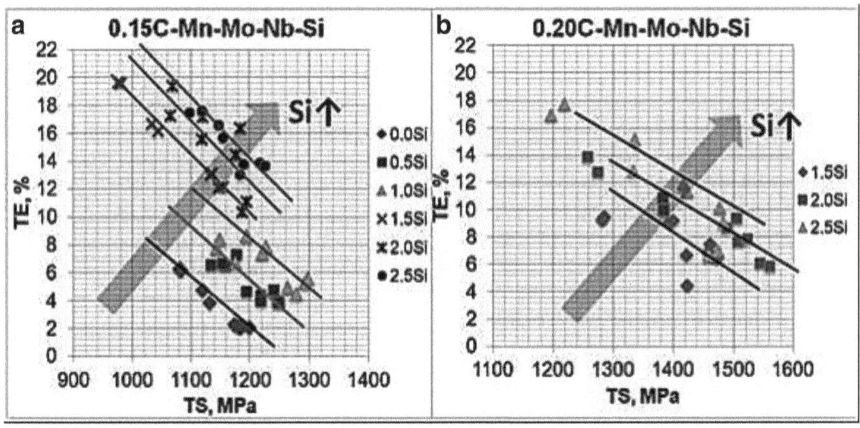

Fig. 4.21 Effect of Si on the balance between strength and ductility for DP steels with (**a**) 0.15 C and (**b**) 0.20 C (Fonstein et al. 2013)

Multiphase steels containing up to 1.5–2.5 % Si are attracting growing practical interest for development of steels targeting properties of 3rd Generation (see Chaps. 8–10).

4.4.1.4   Aluminum

Direct data about Al effect on tensile properties of DP steels are very limited. Since Al significantly raises the A_{c3} temperature, the annealing of steels with increasing Al content at given intercritical temperature is accompanied by progressively decreasing volume fraction of the formed austenite and hence of martensite after cooling.

The data obtained for Fe–Al binary system showed that the strengthening of iron by Al is only about 35 MPa per 1 % Al (Case and Van Horn 1953). The published data do not reveal any strengthening of dual-phase steels by Al, although the microprobe analysis detected some partitioning of Al to ferrite after holding in two-phase temperature range.

Steels with the base composition of 0.09C–2.0Mn–0.6(Cr + Mo)–0.03(Ti + Nb) containing 0.042, 0.40. 0.70, 0.88, and 1.10Al after annealing had the same range of tensile strength, but total elongation increased with higher Al content (Girina and Fonstein 2005). Better balance of TS and TE due to higher Al content, Fig. 4.22, was explained by refined ferrite structure and by smaller sizes of discontinuous martensite islands due to larger portion of new ferrite formation and lower M_s values of remaining austenite.

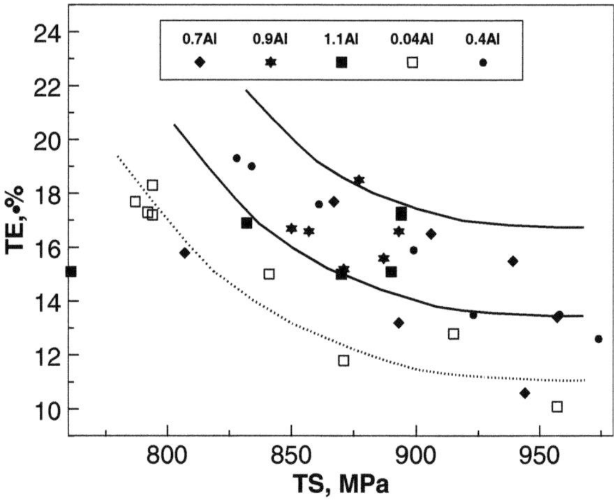

Fig. 4.22 Effect of Al on TS–TE balance (Girina and Fonstein 2005)

4.4.1.5   Chromium

The effects of chromium on the tensile properties are determined primarily by the type of microstructure formed during cooling of austenite, i.e., by the effect of Cr on austenite hardenability. Another contribution is due to lowering the M_s temperature, which reduces the YS/TS ratio to as low as 0.40 (Hashiguchi and Nashida 1980).

Some publications noted refinement of microstructure by Cr additions, but the data are inconsistent and the opposite effect was also reported.

4.4.1.6   Molybdenum

The effect of molybdenum to increase strength of dual-phase steels after relatively slow cooling can be explained by higher hardenability of austenite (Coldren and Eldis 1980; Hashiguchi and Nashida 1980) that results in higher amount of martensite. However, some authors note that the effects of Mo on mechanical properties of DP steels are rather sensitive to both annealing temperature in the $\alpha + \gamma$ range and subsequent cooling rate (Nakagawa et al. 1981).

According to Pottore et al., additions of 0.16–0.30Mo to 0.09C–1.6Mn–0.1Si–0.035Ti–B steel substantially improved the balance between tensile strength and total elongation of water-quenched 980DP steel that was explained by microstructure refinement inherited from hot rolling (Fig. 4.23) (Pottore et al. 2006).

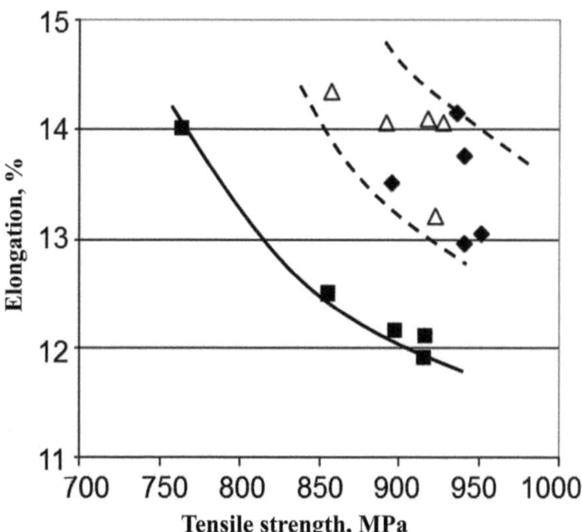

Fig. 4.23 Effect of Mo additions to 0.09C–1.6Mn–0.002B steel on the balance of tensile strength and total elongation (soaking temperature—800 °C, overaging at 200 °C) (Pottore et al. 2006)

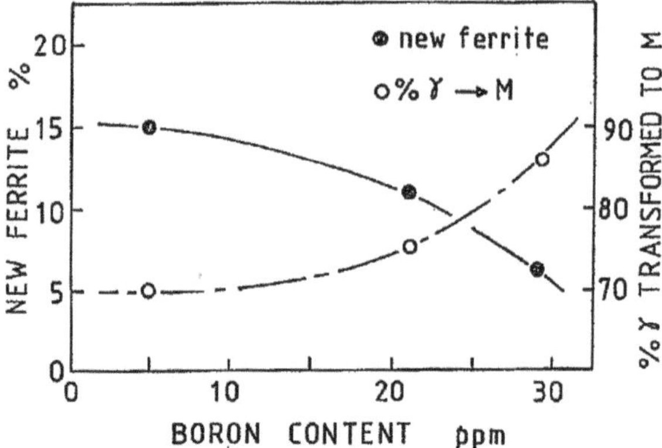

Fig. 4.24 Effect of boron on the volume of new ferrite and the fraction of austenite transformed to martensite at oil quenching from 800 °C; steel of 0.1C–1.35Mn–0.5Si–0.005Ti (Shen and Priestner 1990)

4.4.1.7   Boron

Boron does not directly affect the strength of ferrite or martensite. Its strong influence on properties of DP steels is related exclusively to the above described effect *on austenite hardenability. Therefore effects of boron on mechanical properties of DP steels can be quite different* depending on temperature of annealing in two-phase region (carbon content in austenite) and cooling rate.

As pointed out by Shen and Preistner (Fig. 4.24), the key effect of boron on mechanical properties of DP steels is determined by its corresponding influence on suppression of ferrite formation and effective fraction of martensite formed at austenite transformation of given chemical composition and given parameters of annealing (temperature and cooling rate).

4.4.1.8   Niobium/Niobium Plus Titanium

Effect of microalloying of DP steels by Nb and/or Ti can be related to structure refinement and precipitation hardening.

Theoretical evaluation of precipitation strengthening based on the Orowan model is presented in Chap. 3. The corresponding analysis is possible based on careful TEM measurements of sizes and volume fraction of precipitates (Hoel and Thomas 1981). Effect of strengthening by carbonitride-forming elements is the sum of contribution of precipitation hardening and structure refinement.

It was already shown in 1980s that addition, in particular, of 0.04 % of *niobium* to hot-rolled 0.14C–1.6Mn sheet steel led to higher mechanical properties. Niobium was found to favor the refinement of ferrite grains (Gau et al. 1981). There are numerous data indicating that additions of Nb increase yield strength and tensile strength of DP steels (Hausmann et al. 2013; Song et al. 2014).

After annealing in the lower part of intercritical temperature region, Nb is responsible for the combination of strengthening effects: the shown increase in volume fraction of austenite/martensite with higher Nb content, full or partial

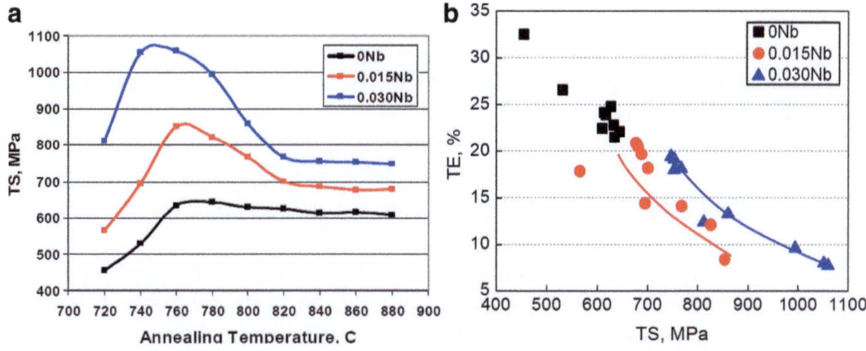

Fig. 4.25 Effect of Nb on (a) tensile strength of 0.085C–2Mn–0.5Cr steel depending on annealing temperature (cooling rate 10 °C/s) and balance of TS and TE (b) (Girina et al. 2008)

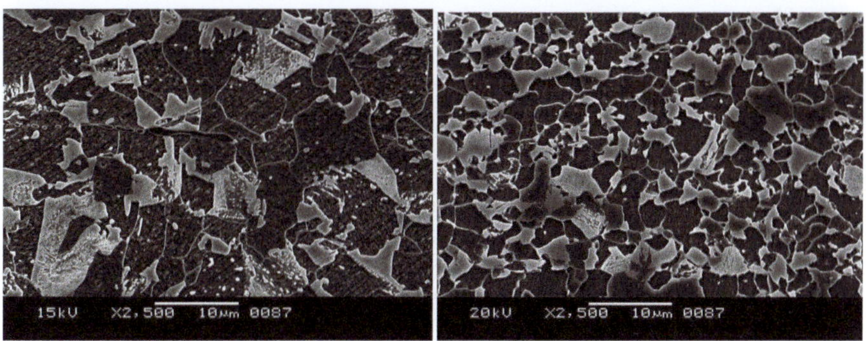

Fig. 4.26 Microstructure of DP steels containing 0.085C–2.0Mn–0.5Cr without Nb (a) and with 0.03 % Nb—modified

inhibition of ferrite recrystallization, general structure grain refinement, and, to a lesser extent, precipitation hardening. As shown in Fig. 4.25a, this combined strengthening effect in the steel of 0.085C–2.0Mn–0.5Cr can be huge: up to 400 MPa increase in strength at lower ductility of steel, of course. At higher temperatures, closer to A_{c3}, the recrystallization of ferrite is complete and the austenite fraction becomes closer to "equilibrium" so that the strengthening effect due to Nb becomes less significant. Nevertheless, grain refinement, slightly bigger portion of martensite (sometimes plus bainite), and higher contribution from precipitation hardening produce together a ~140 MPa increase in strength in steel with 0.03 % Nb when steel without Nb cannot reach necessary TS of 780 MPa. As shown in Fig. 4.25b, strengthening by Nb is featured by better balance of TS and TE that can be explained by dominating role of structure refinement resulting in higher strain hardening and hence in higher uniform elongation.

The comparison of microstructure of steels without Nb and microalloyed by 0.03 % Nb is presented in Fig. 4.26.

As was found, additions of Ti to Nb microalloyed steel facilitated coarsening of precipitates because TiN particles serve as substrate to Nb carbides. As a result, DP steels microalloyed by combined additions of Nb and Ti show lower strength that microalloyed only by Nb.

4.4.1.9   Vanadium

The effect of vanadium is far from being clearly understood. The first reports on dual-phase steels in the USA dealt with a comparison of the properties of ferrite–pearlite and ferrite–martensite microstructures were based on the steel containing, in wt%, 0.12C–1.46–Mn–0.51Si–0.11V. Better combinations of properties of dual-phase steels, which were initially obtained by cooling at moderate rates, were attributed to the beneficial effect of vanadium (Bucher et al. 1979). A series of papers were published in the following years to clarify the vanadium effect. For example, Nakagawa et al. noted that vanadium refines ferrite grains (Nakagawa et al. 1981). This, however, contradicted with the work by Ostrom et al. (1981). Another controversy was related to aging behavior. Vanadium was believed to remove carbon from solid solution thereby weakening aging effects, but in fact vanadium-bearing steels exhibited substantial hardening during aging (Davies 1978).

At the same time, according to Nakagawa et al., fine vanadium carbide precipitates formed in ferrite even during rapid quenching were believed to lower the ductility of DP steels. It should be noted that the consideration of the role of vanadium was often based on generalizations without carrying out parallel experiments using vanadium-free steels or multilevel vanadium concentrations for comparison.

In this context, a dedicated study of four vanadium levels of up to 0.18 % V in 0.08C–1.35Mn–0.35Si steels was conducted (Fonstein et al. 1985). Specimens were heated in the intercritical range at temperature of 730–830 °C, held for 15 min and water quenched or fan cooled at the rates of 600 and 30 °C/s, respectively. To compare with ferrite–pearlite steels, some specimens were normalized at 950 °C.

As shown in Fig. 4.27a, the dual-phase steels were essentially insensitive to vanadium content after water quenching from 750 °C and with 0.18 % V showed slightly lower tensile strength due to a smaller volume fraction of martensite induced by decrease in effective carbon partially bound by vanadium. When fan cooling was used (Fig. 4.27b), the tensile strength increased and the ductility decreased with increasing vanadium, which is similar to the trend observed in normalized ferrite–pearlite steel. This indicated that at relatively low cooling rates from the $\alpha + \gamma$ region, vanadium steels were strengthened by dispersed particles that have had sufficient time to precipitate from supersaturated ferrite during cooling. This conclusion agrees with the results by Nakagawa.

Although the solubility of vanadium in austenite increases with temperature in the $\alpha + \gamma$ range, the biggest strengthening effect of vanadium was observed after heating to lower temperature (750 °C). This can indicate that the amount of precipitates is controlled by supersaturation of carbon in ferrite, which agrees with findings by (Rigsbee and VanderArend 1977).

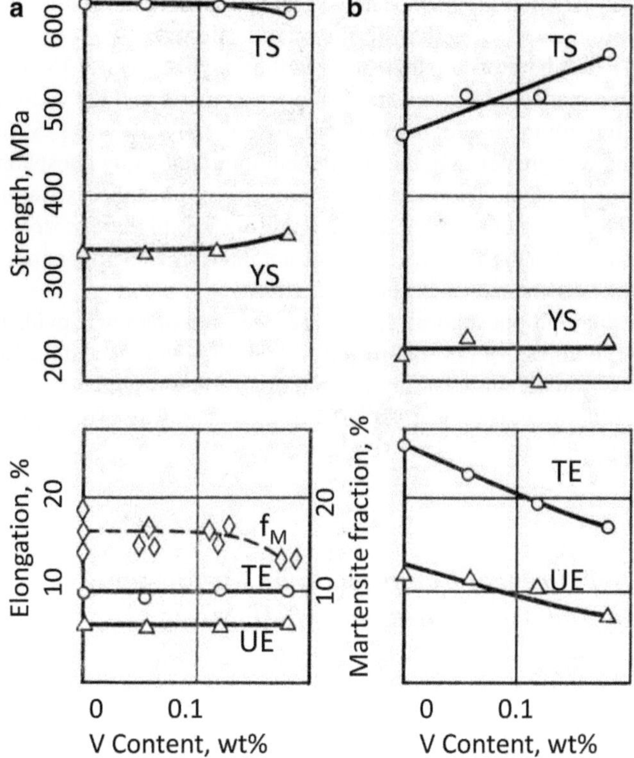

Fig. 4.27 Effect of vanadium on mechanical properties of 0.08C–1.35Mn–0.35Si steel after heating at 750 °C: (**a**) water quenching; (**b**) fan cooled—modified

In general, at a constant holding temperature in the $\alpha+\gamma$ range the effect of increasing vanadium on the tensile properties of dual-phase steels is insignificant, because of the competing opposing interplay of grain refinement and precipitation hardening affected by vanadium, on one hand, and decreasing volume fraction of martensite because of binding of carbon, on the other hand. The latter fact was supported by metallographic observations.

4.4.1.10   Phosphorus

In early developments of DP steels many workers expressed interest in *phosphorus* as a low-cost hardening agent for dual-phase steels. However, the available evidence is ambiguous. In fact, Becker et al. found that 0.1 % P improved ductility of ferrite–martensite steel as a result of substantial increase of the work hardening rate (Becker et al. 1981). The positive influence of additions of phosphorous up to 0.2 % on strain hardening and balance of ductility and tensile strength of DP steel was pointed also by Davies (1979).

In contrast, Araki et al. (1977) showed that phosphorus lowered ductility much more than it improved strength of ferrite–martensite steel.

The study of laboratory-melted steels with 0.06, 0.12, and 0.18 % C and up to 0.1 % P at two manganese levels of 0.5 and 1.5 % found that the yield strength and the tensile strength increased, respectively, by 85–120 MPa and 125–165 MPa per 0.1 % P. This was greater than the strengthening produced by phosphorus addition to ferrite–pearlite steel. The increase in strength by phosphorus was found to be proportional to the amount of ferrite and therefore was higher for steels with lower volume fraction of martensite, i.e., for constant intercritical temperature, with lower carbon contents (Panteleeva and Fonstein 1984)

Commercial high-phosphorus dual-phase steels are unknown, evidently because of the negative impact of phosphorus on weldability (as reflected, in particular, by higher carbon equivalent, CE). Implementation of new welding technologies may make it possible to revise the application of the cheap way of strengthening steels by phosphorus.

4.4.2 Effect of Steel Composition on Strain Aging and Tempering Behavior of DP Steels

Aging of dual-phase ferrite–martensite steels and the factors controlling the aging kinetics and resultant hardening have been discussed above in Chap. 3.

Aging is primarily governed by the interstitial elements' solute in the ferrite and their diffusion to free dislocations.

The effects of alloying elements on aging of dual-phase steels are determined by their impact on key factors controlling aging (a) equilibrium carbon and nitrogen solubility in ferrite; (b) ferrite grain size (diffusion path of interstitials to regions of higher dislocation density); (c) amount of martensite; (d) the M_s temperature (that, along with the amount of martensite, determines the level of residual stresses and density of free dislocations); and the last but not the least, (e) the minimum cooling rate from the $\alpha + \gamma$ region allowing for ferrite–martensite microstructure and therefore the actual amount of interstitials in solution.

Besides, some alloying elements combine with C and N forming carbides/nitrides.

Studies of the effect of Mn (in the range of 1–2 %) revealed the maximum aging resistance at about ~1.5 % Mn (Storozeva et al. 1985). Such an extreme effect of manganese is the result of some competitive influence of Mn on microstructure and propensity for aging. Increase in manganese content in the range of 1.0–1.5 % raises the level of residual stresses due to lower M_s temperature, which should slow the aging. On the other hand, as the manganese content increases above 1.5 %, ferrite grain refinement was observed that should facilitate carbon transport to the areas of higher dislocation density near ferrite grain boundaries. This should accelerate aging. Similar study of 0.03C–0.35Si steels containing 1.03–1.8 % of *manganese* showed that the relaxation time for aging attained a maximum at 1.45 % Mn. This indicates that this steel in the as-quenched condition exhibited the weakest propensity to aging, as confirmed by Takada et al. (1982).

Hashiguchi and Nashida considered the alloy content as an extremely important factor for reducing the rate of cooling from the intercritical temperature during commercial processing of dual-phase steels required to prevent aging (Hashiguchi and Nashida 1980). They plotted nomographs in "alloying element—cooling rate" coordinates, showing ranges of compositions and conditions of cooling from the intercritical region, which require tempering to be used to avoid aging of steel during storage. However, no data are yet available confirming the applicability of this approach to modern high- and ultrahigh-strength DP steels with significantly higher alloy contents.

The effects of *aluminum*, *titanium*, or *niobium* on propensity of DP steels to aging depend on the cooling rate from the intercritical temperature, on the actual ferrite grain size. When the variations in ferrite grain size are small, titanium or aluminum bind interstitials so that increase in their content reduces propensity to aging of non-deformed dual-phase steels.

As shown in Chap. 3, the BH (baking hardenability) effect is enhanced by grain refinement due to larger area of interfaces where carbon can preferably segregate, as well as due to shorter path for carbon diffusion to regions of higher dislocation density. Therefore, microalloying with Nb and/or Ti and the associated grain refinement should induce higher BH effect along with lower propensity to aging at ambient temperatures due to binding of extra nitrogen. Experimental confirmations of this effect are limited (Hashimoto 2007).

In Chap. 3, the important role of nonhomogeneous dislocation distribution as a factor inhibiting aging was pointed out. This is supported by enhanced aging of water-quenched steels containing *aluminum* provided that higher aluminum contributes to ferrite grain refinement (Tanaka et al. 1979).

There are data showing that *vanadium* accelerates aging of DP steels in spite of lowering free carbon content in ferrite by partially binding carbon into carbides (Himmel et al. 1981), as confirmed by the Snoek peak measurements (Storozeva et al. 1985; Pradhan 1985).

Davies suggested that since vanadium combines with interstitial elements such as carbon and nitrogen, adding of vanadium should reduce the propensity to aging of dual-phase steels (Davies 1979). In a more recent study, however, Davies noted considerable hardening as a result of aging in VAN-Q8 steel containing about 0.1 % V, whereas he used aging 100 times shorter than in his previous work. The possible reason for these discrepancies is that in the latter work the specimens were deformed in tension as opposed to rolling employed in the earlier study (Davies 1981).

A possible reason for observed effect, also noted by Davies, was that the vanadium-bearing steel had lower martensite volume fraction. The latter reduces the level of residual stresses in ferrite that normally interfere with the aging. Besides, the vanadium microalloyed steel had substantially finer ferrite grain size and hence shorter interstitial diffusion paths to the nearest area of higher dislocation density.

It is important to conclude that the special study of water-quenched and hot-dipped Zn-coated commercial 590DP, 780DP, and 980DP steels alloyed mostly with various amounts of Mn (sometimes in combination with Cr and/or Mo) did not show any signs of *aging* after storage for 1 year.

There is little data related to the effect of alloy compositions on behavior of DP steels during *tempering*.

Evolution of numerous properties of microalloyed DP steels with TS = 850 and 1023 MPa was studied using Hollomon–Jaffe parameter (HJP) that combines time and temperature of tempering:

$$\text{HJP} = T(\log t + C) \tag{4.7}$$

where T is temperature in K, t is time in hours, and the constant $C = 20$ for studied DP steels (Kamp et al. 2012). It was shown that certain parameters of localized deformation such as bendability and hole expansion continuously increase with HJP, whereas the YS/TS ratio can be increased significantly (from 0.6 to 0.8) after a very short tempering at 350 °C and do not change after.

Effect of alloying elements at *tempering* of dual-phase microstructure is fully determined by their effect on tempered martensite, i.e., by retardation of martensite softening. Grange et al. compared the effects of various elements on low- and medium-carbon martensite (Grange et al. 1977). These authors showed that Mn, Si, Cr, Mo, P, and V not only differed in their ability to retard softening of martensite, but also in the temperature intervals in which the effects of these elements are maximized.

Grange et al. pointed at different mechanisms of apparent "strengthening" of martensite during tempering. With higher *manganese* numerous and finer carbides precipitate; their coalescence is retarded thus preventing recovery of martensite.

It was shown, in particular, in (Baltazar Hernanadez et al. 2011) that fine cementite and slower tempering of martensite in steels alloyed with Cr and/or with higher Mn content confer high resistance to softening compared to leaner chemistries featured by severe decomposition of martensite and forming coarser cementite. DP steels microalloyed with V + Nb also demonstrate significantly lower softening after tempering at up to 600 °C due to strengthening ferrite by precipitation of carbonitrides (Gunduz 2009).

Chromium as a carbide-forming element becomes more efficient after tempering at temperatures above the formation of alloy carbides (>200 °C) providing maximum strength near 425 °C, which then decreases because of coalescence of carbides.

Molybdenum partitions to carbide phase at elevated temperatures and thus keeps the carbides small and numerous; therefore its strengthening effects are especially strong after tempering closer to 500 °C.

Silicon, which increases the hardness of martensite in the wide temperature range due to solid solution strengthening effect, was found to have a much stronger effect after tempering at temperatures around 325 °C that was explained by its well-known effect to inhibit the conversion of ε-carbide to cementite.

Vanadium and niobium. These two elements form stronger carbides than those of Cr and/or Mo. Therefore, these elements can be expected to affect the strength of tempered martensite retarding its softening.

4.5 Summary

It is important to distinguish the direct and indirect effects of steel composition on the properties of dual-phase steels.

The direct effects of alloying elements imply modification of those intrinsic properties of the phases that are essentially independent of the manufacturing process.

Indirectly, the alloying elements influence phase morphologies and their proportions, the required temperatures of heating, the austenitization time sufficient to achieve the desired extent of completion of ferrite recrystallization, as well as the austenite stability in cooling from the $\alpha + \gamma$ range. The effects of composition in this case can be adjusted by a suitable modification of the heat treatment conditions, and vice versa, by suitable selection of composition depending on the manufacturing capabilities.

In conclusion, it should be underscored that the effects of chemical composition on microstructure and properties of dual-phase steels are rather complex and often differ from those during conventional heat treatment after full austenitization.

References

Andrews, K.W. 1965. "Empirical Formulae for Calculation of Some Transformation Temperatures." *Iron and Steel Institute Journal* 203 (Part 7): 721–27.

Araki K., Y. Takada, and K. Nakaoka. 1977. "Work Hardening of Continuously Annealed Dual-Phase Steels." *Trans ISIJ* 17 (12): 710–17.

ASM Handbook; Heat Treating. 1991. Vol. 4. ASM International.

Azizi-Alizamini, H., M. Militzer, and W.J. Poole. 2011. "Austenite Formation in Plain Low-Carbon Steels." *Metallurgical and Materials Transactions A: Physical Metallurgy and Materials Science* 42 (6): 1544–57.

Baltazar Hernanadez, V.H., S.S. Nayak, and Y. Zhou. 2011. "Tempering of Martensite in Dual-Phase Steels and Its Effects on Softening Behavior." *Metallurgical and Materials Transactions* 42 (10): 2011–15.

Becker I., F. Hornbogen, and F. Wendl. 1981. "Mechanical Properties of a Low-Alloyed Mo-Steel with P-Addition up to 1 Wt.% and Dual-Phase Structure." *Zeitschrift für Metallkunde* 72 (2): 89–96.

Bortsov, A.N., and N.M. Fonstein. 1984. "Influence of Carbon Concentration on the Mechanical Properties of the Low-Carbon Ferrite-Martensite Steels." *Physical Metallurgy and Metallography (USSR)* 57 (4): 142–48.

Bucher, J.H., E.G. Hamburg, and J.F. Butler. 1979. "Property Characterization of VAN-QN Dual-Phase Steels." In *Structure and Properties of Dual-Phase Steels*, 346–59. New Orleans.

Calcagnotto, M., D. Ponge, and D. Raabe. 2012. "On the Effect of Manganese on Grain Size Stability and Hardenability in Ultra-Fine-Grained Ferrite-Martensite Dual-Phase Steels." *Metallurgical and Materials Transactions A* 43A (1): 37–42.

Case, S.L., and K.R. Van Horn. 1953. *Aluminium in Iron and Steel.* New York, 345.

Cho, K., C.I. Garcia, M. Hua, J. Lee, Y.S. Ahn, and A.J. De Ardo. 2012. "A New Method to Study the Effect of Cooling Rate on the Decomposition of Austenite in Advanced High Strength Sheet Steels." *Journal of ASTM International* 6 (7): 1–12.

Coldren, A.P., and G.T. Eldis. 1980. "Using CCT Diagrams to Optimize the Composition of an as-Rolled Dual-Phase Steel." *Journal of Metals* 32 (3): 41–48.

Davies, R.G. 1978. "The Deformation Behavior of a Vanadium-Strengthened Dual Phase Steel." *Metallurgical Transactions A* 9 (1): 41–52.

Davies, R.G. 1979. "Influence of Silicon and Phosphorus on the Mechanical Properties of Both HSLA and Dual-Phase Steels." *Metallurgical Transactions* 10a (1): 113–18.

———. 1981. "Tempering of Dual-Phase Steels." In *Fundamentals of Dual-Phase Steels*, 265–78.

Drumond, J., O. Girina, J.F. de Silva Filho, N. Fonstein, and C.A. Silva de Oliveira. 2012. "Effect of Silicon Content on the Microstructure and Mechanical Properties of Dual-Phase Steels." *Metallography, Microstructure, and Analysis* 1 (5): 217–23.

Fonstein, N. 2005. "Effect of Al, C, and Mn on Structure and Mechanical Properties of Dual-Phase and TRIP Steels." In *HSLA '05*. Sanya, China.

Fonstein, N., H.J. Jun, O. Yakubovsky, R. Song, and N. Pottore. 2013. "Evolution of Advanced High Strength Steels (AHSS) to Meet Automotive Challenges." In Vail, CO, USA.

Fonstein, N.M., and A.A. Efimov. 1985. "Study of Austenite Hardenability in the Low Carbon Steels During Cooling from Two-phase Alpha + Gamma Range." *Steel USSR* 15 (3): 137–40.

Fonstein, N.M., L.M. Storozeva, and B.A. Bukreev. 1985. "The Effects of Vanadium on the Properties of Two-Phase Ferrito-Martensitic Steels." *Russian Metal.*, no. 2: 111–16.

Garcia, C.I., K. Cho, K. Redkin, A.J. Deardo, S. Tan, M. Somani, and L.P. Karjalainen. 2011. "Influence of Critical Carbide Dissolution Temperature during Intercritical Annealing on Hardenability of Austenite and Mechanical Properties of DP-980 Steels." *ISIJ International* 51 (6): 969–74.

Garcia, C.I., M. Hua, K. Cho, K. Rodkin, and A.J. De Ardo. 2012. "Metallurgy and Continuous Galvanizing Line Processing of High -Strength Dual-Phase Steels Microalloyed with Niobium and Vanadium." *La Metallurgia Italiana* 6: 3–8.

Gau J.S., J.Y. Koo, A. Nakagawa, and G. Thomas. 1981. "Microstructure and Properties of Dual-Phase Steels Containing Fine Precipitates." In *Fundamentals of Dual-Phase Steels*, 47–58. Chicago.

Girina, O.A., N.M. Fonstein, and D. Bhattacharya. 2008. "Effect of Nb on the Phase Transformation and Mechanical Properties of Advanced High Strength Dual-Phase Steels." In *International Conference of New Developments on Metallurgy and Applications of High Strength Steels*, 1, 29–35. Argentina, Buenos Aires.

Girina, O, and N. Fonstein. 2005. "Influence of Al Additions on Austenite Decomposition in a Continuously Annealed Dual-Phase Steels." In *MS&T '05*, 65–76.

Girina, O., N. Fonstein, D. Bhattacharya, and O. Yakubovsky. 2015. "The Influence of Mo, Cr and B Alloying on Phase Transformations and Mechanical Properties in Nb Added High Strength Dual Phase Steels." In: Proc. HSLA'2015 China.

Granbom, Y., L. Ryde, and J. Jeppsson. 2010. "Simulation of the Soaking and Gas Jet Cooling in a Continuous Annealing Line Using Dilatometry." *Steel Research International* 81 (2): 158–67.

Grange, R.A., C.R. Hribal, and L.F. Porter. 1977. "Hardness of Tempered Martensite in Carbon and Low Alloy Steels." *Metallurgical Transactions* 8A (11): 1775–85.

Gunduz, S. 2009. "Effect of Chemical Composition. Martensite Volume and Tempering on Tensile Behavior of Dual-Phase Steels." *Materails Letters* 63: 2381–83.

Hansen, S.S., and R.R. Pradhan. 1981. "Structure /Properties Relationship and Continuous Yielding Behavior in Dual-Phase Steel." In *Fundamentals of Dual-Phase Steels*, 115–44.

Hara, T., H. Asahi, R. Uemori, and H. Tamehiro. 2004. "Role of Combined Addition of Niobium and Boron and of Molybdenum and Boron on Hardenability in Low Carbon Steels." *ISIJ International* 44 (8): 1431.

Hashiguchi, K., and M. Nashida 1980. "Effect of Alloying Element and Cooling Rate after Annealing on Mechanical Properties of Dual Phase Steel." *Kawasaki Steel Technical Report*, no. 1: 70–78.

Hashimoto, S. 2007. "Effect of Nb on Zn-Coated Dual-Phase Steel Sheet." In *Materials Science Forum*, 539-543:4411–16.

Hausmann, K., D. Krizan, K. Spiradek-Hahn, A. Pichler, and E. Werner. 2013. "The Influence of Nb on Transformation Behavior and Mechanical Properties of TRIP-Assisted Bainitic-Ferritic Sheet Steels." *Material Science and Engineering A* 588: 142–50.

Himmel, L., K. Goodman, and W.L. Haworth. 1981. "Strain-Aging of Adual-Phase Steel Containing Vanadium." In *Fundamentals of Dual-Phase Steels*, 305–14. Chicago, IL: Metal/Society of AIME.

Hironaka, S., H. Tanaka, and T. Matsumoto. 2010. "Effect of Si on Mechanical Property of Galvannealed Dual-Phase Steel." In *Materials Science Forum*, 638-642:3260–65.

Hoel, R.H., and G. Thomas. 1981. "Ferrite Structure and Mechanical Properties of Low-Alloy Duplex Steels." *Scr. Met.* 15 (8): 867–72.

Huang, J., W. Poole, and M. Militzer. 2004. "Austenite Formation during Intercritical Annealing." *Metallurgical and Materials Transactions A* 35 (11): 3363–75.

Irie, T., S. Satoh, K. Hashiguchi, I. Takahashi, and O. Hashimoto. 1981. "Characteristics of Foramble Cold-Rolled High-Strength Steel Shettes for Automotive Use", *Kawasaki Steel Technical Report*, no. 2: 14–22.

Isasti, N., D. Jorge-Badiola, M.L. Taheri, and P. Uranga. 2013. "Phase Transformation Study in Nb-Mo Microalloyed Steels Using Dilatometry and EBSD Quantification." *Metallurgical and Materials Transactions* 44A, 8, 3352-3563.

Kamp A., S. Celottoo, and D.N. Hanlon. 2012. "Effect of Tempering on the Mechanical Properties of High Strength Dual-Phase Steels." *Material Science and Engineering A*, no. 538: 35–41.

Koo, J.Y., and G. Thomas. 1977. "Design of Duplex Fe/X/0.1C Steels for Improved Mechanical Properties." *Metallurgical Transactions A* 8 (3): 525–28.

Krauss, G. 2005. *Steels: Processing, Structure and Performance*. TMS.

Kunitake, T., and H. Ohtani. 1969. "Calculating the Continuous Cooling Transformation Characteristics of Steel from Its Chemical Composition." *The Sumitomo Search*, no. 2: 18–21.

Lanzilotto, C.A., and F.B. Pickering. 1982. "Structure-Property Relationship in Dual-Phase Steels." *Metals Science* 16 (8): 371–82.

Lee, S., D.K. Matlock, and C. Van Tyne. 2011. *ISIJ International* 51 (11): 1903.

Mein, A., G. Fourlaris, D. Growther, and P. Evans. 2012. "Influence of Aluminium Additions upon CCT Diagrams for Hot Rolled TRIP Assited Dual Phase Steel." *Material Science and Technology* 28 (5): 627–33.

Messien, P., J.C. Herman, and T. Greday. 1981. "Phase Transformation and Microstructures of Intercritically Annealed Dual-Phase Steels." In *Conference of Fundamentals of Dual-Phase Steels, Symposium at the 110th AIME Annual Meeting*, 161–80. Chicago, IL, USA

Mohanty, R.R., O.A. Girina, and N.M. Fonstein. 2011. "Effect of Heating Rate on the Austenite Formation in Low-Carbon High-Strength Steels Annealed in the Intercritical Region." *Metallurgical and Materials Transactions A* 42 (12): 3680–90.

Nakagawa, A., I. Koo, and G. Thomas. 1981. "Effect of Vanadium on Structure-Properties of Dula-Phase Fe-Mn-Si-0.1C Steel." *Met Trans* 12A (11): 1865–1972.

Nakaoka, K., K. Araki, and K. Kurihara. 1977. "Strength, Ductility and Ageing Properties of Continuously-Annealed Dual-Phase High Strength Sheet Steels." In *Formable HSLA and Dual-Phase Steels, Metallurgical Society of AIME*, 126–41.

Nakaoka, K., Y. Hosoya, M. Ohmura, and A. Nishimoto. 1979. "Reassessment of the Water-Quenching Process as a Means of Producing Formable Steel Sheets." In *Structure and Properties of Dual-Phase Steels, Symposium at the AIME Annual Meeting*, 330–45.

Nouri, A., H. Saghafian, and Sh. Kheirandish. 2010. "Effects of Silicon Content and Intercritical Annealing on Manganese Partitioning in Dual-Phase Steels." *Journal of Iron and Steel Research Intern.* 17 (5): 44–50.

Ostrom, P., B. Lonnberg, and I. Ludgren 1981. "Role of Vanadium in Dual-Phase Steels." *Metal Technology*, no. 3: 81–93.

Panteleeva, L.A., and N.M. Fonstein 1984. "Influence of Phosphorus on Properties of Steels with Ferrite-Pearlite and Ferrite-Martensite Structures." *Steel in USSR* 14 (3): 144–46.

Pickering, F.B. 1978. *Physical Metallurgy and the Design of Steels*, Applied Science Publishers LTD., London.

Pottore, N., I. Gupta, and R. Pradhan. 2006. "Effect of Composition and Processing in Cold-Rolled Dual-Phase Steels for Automotive Applications." In *MS&T '06*, 653–63. Cincinnati, Ohio.

Pradhan, R. 1983. "Continuously Annealed Cold-Rolled Microalloyed Steels with Different Microstructures." In *International Conference on Technology and Applications of HSLA Steels*, 193–201. Philadelphia, PA, USA

Pradhan, R. 1985. "Metallurgical Aspects of Quenched and Tempered Dual-Phase Steels Produced via Continuous Annealing." In *Technology of Continuously Annealed Cold -Rolled Sheet Steel*, 297–318. Detroit, MI.

Priestner, R., and M. Ajmal. 1987. "Effect of Carbon Content and Microalloying on Martensitic Hardenability of Austenite of Dual-Phase Steels." *Material Science and Technology* 3 (5): 360-364.

Ramos, L.F., D.K. Matlock, and G. Krauss. 1979. "On the Deformation Behavior of Dual-Phase Steels." *Met Trans* 10A (2): 259–61.

Rashid, M.S., and V.N. Rao. 1982. "Tempering Characteristics of a Vanadium Containing Dual Phase Steel." *Metal Trans* 13A (10): 1679–86.

Rigsbee, J.M., and P.J. VanderArend. 1977. "Labaratory Study of Microstructure and Structure Properties Relationships in 'Dual-Phase' HSLA Steels." In *Formable HSLA and Dual-Phase Steels, Metallurgical Society of AIME, Chicago*, 56–86.

Shen, X.P., and R. Priestner. 1990. "Effect of Boron on the Microstructure and Tensile Properties of Dual-Phase Steel." *Metallurgical Transactions* 21A (9): 2547–53.

Song, R., N. Fonstein, H.J. Jun, D.B. Bhattacharya, S. Jansto, and N. Pottore. 2014. "Effects of Nb on Microstructural Evolution and Mechanical Properties of Low-Carbon Cold-Rolled Dual-Phase Steels." *Metallography, Microstructure and Analysis* 3: 174–84.

Speich, G., V. Demarest, and R. Miller. 1981. "Formation of Austenite During Intercritical Annealing of Dual-Phase Steels." *Metallurgical and Materials Transactions A* 12 (8): 1419–28.

Speich, G.R., A.J. Schwoeble, and G.P. Hoffman 1983. "Tempering of Manganese and Mn-Si-V Steels." *Met Trans* 14 (6): 1079–87.

Steven, W., and A.G. Haynes. 1956. "The Temperature of Formation of Martensite and Bainite in Low-Alloy Steels." *Journal of the Iron and Steel Institute* 183 (8): 349–59.

Storozeva, L.M., N.M. Fonstein, and S.A. Golovanenko. 1985. "Examination of Quench and Strain Aging of Low-Carbon Ferritic-Martensitic Steels." *Russian Metallurgy (Metally)*, no. 1: 89–93.

Takada, Y., Y. Hosoya, and K. Nakaoka. 1982. "Possibilities of Achieving Low Yield Ratio with Low Manganese Dual-Phase Steels." In *Metallurgy of Continuously Annealed Steels*, 251–69.

Tanaka, T., M. Nishida, K. Hachiguchi, and T. Kato. 1979. "Formation and Properties of Ferrite plus Martensite Dual-Phase Structures." In *Structure and Properties of Dual-Phase Steels*. Vol. 221–241.

Thomas, G., and J.Y. Koo. 1979. "Developments in Strong, Ductile Duplex Ferritic - Martensitic Steels." In *Conference of Structure and Properties of Dual-Phase Steels, Symposium at the AIME Annual Meeting*, 183–201. New Orleans, LA, USA: Metallurgical Society of AIME, Warrendale, Pa.

Togashi, F., and T. Nishizawa. 1976. *Journal of Japan Institute of Metals* 40 (1): 12.

Toji, Y., T. Yamashita, K. Nakajima, K. Okuda, H. Matsuda, K. Hasegawa, and K. Seto. 2011. "Effect of Mn Partitioning during Intercritical Annealing on Following $\gamma \rightarrow \alpha$ Transformation and Resultant Mechanical Properties of Cold-Rolled Dual Phase Steels." *ISIJ International* 51 (5): 818–25.

Wycliffe, P., G.R. Purdy, and J.D. Embury. 1981. "Austenite Growth in the Intercritical Annealing of Ternary and Quaternary Dual-Phase Steels." In *Conference of Fundamentals of Dual-Phase Steels, Symposium at the 110th AIME Annual Meeting*, 59–83. Chicago, IL, USA

Chapter 5
TRIP Steels

Contents

© Springer International Publishing Switzerland 2015

N. Fonstein, *Advanced High Strength Sheet Steels*,

DOI 10.1007/978-3-319-19165-2_5

5.1   Introduction

Steels with TRIP (*TR*ansformation-*I*nduced *P*lasticity) effect emerged as prospective materials for car body in the 1980s as a result of intensive research of Nippon Steel. However, it took almost two decades to advance dual-phase applications (see Chap. 1), to adjust the mentality of both steel producers and carmakers, to collect more data regarding the benefits of using high-strength steels in general and TRIP steels in particular, and finally to optimize steel compositions that match capabilities of both steel and car producers.

The needs of car industry to meet increasing demand for higher fuel economy and improved crash worthiness, as well as the intention of expanding the number of parts manufactured from high-strength steels kept driving the development of new steels with improved formability and better parameters of mechanical behavior to secure passenger safety. TRIP steels were offering the sought advantages and immediately became very interesting and popular topic for lab research, scientific discussions, and eventually for industrial trials and commercialization.

In fact, real commercialization of TRIP steels began in this millennium, and it took a few years to find a proper place for TRIP steels in the list of automotive grades, as well as to justify the benefits of making certain parts from these advantageous materials.

It is worth to note that, as opposed to dual-phase steels, so far TRIP steels are not produced by every steelmaker, neither are they used by every car manufacturer. There are a few reasons for some "resistance" to produce or apply these steels. First, TRIP steels typically have rather high carbon content (around 0.20 %) that complicates joining of coils or parts by welding. Second, some contradictions still exist regarding mechanical behavior of these steels, particularly their flangeability and delayed fracture.

For the last decade, TRIP steels with tensile strength of mostly 590, 690, and 780 have been developed, but their global production is almost ten times less than that of DP steels.

However, as will become evident in the next Chapters, all the grades of the so-called third-generation AHSS are based on the presence of metastable retained austenite and utilization of the TRIP effect. Besides practical advantages of TRIP steels themselves, this is an additional reason why these steels deserve careful consideration and why it is important to understand the key aspects of their phase transformation, relationship between microstructure and mechanical behavior, as well as specific effects of alloying/microalloying elements.

This chapter aims to analyze and summarize the corresponding data.

5.2   TRIP Phenomenon and the Adaptation of Metallurgical Concept to Low-Alloyed Steels

The discovery of the TRIP phenomenon is generally attributed to the publication by Zackey and Parker in 1960s (Zackey et al. 1967) who worked with highly alloyed 0.25C–9Cr–20Ni–4Mo martensitic stainless steels. Using an artificial heat treatment (quenching in liquid nitrogen after hot rolling), they managed to obtain a significant amount of retained "metastable" austenite. As a result, the steels exhibited low yield strength with significantly delayed plastic instability, showing "extended necking" up to uniform elongation of ~80 %. The authors proposed the term TRIP (TRansformation-Induced Plasticity).

It is known that when austenite is deformed at temperatures above M_S but below the critical M_D temperature, it can transform to martensite if a critical stress or strain have been attained. TRIP phenomenon implies that the local transformation of metastable austenite to martensite hardens the portion of material where this transformation is occurring, thus preventing further strain localization in this region (Fig. 5.1). The same behavior then recurs in the next local region and over the entire specimen ensuring very high strain-hardening rate that facilitates very high uniform elongation in the material.

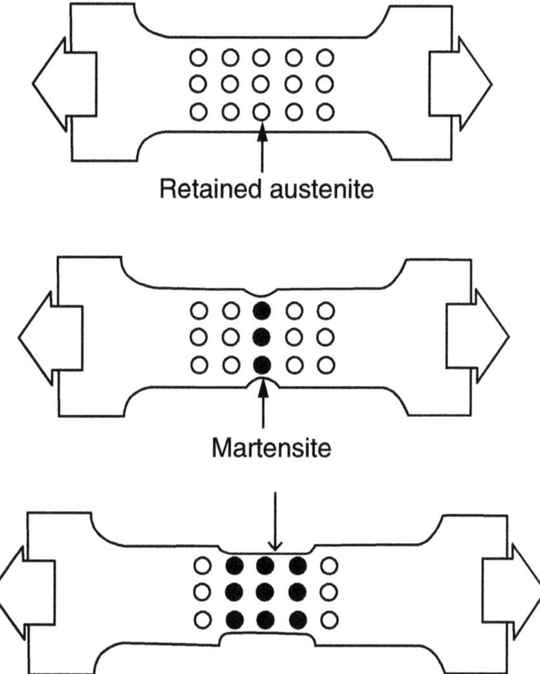

Fig. 5.1 TRIP phenomenon: *dark circles* designate points where austenite has transformed to martensite

Thus, the precondition for the TRIP phenomenon is that the M_S temperature of retained austenite (RA) should be below the room temperature. M_S value can be estimated, in particular, using the following expression (Steven and Haynes 1956)

$$M_S = 539 - 474 \times C - 33 \times Mn - 17 \times Ni - 17 \times Cr - 21 \times Mo \qquad (5.1)$$

Investigations of steels with Al resulted in some corrections related to Al content (Imai et al. 1995):

$$M_S = 539 - 423C - 30.4Mn - 7.5Si + 30Al \qquad (5.2)$$

Taking into account the real composition of low-alloyed steels with ~1.5 % Mn and without Ni, Cr, and Mo, it follows that the carbon content in RA should be about (or above) 1 % although, according to Krauss (2005), small amount of retained austenite can be observed even at carbon content of 0.3 %, probably because of nonhomogeneous distribution of carbon in γ-phase. The M_D temperature depends on chemical composition of RA and its size. It is also implied that M_D should be close to the actual forming temperatures of auto parts.

In comparison with the "classical" high-alloyed TRIP steels, described by Zackey, the retention of austenite in low-alloyed steels requires the occurrence of certain transformation reactions that can lead to artificial "forced" carbon enrichment of the remaining austenite fraction. As shown in Fig. 5.2, the first carbon enrichment

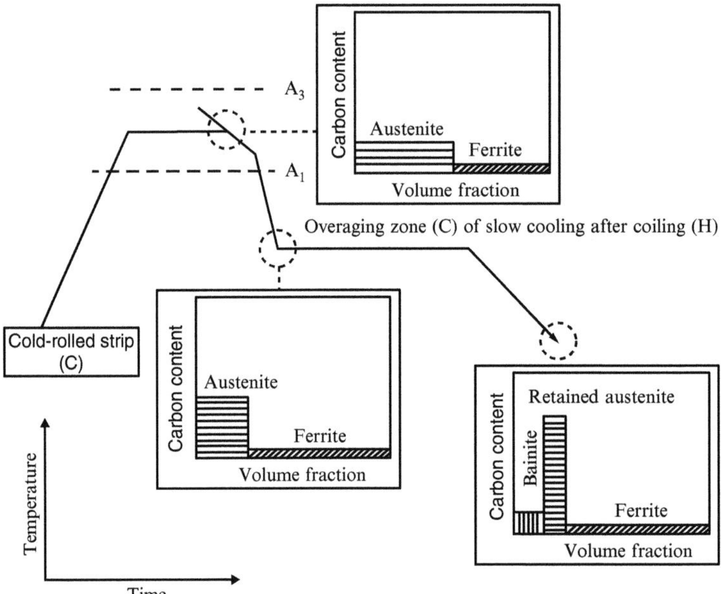

Fig. 5.2 Three stages of potential enrichment of austenite with carbon during processing of TRIP steels (Sakuma et al. 1995)

of austenite can take place during holding steel in the two-phase ferrite + austenite region due to partitioning of carbon to austenite. For example, annealing at temperature that produces only 50 % of austenite doubles the carbon concentration in austenite with respect to the total carbon content in steel.

This is followed by the second enrichment during initial slow cooling of steel when a portion of austenite transforms into new ferrite. Carbon, rejected from the newly formed ferrite, additionally enriches the remainder of austenite. The third step of austenite enrichment with carbon occurs during isothermal soak in the bainite region. Due to the transformation of major part of the available austenite to low-carbon bainite, practically all of the carbon that has been in steel is now concentrated in the remaining retained austenite. The important finding though was the necessity to alloy the steel by strong ferrite stabilizers that facilitate formation of "carbon-free" bainite preventing potential losses of carbon in austenite.

Nippon Steel pioneered the development of TRIP steels. Already during the first studies of DP steel, it was noted that air cooling from lower temperatures of intercritical region down to 400 °C, followed by fast cooling, resulted in increased amount of retained austenite and substantially improved ductility of DP steel due to the TRIP effect. At that time, a 0.1C–0.2Si–2Mn steel was studied (Furukawa et al. 1979; Matsumura et al. 1987).

Starting in the early 1980s with steels containing 0.4 % carbon and increased Si content that made it easier to obtain large portions (~20 %) of high-carbon retained austenite, the researchers demonstrated striking combinations of properties, such as TS of 1000 MPa at 30–35 % elongation. NSC publications inspired numerous researchers to verify whether the promising combination of properties can be reproducible and preserved under external impacts. However, evaluation of spot weldability of the developed steels using the carbon equivalent equation

$$Ceq = C + Si/30 + Mn/20 + 2P + 4S \tag{5.3}$$

resulted in intensive studies of steels with lower content of carbon that had significant advantages compared to conventional HSLA and DP steels (Fig. 5.3).

After intensive laboratory studies aiming to find the optimal composition to ensure sufficient weldability and moderate strength, on the one hand, and to match stamping capabilities of carmakers of that time, on the other, NSC reported very promising formability behavior of 0.11C–1.18Si–1.85Mn steel, which contained about 6 % of retained austenite and had very attractive balance of mechanical properties: YS = 399 MPa and TS = 614 MPa at TE = 35 % (Katayama et al. 1992).

Proposed steel composition and annealing cycle suitable for existing CAPL lines encouraged obtaining uncoated TRIP steels with very high balance of their mechanical properties. However, high silicon content complicated manufacturing of these steels as hot galvanized.

As a result, there was a period of intensive search for alternative steel designs with additions of P, low Si, Al, etc., which will be discussed later.

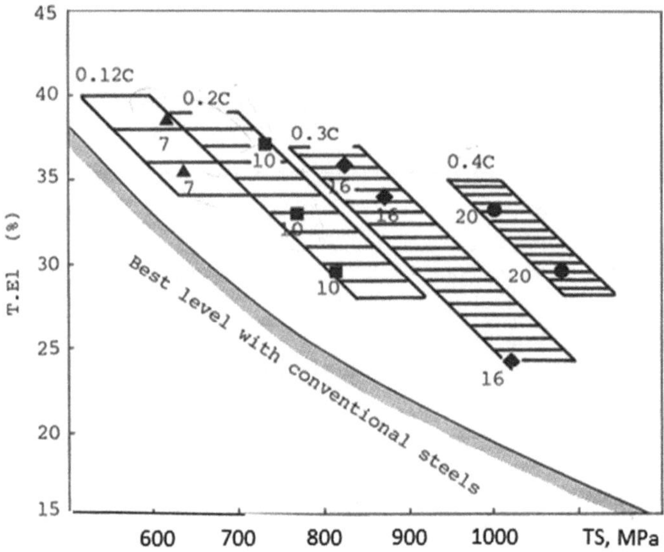

Fig. 5.3 Variations of strength-ductility balance with carbon content in 0.8 mm thick sheet TRIP steel with 1.2Si–1.2Mn ("Steel Sheet with Well-Balanced Strength and Ductility, FORD/NSC Technical Meeting" 1989)

In particular, Thyssen Krupp chose and consistently pursued compositions based on high aluminum content pioneering TRIP steel production in Europe with TS of 600, 700, and 800 MPa (Engl et al. 1998).

Since it was practically impossible to achieve desired length of isothermal holding time for bainite reaction in continuous galvanizing lines, the thermal cycles were modified as to approach pseudo-overaging profile at temperatures near bainite reaction region.

By the beginning of millennium, a general consensus was reached that with equal strength, the TRIP steels offered higher ductility (formability) and strain hardening (energy absorption capability) than the DP steels (Cornette et al. 2001).

At that time, however, the strength of commercial TRIP steels, as well as that of DP steels, was restricted by deterioration of press formability with increasing strength. As shown by the NSC data (Sakuma 2004), up to 2001, the leading high strength grades (total dual-phase and TRIP steels) were steels with TS ~ 590 MPa.

Since then, in response to growing requirements to higher strength of auto parts driven by their weight reduction, considerable experience has been accumulated in cold pressing of HSS and in manufacturing of coated steels with elevated content of Al or/and Si. TRIP steels with tensile strength of 690 and 780 have been developed and commercialized worldwide. Recently, TRIP 980 or TRIP-assisted grade with TS > 980 MPa have been developed as well (POSCO catalog, 2014).

Combination of effects of complicated phase transformations, the chemical composition, and processing on mechanical behavior of TRIP steel evoked deep interest from academia and university researchers worldwide and still continues to be the subject of deep studies using the most modern techniques (Zaefferer et al. 2004; Galan et al. 2012; Barbe et al. 2002; Sakuma et al. 1992).

5.3 Metallurgy of Manufacturing of TRIP Steels

The presence of retained austenite in final microstructure is the key point of the TRIP steel concept. Therefore, it is imperative to achieve maximum possible carbon content in the final portions of austenite to reduce the M_S temperature down to below the ambient temperature in steels that contain only about 0.2 % C. The thermal processing of TRIP steels then must utilize all the possibilities to enrich the γ-phase with carbon during phase transformations occurring at consecutive stages of heat treatment was depicted in Fig. 5.2.

5.3.1 Phase Transformations During Heat Treatment to Produce TRIP Steels

Three processes take place during heating to intercritical temperatures: recrystallization of ferrite, dissolution of cementite, and formation of austenite.

All the considerations of the effects of preliminary structure or heating rate at heating of prior cold-rolled microstructure described in Chap. 2 for dual-phase steels will now be applied to heating of TRIP steels.

Simple calculation assuming the initial content of ~0.2 % C in steel and, for example, ferrite fraction of 50 % to achieve the prescribed strength shows that the (double) enrichment with carbon due to its partitioning to austenite during intercritical annealing is insufficient to approach the necessary carbon content in austenite portion. Partial ferrite formation during slow cooling can increase carbon by another ~10–15 % which is still insufficient.

Thus, to achieve maximum enrichment of austenite with carbon, to prevent martensitic transformation during final cooling to room temperature, an additional stage of carbon enrichment is needed. It is well known that during bainitic transformation, carbon is redistributed and ferrite/cementite (carbide) mixture is formed. Only additions of strong ferrite forming elements that inhibit carbide precipitations can result in the formation of "free carbon bainite" or "bainitic ferrite" and can thus promote further partitioning of carbon to the remaining austenite during bainite reaction.

Due to critical role of alloying elements in retarding carbide formation, the following summary of phase transformation details and effects of annealing parameters on microstructure is inseparable from considerations of chemical composition.

5.3.2   Adaptation of Heat Treatment for TRIP Steel Production to Modern Annealing and Galvanizing Lines

Cold-rolled TRIP steels are produced in continuous annealing lines with immediate transfer of strip to "isothermal" overaging section after cooling from annealing temperature as presented in Fig. 2.2. Coated TRIP steels are produced in hot-dip galvanizing lines equipped by equalizing section of certain length, presented in Fig. 2.3.

Heat treatment of TRIP steel presents serious challenges due to quite different length of overaging zones in annealing lines or in zinc galvanizing lines with holding in equalizing sections at temperatures close to Zn melting temperature. In this respect, it is important to adjust the main factors that can affect phase trans-formations during heat treatment of TRIP steels to match the equipment limitations.

5.3.2.1   Effect of Initial Microstructure

Final properties of TRIP steels depend on the condition of ferrite and the volume fraction of retained austenite, its morphology, and its carbon content that controls the TRIP effect during deformation of steel. All current concepts of heat treatment of TRIP steels aim at complete transformation of cementite into carbon-enriched austenite and at recrystallization or recovery of initial ferrite.

Consequently, there are several different aspects of the influence of the initial microstructure on the formation of final microstructure of TRIP steels.

First, it is the recrystallization of ferrite. As shown, the presence of some portion of martensite in the initial microstructure of relatively high-alloyed hot-rolled steel increase the stored energy and therefore accelerate the recrystallization (Kim et al. 2008).

Second, the initial microstructure influences the kinetics of cementite dissolution during austenitization and hence the saturation of austenite phase with carbon. Thin needle-type cementite dissolves faster, and higher carbon content in γ-phase is also achieved faster. The study conducted by Kim et al. demonstrated that rapid cooling after hot rolling and the resultant refined microstructure accelerated the kinetics of austenitization and saturation of austenite with carbon during subsequent reheating. This raised both tensile strength and elongation of TRIP steel after final processing.

It was observed that higher cold reductions also accelerated the dissolution of cementite (Ehrhardt et al. 2004). This was attributed to stretching of pearlite colonies that increased the specific surface to volume ratio. Low coiling tempera-ture (500 °C compared to 700 °C) was reported to decrease the M_S temperature of retained austenite by formation of homogeneous and very fine microstructure (bainite) of hot-rolled strip and homogeneous distribution of carbon in austenite during intercritical annealing. By contrast, ferrite–pearlite microstructure after high temperature coiling contained coarse cementite that did not dissolve completely or

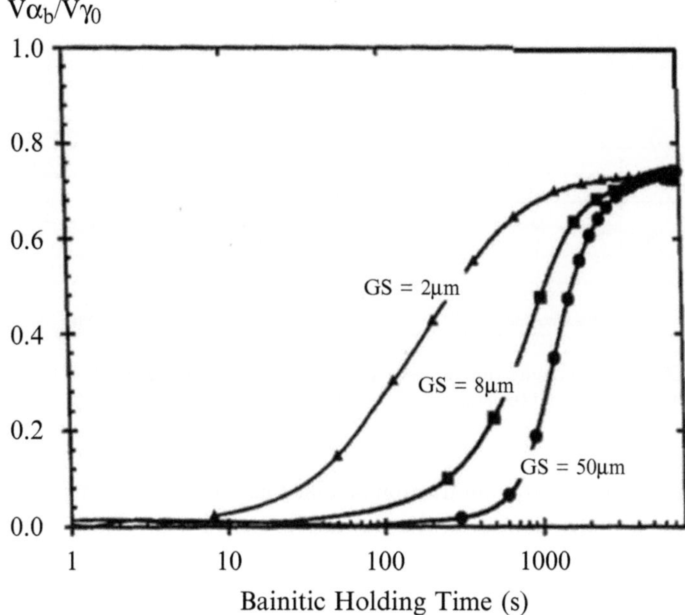

Fig. 5.4 Normalized bainite volume fraction (bainite content with respect to the initial amount of austenite) as a function of isothermal holding time at 360 °C (steel of 0.29C–1.42Mn–1.41Si) and grain sizes achieved by different prior annealing parameters (Godet et al. 2003)

at least entails an inhomogeneous carbon distribution during annealing in two-phase region (Zaefferer et al. 2004).

Third aspect of possible effects of the initial microstructure is related to enrichment of cementite with Mn when higher amounts of Mn are added to increase stability of retained austenite (Jacques et al. 1998). The possibility of substantial enrichment of cementite with manganese was shown by Laquerbe et al., who found an increase in Mn content in cementite up to 15–20 % after high tempering of hot-rolled strip of steels with 1.4–2.0 % Mn (Laquerbe et al. 1999).

While considering the effects of prior microstructure on microstructure evolution, it is worth to point out as well that refinement of the initial austenite after annealing substantially accelerates bainitic reaction (Fig. 5.4).

5.3.2.2   Effect of Annealing Parameters

Due to higher carbon content in TRIP steels than in typical DP steels, there are higher chances to obtain more spheroidized carbides, which, at short annealing times, stay undissolved up to the final cooling. Thus, the annealing parameters, as well as the requirements for initial microstructure should be selected based on the

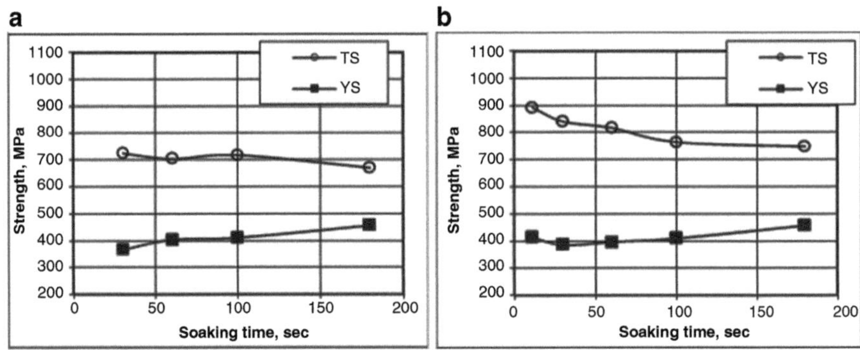

Fig. 5.5 Effect of overaging time at 400 °C on tensile properties of 0.16C–1.6Mn–1.5Si steel annealed at (**a**) 760 °C and (**b**) 805 °C

optimum volume of austenite fraction obtained and its carbon content, on completion of recrystallization, and on dissolution of cementite.

Effect of Annealing Temperature At higher temperatures of annealing, the homogenization of lower carbon concentration in austenite must be accompanied by continuously increasing M_S temperature of the formed austenite and hence by decreasing amount of retained austenite (Samajdar et al. 1998).

Annealing of 0.18C–0.39Si–1.33Mn steel ($A_{C1} = 721$ °C) at 730 and 780 °C for 5 min led to formation of 25 and 75 % of austenite, respectively. Assuming all carbon was in austenite, it was estimated that after annealing at 730 °C, the carbon concentration in austenite was ~0.72 % C and $M_S \approx 200$ °C, whereas after annealing at 780 °C, the carbon concentration in austenite was 0.24 % C and $M_S \approx 400$ °C (Jacques et al. 1998).

Illustration of effects of higher volume fraction of austenite with lower carbon content after austenitization at higher temperature is presented in Fig. 5.5 for 0.16C–1.6Mn–1.5Si steel annealed at 760 °C and 805 °C. In both cases, the tensile strength slightly decreases with overaging time that indirectly indicates growing stability of retained austenite and decrease in martensite fraction formed during final cooling. With annealing at higher temperature, the evolution of microstructure is significantly faster. Slight increase in yield strength is very similar in both cases being indicative of bainite formation at IBT.

Aiming to substantially increase the carbon content already at the first step of heat treatment, some researchers consider annealing at temperatures close to A_{C1}. However, besides some limitations imposed by dissolution of carbides or by completion of recrystallization, the bainitic reaction during isothermal holding can be delayed if the *stability* of initial austenite is *too high*.

The majority of TRIP steels are alloyed with significant amount of Si and/or Al that raise the A_{C1} and especially A_{C3} temperatures. Therefore, obtaining the desired 50–60 % of austenite requires annealing at temperatures that are typically higher than the temperatures of recrystallization and dissolution of cementite. However, the recently developed high-strength TRIP-assisted steels often simultaneously contain Mo and/or Nb. This requires further adjustment of annealing temperatures since the presence of non-recrystallized ferrite can deteriorate ductility of the final products.

Some researchers believe that when cementite is fully dissolved and recrystallization is complete during annealing, the annealing temperature does not have significant impact on transformations in cooling (Pichler et al. 2001). This argument can be supported, in particular, by rather expanded A_{C1}–A_{C3} range in steels with high concentrations of Al that substantially reduces the temperature sensitivity of the amount of austenite formed during annealing.

Effect of Holding Time   The effect of temperature and duration of annealing are interrelated. This is determined by kinetics of austenitization and dissolution of cementite so that at higher annealing temperature, the same effect can be achieved at shorter annealing time.

Since TRIP steels purposely contain elements (Si, Al, and P) that slow down the kinetics of cementite dissolution and partly of recrystallization, the necessary holding time in the intercritical region should be longer than that for DP steels, but it can be adjusted depending on the annealing temperature. According to Samajdar et al., for 0.11C–1.53Mn–1.5Si steel, the holding time of 60 s at 750 °C was required to obtain fully recrystallized microstructure (Samajdar et al. 1998). At higher annealing temperature, it took less time to achieve the required volume fraction of austenite and to complete the recrystallization. For 0.2C–1.67Mn–1.68Si steel, it was found that at any temperature, the austenite volume fraction rapidly increased during first 50 s of soaking. From 50 to 100 s, it continued to grow slowly and finally stabilized (Faral and Hourman 1999). No significant difference after soaking for 30 and 120 s at 825 °C was observed in 590 TRIP 0.093C–1.08Si– 1.66Mn steel (Hosoya 1997).

As shown by the same authors, the growth in the amount of austenite during isothermal holding is accompanied by grain growth. Austenite with coarser grains is expected to be less stable (Rigsbee and VanderArend 1977). Besides, longer duration of isothermal soaking leads to lower carbon content in larger volume of austenite that results hence in higher M_S temperature too.

Effect of Cooling Rate   Initial cooling at low rate ($<$5–7 °C/s) down to the A_{r1} temperature (~650 °C) promotes the formation of new ferrite. This process can be beneficial since the remaining austenite becomes additionally enriched with carbon. As shown in Fig. 5.6, carbon content in austenite grows substantially in slower cooling.

The critical role of initial slow cooling, for example, at 15 °C/s, in facilitating the formation of "new" ferrite, is noted by other authors as well, who considered the formation of new ferrite as an important factor of enrichment of austenite with carbon (Gomez et al. 2007).

On the other hand, the rate of initial cooling was found to have very strong impact on bainitic transformation during isothermal holding. In 0.15C–1.5Mn steel, the bainitic transformation kinetics was slower at lower cooling rate (cooling at

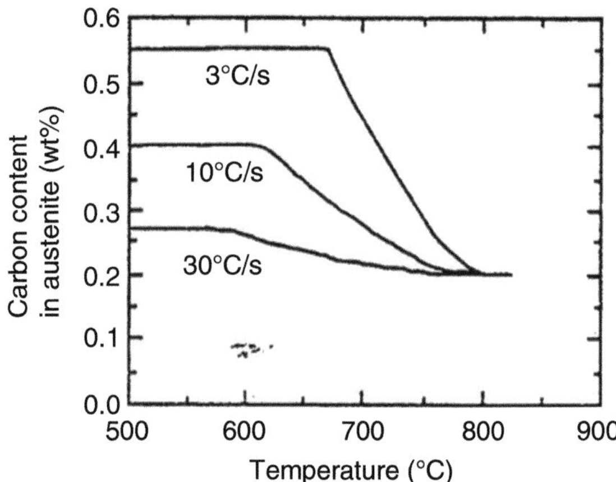

Fig. 5.6 Enrichment of austenite with carbon due to newly formed ferrite during cooling from 825 °C at different cooling rates, 0.2C–1.5Si–1.5Mn steel (Minote et al. 1996)

10 and 70 °C/s to IBT was compared), which was attributed to enhanced ferrite formation during slow cooling that led to *excessive* carbon enrichment of the remaining austenite inducing the delay in transformation (Traint et al. 2004).

The rate of initial cooling rate has to be optimized because relatively "fast" cooling is necessary to prevent the formation of pearlite. At typical ferrite volume fraction of 40–65 %, there is 0.35–0.5 % of carbon content in austenite. Depending on alloying, the cooling rate should be higher than 20–30 °C/s (Pichler et al. 1998). Lower cooling rate can lead to the formation of bainite already during cooling and to carbon deficiency in the remaining austenite.

5.3.2.3   Effect of Isothermal Bainite Transformation (IBT)

The critical step of TRIP steels heat treatment is to hold a steel strip at the temperature within the range of bainitic reaction. The metallurgical term for this procedure is austempering. Depending on processing equipment, it requires overaging or equalizing zones.

Effect of Isothermal Holding Temperature (Austempering Temperature)   The ultimate volume fraction of bainite increases with austempering temperature, which is typically between 325 and 475 °C. Soaking at temperatures near the "bainite nose" (around 400 °C, depending on steel composition) helps shortening the necessary holding time. Vast majority of papers on TRIP steels describe simulations of isothermal holding at 400 °C. Several authors considering HDG operations studied the effects of soaking temperature of up to 465–475 °C (Sugimoto et al. 1998a; Vrieze 1999; Pichler et al. 1998; Yakubovsky et al. 2002).

The effects of holding temperature on the kinetics of bainite transformation in fact strongly depend on chemical composition that will be discussed in Sect. 5.4.

The investigation of a wide range of IBT temperatures typically demonstrates the extremal increase in elongation near the nose of TTT diagrams that can be

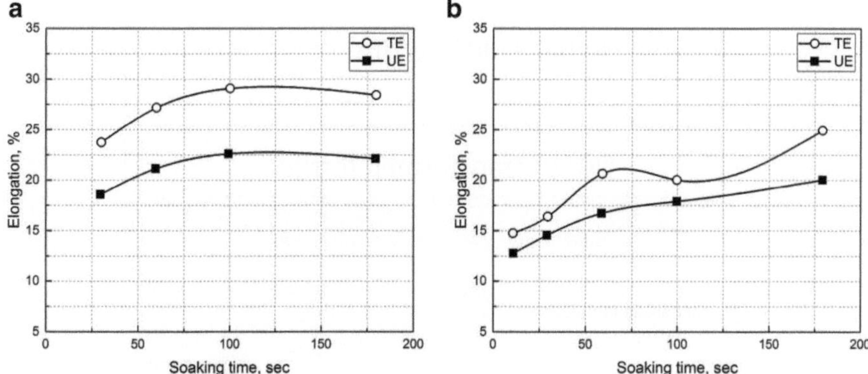

Fig. 5.7 Effect of holding time at 400 °C on elongation of 0.16C–1.6Mn–1.5Si steel annealed at (**a**) 760 °C and (**b**) 805 °C

interpreted as the increased volume of bainite at specified short holding time, which resulted in the necessary enrichment of γ-phase and significant amount of retained austenite. For example, for steels with 1.6 %Si, the position of that nose is close to 400 °C that results in rather fast increase in elongation during holding at this temperature (Fig. 5.7).

Different phase transformation behavior depending on the carbon content in γ-phase is reflected by the evolution of ductility, thanks to retained austenite. After annealing at 760 °C, the stability of lower amount of higher carbon austenite is achieved earlier so that the maximum of uniform and total elongation after annealing at 760 °C is observed after overaging for 100 s. Enrichment of higher volume of low-carbon austenite with carbon after annealing at 805 °C results in relatively low ductility after short IBT and in a continuous increase in elongation with overaging time up to 180 s.

The position of the TTT "nose" strongly depends on the chemical composition of steel. For example, experimental data for TRIP steels with 1.3–1.6 % Al showed relatively slow bainite reaction at 400 °C. At 490 °C, a small driving force for the bainite reaction was determined and therefore almost no enrichment of carbon in austenite was noted. The temperature of 430 °C was proposed as an optimum for hot-dip galvanizing lines since the experimental data showed significant bainite formation and a remarkable amount of retained austenite (Pichler et al. 2001).

Soaking at high temperatures of bainite region negatively impacts TRIP steels by facilitating cementite formation.

Effect of Holding Time in the Bainitic Range Numerous data are available confirming the importance of the duration of isothermal holding.

At the beginning of holding in the bainite region, there is a large amount of remaining austenite. However, the carbon level in this austenite is still too low and therefore the M_S temperature of that austenite is relatively high. Accordingly, the major portion of austenite should transform to martensite during final cooling of

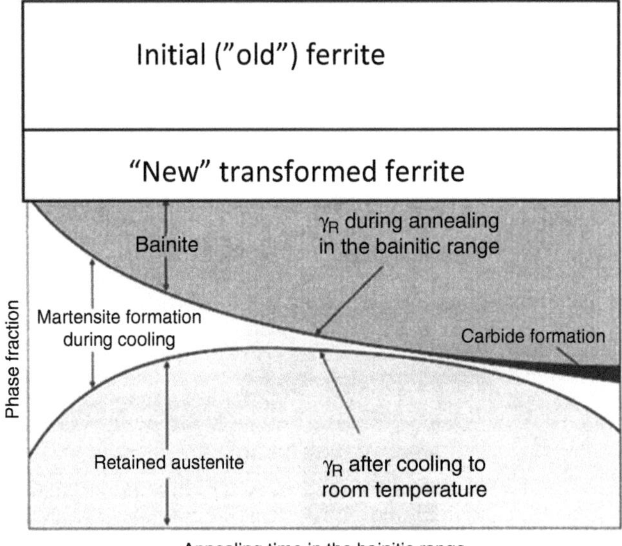

Fig. 5.8 Schematic diagram of microstructural evolution in TRIP steels during austempering

steel to room temperature. This should lead to high tensile strength and relatively low elongation of steel, reproducing the combination of properties of DP steels.

The longer the soaking time, the greater the extent of completion of bainitic transformation and the lower the amount of remaining austenite. If steel contains alloying elements that prevent precipitation of carbides in bainite, the stability of remaining austenite is increased by carbon partitioning. As bainitic transformation progresses, the amount of high-carbon retained austenite increases until it reaches maximum.

Bainitic transformation is accompanied by decrease in the M_S temperature of remaining austenite. "Fresh" martensite progressively disappears from the final microstructure. With sufficient enrichment of austenite with carbon, the M_S temperature decreases below room temperature. This facilitates the formation of microstructure containing ferrite, bainite, and retained austenite stable at ambient temperatures.

As a result of the opposite effects of the potentially available amount of remaining austenite and its stability, some researchers observed the extremal variation of the fraction of retained austenite with time of holding in the bainite region. This is shown in Fig. 5.8 illustrating the general evolution of microstructure during isothermal holding in IBT range.

As shown, ferrite fraction remains unchanged, increase in bainite fraction up to approaching the completion initially leads to a decrease in the martensite content at the expense of growing amount of retained austenite, and its stability but, subsequently, lowers the retained austenite due to formation of carbides at too long holding. The presented scheme is illustrating the potential impact on final

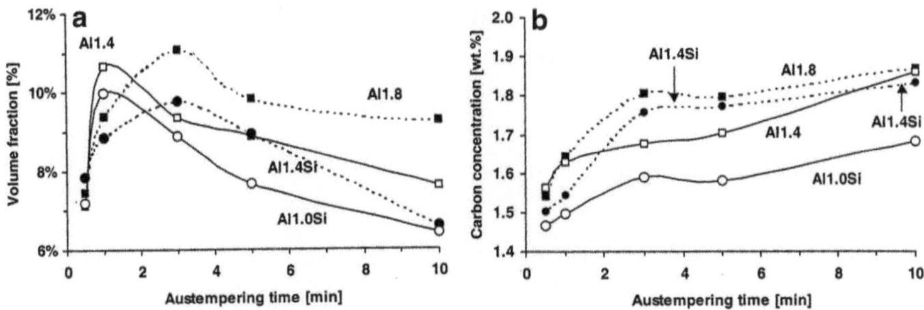

Fig. 5.9 Volume fraction of retained austenite (**a**) and carbon content in retained austenite (**b**) as a function of austempering time (Zhao et al. 1999)

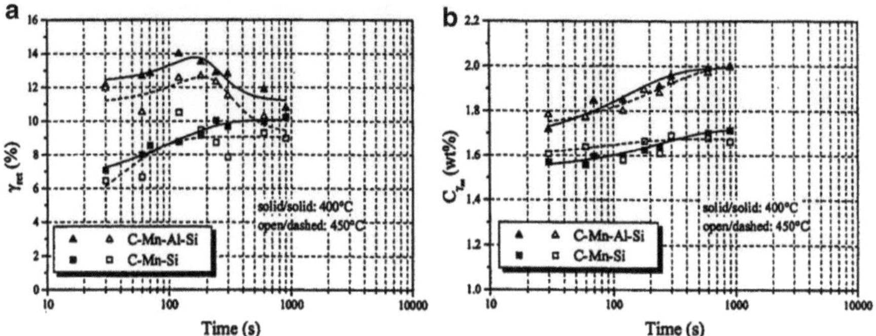

Fig. 5.10 Evolution of amount of retained austenite (**a**) and its carbon content (**b**) during isothermal bainitic reaction (DeMeyer et al. 1999a)

microstructure and properties of TRIP steel manufactured on annealing and galvanizing lines featured by significant difference in possible holding time in bainite area.

The effect of chemical composition and isothermal holding time at 400 °C on the amount of retained austenite was studied, in particular, also by Zhao et al. (1999) when 0.2C–1.52Mn steels with 1.8 %Al or 1.4 %Al were compared with steels containing 1.4 %Al + 0.25 %Si or 0.96 %Al + 0.25 % Si.

As shown in Fig. 5.9, after ~3 min of holding, almost all available carbon partitioned in retained austenite in steels with high Al and without Si. At that moment, the bainitic transformation slowed down and carbon-enriched austenite began to decompose. Higher Al and additions of Si accelerated the enrichment of austenite with carbon.

Whereas the carbon content in retained austenite can grow with time rather long, the existence of maximum in volume of retained austenite or sort of "saturation" points at particular austempering time has been noted in numerous studies, as depicted in Fig. 5.10 where steels with 0.19C–1.57Mn–1.46Si and 0.31C–1.57Mn–0.34Si–1.23Al were compared.

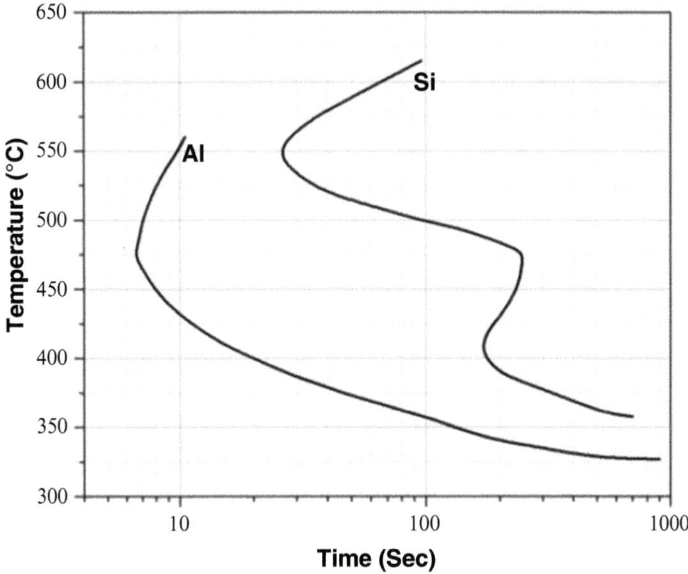

Fig. 5.11 TTT diagrams for 0.16C–1.5Mn steels with additions of 1 % Si or 1.25 % Al (Bhattacharya et al. 2003)

The kinetics of bainitic reaction indicated by TTT diagram depends on the chemical composition of steel. As an example, Fig. 5.11 displays TTT diagrams indicating 25 % transformation for 0.16C–1.5Mn steels with 1 %Si or 1.25 %Al (Bhattacharya et al. 2003). It is evident that Al accelerates bainite reaction in comparison with Si, especially at temperatures close to galvanizing, whereas the lower nose of TTT diagram with Si additions is observed at 400 °C.

During manufacturing, TRIP steels are annealed at intercritical temperatures so that prior to isothermal holding in bainitic region, the microstructure consists of austenite and ferrite. Zhu et al. (2013) studied the kinetics of martensitic and bainitic transformation of austenite depending on the presence of α/γ interfaces in the initial microstructure. They found that the coexisting of ferrite and austenite significantly retards the kinetics of subsequent bainite transformation. The higher the transformation temperature, the more significant the retardation effect was (temperatures of 550, 500, and 430 °C were compared for 0.155C–1.82Mn–0.26Si–0.2Cr steel).

The observed end result is controlled by competition between the accelerating effect of the presence of α/γ interface boundaries themselves and the retardation effect of higher alloy concentration (C, Mn) near the interfaces.

As shown by Traint et al., delay in bainite formation at 400 °C can be expected for steels with 1.3–1.6 %Al, so after 10 s of soaking, almost no bainite was detected. After 60 s, the bainitic reaction progressed only by 50 %. The enrichment with carbon was thus insufficient to prevent the transformation of part of remaining austenite into martensite during final cooling. At the same time, compared to other

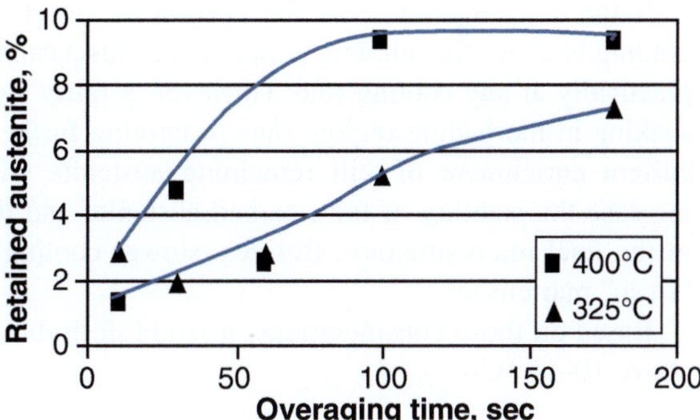

Fig. 5.12 Volume fraction of retained austenite in 0.16C–1.6Mn–1.6Si steel after annealing at 805 °C and overaging at 400 °C and 325 °C

temperatures, longer overaging showed maximum volume fraction of retained austenite just at 400 °C (Traint et al. 2001) that agrees with TTT diagram presented in Fig. 5.11.

In the study of 0.14C–1.94Si–1.66Mn, the increase in holding time at 400 °C from 10 s to 60 s and then to 480 s was accompanied by increase in the amount of retained austenite, γ_R, from 9.5 to 12.7 and then to 13.2 %. However, γ_R grew very slowly between 60 and 480 s. Carbon content in γ_R increased from 0.97 to 1.03 and 1.15 % as bainitic transformation progressed (Ushioda and Yoshinaga 1996). Simultaneous growth in both volume of austenite and its carbon content and observed austenite stability can be explained probably by gradual dissolution of cementite.

The optimal duration of IBT strongly depends on holding temperature and steel composition. As shown in Fig. 5.12, the volume of retained austenite in steel with 0.16C–1.6Mn–1.61 % Si continuously grows with overaging time from 15 to 180 s at 325 °C, whereas holding at 400 °C does not change the volume fraction of retained austenite after holding for 100 s.

Some data point out the possibility to obtain up to 10–15 % of retained austenite after 10.5–20 s holding in the steel of 0.16C–1.5Mn–0.4Si at the temperature of Zn bath (465 °C) (Vrieze 1999), but microstructure contained the mixture of bainite, retained austenite, and martensite resulting in very low YS/TS, and stability of the retained austenite was not investigated. Based on TTT diagram presented in Fig. 5.11, it is more realistic to reach necessary amount of retained austenite and properties of TRIP steel using additions of Al.

When isothermal holding is *too* long, the volume fraction of retained austenite began to decrease due to precipitation of cementite in bainite, which sucks carbon out from austenite. This was observed at high IBT temperatures close to the temperatures of Zn bath or galvannealing furnace (Gallagher et al. 2003) or at holding time over 480 s (Ushioda and Yoshinaga 1996).

Effect of the Rate of Final Cooling Stable retained austenite should not be sensitive to the rate of final cooling. Nevertheless, it is reasonable to assume that carbon distribution in the remaining austenite is not very homogeneous.

If the M_S temperature of the portion of austenite remaining after isothermal holding is above the ambient temperature, this austenite transforms into martensite practically at any cooling rate. However, a fairly slow final cooling adds time to soaking in the bainite region, thus promoting further bainite formation with concurrent enrichment of still remaining austenite with carbon. The latter would increase the stability of the retained austenite and thereby of its volume fraction in the final microstructure. Besides, slower cooling facilitates auto-tempering of "fresh" martensite.

Based on those considerations, it is not desirable to have the final cooling rate above 10–15 °C/s.

5.3.3   Hot-Rolled TRIP Steels

Applications and hence production of as-hot-rolled TRIP steels are limited. The general approach to the design of hot-rolled TRIP steels combines the manufacturing concepts for as-hot-rolled DP steels and annealed TRIP steels. Hot-rolled steels are supposed to undergo heavy reductions at possible low finishing temperatures close to A_{r3} to accelerate ferrite transformation during initial relatively slow cooling followed by rapidly cooling to bainite region and coiling at temperatures of 400–450 °C, as presented in Fig. 5.13. Additions of Si in high concentrations facilitate raising the finishing temperature so that to finish rolling close or below A_{r3}.

Some aspects of manufacturing hot-rolled TRIP steels were clarified by Tsukatani and coworkers (Tsukatani et al. 1991). Comparing seven compositions with 0.2 C and various Si and Mn contents, the authors found the best combination of strength and ductility (TS = 800 MPa at TE ~ 35 %) for steel containing 1.5 %Si

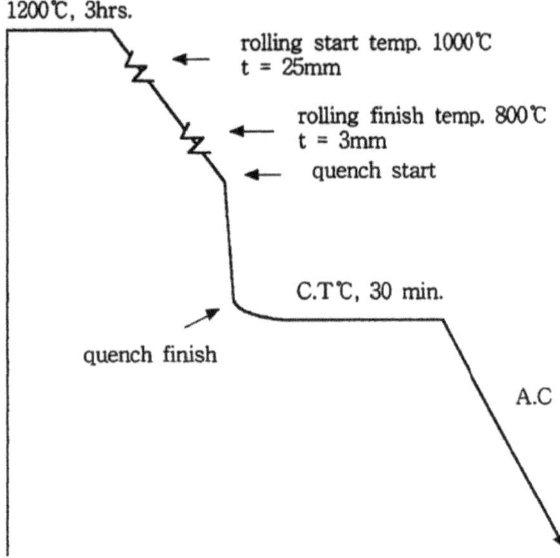

Fig. 5.13 General manufacturing concept for as-hot-rolled TRIP steels (Kim et al. 2008)

and 1.5 %Mn coiled at 400 °C and held before final cooling for 30 min. The properties were in good correlation with the amount of retained austenite. It was also shown that decrease in finishing hot-rolling temperatures promotes the increase in the volume fraction of retained austenite with the latter being maximum at finishing temperatures around 780 °C.

Studies of the effects of isothermal holding temperature and time performed by Jun et al. (2004) using 0.20C–1.96Si–1.51Mn steel showed that the difference in decomposition behavior of retained austenite correlated with carbon diffusivity. At low coiling temperature (350 °C), the amount of retained austenite (7–10 %) was almost insensitive to holding time for up to 8 h because of low carbon diffusivity. At temperatures above 370 °C, austenite decomposed into ferrite and cementite due to redistribution of solute carbon. At 450 °C, carbon diffusivity is ten times higher than that at 350 °C, and the fraction of retained austenite decreased with increasing holding time from 14 % to zero due to austenite decomposition after 2 h. At 400 °C, the volume fraction of retained austenite changed from 16 % after ~60 min of holding to zero after 8 h.

Kim et al. studied effects of small (~0.3–0.4 %) additions of Ni or Ni + Cu to the base 0.20C–1.41Si–1.53Mn composition along with the effects of coiling temperature (400 and 450 °C) with 30 min holding time. Tensile strength from 750 to 1200 MPa was obtained at elongation ranging from 40 to 20 %, respectively. The best TS–TE balance was found for the steels with 0.41Ni or 0.4Cu–0.42Ni coiled at 450 °C. Both compositions contained 16–18 % of retained austenite (Kim et al. 2008).

Hashimoto et al. studied the effects of 0.02–0.04 %Nb and 0.1–0.2 %Mo additions to base 0.20C–1.5Si–1.50Mn composition after coiling at 400 °C and holding for 10 min (Hashimoto et al. 2004). Additions of 0.05 % Nb resulted in higher elongation (32 % as opposed to.27 % without Nb) and higher TS (832 vs. 762 MPa). While strengthening of steel was attributed to precipitation of fine NbC carbides mainly during coiling, good ductility of Nb bearing steel was explained mainly by high volume fraction of retained austenite and its dispersed morphology that facilitated its higher stability and contributed to higher ductility through high strain hardening.

Additions of Mo promoted further strengthening by ~70 MPa due to larger amount of fine NbMoC precipitates.

5.4   Relationship of Structure and Static Mechanical Behavior of TRIP Steels

The microstructure of TRIP steels comprises ferrite, bainite of various morphologies (with dominating so-called bainitic ferrite or carbon-free bainite), retained austenite of different dispersion, and fresh martensite. Significant problems of correlating the microstructure and properties of TRIP steels are related to difficulties in distinguishing individual microconstituents. Whereas the amount of retained austenite and its carbon content can be estimated by XRD, it is almost impossible to separate ferrite and carbon-free bainite, as well as martensite and austenite by etching in microstructure observations.

There is a lack of studies focusing on the relationship between microstructure and mechanical behavior of TRIP steels, which should serve as a basis for further improvement of steels of this class. Below an attempt is presented to outline and highlight the most important parameters of microstructure of TRIP steels responsible for specific features of their mechanical behavior.

5.4.1 Effect of Structure Parameters of TRIP Steels on the Balance of Tensile Strength and Ductility

Austenite The common consensus is that higher RA amount is responsible for higher ductility of TRIP steels. In particular, some publications report fairly strong correlation between the volume fraction of retained austenite (up to $V_\mathrm{M} \approx 0.2$) and such formability indicator as the $\mathrm{H_o}$ position in FLD diagram (Matsumura et al. 1992, b). At the same time, numerous data indicate that the volume fraction of retained austenite does not correlate directly with ductility of these steels. This contradiction could be attributed to a discrepancy as to which type of retained austenite is efficient in improving ductility of TRIP steels. Some reports concluded that the film-shaped austenite in bainite lath is efficient but the granular austenite is not. On the contrary, others recognized the benefit of granular austenite. There are many other reasons for discrepancy in evaluation of the role of the amount of retained austenite, among which the most important is the differences in carbon content of retained austenite in compared steels.

The key factor is the austenite stability that depends on both carbon content in RA and size of RA islands (often less than 1 µm) because TRIP effect implies the austenite-to-martensite transformation occurring within fairly large span of tensile strains to delay necking. If RA is unstable, it transforms to martensite early during deformation without any sensible effect on ductility regardless of its amount. On the other hand, if retained austenite islands are too small and too stable to transform to martensite under strain, they are not involved in TRIP effect.

Such parameters of ductility as flangeability and hole expansion of TRIP steels are often considered as their weak points since hole punching results in a wide damage zone due to highly localized stresses that induce austenite-to-martensite transformation. As shown in numerous studies by Sugimoto, that strength–stretch flangeability balance (TS $\times$ λ) increases linearly with volume fraction of retained austenite and carbon concentration in it. It was concluded that the best combination of elongation and stretch-flangeability can be achieved when austempering is performed at 325–350 °C (Sugimoto et al. 2007), i.e., below the M_S temperature. Such treatment closely resembles the Q&P process discussed in Chap. 10.

Ferrite For optimal ductility of TRIP steels, it is important to have ductile, recrystallized, or fully recovered ferritic matrix. With all other factors being constant, the increase in ferrite fraction, if strength requirements permit, results in higher carbon content in austenite and consequently in its higher stability.

Coexistence of pure and highly deformable ferrite typical for intercritical annealing, on the one hand, and ductile retained austenite stable up to certain strain, on the other hand, can also be important for mechanical behavior of TRIP steels because this combination can facilitate avoiding strain localization near interfaces, as opposed to DP microstructures.

Bainite Lower bainite obtained at low temperatures or during short holding in IBT region contains high densities of defects due to high carbon supersaturation and insufficient recovery. A few quantitative estimates show that TRIP steels usually contain a few percent of low bainite.

Fresh Martensite Yield strength of TRIP steels depends on the presence of fresh martensite in its final microstructure. After short time holding in IBT region, and/or during sluggish bainitic reaction due to high alloying, a part of remaining austenite does not have sufficient amount of carbon to prevent transformation to martensite during final cooling. In this case, TRIP steels display the same low YS/TS ratio as DP steels also demonstrating continuous yielding. As the holding time increases and the bainitic reaction becomes more complete, YS grows and yield point elongation is observed (Yakubovsky et al. 2002). Sakuma et al., while studying 0.15C–1.2Si–1.5Mn steel, pointed out that YPE in TRIP steels was observed due to low dislocation density in ferrite grains when limited or no localized martensite transformation took place (Sakuma et al. 1992).

Post annealing (low tempering) of TRIP steels as the critical remedy to improve properties of TRIP steels was first mentioned in the paper by Tomita and Morioka. All the compared heat treatment cycles of 0.6C–1.5Si–1.8Mn steel included final "light tempering" at 200 °C to temper martensite (Tomita and Morioka 1997). In fact, post tempering of TRIP steels at 200 °C is insufficient to significantly improve the behavior in localized deformation and to achieve high hole expansion, but in general, tempering of TRIP steels is found to be accompanied by increasing carbon content in the remaining portion of retained austenite resulting in both improving elongation and hole expansion.

Tempered Martensite Tomita and Morioka tried post tempering treatment as the final step of heating at 200 °C after isothermal holding at 400 °C for 300 s (Tomita and Morioka 1997). Improvement of properties of TRIP steels (and, in particular, increase in YS) is expected due to tempering of fresh martensite and relaxation of residual stresses.

5.4.2   Effect of Amount and Stability of Austenite on Strain Hardening of TRIP Steels

Very high strain hardening up to the ultimate uniform elongation is the main feature of TRIP steels, defining their enhanced ductility, formability, and high absorbed energy at fracture.

In general, deformation of TRIP steels is controlled by two factors. First, there are long-range internal stresses in the matrix that contain not only untransformed (and perhaps partially non-transformable) retained austenite but also other hard "second" phases. Second, it is the strain-induced transformation of retained austenite to martensite accompanied by stress relaxation. The first factor contributes to high strain-hardening rate during early stages of deformation. The second factor is dominating and results in relatively high strain-hardening rate within a wide strain range thus suppressing the onset of diffuse necking (Sugimoto et al. 2002).

Figure 5.14 displays the "instantaneous n_i-values—true strain" curves. TRIP steels (a, b) show n_i-values up to 0.25. Compared to a dual-phase steel of the same tensile strength of ~650 MPa (c), TRIP steels demonstrate an important advantage of high n_i-values up to the limit of uniform elongation, whereas dual-phase steel is featured by typical monotonously decreasing n-values. As-annealed TRIP steel (Fig. 5.14a) exhibits a very high initial n_i, then drops to zero during YPE, and increases beyond the YPE strain up to 0.25. Temper rolling suppresses YPE (b) resulting in a very short strain range of initially low n_i; then the maximum value of n_i decreases from 0.25 to 0.21 but stays high at strains up to 0.20.

The observations showed that the pace of increase of the n-value to maximum depends on the pace of transformation of RA. When strain-induced transformation to martensite occurs gradually, a slower increase to a maximum n-value was observed (Evans et al. 1997). Similar observations brought about the understanding of important or even dominating role not of the amount of RA but rather of its stability. Highly unstable RA very rapidly transforms to martensite on the early stages of deformation, while it is desirable to keep a part of austenite in car components even after stamping to increase energy absorption during a potential crush.

Therefore, it is essential to find the appropriate compromising "metastability" of retained austenite in steel in order to maximize practical benefits of the TRIP effect.

Figure 5.15 compares the "n_i-strain" plots of as-annealed of 1 % Si steel and 1.6 % Al TRIP steels annealed at different temperatures (760 and 830 °C, respectively) to reach approximately the same (35–40 %) initial volume fraction of austenite. Steels were overaged at 325, 400, and 475 °C for 10, 30, 60, 100, and 180 s. Dilatometry detected significant amount of "fresh" martensite forming during final cooling after isothermal holding at 325 °C, especially after short soaking. There was no evident difference in behavior of Si- (a) and Al-bearing (b) steels at this temperature. Longer soaking produced less "fresh" martensite that, in turn, resulted in lower initial n_i-value and increasing n_i-value at higher strains. After short holding at 400 °C, the 1 % Si steel also showed rapid decrease in n_i-value with increasing true strain, which is typical for dual-phase structure. This

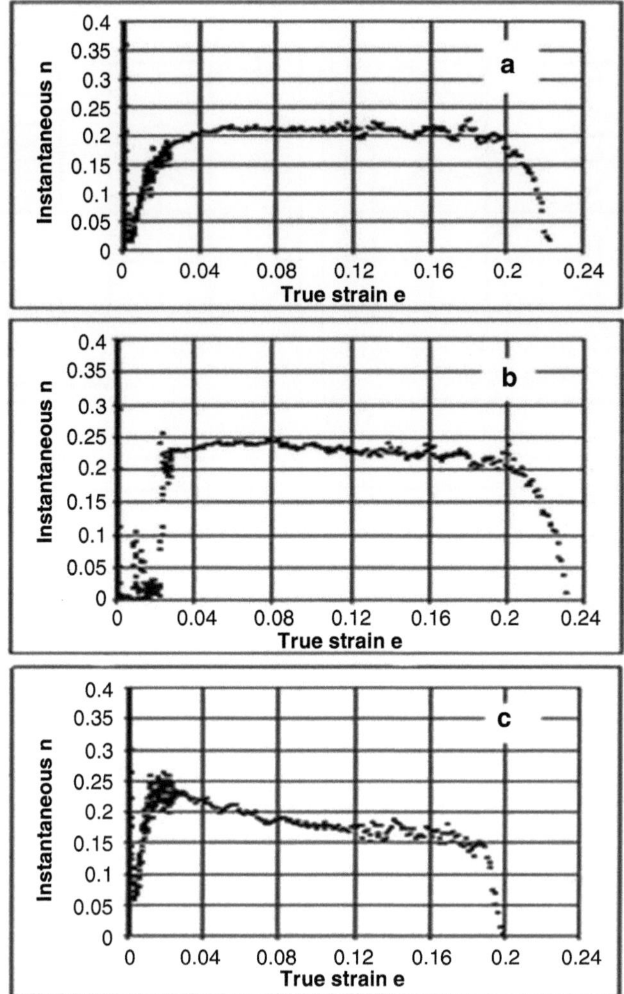

Fig. 5.14 True strain dependence of instantaneous n_i-value of (**a**) as-annealed Al(Si-P) TRIP steels, (**b**) the same steel after temper rolling at 0.6 % reduction and (**c**) dual-phase steel of the same strength (~650 MPa) (Yakubovsky et al. 2002)

took place after isothermal holding in bainite region time for 10 s when dilatometry detected 29 % of martensitic transformation during final cooling of the remaining austenite. This decrease in n_i-value becomes less significant with longer isothermal holding, showing only 6 % of martensite after 60 s of holding, and at last, after 180 s, the n_i-value plot becomes almost flat, which is a manifestation of "true" TRIP steels. No martensite was found in this microstructure.

In spite of some contradictions, the majority of researchers pointed out a higher rate of bainitic reaction in Al-bearing TRIP steels that accelerates approaching of

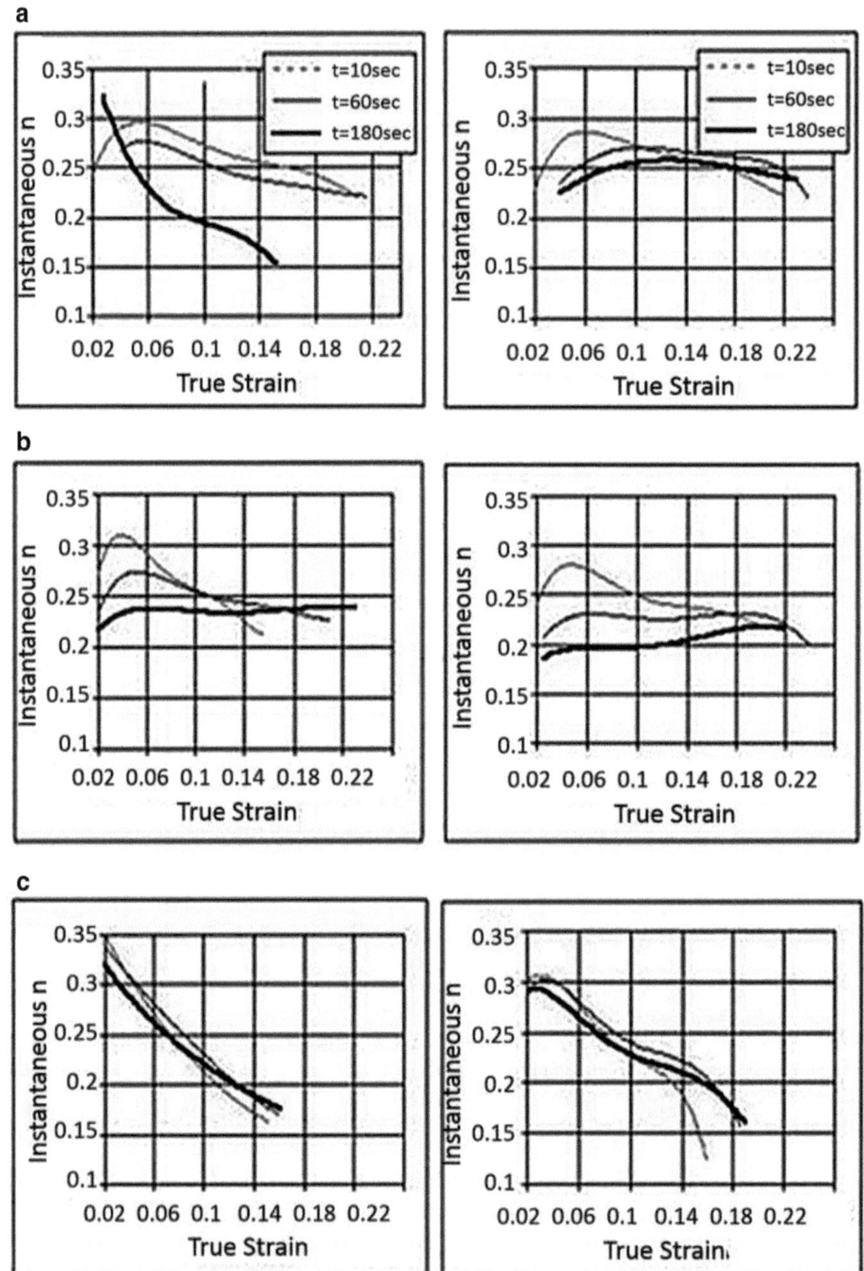

Fig. 5.15 "Instantaneous n_i-values—true strain" plots for 1 % Si (*left column*) and 1.6 % Al (*right*) TRIP steels as a function of temperature and time of isothermal holding treatment in bainite region: (**a**) IBT = 475 °C; (**b**) 400 °C; (**c**) 325 °C (Yakubovsky et al. 2002)

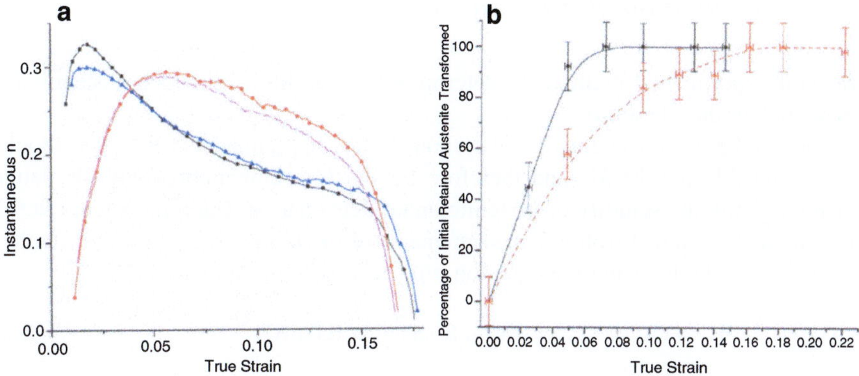

Fig. 5.16 Variation in instantaneous *n*-value with true strain for equiaxed and lamellar microstructures of steel held for 100 s at 450 °C (two samples per microstructure were tested) (**a**) and transformation of austenite in these two microstructure as a function of strain (**b**); *blue lines—*equiaxed microstructure, *red—*lamellar (Chiang et al. 2011)

the "completion" of bainitic reaction with higher carbon content in the remaining portion of austenite. As a result, 1.6 % Al TRIP steel exhibited a flat n_i-curve and significant amount of retained austenite (10–12 %) after shorter IBT time.

Nevertheless, after holding at IBT = 475 °C, when no martensite was found, n_i tended to decrease with increasing true strain. This took place after short soaking of both 1 % Si and HiAl steels, but after longer holding, it was more significant for Si-containing steel. Two important conclusions can be made. First, this means that "flat" or increasing n_i with true strain should not be explained only by the absence of "fresh" martensite emerging during final cooling but also by continuing transformation of retained austenite into martensite during tensile test. This kind of evolution of n_i with strain can be considered as attributable to steel with stable RA and profound TRIP effect. Second, 1.6 %Al steels likely preserve their TRIP features at higher temperatures and shorter holding at IBT than 1 % Si-TRIP steel.

Chiang et al. studied the effect of retained austenite morphology on strain-hardening behavior of TRIP 0.17C–1.53Si–1.5Mn–0.02Ti steel after annealing in the intercritical region and after two-stage treatment including water quenching from the full austenitization prior to intercritical annealing. The heat treatment of these two types produced the so-called "equiaxed" and "lamellar" microstructures. As shown in Fig. 5.16, the equiaxed microstructure was featured by higher strain-hardening rate at low strains with subsequent rapid decay, while the lamellar microstructure demonstrated superior sustained strain hardening up to higher strain. The authors explained their findings by higher stability of lamellar RA that prevents it from transformation to martensite up to higher strains (Chiang et al. 2011).

5.4.3   Evaluation of Austenite Stability

Several experimental methods to determine the stability of retained austenite are described in the literature.

One of them is the evaluation of carbon content in retained austenite that should be responsible for the M_S and therefore for the M_D temperatures with the latter characterizing the stability of austenite under deformation. Carbon content (wt.%) C_γ can be determined from the measurements of the lattice parameter of austenite, a_γ, for example, by using the equation proposed by Sugimoto et al. (2002):

$$C_\gamma = \left(a_\gamma - 0.35467\right)/ 0.04.67 \tag{5.4}$$

Several other equations can be found in the literature that also take into account the effects of alloying elements, as, for example, one proposed by Iung et al. (2002)

$$a_\gamma = 3.553 + 0.0474C_\gamma + 0.00095\text{Mn} + 0.0056\text{Al} + 0.0006\text{Cr} \tag{5.5}$$

For example, higher and more stable n-values up to higher true strains resulting from partial substitution of Si with Al, presented in Fig. 5.17, appeared to be in good agreement with larger volume fraction of retained austenite and its higher carbon content in Al-bearing TRIP steel (DeMeyer et al. 1999a, b).

Some authors evaluate different stability of austenite and rate of austenite-to-martensite transformation by direct measurements of the amount of retained

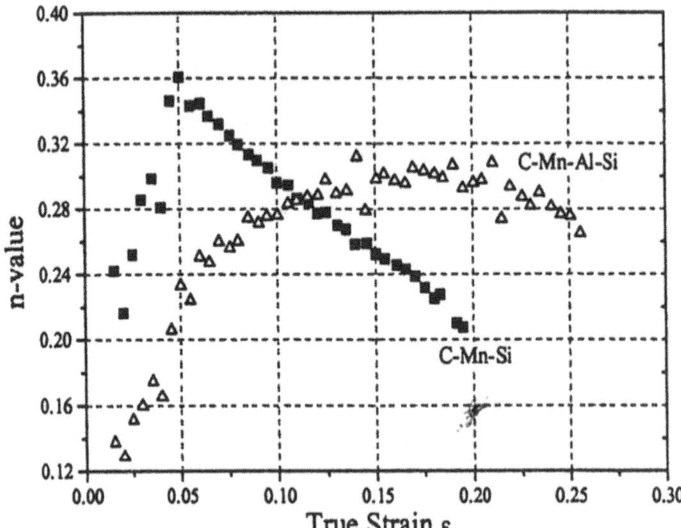

Fig. 5.17 Comparison of variations of strain-hardening rate n with true strain for C–1.5Mn–1.5Si and C–1.5Mn–0.3Si–1.2Al steels (annealing during 4 min at 770 °C followed by holding at 450 °C for 2 min) (DeMeyer et al. 1999a)

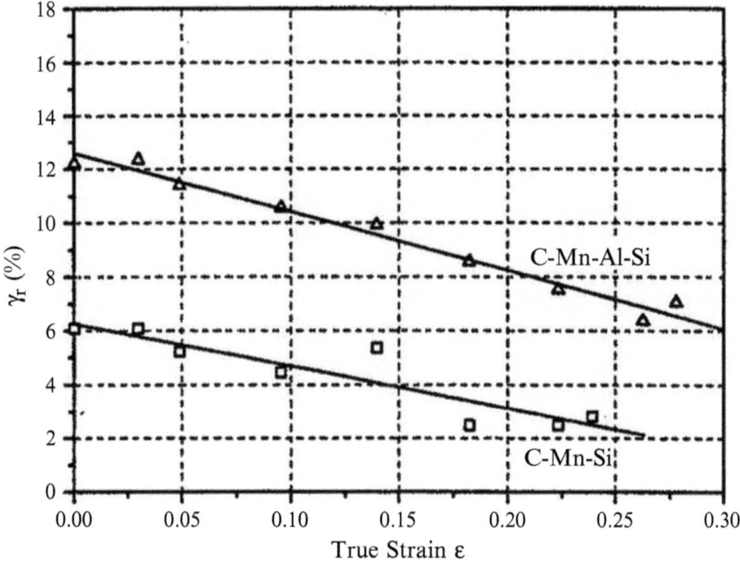

Fig. 5.18 Changes in volume fraction of retained austenite during tensile testing as a function of chemical composition of steel (DeMeyer et al. 1999a, b)

Fig. 5.19 Effect of deformation mode on austenite-to-martensite transformation rate during deformation (Sakuma et al. 1995)

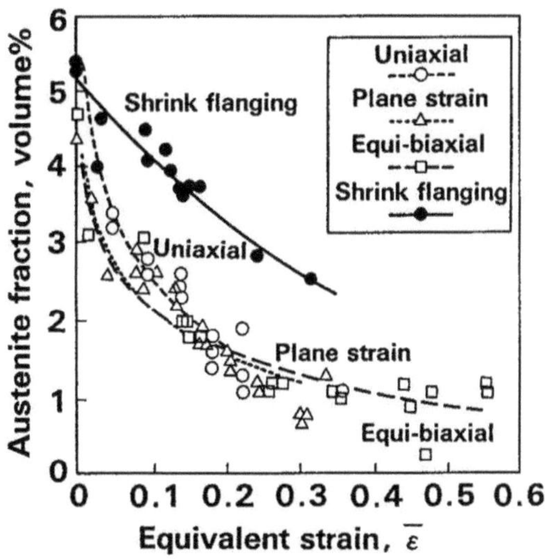

austenite during interrupted deformation. Figure 5.18 depicts the different resistance of RA to martensitic transformation for the same Si and Si–Al steels as in Fig. 5.17.

However, as shown in several publications (Sakuma et al. 1995; Stretcher et al. 2002), austenite-to-martensite transformation rate per strain can be different depending on the deformation mode (Fig. 5.19).

K.-I. Sugimoto used the coefficient k as the parameter for stability of retained austenite against strain-induced martensitic transformation. The coefficient k is defined from the equation

$$\log f_\gamma = \log \gamma_{f0} - k\varepsilon_p \tag{5.6}$$

where ε_p is a plastic strain. According to Sugimoto, the austempering temperature determines the initial carbon content in retained austenite rather than its amount. The amount of RA depends more on the duration of isothermal holding in bainite region. However, as exemplified in Fig. 5.20 for 0.2C–1.5Si–1.5Mn steel, the value of k-value seems to be controlled by both austempering time and temperature.

It has been recognized that the most important parameter of TRIP steel behavior is the thermodynamic stability of the retained austenite that can be characterized by single parameter, the M_S^σ temperature described schematically in Fig. 5.21 that illustrates austenite transformation mechanisms at various temperatures (Haidemenopoulos and Vasilakos 1996). When stress (below the yield stress of the parent phase) is applied at temperatures higher than M_S, the stress-assisted nucleation of martensite occurs at preexisting nucleation sites. Thus, spontaneous

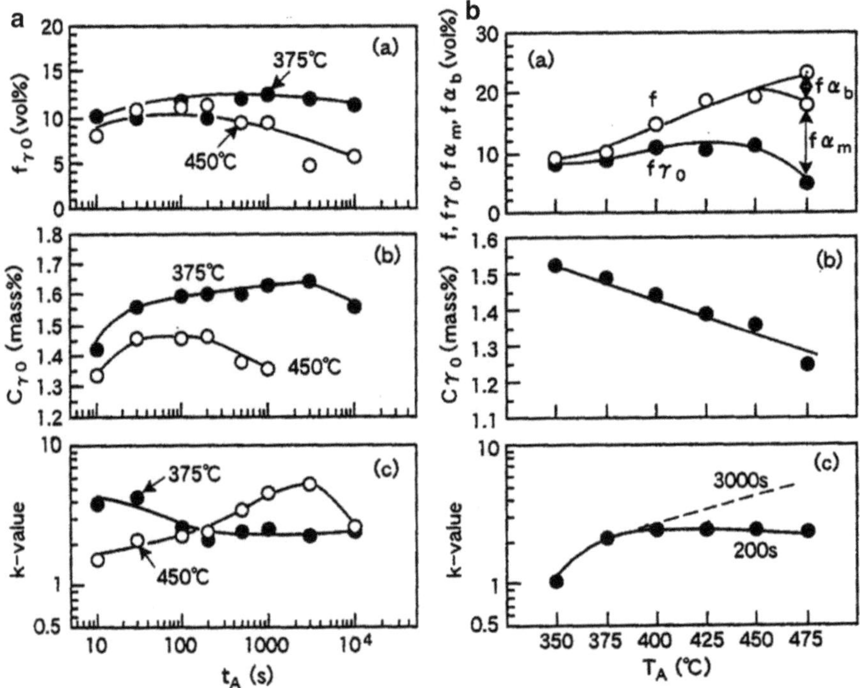

Fig. 5.20 Effects of austempering time (a) and temperature (b) on the initial volume fraction of retained austenite $f_{\gamma 0}$, its initial carbon content $C_{\gamma 0}$, and the k-value (Sugimoto et al. 2000)

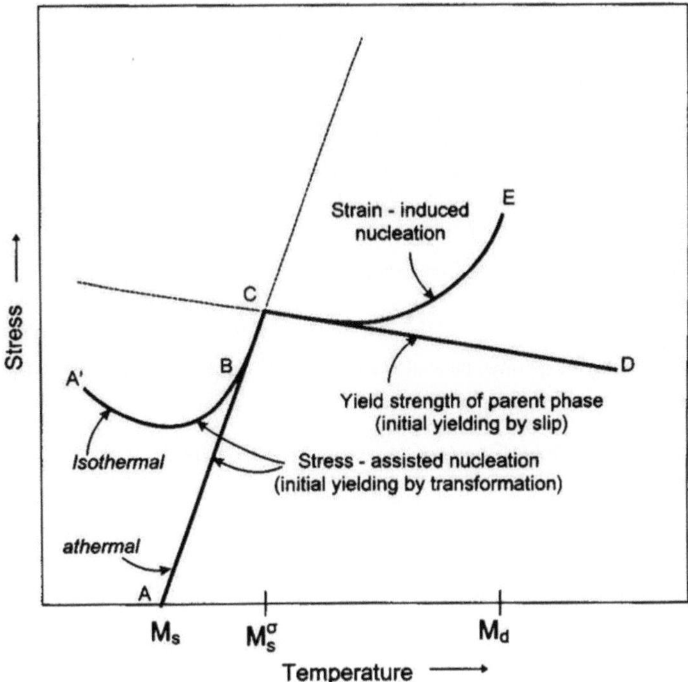

Fig. 5.21 Determination of M^σ temperature—common scheme

transformation would have occurred on cooling below M_S. Above the M_S^σ temperature, when the stress reaches the yield stress for slip in austenite, new potent nucleation sites are generated triggering the strain-induced transformation of austenite to martensite. Near the M_S^σ temperature, both transformation modes operate concurrently. At temperatures above M_D, martensite can no longer be induced by deformation.

Lower carbon content in retained austenite (or other factors that reduce its stability) leads to higher M_S, M_S^σ, and M_D temperatures.

Based on the data on yield point behavior and the associated strength differential effect as a function of temperature, Olsen et al. concluded that it is possible to determine the M_S^σ temperature from the transition between smooth yielding to discontinuous yielding using the so-called SS–TV–TT (single specimen–temperature variable–tension test) technique. Transformation leads to load relaxation resulting in lowering the yield point in stress–strain curve. Thus, the first moment the yield point is detected at decreasing temperature is indicative of the mechanically induced martensitic transformation of retained austenite and consequently the corresponding temperature is equal to M_S^σ, as illustrated in Fig. 5.22.

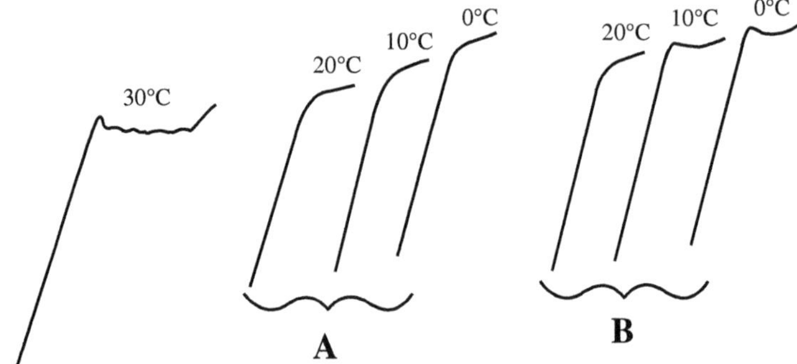

Fig. 5.22 Example of experimental evaluation of the M^σ temperature (Vasilakos et al. 1999)

5.4.4 Factors that Control Austenite Stability

The factors, which determine the stability of austenite after austempering, are typically listed in the following order: chemical stabilization of austenite by enrichment with carbon as the most important for austenite retention, then it is partitioning of alloying elements, and austenite size.

Imai et al. compared 0.2C–2Si–1.5Mn and 0.2C–2Al–1.5Mn TRIP steels. These authors showed that with the same volume fraction of austenite, the carbon content in austenite, C_γ, of Al-alloyed steel was significantly higher, and austenite demonstrated more stable behavior during tensile test when at the same strain, the loss of RA in Al-alloyed steel was twice as low as the loss in Si-alloyed steel (Imai et al. 1995).

In the study performed by Itami and Takahashi, only the time of austempering of 0.14C–1.9Si–1.7Mn steel was varied with the intention to change carbon content in retained austenite. While the observed volume fraction of retained austenite increased from 9.5 % to 12.7 % and then to 13.2 % with transformation time increasing from 10 to 60 and then 480 s, respectively, higher austenite stability after longer holding was observed in subsequent tensile tests (Fig. 5.23). This was considered as being the result of higher carbon content in RA (Itami et al. 1995). The same observations were reported by Traint et al. (2003) in studying 0.2C–1.% Mn steels with additions of 1.5Si or 2.0 %Al. Whereas in retained austenite of Si-added steel, after holding at 400 °C, the maximum carbon content of ~1 % was achieved, in Al-added steel, this approached ~1.6 % inducing higher stability of RA under deformation.

Utilization of high alloying to stabilize austenite against ferrite and pearlite transformations during initial cooling is typically limited for TRIP steel because alloying also retards the bainitic reaction.

An attempt to quantify various mechanisms of stabilization of retained austenite was made by Wang and Van der Zwaag (2001). They also confirmed that most important is the chemical stabilization due to enrichment of retained austenite with

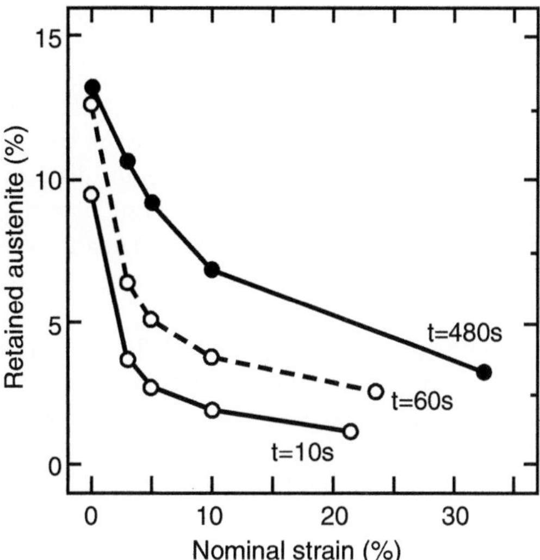

Fig. 5.23 Effect of bainite transformation time (t) on the changes in volume fraction of retained austenite with strain (Itami et al. 1995)

carbon. The M_S temperature of 1.5Mn–1.5Si steel was described by exponential relationship with carbon concentration in retained austenite as

$$M_S\ (K) = 273 + 545.8\ e^{-1.362\ C} \tag{5.7}$$

M_S decreased from 473 to 308 K as carbon in austenite increased from 1.0 to 2.0 mass %.

Athermal transformation of metastable austenite into martensite is dictated by the density of potential nuclei and austenite grain (Olson and Cohen 1976). For example, theoretical calculations give the required nuclei density as high as 2.5×10^{17} m^{-3} for austenite grain size of 1 μm. If the necessary nuclei density cannot be achieved, the shortage of nuclei can induce the austenite retention.

Strong influence of austenite grain size on the M_S temperature was estimated by Wang and Van der Zwaag. Refining RA grain size led to significant decrease in the M_S temperature, so that substantially higher stresses for austenite transformation to martensite were required. Similar findings were reported in other publications (Yang and Bhadeshia 2009; Lee and Lee 2005).

On the other hand, very fine austenite grains can be too stable, thus inhibiting the TRIP effect. As was demonstrated by Park et al. in their in situ EBSD analysis of changes in volume fraction of austenite with strain, volume fraction of tiny film-type austenite in the amount ~8 % or 50 % of the total RA did not change with strain, i.e., did not contributed to transformation-induced plasticity (Park et al. 2002).

5.4.5 Bake Hardening of TRIP Steels

Bake-Hardening properties of TRIP steels, like those observed in dual-phase steels, are determined by interactions between carbon atoms and dislocations. The dislocation network is very heterogeneous: dislocations are preferentially concentrated near the grains boundaries of prior austenite grains transformed to bainite or martensite, i.e., near ferrite–bainite/martensite interfaces. Pinning of dislocations by carbon atoms requires long-range diffusion, which is very weak at room temperature.

Consequently, as-annealed TRIP steels are resistant to natural aging.

The amount of retained austenite obtained after heat treatment does not change not only during holding at room temperature but also during heating up to 170 °C. Only at 220 °C, the volume fraction of RA in TRIP 700 steels began to decrease after holding during 700 min (Bleck and Bruhl 2008).

Pre-straining, especially by unidirectional tension, creates homogeneous dislocation network, when only short-distance carbon atom diffusion is necessary for dislocation pinning. Consequently, similarly to dual-phase steels, TRIP steels show substantial increase in yield strength after paint baking treatment. For example, simulation of paint baking of TRIP800 steel (0.2C–1.67Mn–1.68Si) at 170 °C for 20 min produced after 2 % pre-strain (BH_2) yield strength increase of as high as 97 MPa (Cornette et al. 2001).

The study by Yakubovsky et al. included a variety of laboratory heats (Yakubovsky et al. 2002). The base chemical compositions of steels with 0.16C–1.5Mn, differed mostly in Si and Al contents, are given in Table 5.1, which were heat treated to get dual-phase and ferrite–bainite microstructures for low Si composition, as well as TRIP and multiphase steels of various compositions. The data presented in Table 5.2 show no evident correlation between BH effect and steel chemistry or microstructure but only some trends of changes in the critical parameters that can affect BH of TRIP and multiphase steels.

In particular, this table describes the influence of annealing temperature on BH_2 effect in low-Si steel that contains almost no RA and can hardly be classified as a TRIP steel. The increase in BH_2 with higher annealing temperature can be explained here by increase in volume fraction of bainite (or martensite in water

Table 5.1 Chemical composition of studied lab heats

Symbol	Si	Al	Other
LoSi	0.25–0.45	0.04–0.06	–
1 % Si	0.9–1.2 %	0.04–0.06	–
HiSi	1.5–1.7	0.04–0.06	–
LoSi-MA	0.25–0.45	0.04–0.06	Nb + Mo or Ti + Mo
MidAl	0.10	0.8	
MidAl (Si and/or P)	0.25–0.45	0.8	0.10 P
HiAl	0.10	1.5–2.0	

Table 5.2 Microstructural parameters and BH_2 values of investigated steels (IBT at 400 °C for 90–120 s)

Steel	Annealing temperature (°C)	Structure	RA (%)	BH_2 (MPa)
LoSi	760	F+45 % B	<1	62
	760	F+45 % M	1–2	55
	850	F+90 % B	<1	80
	850	F+90 % M	<1	95
LoSi-MA Ti+Mo (Nb+Mo)[a]	760	F+45 % B	<1	60; (112)[a]
	850	F+95 % B	<1	100
1Si-TRIP	760	Fe+35 %(B+RA)	8	64
MidAl (Si+P)-TRIP	805	Fe+45 %(B+RA)	12	85
	805	Fe+45 % M	1–2	88

[a]Steel exhibited 10–15 % unrecrystallized structure

quench cycle) and by corresponding increase in dislocation density that induced more significant strengthening effect due to their pinning.

Similar increase in BH effect with higher annealing temperatures can be observed in TRIP steels. The important role of dislocation density is confirmed by significant increase in BH in microalloyed (MA) steels with additions of Nb +Mo. When annealing temperature does not ensure full recrystallization of MA steel with elevated recrystallization temperature, much higher BH_2 effect (up to 112 MPa) due to increase in total dislocation density can be immediately observed. Similar behavior was shown earlier for dual-phase steels.

1 %Si-TRIP steel containing about 6–8 % of RA demonstrated approximately the same level of BH effect as low-Si steel that had ferrite–bainite microstructure after holding at the same IBT temperature (in this case, 400 °C). Thus, the contribution of RA itself to BH effect of TRIP steels is insignificant. One can expect, though, some variations attributable to the presence of Al or Si that prevent carbide formation in bainite during IBT holding and consequently promote a higher solute carbon content not only in austenite but also in ferrite and bainitic ferrite. This can affect diffusion rate of free carbon atoms as well.

It was confirmed that bainite in TRIP steels is highly supersaturated with carbon. Results of De Meyer et al. also pointed at some decrease in carbon content in austenite during paint baking simulations suggesting that at 170 °C carbon atoms could diffuse from austenite to dislocations (De Meyer et al. 2000).

As was mentioned earlier, TRIP steels often exhibit YPE. Some researchers consider this fact as an indication of relatively low dislocation density. Apparently, when YPE is close to or higher than 2 %, the 2 % pre-strain for BH evaluation can be insufficient to create the amount of free dislocations necessary to achieve significant BH effects.

Small amounts of "fresh" martensite (even 2–3 %) that can be formed by partial transformation of lower carbon portion of remaining austenite during final cooling are sufficient to suppress YPE. As shown in Fig. 5.24, the actual position of IBT

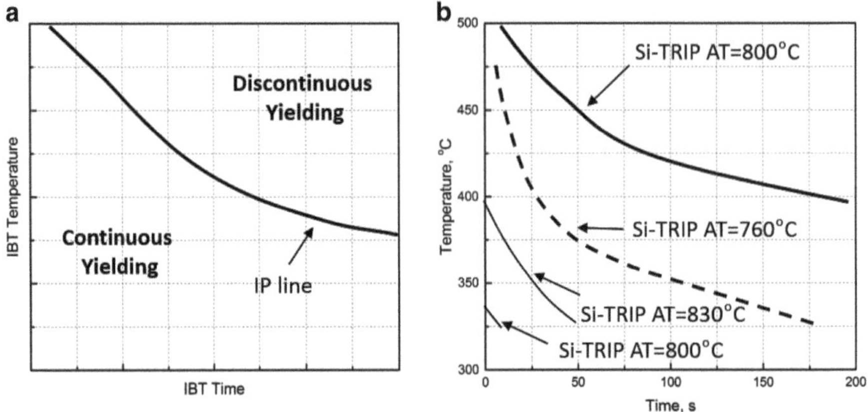

Fig. 5.24 Effect of temperature and IBT time in bainite zone and steel composition on stress–strain curves of TRIP steels: IP curve designates the boundary where *Inflection Points* on S–S curves appear: (**a**) schematic, (**b**) examples of chemical composition effects (Yakubovsky et al. 2002)

"Temperature—Time" borderline between continuous and discontinuous stress–strain curves depends on chemical composition that controls the temperature range and the rate of bainitic reaction and consequently the enrichment of retained austenite with carbon that can prevent martensite formation. Naturally, TRIP steels containing some amount of martensite demonstrate higher baking hardenability.

The role of necessary amount of solute carbon is confirmed by comparative measurements of BH effects and internal friction in steels that were air- and water-cooled after holding in IBT range. It was shown that after final water cooling, the Snoek peak was higher (i.e., higher concentration of solute atoms) that, in turn, resulted in higher BH effect.

As mentioned above, baking hardenability of as-annealed steel, exhibiting YPE above 2 %, can be very low due to insufficient amount of dislocations in TRIP steels with ferrite–bainite matrix. However, temper rolling at 0.4–0.6 % reduction can be sufficient to create enough dislocations to ensure $BH_o = 45$ MPa, and starting from pre-strain of 0.5 %, the BH effect reaches 80 MPa.

At smaller YPE, the pre-strain of 2 % should be sufficient to reach necessary $BH_2 > 40$ MPa even without temper rolling.

Compared to 0.15C–1.5Mn steel with 0.9Al–0.3Si–0.06P additions, 1Si-TRIP steel exhibits a little lower BH_2 of about 60 MPa. Partially it is because at the same strength level, due to the significant strengthening of ferrite by Si, the treatment of Si steel was performed at higher IBT in area of softer bainite and therefore the 1 % Si steel had lower dislocation density (Fig. 5.25). Anyway, the total sum of bake and strain hardening during forming operations provide substantially higher final strength of parts after stamping and paint baking.

Studying of various pre-strain effects on BH values of TRIP-800 steels after holding at 170 °C for 20 min revealed not only the high level of bake-hardening

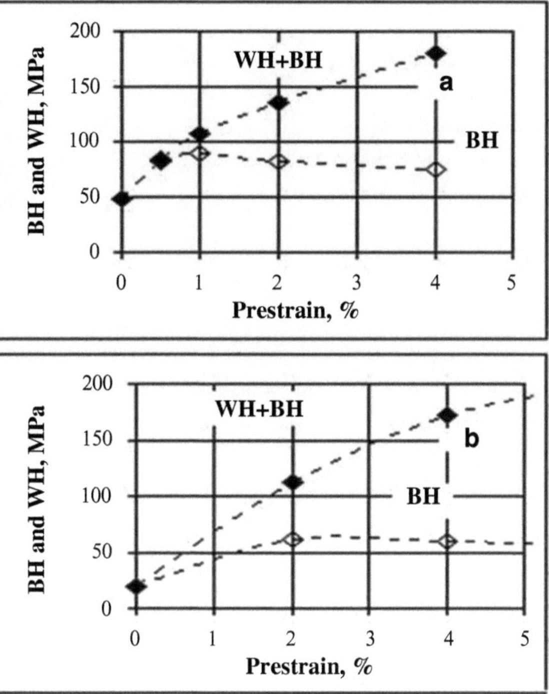

Fig. 5.25 Effect of pre-strain on bake and strain hardening (BH and SH) of temper-rolled (**a**) Al + Si + P and (**b**) 1Si steel

effect that exceeded the level for DP-600 steel but also their different strain dependence. Whereas DP steels often demonstrate the maximum BH value at 1.5–2 % pre-strain (see Chap. 3), TRIP steels show increasing BH effect up to 4 % pre-strain, and it stays high up to 8–12 % pre-strain (Fig. 5.26) (Hance et al. 2003).

5.5   Effect of Steel Composition on Phase Transformations, Final Microstructure, and Mechanical Properties of TRIP Steels

Effects of the steel composition on microstructure and properties of TRIP steels include all features discussed with regard to dual-phase steels: the effects of alloying on A_{C1}–A_{C3} range and therefore on kinetics of austenitization, as well as the effects on recrystallization and overlapping of recrystallization and austenitization. Since the annealing temperatures in processing of TRIP steels are rather high due to alloying with elements that raise A_{C1} temperatures, the recrystallization may not be a very important limitation in heat treatment of TRIP steels. However, the same factors (raising A_{C1} and expanding A_{C1}–A_{C3} temperature range, Table 5.3) produce strong effects of chemical composition on kinetics of

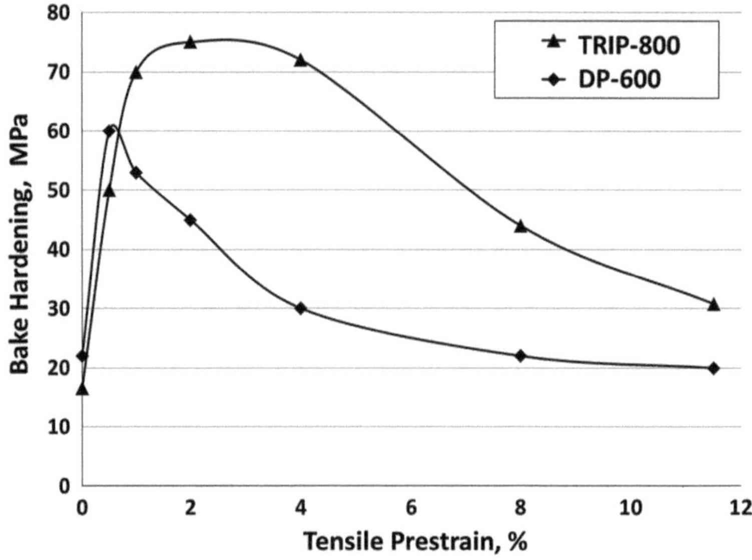

Fig. 5.26 Bake hardenability (BH) measured after holding at 175 °C for 30 min as a function of tensile pre-strain for DP and TRIP steels

Table 5.3 Parameters of transformations of investigated steels

Logo	A_{C1}	A_{C3}	$T_{50\%\gamma}$	%M
LoSi	710	880	785	58
1Si	725	900	810	
HiSi	735	915	830	
0.4Al	720	900		47
0.8Al	715	950	825	56
1.5 Al	725	1130		
1.9Al	735			

austenitization, as shown in Fig. 5.27. Table 5.3 lists the temperatures of formation of 50 % austenite during 60 s holding that were determined by dilatometry and verified by quantitative metallographic analysis of microstructures of quenched samples.

Since the main features of TRIP steels are based on the presence of retained austenite, the most important role of alloying is to enhance hardenability of γ-phase, to prevent pearlite formation during initial cooling from intercritical temperature down to IBT so that to preserve all carbon in the remaining austenite, and to ensure the necessary kinetics of bainite formation and proper enrichment of remaining austenite with carbon to decrease M_S below room temperature.

The discussion below consider as well the role of ferrite forming and microalloying elements, which can influence the transformations, final microstructure, and properties through affecting the carbon partitioning, carbide formation, microstructure refinement, and contribution to precipitation strengthening of ferrite as well.

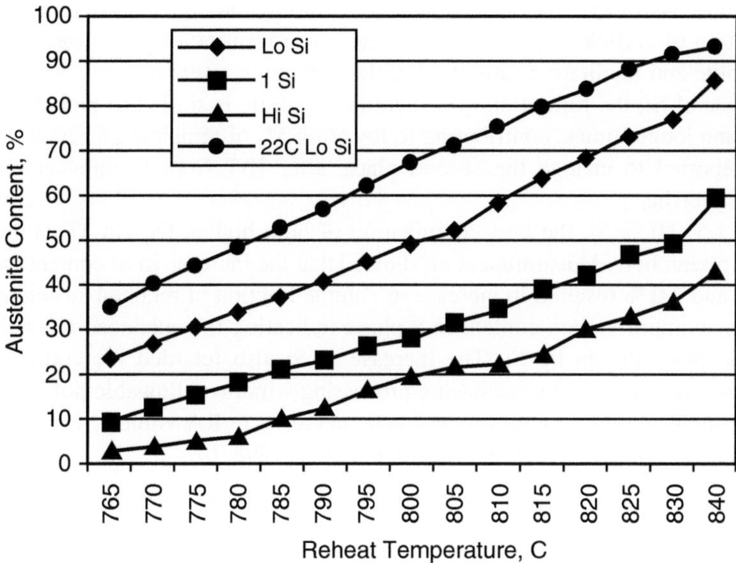

Fig. 5.27 Effect of chemical composition on austenitization kinetics of 0.16C–1.5Mn steels with 0.35 (LoSi), 1.0, and 1.6 % Si (HiSi) and 0.22C–1.5Mn–0.35Si steel during heating at 5 °C/s (dilatometry)

5.5.1   Effect of Ferrite-Stabilizing Elements

The presence of Si, Al, and other ferrite stabilizers in TRIP steel composition is necessary to suppress the formation of carbides during bainitic reaction, thus preserving all carbon in the remaining austenite and facilitating the formation of bainitic ferrite. All of the initial TRIP designs were based on alloying with 1.0–2.0 % Si. Since high Si is not desirable for galvanizing process, other ferrite forming elements began to draw attention, including Al, P, V, Cr, Mo, Ti, and Nb.

The most successful substitutes of Si turned out to be Al and P individually and/or in various combinations with Si.

Silicon   Silicon is the key element in TRIP steels that makes the retention of austenite feasible at relatively low carbon content. The additions of Si facilitate enrichment of austenite with carbon and thus impact the stability of remaining austenite leading to higher volume fraction of retained austenite.

Alloying with ferrite stabilizers and favorable thermal cycles of existing CAPL lines significantly facilitated manufacturing of uncoated TRIP steels with very high balance of their mechanical properties. Lab results and industrial trials of steel with ~0.4 % Si showed the possibility of obtaining up to 6–8 % of austenite after annealing at temperatures close to A_{C1} and long overaging, but that austenite was not very stable and transformed to martensite on early stages of deformation (Jacques et al. 2001).

Additions of more than 1 % Si strongly retard the precipitation of carbides. This inhibition of carbides by Si is responsible for prevention of formation of bainitic cementite and facilitating saturation of the remaining austenite with carbon. In the presence of Si, the formation of cementite in bainite is shifted to higher temperatures and longer times, contributing to lowering M_S of remaining austenite. 1 % Si was reported to change the second phase after IBT from bainite–cementite to bainitic ferrite.

At 1.5–2.0 % Si, the best combination of both higher TS and TE is noted by many researchers. Matsumura et al. showed that the increase in Si content from 1.2 to 1.5 and 2.0 % resulted in increase in volume fraction of retained austenite at the same maximum carbon content in γ-phase indicating the enhanced retardation of carbide precipitation by Si. The increase in Si also retarded the completion of bainitic reaction, thereby increasing processing window: allowable holding at IBT was shifted to longer times where drastic decrease in RA volume fraction can be observed at lower Si contents (Matsumura et al. 1992, b).

Aluminum Numerous comparisons of Si- and Al-added TRIP steels showed that by adjustment of chemical composition (C, Mn) and isothermal holding temperature/time, the similar combinations of strength and ductility can be achieved in Al-bearing steels at higher uniform elongation and carbon content in austenite.

The effects of Si and Al are illustrated in (Zhao et al. 1999) showing the product of the amount of retained austenite and its carbon content against austempering time at 400 °C. As was shown in Fig. 5.9, that in Al-alloyed steels, almost all carbon of steel was in the retained austenite, indicating strong effect of aluminum on suppression of carbide formation during bainitic reaction. The observed lower values C_γ,%·V product due to replacement of 0.4 %Al by 0.25 %Si reflect the facts that Si additions reduce both volume fraction of austenite and its carbon content being less effective on suppression of carbon formation during the bainitic transformation (Fig. 5.28).

By contrast, Girault et al. concluded that Al as strong carbide inhibitor is less efficient than Si in the same concentration, although this contradiction with numerous literature data can be in part related to low carbon content (0.11 %) in steel used for this study (Girault et al. 2001).

Gomez et al. performed a systematic analysis of phase transformation parameters of 0.15C–1.5Mn steels with 0–1.5 % of both Al and Si using JMatPro software. These authors pointed at higher volume fraction of austenite at its lower carbon content in Si-alloyed steel than in Al-alloyed steel, annealed at the same temperatures. Computed CCT diagrams for 1 % Al showed that ferrite formation was shifted to significantly shorter times (higher cooling rates) than in case of 1 % Si steel. At initial cooling rate of 10–15 °C/s, it was almost impossible to avoid new ferrite in Al-added steel, whereas in Si-alloyed steel, there was a risk of pearlite formation. As was predicted, Al slightly accelerates bainitic reaction at 400 °C but especially strong at the Zn pot temperature (460 °C) (Gomez et al. 2009) that qualitatively agrees with TTT diagram presented in Fig. 5.11.

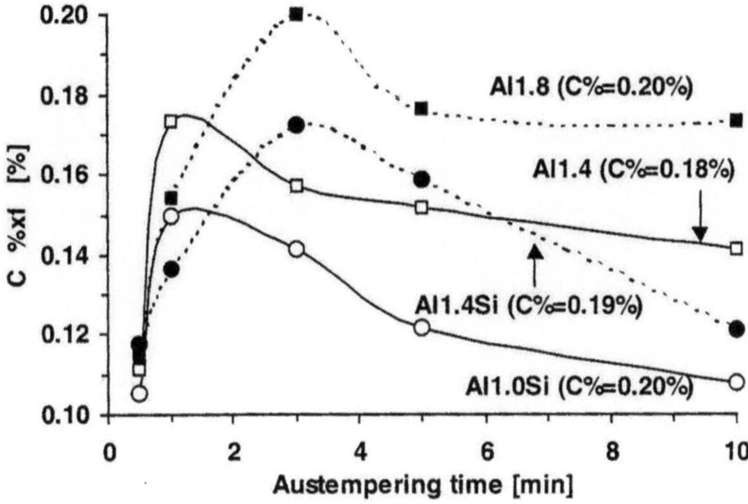

Fig. 5.28 Effect of alloying by Al and Si on product of carbon content and volume fraction of retained austenite $C_\gamma,\%\cdot V_\gamma$ as a function of austempering time (Zhao et al. 1999)

Phosphorus Additions of phosphorus were considered to partially substitute Si and/or Al but mostly for application together with other ferrite-stabilizing elements.

It is known that phosphorous, like other ferrite stabilizers, increases C activity in ferrite and cementite and retards kinetics of cementite precipitation. Phosphorous is also known as one of the strongest solid solution strengthening elements.

The beneficial effect of P in the amount of 0.07 % together with 0.5 % Si in (0.14–0.18C)–1.5Mn steels was found to produce higher amount of RA and combination of higher strength and ductility. The formation of very stable portion of austenite in high-P steel was attributed to both small size of austenite particles and to higher matrix constraint resulting from solid solution strengthening of ferrite by phosphorus. On the other hand, significant portion of very small austenite particles that appeared to be of very high mechanical stability had a negligible effect on ductility (Chen et al. 1989).

In the other study, the biggest amount of austenite and the best combination of properties were found for 0.19C–1.6Mn–0.4Si–0.6Al–0.07P steel. It was then concluded that Al and P can only partially replace Si (Barbe et al. 2002).

Comparison of 1.6 %Si additions with additions of 1.9 % Al or 0.1 %P + 0.3 %Si to 0.15C–1.5Mn steels aiming to find the opportunity to decrease Si content in TRIP steels showed that after overaging at 400 °C, the steel with Al contained 8–11 % of retained austenite, while steel with Si showed 8–9 %, and the steel with P and low Si had only 4 % (Gallagher et al. 2002). Figure 5.29 displays microstructures of these steels annealed at different intercritical temperatures corresponding to 50 % of initial austenite, after overaging at 400 °C for 100 s (Gallagher et al. 2002).

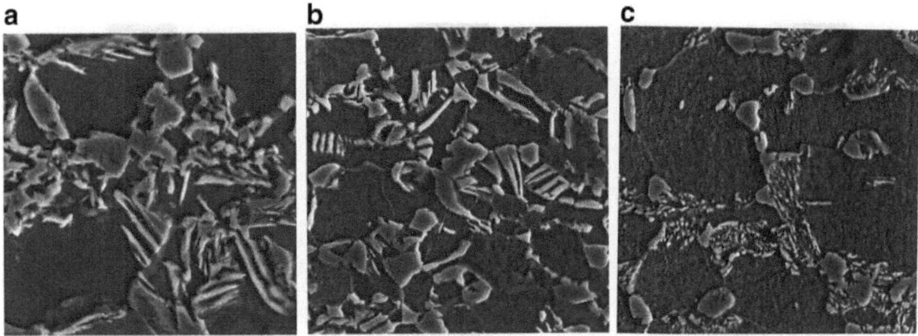

Fig. 5.29 Comparison of microstructure of 0.15C–1.5Mn steels alloyed by (**a**) 1.6 % Si, (**b**) 1.9 % Al, and (**c**) 0.1P + 0.3Si after overaging at 400 °C for 100 s, SEM ×5000 (Gallagher et al. 2002)

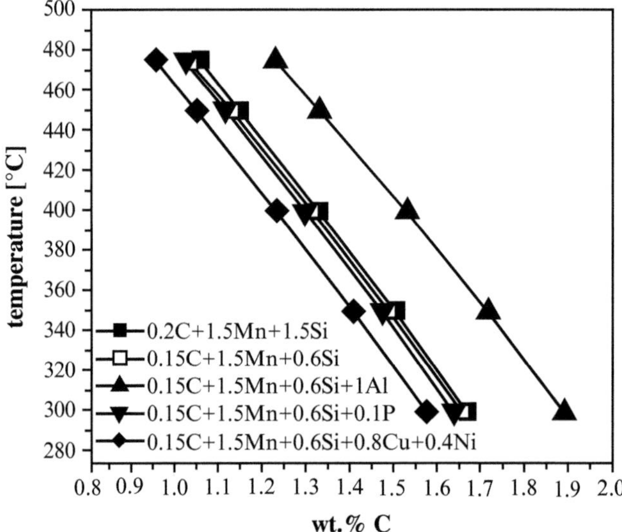

Fig. 5.30 Influence of Si, Al, P, and Cu on the maximum enhancement of carbon in the austenite calculated according to the T_0 concept using ThermoCalc (para-equilibrium) computations (Traint et al. 2004)

Traint et al. reported that additions of the same 0.1 % P to 0.15C–1.5Mn–0.6Si (at slightly higher Si content) steel resulted in high amount of RA and satisfactory mechanical properties. Although possible embrittlement or problems with welding due to high P content cannot be ruled out, additional benefits of alloying with P were considered as a way to reduce surface defects characteristic for Si steels during galvanizing.

The effects of ferrite stabilizers on maximum carbon content in retained austenite as a function of overaging temperature calculated using Bhadeshia's T_0 concept (Bhadeshia 1992) are shown in Fig. 5.30. It is seen that such elements as Si and P

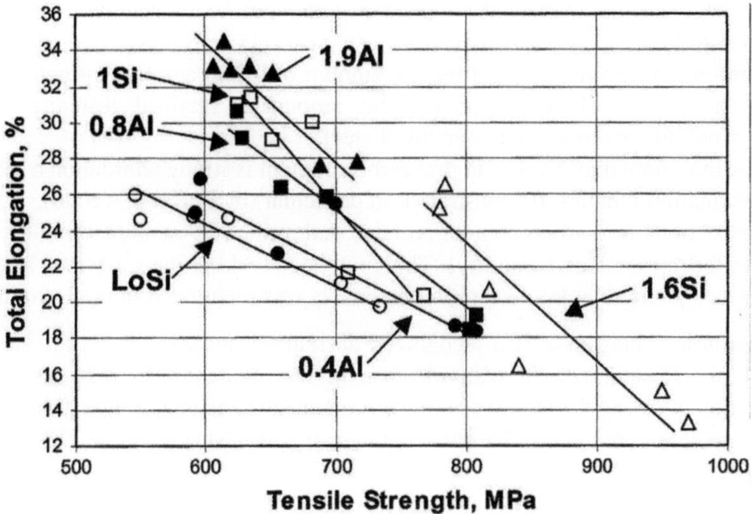

Fig. 5.31 Comparison of balance of total elongation and tensile strength of 0.16C–1.5Mn steels with various amounts of Al and Si

have minor impacts on carbon content, while additions of Al significantly increase maximum carbon concentration in RA (Traint et al. 2004).

These computations are in a good agreement with the published data, indicating that the additions of Al (around 1.5 %) after overaging time at 400 °C produced higher amount of RA, higher carbon content in austenite, and higher elongation than the additions of equivalent amount of Si. However, due to substantial strengthening by Si (up to 150–200 MPa), tensile strength of Si-alloyed steel is higher and so are the products TS × TE or TS × UE, because the difference in elongation does not allow to offset the difference in tensile strength.

Some comparative data on balance of total elongation and tensile strength for TRIP steels with 0.16C–1.5Mn and various amounts of Al and Si (Yakubovsky et al. 2004) are presented in Fig. 5.31.

5.5.2 Effects of Alloying Elements Enhancing Hardenability of Austenite

The number of studies considering the influence of alloying elements in TRIP steel composition that affect transformations of austenite during initial cooling and further processing is quite low.

Carbon It is generally accepted that carbon is the most powerful element that enhances the hardenability of austenite.

The increase in C content also lowers the temperature of transition from upper to lower bainite.

Carbon content in steel controls the amount of initial austenite at annealing of TRIP steels, its carbon content, possible amount of retained austenite in final microstructure, and potential strength of steel.

The important drawback of higher carbon content is strong retardation of bainitic transformation kinetics, demonstrated, in particular, by Lee et al. (2008).

Restrictions for carbon content are also determined by manufacturability limitations (such as, e.g., weldability). This motivates seeking the ways to achieve the prescribed strength at lower carbon content, as was mentioned in the Introduction.

Manganese Manganese is a recognized austenite stabilizer so that the austenite hardenability is increased with increase in Mn content. Mn prevents pearlite formation allowing for slower cooling from annealing temperatures, decreases the M_S temperature, but retards bainite formation. Typical Mn content in TRIP steels is ~1.3–1.7 %.

Manganese increases the strength of steel by solid solution strengthening. However, excessive manganese should be avoided because the inhibition of ferrite formation by Mn can reduce carbon enrichment of austenite during initial cooling.

As shown in Fig. 5.32, with increasing Mn content, longer holding in IBT range is required to achieve maximum volume fraction of retained austenite and the best ductility of TRIP steels. As shown, the increase in Mn content facilitates higher

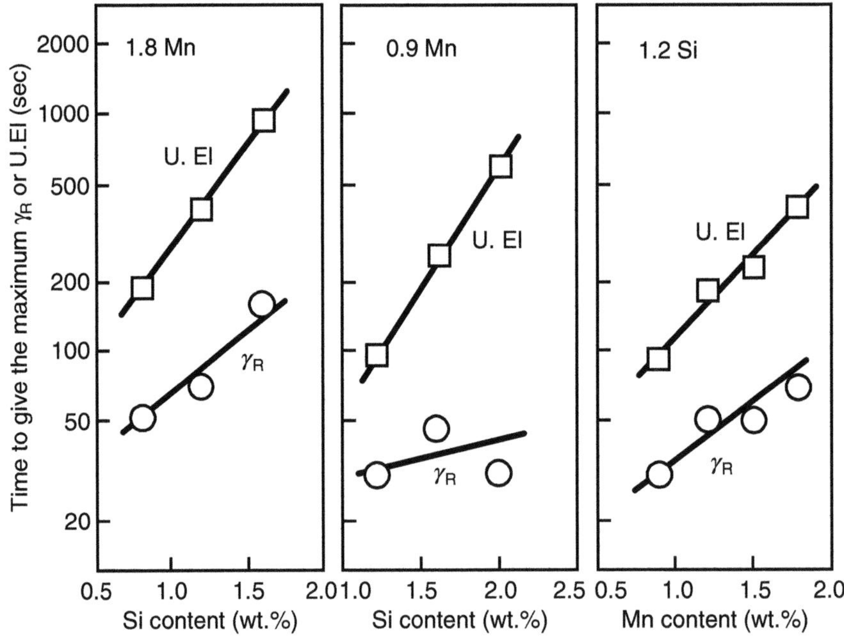

Fig. 5.32 Changes in holding time at 400 °C to attain maximum content of retained austenite or UE depending on Si and Mn content in 0.2 % C steels (Sakuma et al. 1991)

amount of RA, but without enhancing its stability due to the dominant role of decreasing its carbon content.

Molybdenum Mo acts as an austenite stabilizer, when added in small quantities, but as a ferrite stabilizer, when it is present in greater amounts.

Mo retards ferritic and pearlitic transformations due to its effect on both kinetics and thermodynamics of carbide formation in steel, lowering A_{r1} by roughly 35 °C at concentration of 0.15 % Mo.

As an element that increases austenite stability, Mo retards the transformation to ferrite beyond realistic annealing parameters. Therefore, steels with Mo contain only the initial, "old" ferrite. Mo lowers the bainitic start temperature and retards bainite formation. On the other hand, Mo weakens the effect of overaging time and temperature on the amount of retained austenite, and fairly high volume fraction of RA (~8 %) was observed in 0.12C–1.5Mn–1.2Si–0.14Mo steel that is rather constant from 1 up to 15 min holding at 400 and 450 °C (Sakuma et al. 1993).

As was mentioned in Chap. 4, a number of publications have suggested that Mo imposes strong solute drag on grain boundary migration that leads to substantial delays in recrystallization and precipitation. The known retardation of precipitation of Nb(C,N) by Mo is explained by increased solubility of carbonitrides stemming from decrease in carbon activity and thereby decreased driving force for precipitation. Mo also affects diffusivities of precipitating elements (i.e., Nb, C, and N). Thus, Mo can enhance the efficiency of both Nb and C in retaining and stabilizing austenite by delaying the precipitation of Nb(C,N) which could reduce austenite stability (Mohrbacher 2007).

In some publications, the strengthening by Mo is noted, but it should be rather related to microstructure refinement or accompanying factors that increase the volume fraction of fresh martensite during final cooling due to incomplete bainite reaction since the strengthening of ferrite by Mo itself is negligible.

Chromium At high concentrations (more than 9 %), Cr makes steel ferritic and is considered as a ferrite stabilizer. During continuous cooling of low-Cr steel, the prevailing role of Cr is the enhancement of austenite hardenability by reducing the formation of new ferrite. As to IBT, it is known that Cr slows bainitic reaction. As a result, dual-phase steel type behavior was noted in TRIP steels with ~0.4 % Cr additions due to incomplete bainitic reaction and consequently significant amount of fresh martensite formed during the final cooling (Kim et al. 2003; Suh et al. 2008; Gomez et al. 2010). At the same time, a number of steel grades containing 0.2–0.5 % Cr were commercialized, where the negative effects of Cr are compensated by reducing the concentrations of Mn or Mo and by high Al content (1.2–1.7 %).

Copper and Nickel Cu and Ni are austenite stabilizers. Their additions were found beneficial for promoting higher RA and therefore higher ductility of TRIP with 0.15C–1.5Mn–1.5Si steels when individual and combined effects of 0.5 % Cu and 0.4 % Ni were investigated after isothermal holding at 430 °C. The known strengthening effect of Cu is usually considered to be related to very fine ε-Cu precipitation in ferrite grains, but it has been reported only for Cu content above 1 % for longer than tens of minutes holding. Thus, strengthening by 0.5 % Cu that allowed for TS of 817 MPa compared to 727 MPa in Cu-free steel at total elongation 36.4 % and 29.2 %, respectively, is explained by higher fraction of RA and solid solution strengthening by Cu. Additions of 1 % Si or 1 % Al to 0.15C–1.5Mn–0.8Cu–0.4Ni steel resulted in the increase in strength or elongation, respectively (Traint et al. 2001).

5.5.3   Effect of Microalloying

In search for efficient resources to increase strength of TRIP steels, particularly without deterioration of their ductility, numerous studies are focusing on potential contributions of microalloying elements.

Niobium Depending on coiling temperature applied, Nb can be partly in solid solution or to form carbonitrides. Combined effects of Nb additions on recrystallization during hot rolling and annealing of cold-rolled material result in significant microstructure refinement. That, in turn, accelerates transformation of austenite facilitating formation of new ferrite that should promote initial stage of carbon enrichment of γ-phase.

As presented in Fig. 5.33, additions of Nb delays bainite formation, which can be attributed to blocking of the nucleation sites by fine dispersed carbonitrides and enrichment of austenite with carbon due to enhanced formation of ferrite during initial cooling (Ohlert et al. 2003).

Smaller austenite particles in Nb microalloyed steels contain less potential nucleation sites for transformation to martensite and hence require greater total driving force for martensite nucleation. The smaller they are, the greater is their geometric stability against transformation to austenite (Olson and Cohen 1976). Thus precipitates not only control the grain size and formation of highly

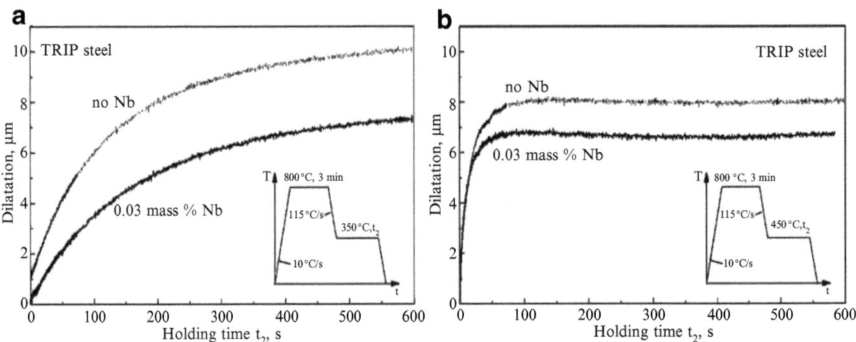

Fig. 5.33 Effect of Nb on isothermal bainite formation in TRIP steels at (**a**) 350 °C and (**b**) 450 °C; 0.2C–1.4Mn–0.5Si–0.75Al–0.04P steel (Ohlert et al. 2002)

homogeneous microstructure, which enhances carbon diffusion, but also considerably influence the martensite/bainite nucleation. As noted by Bleck et al. (2001), microalloying of TRIP steels with Nb lowers the M_S temperature. In particular, the addition of 0.03 %Nb to 0.2C–1.4Mn–0.5Si–0.75Al–0.04P steel in combination with coiling at low temperature reduces the M_S temperature by more than 100 °C.

However, this effect of Nb is restricted to low temperature coiling because at high coiling temperatures, the precipitates become coarser and less efficient in inhibiting austenite grain growth during annealing.

Additions of Nb (up to 0.04 %) in 0.17C–1.4Mn–1.5Si (Hulka 1999) or 0.2C–1.55Si–1.55Mn (Zarei-Hanzaki et al. 1995) TRIP steels significantly increased the amount of retained austenite. This was attributed to lowering the M_S temperature first because of increased hardenability of austenite due to inhibiting of carbide formation and, second, because of solid solution strengthening in the presence of dissolved Nb and C that raises the resistance of matrix against the volume expansion needed for transformation.

As shown in Chap. 4, regardless of the coiling temperature, after annealing, the precipitated fraction of Nb is over 90 %, but in case of low coiling, a much larger portion of Nb precipitates as very small particles affecting strength of steels. As shown by Traint et al., microalloying of Al-bearing TRIP steel with 0.045 % Nb adds ~30–50 MPa to TS which is accompanied by some deterioration of ductility. The authors observed some intensification of ferrite formation due to Nb additions to 0.20C–1.5Mn steel alloyed with 1.5Si or 2.0Al. However, they did not detect bigger amount of austenite or its higher stability under deformation (Traint et al. 2003).

According to (Heller et al. 2005), additions of 0.03–0.04 % Nb raised yield and tensile strengths of 0.19C–Mn–Al–Si steel up to 100 MPa due to grain refinement and precipitation hardening and increased the TS × TE product. Further addition of Nb up to 0.07 % did not enhance strength but improved elongation. One explanation is related to finer grain sizes of retained austenite and their higher resistance to martensitic transformation up to higher applied stresses, shifting the onset of strain localization (necking) to higher elongation.

Other explanations employ strengthening by precipitation hardening and possible compensation of deteriorated elongation of TRIP steels by higher volume fractions of ferrite and retained austenite (Ohlert et al. 2002).

Investigation of 0.20C–1.78Mn–0.27Si–1.40Al–0.006P with 0.016 and 0.03 % Nb additions had shown that additions of 0.016 % Nb increase both strength and elongation of TRIP steels, but further increase in Nb content did not change properties as shown in Table 5.4. Obtained results should be explained by structure refinement presented in Fig. 5.34.

Table 5.4 Effect of Nb on mechanical properties of TRIP 780 steel

% Nb	TS (MPa)	YS (MPa)	YS/TS	TE (%)	UE (%)	½ (TS + YS) × UE
<0.003	802	357	0.45	27	19	11,010
0.016	844	547	0.65	28	21	14,735
0.032	849	567	0.67	26.5	20.5	14,480

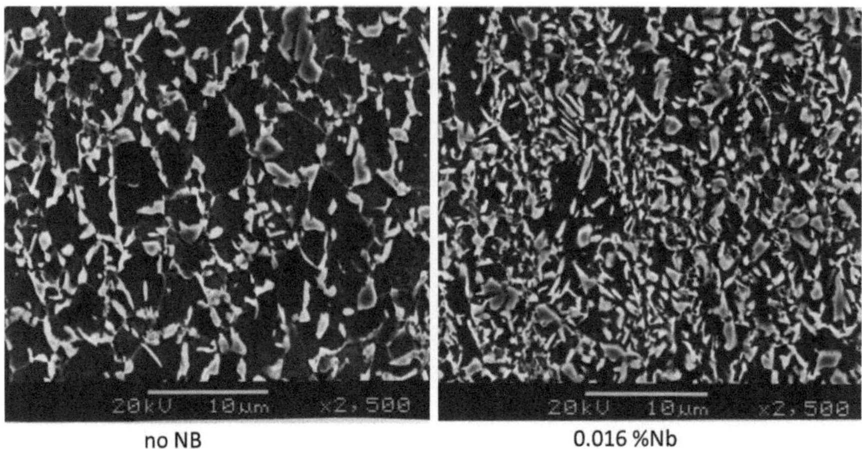

Fig. 5.34 Effect of Nb on microstructure of TRIP780 steel

As a summary, it is worth pointing out that the overall effects of Nb on the properties of TRIP steels should be categorized as the direct effects (grain refinement, potential precipitation hardening, influence on ferrite, bainite, and martensite formation) and indirect effects related to enhancing the enrichment of austenite with carbon and raising austenite stability.

Vanadium Very few publications can be found in the literature that discussed the role of vanadium in TRIP steels. For example, Scott et al. studied the effects of additions of vanadium (up to 0.2 %) and nitrogen (up to 0.015 %) to 0.19C–1.5Si–1.5Mn steel. It was found that microalloying with 0.1 V–0.007 N produced strengthening by 60–100 MPa, while preserving good total elongation (25–27 %) after overaging at 400 °C. In case of 0.2 V–0.015 N, the increase in strength was much higher (214 MPa); however, it was accompanied by a dramatic decrease in ductility. It was established that V additions did not affect the amount or stability of retained austenite (its carbon content), and the evolution of strength should be entirely related to precipitation hardening, which is very sensitive to hot-rolling conditions and the temperature of annealing in the intercritical region (Scott et al. 2004).

5.6   Fracture of TRIP Steels

As was pointed out in Chap. 1, one of the main reasons of interest the carmakers have for TRIP steels is their unique high energy absorption potential that can be the decisive in choosing these materials for safety parts intended for passenger cage protection by accordion-type failure.

In the field of crashworthiness, the high rate of work hardening that leads to substantial increase in strength of steel during forming and subsequent bake hardening should be considered as an additional contribution to energy absorption.

Toughness tests using thin sheet specimens with thickness below 2 mm is problematic, so different studies used different approaches and consequently the corresponding estimates discussed below should be rather considered as comparative data.

5.6.1   Energy Absorption of TRIP Steels

Fracture parameters of TRIP steels are excellent in spite of transformation of retained austenite into high carbon and hence brittle martensite. It was shown that the propensity of martensite to cracking in mixed austenite–martensite microstructure depends on the absolute sizes of martensite islands. To crack fine martensite grains is more difficult, as was demonstrated using 0.98C–1.46Si–1.89Mn–1.26Cr–0.2Mo steel, which is relatively close to composition of austenite in commercial TRIP steels in carbon and total alloying content. Therefore, fine retained austenite in TRIP-assisted steels prevents the formed martensite from deteriorating their mechanical properties (Chatterjee and Bhadeshia 2006).

Carmakers need steels with the best relationship between sheet thickness and maximum static and dynamic energy absorption. At low strains, high strain rates nearly double YS of mild steels. At higher strain levels, the dynamic hardening effect is reduced, and the difference between dynamic and static stress–strain curves diminishes. TRIP steels, however, show different strain rate hardening effect. Yield strength of TRIP steel increases by up to 30 % under dynamic loading compared to static behavior. What is important is that TRIP steels also demonstrate high strain rate hardening effect at higher strain level as well, and the magnitude of the effect is close to normal hardening under static load. Therefore, TRIP steels absorb more energy regardless of strain rate and strain.

Figure 5.35 shows the experimental results of testing spot-welded box beam made from mild steel DC 04 and TRIP 800 sheet steels with thickness of 1.5 mm and 1 mm, respectively. The box beam of original height of 270 mm was reduced to 112 mm for mild steel and to 185 mm for TRIP steels in spite of thinner walls. This illustrates that using TRIP steels in crash-critical parts allows to absorb more energy and extremely useful for parts that have to bear high rates of strain (Engl et al. 1998).

The evaluation of toughness of thin sheets using the cup impact test has become the most used in the last years. It provides information on toughness properties of cold-drawn materials by simulating their behavior in car parts.

Due to complexity of dynamic tests, the area under stress–strain curve can be used as a comparative estimate for the absorbed energy, AE, or crashworthiness. Properties of final parts were simulated by BH_2 treatment when 2 % of pre-strain

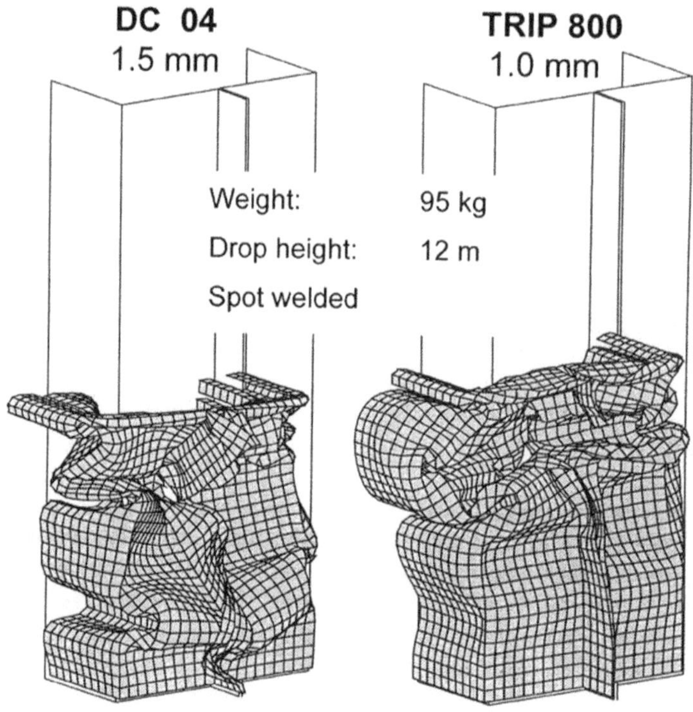

Fig. 5.35 Crash test results for a box beam (Engl et al. 1998)

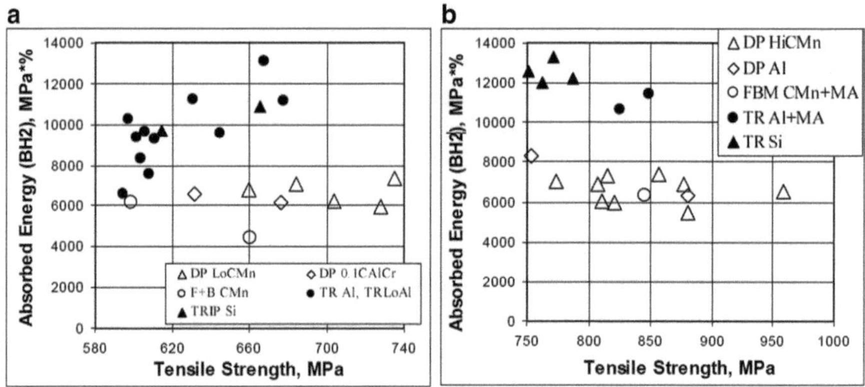

Fig. 5.36 Absorbed energy after BH_2 treatment of different cold-rolled steels; DP, ferrite–bainitic steels (FB), and TRIP steels of various composition including microalloyed (MA): (**a**) TS = 580–740 MPa; (**b**) TS = 750–950 MPa (Yakubovsky et al. 2004)

were considered as simulation of stamping, while aging at 170 °C for 20 min was simulating bake hardening.

The balance between AE, designated as AE (BH_2) after BH treatment, and TS of cold-rolled steels is shown in Fig. 5.36. Both DP steels with ferrite–martensite microstructure and steels with ferrite–bainite microstructure demonstrate approximately the same levels of AE $(BH_2) \approx 6000$–8000 MPa $\times$ % within the entire tensile strength range (580–950 MPa). TRIP steels demonstrate substantial increase in AE (BH_2) values that at TS higher than 640 MPa become 1.5–2.0 times higher compared to $AE(BH_2)$ of F + M and F + B steels (Yakubovsky et al. 2004)

Absorbed energy after BH_2 treatment depends on three parameters: work hardening during 2 % pre-strain (WH_2), bake hardening (BH_2), and residual uniform elongation of steel after bake hardening (UE_{BH2}). It is well known that DP steels have the highest initial WH_2 rate. TRIP steels, on the contrary, can sometimes show low initial strain hardening due to yield point elongation. As for bake hardening (BH_2), the values for both DP and TRIP steels are within rather wide range (50–120 MPa) and depend on various factors discussed above. On the other hand, TRIP steels show the highest UE in both as-processed condition and after bake hardening.

Thus, the higher levels of $AE(BH_2)$ of TRIP steels can be mostly attributed to higher UE preserved after BH_2 treatment.

5.6.2 Fatigue Behavior

According to the NSC data, fatigue limits at stress-controlled 500,000 cycles of TRIP steels are superior to those of DP and HSLA steels at comparable tensile strength of about 600 MPa. Insignificant decrease in volume of RA was observed after fatigue test. It was suggested that high fatigue limit of TRIP steel could be explained by retardation of micro-crack propagation due to presence of compressive subsurface residual stresses induced by the transformation of RA to martensite under cyclic load (Yokoi et al. 1996).

Fatigue strength and crack initiation/propagation were studied using 0.17C–1.4Si–2.02Mn TRIP-aided bainitic steels (Sugimoto et al. 1998b). TRIP steels demonstrated the highest fatigue limit compared to DP steels with the same strength of ~900 MPa. The threshold strain intensity value (ΔK_{th}) turned out to be lower for TRIP steels than for dual-phase steels, but the crack propagation rate at a high stress intensity range was very close to other steels. It was concluded that the retained austenite suppressed propagation of micro-cracks due to stress relaxation by strain-induced transformation and also because the average spacing between fine islands/films of retained austenite was smaller than the plastic zone ahead of crack tip.

5.6.3 Resistance to Hydrogen Embrittlement

Experimental data regarding hydrogen resistance of TRIP steels, as of the majority of AHSS, are inconsistent, in particular, mostly because of the different methods employed in comparative tests.

Hydrogen uptake and its changes with time were studied using CMnSi and CMnAlSi eloctrogalvanized TRIP 800 steels. Slight increase in H content was observed after pickling, although within standard deviation. No significant increase was also induced by elctrogalvanizing of samples. Using special charging, H concentrations of up to 48 ppm were measured immediately after saturation. However, the diffusivity of hydrogen in TRIP steels was high and its concentration returned to initial level already after 60 min (De Meyer et al. 2000).

Hojo et al. studied the effect of aluminum on hydrogen absorption behavior and delayed fracture of 0.2C–Si–Al–1.5Mn steel with constant sum of Si + Al concentrations of about 1.5 %. Delayed fracture was tested under constant load in 4-point bending of materials in an H_2SO_4 + KSCN solution with the maximum fracture strength enduring 5 h used as criterion for the resistance to delayed fracture. With addition of aluminum, the total charged hydrogen did not change but the diffusible hydrogen increased. At the same time, the delayed fracture strength at 0.2–1.0 % Al increased significantly. It was suggested that this was mainly caused by suppression of the stress-assisted martensite transformation due to stabilization of carbon-enriched RA by Al additions. As the second important factor, it was suggested that the diffusible hydrogen in Al-added TRIP steel was trapped in refined retained austenite and at the extended surface of martensite lath boundaries due to martensite refinement with additions of Al. Finally, the resistance to delayed fracture was facilitated by relaxation of local stress concentration due to strain-induced plasticity (Hojo et al. 2008).

In the other publication, Hojo et al. studied the effects of diffusible hydrogen and alloy content (Cr, in particular) on hydrogen embrittlement of TRIP-aided steels with bainitic and martensitic matrixes. Steels employed in the work contained ~0.20C–1.5Si–1.5Mn–0.05Nb and variable amount of Cr; some of the heats contained Ni, Mo, Ti, and B. The loss of elongation measured before and after hydrogen charging was used as an indicator for hydrogen embrittlement sensitivity (HES). In general, the higher the volume fraction of retained austenite and carbon content in it due to different alloying was, the lower the HES was obtained. The explanation implied larger amount of hydrogen soluble in austenite than in martensite or bainite. Martensitic transformation should, in general, lead to diffusion of undissolved hydrogen towards grain and lath boundaries, and this can cause hydrogen embrittlement. However, stable RA in TRIP steels restrains martensitic transformation and therefore improves resistance to hydrogen embrittlement. In case of martensitic matrix, steels with 1 %Cr were featured by particularly low HES due to refinement of martensite lathes and restraining precipitation of cementite in the martensitic matrix (Hojo et al. 2013).

5.7 Summary

The combination of high strength, high uniform elongation, and high strain hardening of TRIP steels allows for manufacturing of many complex autobody parts by deep drawing despite high strength and relatively low anisotropy coefficient inherent to high-strength steels.

All researchers note that ductility of TRIP steels is negatively impacted by the presence of "fresh" martensite. The latter is formed due to delayed bainite reaction induced by extra alloying and consequently due to insufficiently long holding in IBT range.

Post annealing (low tempering) of TRIP steels is found to be accompanied by increasing carbon content in the remaining portion of retained austenite resulting in both improving elongation and hole expansion.

Intensive studies of the TRIP effect and properties of retained austenite are being continued. Presently, they are strongly motivated by search for the best solution in developing the so-called third-generation advanced high-strength steels.

References

Barbe, L., L. Tosai-Martinez, and B.C. DeCooman. 2002. "Effect of Phosphorous on Tensile Properties of Cold Rolled and Intercritical Annealed TRIP-Aided Steel." In *TRIP_aided Ferrous Alloys*. Gent.

Bhadeshia, H.K.D.H. 1992. *Bainite in Steels*. Cambridge University Press.

Bhattacharya, D., N. Fonstein, O. Girina, I. Gupta, and O. Yakubovsky. 2003. "A Family of 590 MPa Advanced High Strength Steels with Various Microstructures." In *45th MWSP Conference*, 173–86.

Bleck, W., and S. Bruhl. 2008. "Bake-Hardening Effects in Advanced High Strength Steels." In *New Development on Metallurgy and Applications of High Strength Steels*. Buenos Aires, Argentina.

Bleck, W., A. Frehn, and J. Ohlert. 2001. "Niobium in Dual-Phase Steels and TRIP Steels." In , 727–52. Orlando.

Chatterjee, S., and H.K.D.H. Bhadeshia. 2006. "TRIP-Assisted Steels: Cracking of High-Carbon Martensite." *Material Science and Technology* 22 (6): 646–649.

Chen, H., H. Era, and M. Shimizu. 1989. "Effect of Phosphorous on the Formation of Retained Austenite and Mechanical Properties in Si-Containing Low-Carbon Steel Sheet." *Metallurgical Transactions A* 20A (3): 437–45.

Chiang, J., B. Lawrence, J.D. Boyd, and A.K. Pilkey. 2011. "Effect of Microstructure on Retained Austenite Stability and Work Hardening of TRIP Steels." *Material Science and Engineering A* 528: 4516–21.

Cornette, D., T. Hourman, O. Hudin, J.P. Laurent, and A. Reynaert. 2001. "High Strength Steels for Automotive Safety Parts." SAE International – 2001-01-0078.

DeMeyer, M., K. De Wit, and B.C. De Cooman. 2000. "The Baking Hardening Behavior of Electro-Galvanized Cold Rolled CMnSi and CMnAlSi TRIP Steels." *Steel Research* 71 (12): 511–18.

DeMeyer, M., D. Vanderschueren, and B.C. DeCooman. 1999. "The Influence of Al on the Properties of Cold-Rolled C-Mn-Si TRIP Steels." In *41st MSWP Conference*, 265–76.

DeMeyer, M., D. Vanderschueren, and B.C. De Cooman. 1999. "The Influence of the Substitution of Si by Al on the Properties of Cold rolled C-Mn-Si TRP Steels." *ISIJ International* 39 (8): 813–32.

Ehrhardt, B., T. Gerber, and T.W. Schaumann. 2004. "Approaches to Microstructural Design of TRIP and TRI-Aided Cold-Rolled High Strength Steel." In *AHSSS*, 39–50.

Engl, B., L. Kessler, F-J. Lenze, and T.W. Schaumann. 1998. "Recent Experience with the Application of TRIP and Other Advanced Multiphase Steels." In *IBEC 98*, 1–8.

Evans, P., L. Crawford, and A. Jones. 1997. "High Strength C-Mn Steels for Automotive Applications." *Ironmaking Steelmaking* 24 (5): 361.

Faral, O., and T. Hourman. 1999. "Influence of Continuous Annealing Conditions on Dual-Phase and TRIP Steels for Automotive Application." In 253–64.

Furukawa, T., H. Morikawa, H. Takechi, and K. Koyama. 1979. "Process Factors for Highly Ductile Dual-Phase Steels." In *Structure and Properties of Dual-Phase Steels*, 281–303. New Orleans, LA, USA

Galan, J., L. Samek, P. Verleysen, and et al. 2012. "Advanced High Strength Steels for Automotive Industry." *Revista De Metalurgia* 48 (2): 118–31.

Gallagher, M., J.G. Speer, and D. Matlock. 2003. "Effect of Annealing Temperature on Austenite Decomposition in a Si-Alloyed TRIP Steel." In *Symposium of Austenite Formation and Decomposition*, ISS, Chicago, IL, USA, 536–76.

Gallagher, M., J.G. Speer, D.K. Matlock, and N. Fonstein. 2002. "Microstructure Development in TRIP-Stet Steels Containing Si, Al. P." In *44 MSWP Conference*, 153–72.

Girault, E., A. Mertans, P. Jacques, Y. Houbaert, B. Verlinden, and J.-V.Y. Hambeeck. 2001. "Comparison of the Effect of Silicon and Aluminium on the Tensile Behavior of Multiphase TRIP-Assisted Steels." *Scripta Materialia* 44: 885–92.

Godet, S., C. Georges, and P.J. Jacques. 2003. "On the Austenite Retention in Low Alloy Steels." In *TMS-ISS Conference*, 523–36. Chicago.

Gomez, M., C. Garcia, and A. DeArdo. 2007. "Microstructural Evolution during Continuous Galvanizing and Final Mechanical Properties of High Al-Low Si TRIP Steels." In *MS&T'07*, 1–14.

Gomez, M., C.I. Garcia, and A.J. DeArdo. 2010. "The Role of New Ferrite on Retained Austenite Stabilization in Al-TRIP Steels." *ISIJ International* 50 (1): 139–46.

Gomez, M., C.I. Garcia, D.M. Haezenbrouck, and A.J. DeArdo. 2009. "Design of Composition in (Al/Si) Alloyed TRIP Steels." *ISIJ International* 49 (2): 302–11.

Haidemenopoulos, G., and A. Vasilakos. 1996. *Steel Research* 67 (11): 513–19.

Hance, B.M., T.M. Link, and D.P. Hoydick. 2003. "Bake Hardenability of Multiphase High Strength Sheet Steels." In *45 Conference MWSP, 195–206*.

Hashimoto, S., S. Ikeda, K.-I. Sugimoto, and S. Miyake. 2004. "Effect of Nb and Mo Additions to 0.2%C-1.5%Si-1.5%mn Steel on Mechanical Properties of Hot Rolled TRIP-Aided Steel Sheets." *ISIJ International* 44 (9): 1590–98.

Heller, T., I. Heckelmann, T. Gerber, and T.W. Schaumann. 2005. "Potential of Niobium in Sheet Steels for the Automotive Industry." In *Recent Advances of Niobium Containing Materials in Europe*, 21–28. Dusseldorf.

Hojo, T., and et al. 2013. "Hydrogen Embrittlement Properties of Cr Added Ultra-Strength TRIP-Aided Martensitic Steel." *CAMP-ISIJ* 26 (2): 624–27.

Hojo, T., K.-I. Sugimoto, Y. Mulai, and S. Ikeda. 2008. "Effect of Aluminium on Delayed Fracture Properties of Ultra-High Strength Low-Alloy TRIP-Aided Steels." *ISIJ International* 48 (6): 824–29.

Hosoya, Y. 1997. "Metallurgy of Continuously Annealed TRIP Steel Sheet." In 148–54. Turkey.

Hulka, K. 1999. "Relationship between Heat Treatment Conditions, Microstructure and Properties of Nioboum Microalloyed TRIP Steel." In *41st MSWP Conference*, XXXVII:67–77.

Imai, N., N. Komatsubara, and K. Kunishige. 1995. "Effect of Alloying Element and Microstructure on Mechanical Properties of Low-Alloy TRIP-Steels." *CAMP-ISIJ* 8: 572–75.

Itami, A., M. Takahashi, and K. Ushioda. 1995. "Plastic Stability of Retained Austenite in the Cold-Rolled 0.14C-1.9Si-1.7Mn Sheet Steel." *ISIJ International* 35 (9): 1121–27.

Iung, T., J. Drillet, A. Couturier, and C. Olier. 2002. "Detailed Study of the Transformation Mechanism in Ferrous TRIP Aided Steels." In 31–37. Gent, Belgium.

Jacques, P.J., X. Cornet, Ph. Harlet, J. Ladriere, and F. Delanney. 1998. "Enhancement of the Mechanical Properties of a Low-Carbon, Low-Si Steel by Formation of a Multiphased Microstructure Containing Retained Austenite." *Metallurgical and Materials Transactions A* 29A (9): 2383–93.

Jacques, P.J., E. Girault, Ph. Harlet, and F. Delanney. 2001. "The Development of Cold-Rolled TRIP-Assisted Multiphase Steels. Low Silicon TRIP-Assisted Multiphase Steels." *ISIJ International* 41 (9): 1061–67.

Jun, H.J., S.H. Park, S.D. Choi, and C.G. Park. 2004. "Decomposition of Retained Austenite during Coiling Process of Hot-Rolled TRIP Aided Steels." *Material Science and Engineering A* 379: 204–9.

Katayama, T., M. Takahashi, M. Usuda, and O. Akisie. 1992. "Study of Formable High Strength Steel Sheets for Automotive Panels." SAE International 920247.

Kim, S.I., Y.H. Jin, J.N. Kwak, and K.G. Chin. 2008. "Influence of Cooling Process after Hot Rolling on Mechanical Properties of Cold Rolled TRIP Steel." In *MS&T'08* , 1784–1802. Pittsburgh.

Kim, S.-J., C.G. Lee, T.-H. Lee, and C.-S. Oh. 2003. "Effect of Cu, Cr and Ni on Mechanical Properties of 0.15 Wt.% C TRIP-Aided Cold Rolled Steels." *Scripta Materialia* 48: 539–44.

Krauss, G. 2005. *Steels: Processing, Structure and Performance.* TMS.

Laquerbe, L., J. Neutjens, Ph. Harlet, F. Caroff, and P. Cantinieaux. 1999. "New Processing Route for the Production of Silicon-Free TRIP-Assisted Cold-Rolled and Galvanized Steels." In *Proceedings of 41st MWSP Conference ISS*, XXVII:89–99.

Lee, K., Y.-R. Im, and K.G. Chin. 2008. "Effect of Carbon Content on the Microstructure and the Transformation Kinetics of Super Bainitic TRIP Steels." In *MS&T*, 1785–93. Pittsburgh.

Lee, S.-J., and Y.-K. Lee. 2005. "Effect of Austenite Grain Size on Martensite Transformation of the Low Alloy Steel." *Material Science Forum* 475–479: 3169–72.

Matsumura O., Y. Sakuma, and H Takechi. 1987. "Enhancement of Elongation by Retained Austenite in Intercritical Annealed 0.4C-1.5Si-0.8Mn." *Trans ISIJ* 27: 570–79.

Matsumura, O., Y. Sakuma, Y. Ishii, and J. Zhao. 1992. "Effect of Retained Austenite on Formability of High Strength Sheet Steels." *ISIJ International* 32 (10): 1110–16.

———. 1992. "Retained Austenite in 0.4C-Si-1.2 Mn Steel Sheet Heated and Austempered." *ISIJ International* 32 (9): 1014–20.

Minote, T., S. Torizuka, A. Ogawa, and M. Nijura. 1996. "Modeling of Transformation Behavior and Compositional Partitioning in TRIP Steel." *ISIJ International* 36 (2): 201–7.

Mohrbacher, H. 2007. "Microalloying with Niobium in TRIP Steels." In Proc, of Conf. *Metal 2007*, 1–8. Hradec nad Moravici, Cz.

Ohlert, J., W. Bleck, and K. Hulka. 2002. "Control of Microstructure in TRIP Steels by Niobium." In *International Conference: TRIP Assisted High Strength Ferrous Alloys*, 199–206. Gent.

Olson, G.B., and M. Cohen. 1976. "A General Mechanism of Martensite Nucleation." *Metallurgical Transactions A* 7A: 1897–1904.

Park, K.K., S.T. Oh, S.M. Baeck, and D.I. Kim. 2002. "In-Situ Deformation Behavior of Retained Austenite on TRIP Steel." In *Materials Science Forum*, 408–412:571–76.

Pichler, A., P. Stiaszny, R. Potzinger, R. Tikal, and E. Werner. 1998. "TRIP Steels with Reduced Si Content." In *Proceedings of 40th MWSP Conference ISS*, 259–74.

Pichler, A., S. Traint, H. Pauli, H. Mildnier, and et al. 2001. "Processing and Properties of Cold-Rolled TRIP Steels." In *43rd MSWP Conference -ISS*, 411–34.

Rigsbee, J.M., and P.J. VanderArend. 1977. "Laboratory Studies of Microstructure and Structure-Properties Relationship in 'Dual-Phase' HSLA Steels." In *Formable HSLA and Dual-Phase Steels, Metallurgical Society of AIME*, 56–86.

Sakuma, Y. 2004. "Recent Achievement in Manufacturing and Application of High-Strength Steel Sheets for Automotive Body Structure." In *Proceedings of AHSS-2004*, 11–18.

Sakuma, Y., A. Itami, O. Kawano, and N. Kimura. 1995. "Next Generation High Strength Sheet Steel Utilizing Transformation Induced Plasticity." *Nippon Steel Technical Report*, no. 64: 20–25.

Sakuma, Y., D. Matlock, and G. Krauss 1992. "Intercritically Annealed and Isothermally Transformed 0.15 Pct C Steel Containing 1.2Si-1.5MN and 4 %Ni." *Metallurgical Transactions A* 23A: 1221–32.

———. 1993. "Effect of Molybdenum on Microstructure and Mechanical Properties of Intercritically Annealed and Isothermally Transformed Low Carbon Steel." *Material Science and Technology* 9 (4): 718–24.

Sakuma, Y., O. Matsumura, and H. Takechi. 1991. "Mechanical Properties and Retained Austenite in Intercritically Heat-Treated Bainite-Transformed Steel and Their Variation with Si and Mn Additions." *Metallurgical Transactions A* 22A (2): 489–99.

Samajdar, I., E. Girault, B. Verlinden, E. Aernoudt, and J. Van Humbeeck. 1998. "Transformation during Intercritical Annealing of TRIP-assisted Steel." *Transactions of ISIJ International* 38 (9): 998–1006.

Scott, C., D. Maugus, P. Barges, and M. Gouné. 2004. "Microalloying with Vanadium for Improved Cold Rolled TRIP Steels." In *AHSS for Automotive Application*, 181–181. Winter Park, CO.

"Steel Sheet with Well-Balanced Strength and Ductility, FORD/NSC Technical Meeting." 1989. NSC.

Steven, W., and A.G. Haynes. 1956. "The Temperature of Formation of Martensite and Bainite in Low-Alloy Steels." *Journal of the Iron and Steel Institute* 183 (8): 349–59.

Stretcher, A., J.G. Speer, and D.K. Matlock. 2002. "Forming Response of Retained Austenite in C-Si-Mn High Strength TRIP Sheet Steel." *Steel Research* 73 (6+7): 287–93.

Sugimoto, K.-I., T. Iida, J. Sakaguchi, and T. Kashima. 2000. "Retained Austenite Characteristics and Tensile Properties in a TRIP Type Bainitic Sheet Steel." *ISIJ International* 40 (9): 902–8.

Sugimoto, K.-I., M. Murata, T. Muramutso, and Y. Mukai. 2007. "Formability of C-Si-Mn-Al-Nb-Mo Ultra High-Strength TRI-Aided Sheet Steels." *ISIJ International* 47 (9): 1357–82.

Sugimoto, K.-I., Y. Shimizu, and J. Sakaguchi. 1998a. "Microstructure and Formability of High Strength TRIP Aided Bainitic Sheet Steels." In 275–81.

Sugimoto, K.-i., M. Kobayashi, K. Inoue, X. Sun, and T. Soshiroda. 1998b. "Fatigue Strength of TRIP-Aided Bainitic Sheet Steels." *Tetsu-to-Hagane* 84 (8): 559–65.

Sugimoto, K.-i., K. Nakano, S.-M. Song, and T. Kashima. 2002. "Retained Austenite Characteristics and Stretch-Flangeability of High-Strength Low-Alloy TRIP Type Bainitic Sheet Steels." *ISIJ International* 42 (4): 450–55.

Suh, D.-W., S.-J. Park, and S.-J. Kim. 2008. "Influence of Cr and Ni on Microstructural Evolution during Heat Treatment of Low Carbon Transformation Induced Plasticity Steels." *Metallurgical and Materials Transactions* 39A (9): 2015–19.

Tomita, Y., and K. Morioka. 1997. "Effect of Microstructure on Transformation-Induced Plasticity of Silicon-Containing Low-Alloy Steel." *Materials Characterization* 38: 243–50.

Traint, S., A. Pichler, M. Blaimschein, B. Bohler, C. Krempazsky, and E. Werner. 2004. "Alloy Design, Processing and Properties of TRIP Steels: A Critical Comparison." In *AHSS*, 79–98.

Traint, S., A. Pichler, K. Spiradek-Hahn, K. Hulka, and E. Werner. 2003. "The Influence of Nb on the Phase Transformation and Mechanical Properties in Al- and Si-Alloyed TRIP Steels." In *Symposium of Austenite Formation and Decomposition*, 577–94. Chicago, IL, USA.

Traint, S., A. Pichler, P. Stiaszny, and E. Werner. 2001. "Mechanical Behavior and Phase Transformation of an Aluminium Alloyed TRIP Steel." In *43rd MSWP Conference -ISS*.

Tsukatani, I., S. Hashimoto, and T. Inque. 1991. "Effect of Silicon and Manganese Additions on Mechanical Properties of High Strength Sheet Steel Containing Retained Austenite." *ISIJ International* 31 (9): 992–1000.

Ushioda, K., and N. Yoshinaga. 1996. "Recent Progress in Physical Metallurgy of Cold Rolled Sheet Steels." In *Thermomechanical Processing in Theory, Modeling & Practice*, 162–88. Stockholm, Sweden.

Vasilakos, A., K. Papamantellos, G. Haidemenopoulos, and W. Bleck. 1999. "Experimental Determination of the Stability of Retained Austenite in Low Alloy TRIP Steels." *Steel Research* 70 (11): 466–471

Vrieze, J. 1999. "Annealing Treatment for Producing Cold-Rolled Dual-Phase and TRIP-Steels for Automotive Applications." In *Proceedings of 41th MWSP Conference, ISS*, 277–94.

Wang, J., and S. Van der Zwaag. 2001. "Stabilization Mechanisms of Retained Austenite in Transformation Induced Plasticity Steels." *Metallurgical and Materials Transactions A: Physical Metallurgy and Materials Science* 32A (6): 1521–39.

Yakubovsky, O., N. Fonstein, and D. Bhattacharya. 2002. "Stress-Strain Behavior and Bake-Hardening of TRIP and TRIP-aided Multiphase Steels." In 263–70. Gent.

Yakubovsky, O., N. Fonstein, and D. Bhattacharya. 2004. "Effect of Composition and Microstructure on the Stress-Stress Behavior of TRIP and Dual-Phase Steels." In *International Conference on Advanced High Strength Sheet Steels for Automotive Applications*, 295–306. Winter Park, CO.

Yang, H.-S., and H.K.D.H. Bhadeshia. 2009. "Austenite Grain Size and the Martensite-Start Temperature." *Scripta Materialia* 60: 493–95.

Yokoi, T., K. Kawasaki, M. Takahashi, K. Koyama, and M. Mizui. 1996. "Fatigue Properties of High Strength Steels Containing Retained Austenite." *Technical notes/JSAE REview* 17: 210–12.

Zackey, V., E. Parker, D. Fahr, and R. Busch. 1967. "The Enhancement of Ductility in High-Strength Steels." *Transactions of ASM* 60: 252–59.

Zaefferer, S., J. Ohlert, and W. Bleck. 2004. "A Study of Microstructure, Transformation Mechanisms and Correlation between Microstructure and Mechanical Properties of a Low-Alloyed TRIP-Steel." *Acta Materaila* 52: 2765–78.

Zarei-Hanzaki, A., P.D. Hidgson, and S. Yue. 1995. *ISIJ International* 35 (3): 324.

Zhao, L., J. Sietsma, and S. van der Zwaag. 1999. "Phase Transformation and Microstructure Evolution in Al-Containing TRIP Steels." In 77–82.

Zhu, K., H. Chen, J.-Ph. Masse, O. Bouaziz, and G. Gachet. 2013. "The Effect of Prior Ferrite Formation on Bainite and Martensite Transformation Kinetics in Advanced High Strength Steel." *Acta Materalia* 61: 6025–36.

Chapter 6
Complex Phase Steels

Contents

6.1 Introduction

Wider usage of AHSS requires application of high-strength steels for forming complicated parts. Consequently, AHSS steels undergo bending at lower bend radius, while part designs include multiple flange stretching so that the localized stress–strain behavior of steel becomes very important.

The general trend of increasing tensile strength of automotive steels has revealed some shortcomings of AHSS. Such microstructure designs as DP, TRIP steels, and TRIP-aided DP steels can achieve very high combination of strength and elongation; however, the higher the strength, the higher the sensitivity to edge fracture and the lower the hole expansion, which hinders efficient part design and stamping technology.

As shown earlier, the balance of high strength and elongation of DP and TRIP steels is to a considerable extent based on high strain hardening rate, which is the main factor of damaging local area near edges during punching holes. The emerged concept of "Complex phase" (CP) steels containing the mixture of ferrite, bainite, martensite, and precipitates is intended to address this problem specifically for parts

© Springer International Publishing Switzerland 2015   241
N. Fonstein, *Advanced High Strength Sheet Steels*,
DOI 10.1007/978-3-319-19165-2_6

that require higher flangeability, sometimes sacrificing high elongation that can be achieved on account of high strain hardening rates.

In parallel with that, the CP microstructure design is expected to improve another drawback of existing AHSS such as low YS/TS (yield strength/tensile strength) ratio because safety critical parts should resist intrusion and higher YS of material is crucial.

The developed complex phase steels contain the same alloying elements as DP or TRIP steels, but the combinations of modified chemistry and thermal cycles focus on obtaining certain amounts of bainite; small additions of microalloying are usually present to increase yield strength by precipitation hardening.

This chapter is intended to summarize the existing data on the subject and describes the approaches to achieve better steel resistance to localized deformation and stress concentration that are the major mechanical factors responsible for edge cracking and hole expansion (HE).

6.2 Effect of Bainite on Hole Expansion of AHSS and "Complex Phase" Steels Concept

As shown in several publications, there exists an inverse correlation between elongation of low YS/TS ratio type DP steel and flangeability (hole expansion) (Matsuoka et al. 2007).

The usual recommendations to improve flangeability of DP steels include all possible means to minimize the difference in strength between martensite and ferritic matrix:

- Increase tempering temperature to soften martensite
- Reduce carbon content in martensite
- General microstructure refinement, e.g., by microalloying with Nb (Lee et al. 2012)
- Eliminate or reduce the soft ferrite fraction or increase its strength by Si, precipitates, or grain refinement.

In particular, Hasegawa et al. pointed out the clear correlation of hole expansion and difference in hardness between ferrite and martensite as presented in Fig. 6.1 (Hasegawa et al. 2004).

Observed advantages of ferrite–bainite microstructure stem also from smaller differences in strength properties between phases (compared to ferrite-martensite) and lower internal stresses. These factors can play a key role for improvement of hole expansion, as shown in many publications (Kagechi 2003; Sakata et al. 2003; Shimuzi et al. 2004).

Along the same lines, a number of publications demonstrated the advantages of partial or full replacement of martensite in DP steels by bainite as a less hard microstructure constituent. Shirasawa et al. clarify this statement by pointing at the

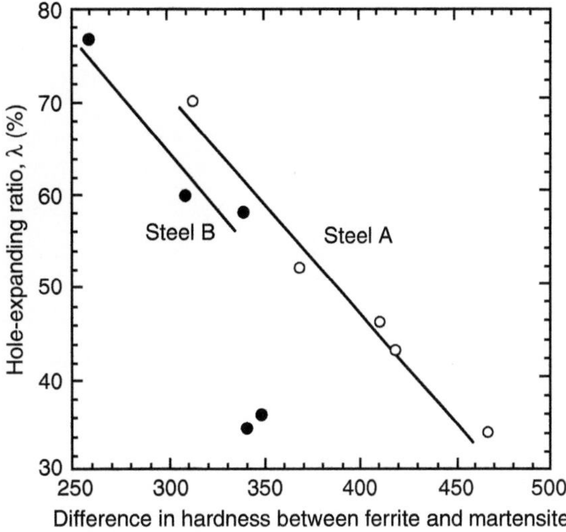

Fig. 6.1 Relationship between the difference in hardness between ferrite and martensite and the hole expansion (Hasegawa et al. 2004)

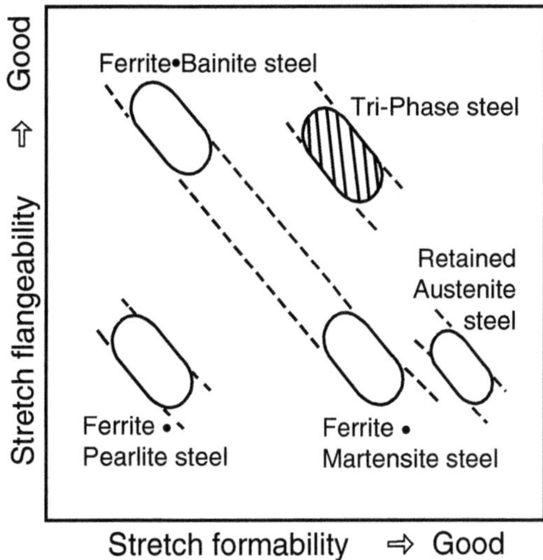

Fig. 6.2 Relationship between hole expansion/stretch flangeability and elongation/stretch formability for different types of microstructure (Shirasawa et al. 1993)

advantages of specific types of ferrite–bainite or multiphase microstructures (Shirasawa et al. 1993) (Fig. 6.2).

Discovering the advantages of partial or full replacement of martensite with bainite resulted in creation of requirements to the so-called complex phase steels with significant fraction of bainite microstructure constituent (Bode et al. 2004; Karelova et al. 2009). This was reflected in corresponding standards with requirements for guaranteed elevated values of hole expansion.

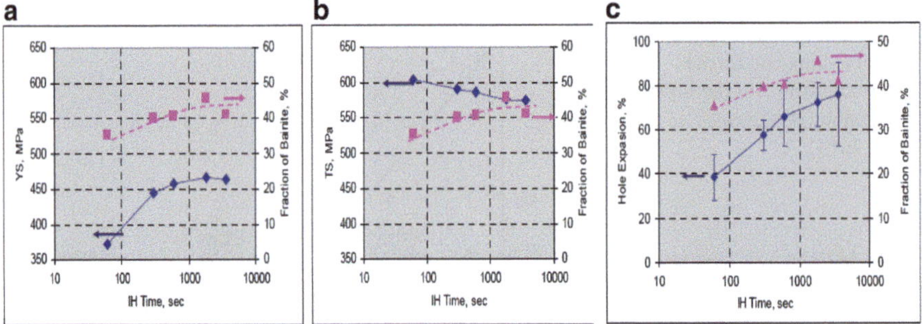

Fig. 6.3 Effect of IH time on tensile properties, bainite fraction, and hole expansion (Fonstein et al. 2011)

Experience with hot-rolled steels of high flangeability (usually called as multiphase steels) with TS > 800 MPa is also a useful proof that bainitic matrix or bainite–ferrite mixture are the best microstructures for AHSS with high hole expansion.

Aiming to get quantitative data regarding the effects of replacement of martensite with bainite on the mechanical properties including hole expansion, Fonstein et al. studied the microstructure of mill cold-rolled 1.5 mm thick 0.1 C–1.8 Mn–0.3 Si (weight%) steel strips after two heat treatment cycles (Fonstein et al. 2011).

The selected annealing cycle aimed at obtaining various amounts of bainite of the same morphology in the final multiphase microstructure that was achieved by holding isothermally the preliminary austenitized samples at the same temperature (400 °C) for various periods of time ranging from 60 to 3600 s. The increase in isothermal holding (IH) time resulted in gradual growth of bainite fraction at the expense of martensite.

As shown in Fig. 6.3a, b, YS gradually increases, and tensile strength decreases as the amount of bainite grows. As a consequence, YS/TS ratio changes from 0.6 to 0.8 with IH time increasing from 60 to 600 s. Starting from 300 s of holding, with ~40 % of bainite and only ~5 % of martensite present in multiphase structure, the yield point elongation (YPE) appears.

Based on these results, it was concluded that the replacement of 10 % of martensite with bainite induces a 40 MPa loss in TS, i.e., strengthening effect of bainite is 4 MPa/% less than that of martensite. Figure 6.3c shows the effect of bainite on hole expansion (HE). The above-mentioned 40 MPa loss in tensile strength due to gradual replacement of martensite with bainite is accompanied by almost double increase in HE.

Growing demand for steels with higher hole expansion has motivated numerous studies of mechanical parameters responsible for resistance to localized deformation. For example, Fig. 6.4 illustrates very strong correlation between hole expansion (here TSxλ) and reduction of area (a) or with elongation at tensile testing of short gauge length (5 mm) specimens (Sugimoto et al. 2000). Similar correlation

Fig. 6.4 Correlation
between hole expansion and
reduction of area (**a**) and
with elongation measured
using tensile specimens
with 5 mm gauge length (**b**)
(Sugimoto et al. 2000)

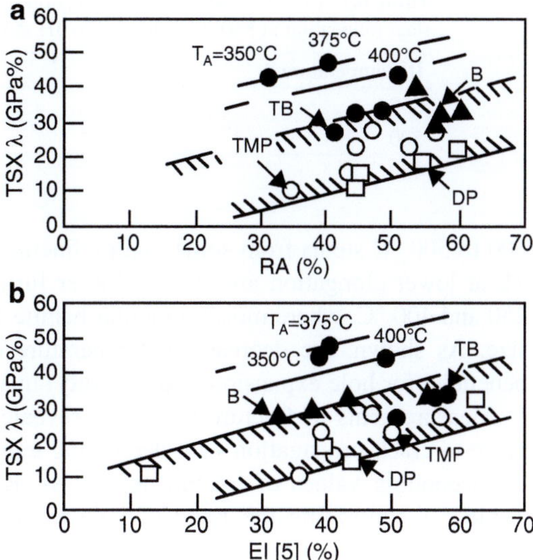

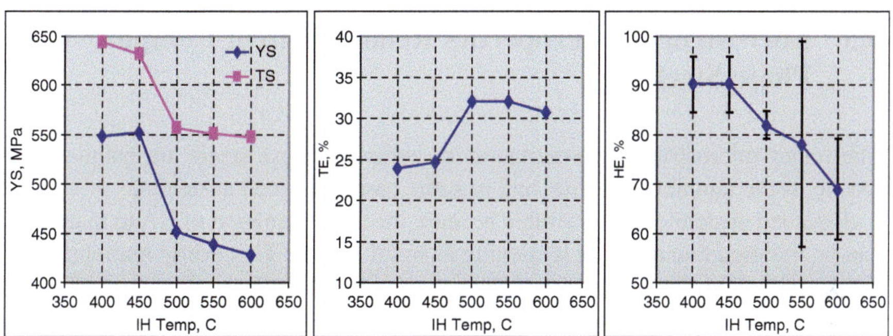

Fig. 6.5 Effect of IH temperature on tensile properties and hole expansion

was found earlier (Bansal 1977). These data emphasize the role of post-necking deformation that is typically higher at lower strain hardening.

Careful investigations of commercial steels did not reveal any direct dependences of hole expansion on the presence of such surface imperfections as micro-tearing, micro-cracks, and/or inclusions that could serve as crack initiators. Therefore, it was reasonably assumed that hole expansion (flangeability) of AHSS should be controlled not by crack initiation but mostly by the resistance to crack propagation that can be sensitive to morphology of strengthening phase (Fonstein et al. 2011).

The effect of bainite morphology on tensile properties and hole expansion is described in Fig. 6.5. Small increase in YS and TS at lowering IH temperature from

Table 6.1 Effect of bainite morphology on fracture toughness of multiphase steel (annealing at 880 °C, followed by IH at various temperatures for 150 s)

IH temperature	K_{EE} MPa√m
550	167
500	188
450	227

600 to 500 °C stems from some grain refinement. Significant increases in YS and TS at lower elongation are observed after further lowering of IH temperature to 450 and 400 °C, where mostly granular bainite forms with significantly finer grain size. As shown, the decrease in temperature of bainite formation is obviously beneficial for hole expansion (flangeability) in spite of 70 MPa increase in TS.

Measurements of fracture toughness revealed significant changes in the resistance to crack propagation with decreasing temperature of bainite formation. The corresponding values of fracture toughness, K_{EE} (MPa m$^{-1/2}$), are presented in Table 6.1 confirming the trend of hole expansion improvement with increased resistance to crack propagation.

6.3 Microstructure–Properties Relationship of Complex Phase Steels

The major microstructure constituents of complex phase steels are bainite, martensite, some amount of ferrite, and in some cases, retained austenite.

Retained austenite is undesirable because the transformation of RA to martensite can be initiated due to strain hardening at small strains. This brittle martensite can crack and potentially damage the edges during hole punching.

As presented above, the replacement of martensite with bainite improves hole expansion.

Conversely, with increasing volume fraction of martensite in microstructure, the hole expansion gradually decreases until martensite becomes the dominant phase with the volume fraction over 80 % (Fig. 6.6). This result confirms the importance of microstructure uniformity to achieve better flangeability (Ryde et al. 2012).

Similar effects were found with increasing amount of ferrite (Fig. 6.7).

These results demonstrate the critical role of microstructure homogeneity in achieving higher hole expansion. Therefore, the processing parameters along with appropriate selection of chemical composition should aim at reducing the amounts of martensite and ferrite as much as possible, as well as at eliminating the retained austenite.

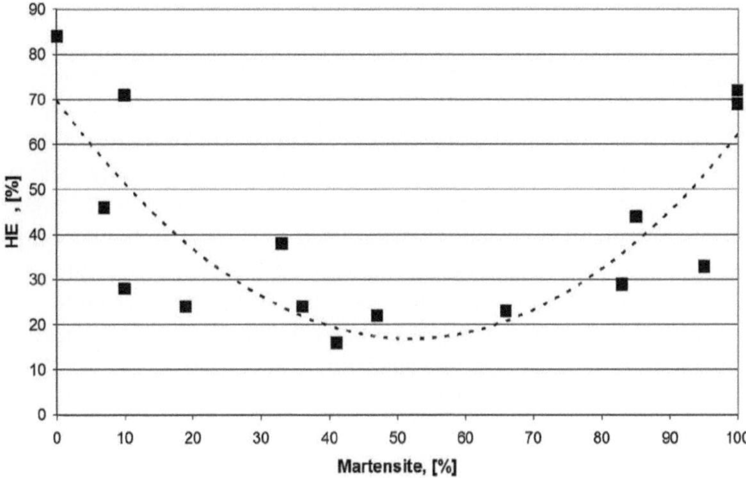

Fig. 6.6 Effect of martensite volume fraction on hole expansion (Ryde et al. 2012)

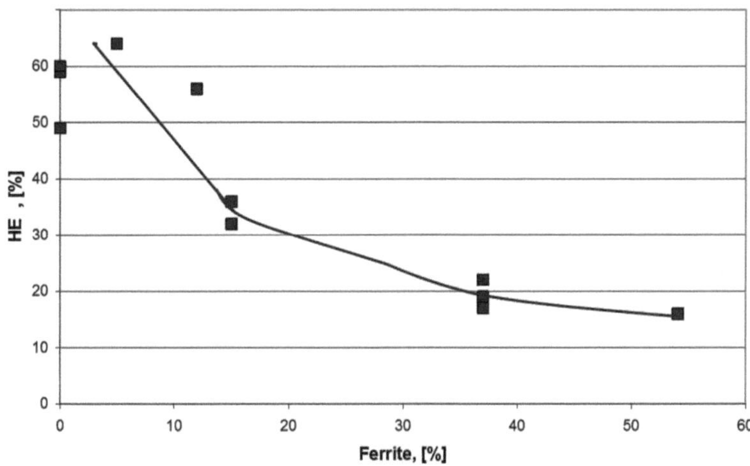

Fig. 6.7 Effect of ferrite volume fraction on hole expansion of CP steels (Ryde et al. 2012)

6.4   Adaptation of Processing Parameters to Obtain Bainite Fraction

Utilization of special cooling profile after hot rolling in combination with proper chemical composition of steel can ensure prevailing of bainite constituent in the microstructure.

Annealing of bare steels using austempering cycle with isothermal holding in bainitic region can also produce some amount of bainite constituent. However, if annealing begins with an intercritical temperature, at least the initial fraction of ferrite is unavoidable, as well as the enrichment of austenite with carbon. This can result in substantial portion of retained austenite in the final microstructure that can prevent high hole expansion owing to microstructure nonuniformity. Annealing at supercritical temperatures significantly reduces the hardenability of lower carbon austenite, and it therefore requires special ways (e.g., high initial cooling rate or alloying) to avoid ferrite formation during cooling to isothermal holding temperature. Some continuous annealing lines, such as CAPL, allow for up to 6–10 min isothermal holding that, together with properly selected chemistry, makes it possible to minimize the amount of RA and probably render the best way of obtaining the microstructure with dominant bainitic matrix in bare steels, provided that the initial high cooling rates are available.

With regard to cold-rolled galvannealed/galvanized steels, two additional challenges are to be faced. The first is related to the limitations in the annealing parameters: typical initial slow cooling that can facilitate the formation of undesirable ferrite, as well as the limited time for isothermal holding in equalizing zones at the relatively high temperatures of Zn melting in bainite region. This requires to utilize all the means to accelerate the bainitic reaction (low carbon, minimum alloying). The second challenge is to achieve the necessary strength of bainite that can call for higher carbon content and higher alloying.

To find applicable compromise to meet these challenges, it is useful to carefully summarize in detail the effects of chemical composition, initial microstructure, and annealing parameters on bainite transformation.

6.4.1 Effect of Chemical Composition on Phase Transformation of Low-Carbon Steels During Isothermal Holding and Continuous Cooling

If the initial cooling rate after *full austenitization* is high enough to avoid ferrite formation but is lower than the critical cooling rate for martensite transformation, low-carbon steels usually undergo the bainite transformation. Knowing the bainite transformation start temperature (B_s) is important for appropriate selection of isothermal holding temperature in austempering processing. The B_s temperature, like the Ms and Fs temperatures, depends on chemical composition. Several empirical equations listed in Table 6.2 show the effect of alloying elements on the B_s temperature. In low-alloy steels, carbon has the strongest effect on the B_s temperature, which is followed by Mn (Steven and Haynes 1956; Bodnar et al. 1989).

It is difficult to obtain a fully bainitic microstructure in low-alloy steels because of the proximity of the temperature range for bainite transformation to that of martensite transformation. Hence it is desirable to separate the bainite and

Table 6.2 Empirical equations for B_s temperature (input number in wt%)

No.	Equation	References
1	B_s (C) = 830–270 C–90 Mn–37 Ni–70 Cr–83 Mo	Steven and Haynes (1956)
2	B_s (C) = 630–45 Mn–40 V–35 Si–30 Cr–25 Mo–20 Ni–15 W	Zhao et al. (2001)
3	B_s (C) = 637–58 C–35 Mn–15 Ni–34 Cr–41 Mo	Li et al. (1998)
4	B_s (C) = 844–597 C–63 Mn–16 Ni–70 Cr	Bodnar et al. (1989)

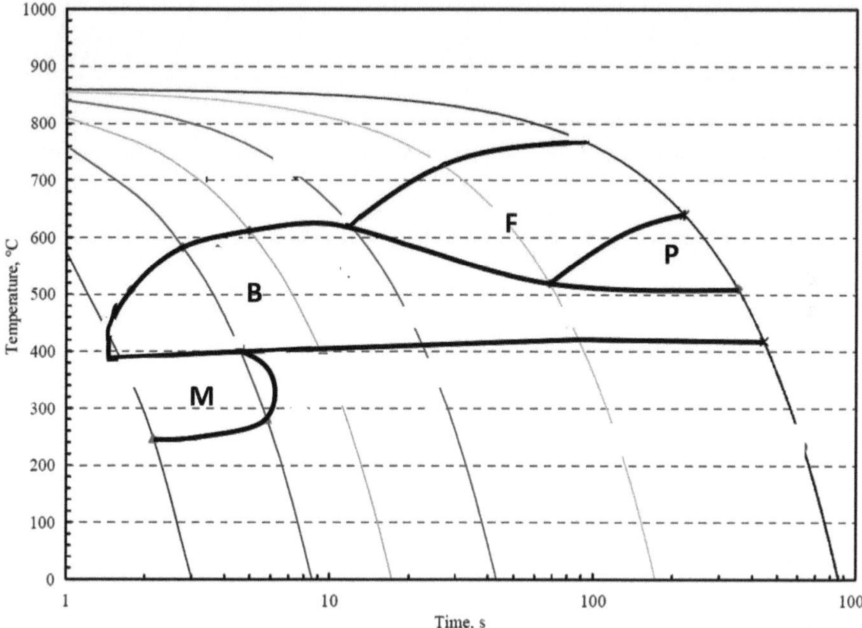

Fig. 6.8 Ideal CCT diagrams indicating the ranges of cooling rates optimal for producing complex phase steels—original

martensite regions as much as possible. During continuous cooling in CGL lines, the best type of CCT diagrams, suitable for obtaining preferably bainite in a wide range of cooling rates with minimum amount of martensite, that will form after galvanizing at final cooling and will be therefore untempered, is presented in Fig. 6.8.

Reports on several comprehensive studies of the effects of alloying elements on bainite transformation of low-alloy steels can be found in the literature, e.g., (Cias 1973; Wang et al. 2000; Mesplont et al. 2001; Mesplont 2002), as well as the reports on several cooperative European projects aiming at optimization of bainite containing steels and CP steels in particular ("Bainhard Report" 2012; *Cold-Rolled Complex Phase (CP) Steel Grades with Optimized Bendability, Stretch*

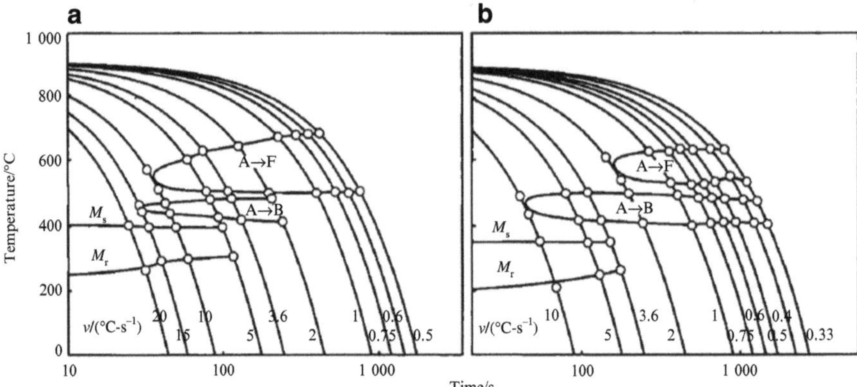

Fig. 6.9 CCT diagrams for steels of 1.4–2.3 Mn and 0.5 Si with 0.1 C (**a**) and 0.17 C (**b**) (Xu et al. 2010)

Flangeability and Anisotropy (CP Steels) 2012). Some of the important conclusions of these studies are:

Increase in C content increases hardenability of steel and decreases B_s temperature, but facilitates excessive martensite, so CP steels typically do not contain more than 0.1–0.17%C. At the same time, as shown by Xu et al., the increase in carbon content from 0.10 to 0.17 in the steel with 1.4–2.3 Mn and 0.5 Si allows for suppressing ferrite formation more significantly than facilitating martensite transformation (Xu et al. 2010) (Fig. 6.9).

Alloying additions of Mn or Cr delay both pearlite and bainite transformation but result in slightly higher proportion of the bainitic transformation, which may occur over wider range of cooling rates. The concentration of Mn must be high enough (around 1.6–2 %) to avoid ferrite formation. On the other hand, Mn increases incubation time and decreases the temperature of bainite transformation.

As shown in Fig. 6.10, the retarding effect of Cr on isothermal bainite transformation in 0.2 C–1.5 Mn–Cr steel at 480 °C is very pronounced (Quidort and Brechet 2002).

Molybdenum substantially delays the decomposition of austenite in the upper temperature range and extends the range of cooling rates for bainite transformation (Fig. 6.11) (Dacker et al. 2012).

It is known that additions of boron markedly retard the ferritic reaction. The effect of B stems from preferential segregation of B to the prior austenite boundaries. This permits the bainitic reaction in low-carbon austenite at shorter times.

Introduction of molybdenum into a boron-bearing steel strongly delays the decomposition of austenite into ferrite and pearlite and vastly expands the range of cooling rates for the formation of finer and stronger bainitic microstructures. B + Mo effect has been clearly observed in 0.17 wt% C–1.6 wt% Mn steel containing 25 ppm solute B and 0.2 wt% Mo.

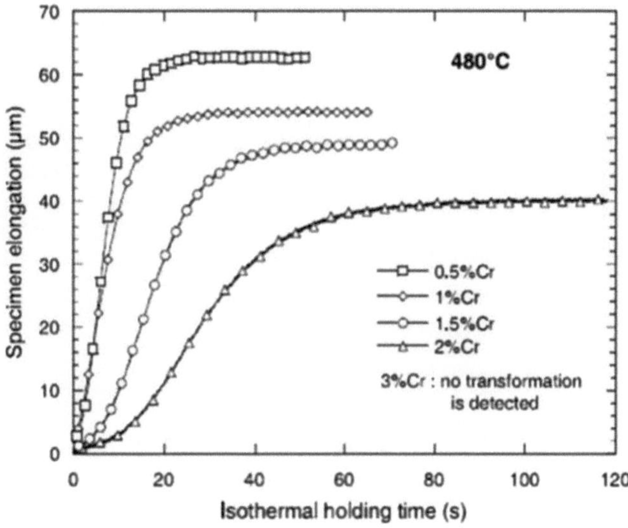

Fig. 6.10 The effect of Cr on isothermal bainite transformation in 0.2 C–1.5 Mn–Cr steel (Quidort and Brechet 2002) and CCT diagram of 0.2 C–1.5 Mn steel (Quidort and Brechet 2002)

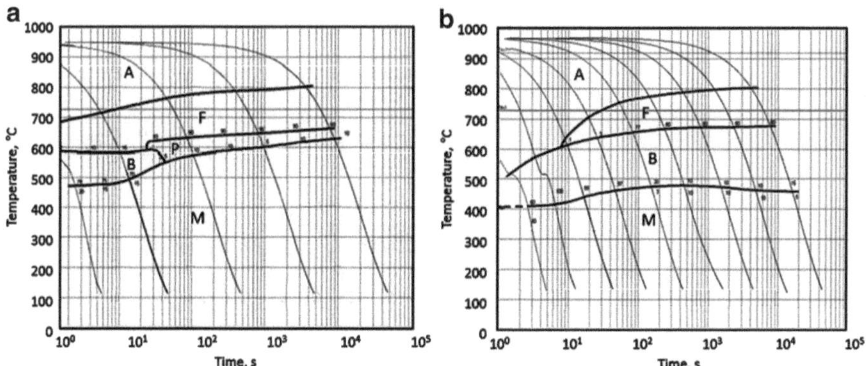

Fig. 6.11 The CCT diagrams for 0.05 C, 1.5 Mn steel without (*left*) and with (*right*) addition of 0.5 % Mo (Dacker et al. 2012)

CCT diagram of Fig. 6.12 shows that fully bainitic microstructure can be obtained within a wide range of cooling rates.

As a result of Mo addition, the critical cooling rate for bainite transformation decreases, the bainitic region is expanded dramatically, the ferrite and pearlite transformations are greatly suppressed, and the bainite start temperature increases at low cooling rates and decreases at high cooling rates, although the formation of martensite can be avoided completely at cooling rates below 20 °C/s.

Sometimes CP steels contain certain amounts of Si for ferrite strengthening. The consumption of carbon by austenite in front of bainitic ferrite is an essential step

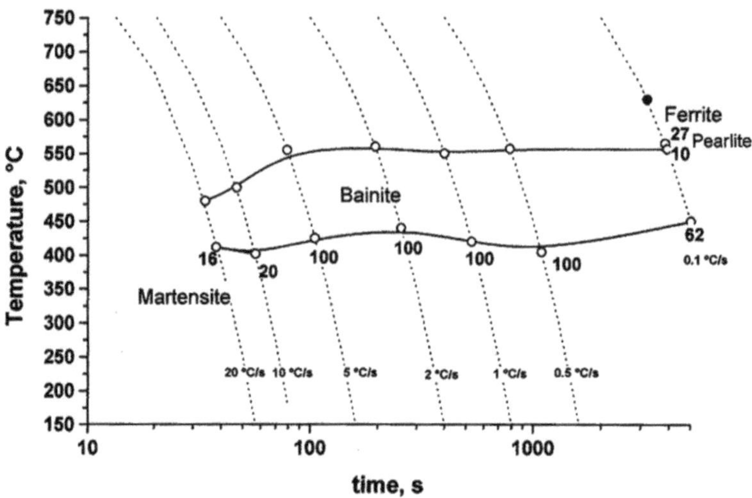

Fig. 6.12 CCT diagram of 0.17 C–1.6 Mn–0.2 Mo–0.0025 B–0.02 Ti steel (Mesplont 2002)

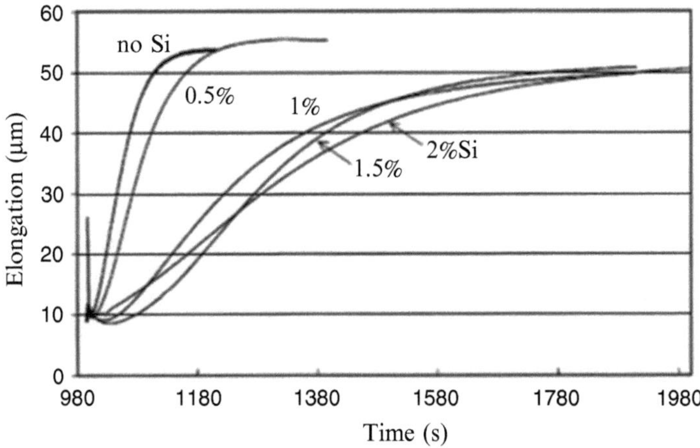

Fig. 6.13 Effect of Si on bainite transformation (dilation elongation) in 0.5 C–5 Ni–Si steel at 400 °C (Quidort and Brechet 2002)

that precedes further progress of bainite transformation. Therefore, the addition of Si and Al, which retard the formation of cementite, should slow down the overall bainite transformation (Quidort and Brechet 2001), when sufficiently high Si and Al can completely cease the formation of cementite producing the so-called Carbide-Free-Bainite discussed later in Chap. 8.

Quidort studied the effect of Si on isothermal bainite transformation in 0.5 C–5 Ni–Si steel. As shown in Fig. 6.13, Si significantly retards the formation of carbides and bainite (Quidort and Brechet 2002).

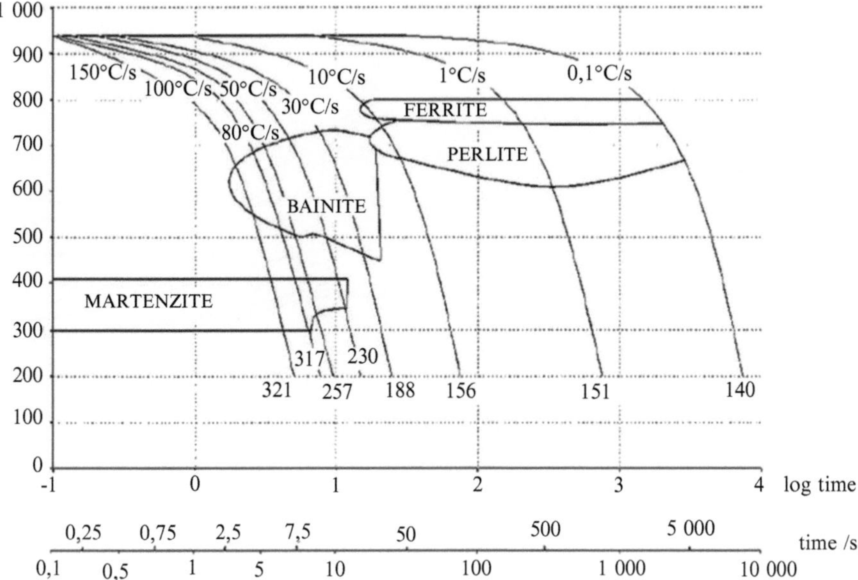

Fig. 6.14 CCT diagram for 0.08 C–1.5 Mn–0.4 Si–0.3 Cr–0.2 Ni–0.1 Ti–0.2 Cu steel (Kawalek et al. 2013)

As shown in the study of the combined effects of Mn and Si (Liu and Zhang 1990), Si enhances the segregation of Mn to austenite grain boundaries and inhibits the Fe_3C precipitation, thereby increasing the carbon content in untransformed austenite and hence suppressing the nucleation of bainitic ferrite.

Therefore, silicon content is typically limited to 0.25 wt% to maximize the incubation time for ferrite and minimize it for bainite.

There are numerous confirmations that Nb retards the formation of proeutectoid ferrite, although explanations can be different, whether it is related to Nb segregations on interphase boundaries or solute drag effect. According to Rees et al., the effect of Nb on bainite formation is due to precipitates. These authors showed that precipitates strongly influenced the critical cooling rates defining boundaries for specific transformations. In accordance with observed data, more precipitates (after low heating at 900 °C and 950 °C for 460 s) accelerated the bainite formation as compared with steel with less precipitates or with Nb-free steel. The reason was not explained, but probably the effect was due to increasing number of nucleation sites (Rees et al. 1995).

Keeping in mind different processing parameters available for specific equipment, different authors demonstrate different CCT, which indicate wider or narrower processing windows where dominant bainitic microstructure can be obtained. This, for example, is depicted in Fig. 6.14 for 0.08 C–1.5 Mn–0.4 Si–0.3 Cr–0.2 Ni–0.1 Ti–0.2 Cu steel (Kawalek et al. 2013).

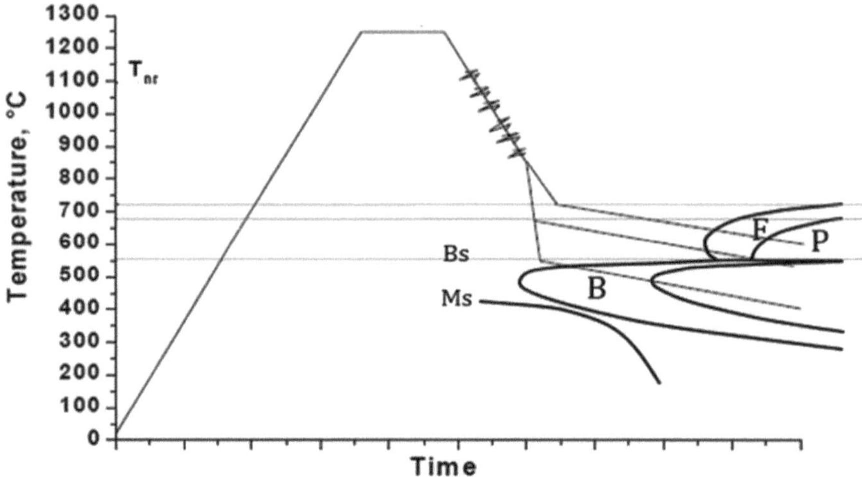

Fig. 6.15 Scheme illustrating the hot rolling parameters to obtain complex phase structure—original

6.4.2 Effect of Processing Parameters

Hot-rolled CP (or MP) steels are produced using thermomechanical processing focusing on prevention of ferrite formation during initial cooling. The most important parameter is the coiling temperature (CT) that determines the type of transformation and the morphology of products. As shown it should be higher than Ms temperature (Fig. 6.15).

The annealing cycle for bare steels is based on using continuous annealing line thermal cycle with direct transfer to overaging zone with the temperature in bainite region.

As mentioned above, for both CAL and CGL, it is desirable to suppress the formation of ferrite during initial cooling not only to prevent the development of microstructure nonuniformity but also to avoid enrichment of austenite with carbon that would delay the bainite transformation and increase the amount of martensite. Minimizing the carbon enrichment of austenite raises the Ms temperature that facilitates self-tempering of martensite formed during final cooling, as well as minimizes the amount of retained austenite.

Decreasing the austenitization temperature (at the same grain size) before fast cooling to the temperatures of isothermal holding appeared to reduce the time needed to initiate transformation, which was explained by reducing the amount of nonequilibrium segregation of Mn to austenite grain boundaries (Liu and Zhang 1990).

Since the selection of cooling rate is limited by the equipment configuration, the design of chemical composition should aim at expanding the bainite region.

6.5 Effect of Alloying and Microalloying Elements on the Mechanical Properties of Bainite and the Contribution from Precipitation Hardening

Since the complex phase steels are high-strength steels typically with 800–1000 MPa, it is necessary to reduce the amount of martensite in microstructure and to substitute it with higher strength bainite to meet the steel strength requirements.

Bainite as a microstructure constituent can significantly differ in strength which is determined by such parameters as ferrite lath size, solid solution strengthening due to carbon, substitutional solute strengthening, and dislocation density.

At lower B_s temperature, both the soluble carbon content and the dislocation density increase, which results in higher strength of bainite with temperature decrease. The correlation between the strength of bainite and the B_s temperature is presented in Fig. 6.16 (DeArdo et al. 2009). Depending on equations used for calculations of strength of bainite, authors found that increase in 10 MPa tensile/ yield strength of bainite can be achieved on account of the decrease in bainite start temperature in 3–5.5 °C.

Bainite is a less strong microstructure constituent than martensite. Therefore, to reach the same tensile strength as DP steels (800–1000 MPa), various means of strengthening should be utilized. Some contribution to the strength of CP steels

Fig. 6.16 The correlation between strength of bainite and the B_s temperature (DeArdo et al. 2009)

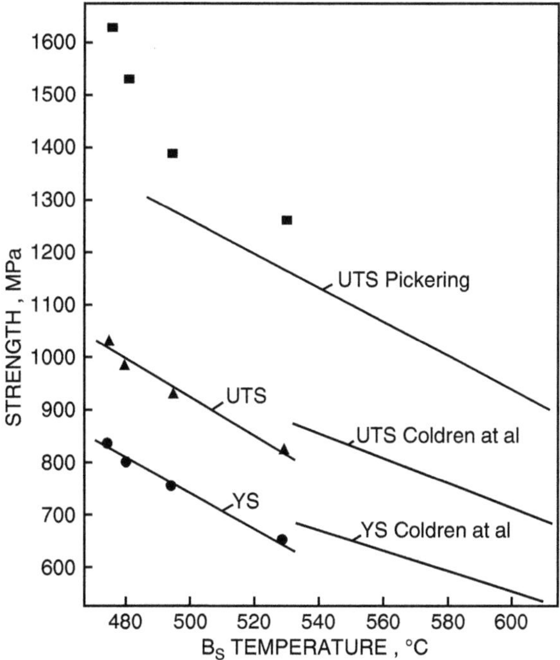

comes from martensite, whose portion depends on annealing cycle and the chemical composition of steel.

In the bainite transformation temperature range (600–450 °C), the strength increases with the increase in Mo addition. By using the EBSD technique, it was shown that molybdenum addition led to increase in average grain misorientation. This can be related to higher dislocation density and/or finer subgrain structure in Mo-containing steels.

It was found that hardness of B-bearing steel is always slightly higher than that of B-free steel. According to the microstructure quantification results, this is attributed mainly to microstructure refinement observed with additions of B.

Si can make a small contribution to strength because of limitation in its content. Microalloying elements that are considered for increasing the strength of CP steels include Nb and V.

The strength of hot-rolled CP or multiphase steels is determined by additions of Mo or Cr and by very high content (above 0.1 %) of V or Ti. Since the precipitations' temperature of Ti or V is higher than the optimum coiling temperature required to obtain the best morphology and the highest strength of bainite, the compromise should be found based on specified requirements for tensile properties and hole expansion.

For cold-rolled and annealed CP steels, microalloying with Nb allows for higher hardness (strength) compared to compositions without Nb. Hardness increases with lower isothermal holding temperature. The amount of Nb that precipitates at lower holding temperatures is less than that at higher temperatures, and this leads to a decrease in hardness. In this case, the lower hardness is compensated by refinement of the microstructure in whole and particularly of bainite at lower holding temperatures.

Nb strengthens the steel by fine NbC or Nb(C,N) precipitates, which also retard recrystallization and grain growth, thus facilitating grain refinement. In combination with B, the growth of coarse $Fe_{23}(CB)_6$ precipitates at grain boundaries is avoided because carbon is tied up by Nb.

As demonstrated by Sugimoto, Nb microalloying increases both yield and tensile strength of bainite-containing steels due to microstructure refinement and Nb precipitates. The total elongation was also improved (Sugimoto and Murata 2007).

It was shown that V affects the final mechanical properties of bainitic steels due to both strong solid solution hardening and precipitation hardening.

6.6 Summary

CP steels with relatively low elongation are used to produce safety parts where improved hole expansion and bending radius as well as the increased yield strength are critical.

Existing CP steels with tensile strength of mostly 800 and 1000 MPa have different chemical compositions, which should match the available annealing equipment and thermal cycles. For example, CP 800 (TS > 800 MPa) with HE > 35 % were developed using quite different chemistries with 0.11 or 0.14 % C, 1.5–2.1 % Mn, 0.15–0.5 % Si, without or with 0.3 Cr, without or with 0.15–0.2 Mo, and without or with 0.015–0.02 Nb.

Recently, CP 1000 steels have become commercial, and CP 1180 steels are under development.

The best composition should ensure the widest possible range of cooling rates for bainite transformation and maximum possible cooling rates for martensite transformation. Eliminating or minimizing the formation of ferrite is important not only for the uniformity of microstructure but also in view of hindering carbon enrichment of austenite to lower the carbon content in martensite and raise the Ms temperature, as well as to reduce the amount or to fully eliminate the retained austenite.

References

Bansal, A.K. 1977. "Stretch Bend Testing of High Strength Low Alloy Sheet Steels." Hamilton, Canada: McMaster.

Bode, R., M. Meurer, T.W. Schaumann, and W. Warnecke. 2004. "Selection and Use of Coated Advanced High-Strength Steels for Automotive Applications." *Stahl Und Eisen* 101 (7–8): 551–56.

Bodnar, R., T. Ohhashi, and R. Jaffee. 1989. "Effects of Mn, Si, and Purity on the Design of 3.5NiCrMoV, 1CrMoV, and 2.25Cr-1Mo Bainitic Alloy Steels." *Metallurgical and Materials Transactions A* 20 (8): 1445–60.

Cias, W.W. 1973. "Phase Transformation Kinetics and Hardenability of Low-Carbon Boron Treated Steels." *Metallurgical Transactions* 4 (2): 603–14.

Cold-Rolled Complex Phase (CP) Steel Grades with Optimized Bendability, Stretch Flangeability and Anisotropy (CP Steels). 2012. EUR 25041. European Commission.

Dacker, C.-A., O. Karlsson, C. Luo, K. Zhu, J.-L. Collet, M. Green, P. Morris, et al. 2012. *Bainitic Hardenability – Effective Use of Expensive and Strategically Sensitivealloying Elements in High Strength Steels (BainHard)*. European Commission, Directorate-General for Research and Innovation.

DeArdo, A.J., M.J. Hua, K.G. Cho, and C.I. Garcia. 2009. "On Strength of Microalloyed Steels: An Interpretive Review." *Materials Science and Technology* 25 (9): 1074–82.

Fonstein, N., H.J. Jun, G. Huang, and et al. 2011. "Effect of Bainite on Mechanical Properties of Multi-Phase Ferrite-Bainite-Martensite Steels." In *MST'11*.

Hasegawa, K., K. Kawamura, T. Urabe, and Y. Hosoya. 2004. "Effects of Microstructure on Stretch-Flange-Formability of 980 MPa Grade Cold-Rolled Ultra-High Strength Steel Sheets." *ISIJ International* 44 (3): 603–9.

Kagechi, H. 2003. "Recent Progress and Future Trends in the Research and Development of Steels." *NKK Technical Review* 88: 6–9.

Karelova, A., C. Krempaszky, M. Dünckelmeyer, E. Werner, T. Hebesberger, and A. Pichler. 2009. "Formability of Advanced High Strength Steels Determined by Instrumented Hole Expansion Testing." In *MS&T2009*, 1358–68. Pittsburgh, PA

Kawalek, A., J. Rapalska-Nowakowska, H. Dyja, and R. Koczurkiewicz. 2013. "Physical and Numerical Modelling of Heat Treatment the Precipitation-Hardening Complex-Phase Steel (CP)." *Metallurgia* 52 (1): 23–26.

Lee, J., S.-J. Lee, and B.C. De Cooman. 2012. "Effect of Micro-Alloying Elements on the Stretch-Flangeability of Dual Phase Steel." *Materials Science and Engineering: A* 536 (0): 231–38.

Li, M., D. Niebuhr, L. Meekisho, and D. Atteridge. 1998. "A Computational Model for the Prediction of Steel Hardenability." *Metallurgical and Materials Transactions B* 29 (3): 661–72.

Liu, S.K., and J. Zhang. 1990. "The Influence of the Si and Mn Concentrations on the Kinetics of the Bainite Transformation in Fe-C-Si-Mn Alloys." *Metallurgical Transactions A* 21 (6): 1517–25.

Matsuoka, S., K. Hasegawa, and Y. Tanaka. 2007. "Newly-Developed Ultra-High Tensile Strength Steels with Excellent Formability and Weldability." *JFE Technical Report*, no. 10: 13–18.

Mesplont, C. 2002. "Phase Transformations and Microstructure-Mechanical Properties Relations in Complex Phase High Strength Steels." Ph.D., Gent: Gent University.

Mesplont, C., S. Vandeputte, and B.C. De Cooman. 2001. "Microstructure-Properties Relationships in Complex-Phase Cold-Rolled High Strength Steels." In *MWSP*, 39:359–71.

Quidort, D., and Y. Brechet. 2002. "The Role of Carbon on the Kinetics of Bainite Transformation in Steels." *Scripta Materialia* 47 (3): 151–56.

Quidort, D., and Y.J.M. Brechet. 2001. "Isothermal Growth Kinetics of Bainite in 0.5% C Steels." *Acta Materialia* 49 (20): 4161–70.

Rees, G.I., J. Perdrix, T. Maurickx, and H.K.D.H. Bhadeshia. 1995. "The Effect of Niobium in Solid Solution on the Transformation Kinetics of Bainite." *Materials Science and Engineering: A* 194 (2): 179–86.

Ryde, L., O. Lyytinen, P. Peura, and M. Titova. 2012. "Cold-Rolled Complex-Phase (CP) Steel Grades with Optimised Bendability, Stretch-Flangeability and Anisotropy (CP-Steels)." European Commission.

Sakata, K., S. Matsuoka, and K. Sato. 2003. "Highly Formable Sheet Steels for Automobile through Advanced Microstructure Control Technology." *Kawasaki Technical Report*, no. 48: 3–8.

Shimuzi, T., Y. Funakawas, and S. Kanaka. 2004. "High Strength Steel Sheets for Automotive Suspension and Chassis Use: High Strength Hot Rolled Steel Sheets with Excellent Press Formability and Durability for Critical Safety Parts." *JFE Technical Report*, no. 4: 25–31.

Shirasawa, H., K. Mimura, T. Yokoi, Z. Shibata, and N. Inoue. 1993. "Formability of TS590 N/mm2 'Tri-Phase' Hot Rolled Steel." *SAE Technical* 930284.

Steven, W., and A.G. Haynes. 1956. "The Temperature of Formation of Martensite and Bainite in Low-Alloy Steels." *Journal of the Iron and Steel Institute* 183 (8): 349–59.

Sugimoto, K.I., and M. Murata. 2007. "Applications of Niobium to Automotive Ultra High-Strength TRIP-Aided Steels with Bainitic Ferrite And/or Martensite Matrix." In *MS&T2007*, 1:15–26. Detroit, MI.

Sugimoto, K.-I., J. Sakaguchi, T. Ida, and T. Kashima. 2000. "Stretch-Flangeability of a High-Strength TRIP Type Bainitic Sheet Steel." *ISIJ International* 40 (9): 920–26.

Wang, J., P.J. Van Der Wolk, and S. van der Zwaag. 2000. "On the Influence of Alloying Elements on the Bainite Reaction in Low Alloy Steels during Continuous Cooling." *Journal of Materials Science* 35: 4393–4404.

Xu, F.-y., Y.-w. Wang, B.-z. Bai, and H.-s. Fang. 2010. "CCT Curves of Low-Carbon Mn-Si Steels and Development of Water-Cooled Bainitic Steels." *Journal of Iron and Steel Research, International* 17 (3): 46–50.

Zhao, Z., C. Liu, Y. Liu, and D.O. Northwood. 2001. "A New Empirical Formula for the Bainite Upper Temperature Limit of Steel." *Journal of Materials Science* 36 (20): 5045–56.

Chapter 7
Martensitic Sheet Steels

Contents

7.1 Introduction

Martensitic steels are the hardest type of steels featured simultaneously by high tensile strength and very high YS/TS ratio that is very important for the safety parts of cars that are intended to protect the driver and the passengers from intrusions during collisions. Because of the importance of this function, the production of martensitic grades of sheet steels historically started far earlier than that of AHSS, in particular, in the early 80s in the USA.

Microstructure of low-carbon martensitic steels is mainly composed of lath martensite as a result of austenite transformation during quenching after hot rolling or annealing. These steels are often subjected to post-quench tempering that improves ductility and toughness, as well as provides good formability even at very high yield strength.

Strength of martensitic grades is practically controlled by the carbon content; however, the alloying elements are added to achieve the necessary hardenability during processing and to affect other properties such as ductility, bendability, and delayed fracture resistance.

© Springer International Publishing Switzerland 2015

N. Fonstein, *Advanced High Strength Sheet Steels*,

DOI 10.1007/978-3-319-19165-2_7

7.2 Martensitic Grades for Roll Forming

Until recently, the majority of parts in modern cars with martensitic structure were produced using annealed martensitic steels. The existing group of martensitic grades includes steels with TS = 900, 1100, 1300, and 1500 MPa. With high YS/TS ratio, typically above 0.85, these materials are used preferably for roll forming that has proved to be very effective method to manufacture numerous car parts critical for crash safety. In contrast to steels intended for forming using other methods, steels for roll forming should have high yield to tensile strength ratio to minimize strength gradient and confine the plastic strain to bent corners. High YS/TS ratio also imparts high initial and therefore high retained strength of nondeformed portions of the profile.

7.2.1 Processing and Compositions of Annealed Martensitic Grades

In fact, any continuous annealing line with water quenching capability can produce martensitic grade.

Composition of annealed martensitic grades depends on the design of cooling/ water quenching sections. As shown in Fig. 7.1, continuous annealing lines differ in capability to use water quenching directly from soaking at temperatures of full austenitization (solid line) and those that have unavoidable temperature drop in the pre-quenching cooling section (dashed line) with some restrictions for maximum annealing temperatures.

Therefore, in the first case, the complete austenite-to-martensite transformation is possible with minimum alloying, Table 7.1. As shown in Fig. 7.2, the strength of martensite is mostly controlled by carbon content. With up to 0.5 % C, the effect of carbon on steel hardness is practically linear (Krauss 2005). Consequently,

Fig. 7.1 Typical thermal cycles for CAL equipped with water quenching

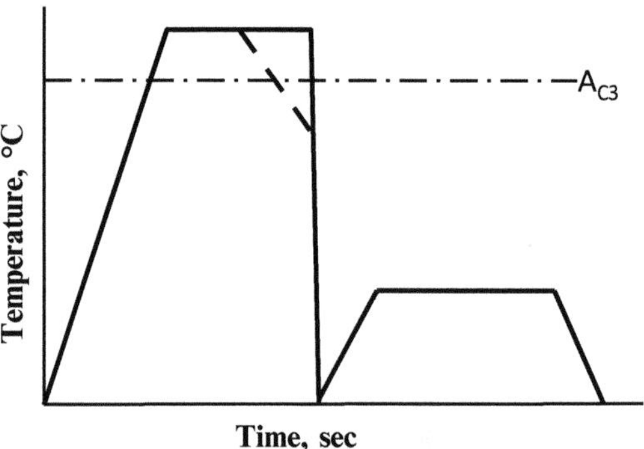

Table 7.1 Chemical compositions of lean martensite grades (ArcelorMittal 2014)

	C	Mn	P	S	Other	Form
M900	0.08	0.45	0.01	0.015	B, Ti	CR, EG
M1100	0.12	0.45	0.01	0.015	B, Ti	CR, EG
M1300	0.19	0.45	0.01	0.015	B, Ti	CR, EG
M1500	0.25	0.45	0.01	0.015	B, Ti	CR, EG
M1700	0.30	0.45	0.01	0.015	B, Ti	CR

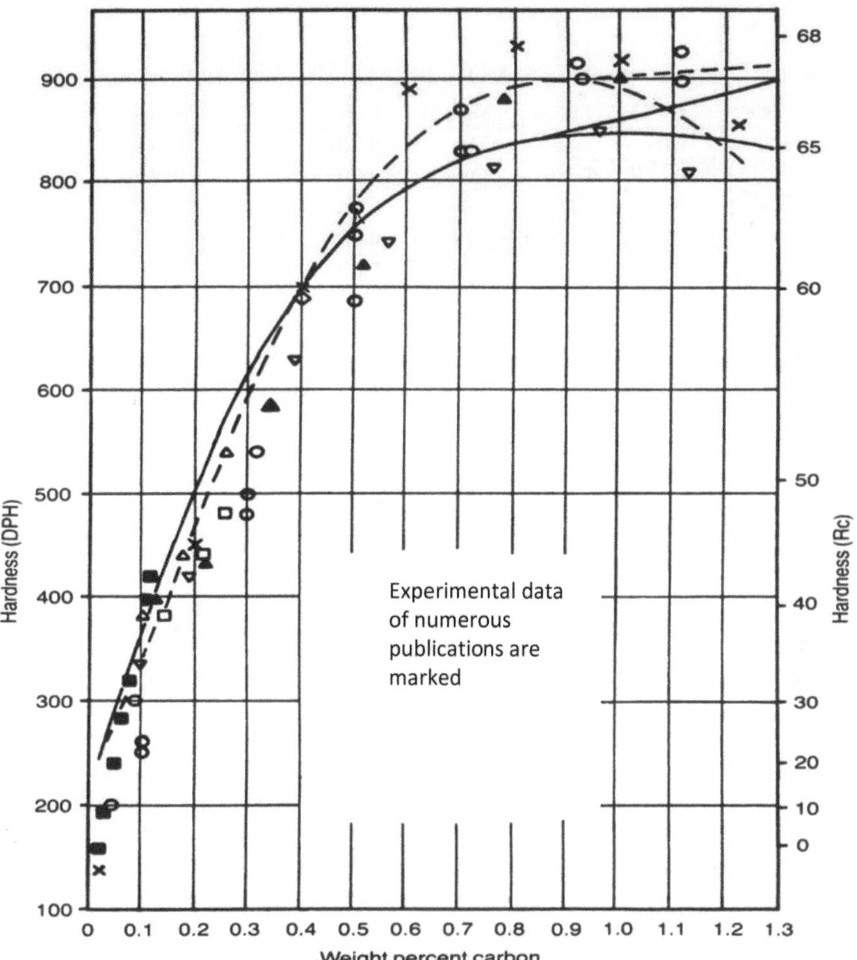

Fig. 7.2 Effect of carbon on strength of martensite (Krauss 2005)

martensite grades with TS = 900, 1100, 1300, and 1500 MPa differ only in carbon content with very lean alloy composition of 0.45 % Mn and small additions of boron protected from interaction with nitrogen by microalloying with Ti. The mechanical properties of such steels are presented in Table 7.2.

Table 7.2 Mechanical properties of lean martensite grades (ArcelorMittal 2014)

	Test—direction	Yield strength (MPa)	Ultimate tensile strength (MPa)	Total elongation	Recommended bend ratio
M900	ASTM-L	877	1015	6	4T
M1100	ASTM-L	1018	1179	6	4T
M1300	ASTM-L	1212	1423	6	4T
M1500	ASTM-L	1370	1629	6	4T
M1700	ASTM-L	1520	1820	5	4T

Table 7.3 Chemical composition of DOCOL grades (SSAB Catalog 2015)

Steel grade	C (%)	Si (%)	Mn (%)	P (%)	S (%)	Al_{tot} (%)	Nb (%)
Docol 130 M	0.05	0.20	2.00	0.010	0.002	0.040	0.015
Docol 175 M	0.11	0.20	1.60	0.010	0.002	0.040	0.015
Docol 190 M	0.14	0.20	1.50	0.010	0.002	0.040	0.015
Docol 205 M	0.18	0.20	1.20	0.010	0.002	0.040	0.015
Docol 220 M	0.20	0.20	1.00	0.010	0.002	0.040	0.015

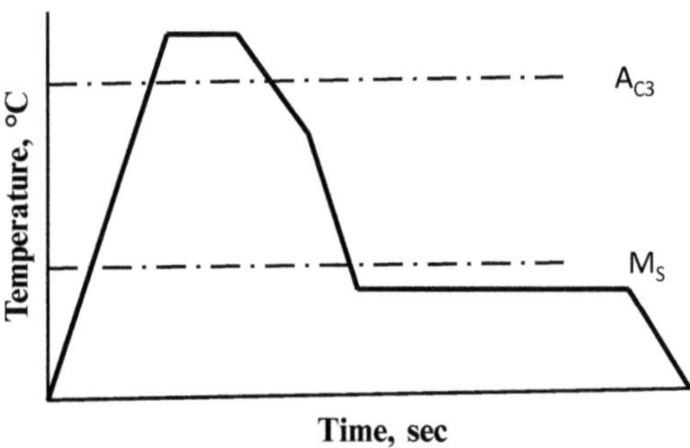

Fig. 7.3 Thermal cycle to produce martensitic grades based on direct transfer from cooling to overaging (tempering) section

In the case of initial temperature drop (dashed line in Fig. 7.1), higher alloying is necessary to prevent the formation of ferrite and pearlite during initial slow cooling. Consequently, the corresponding martensitic grades produced, in particular, by SSAB (Olsson and Sperle 2006), contain additions of 1–2 % Mn (the higher the lower carbon content) to ensure sufficient hardenability of austenite, Table 7.3. Increased Mn content also helps to decrease the A_{c3} and therefore annealing temperatures of martensitic grades.

In the case of equipment with direct transfer to overaging zone shown in Fig. 7.3, alloying of steel should be even higher due to slower cooling rate (typically not higher than 70–80 °C/s) from soaking temperatures down to the overaging/tempering section. Depending on the target strength, the martensitic grades should have, besides boron additions, higher amount of Mn, Cr, and sometimes Mo.

Strong demand to further reduce the weight of automobiles and simultaneously achieve higher reliability of safety parts motivated steel producers to develop

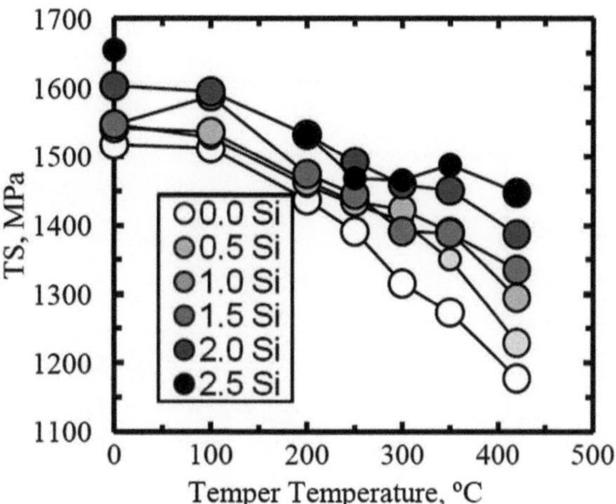

Fig. 7.4 Effect of Si additions and tempering temperatures (150 s holding time) on tensile strength of martensite (Johnson et al. 2013)

martensite grades with TS > 1700 MPa and >2000 MPa. There is a possibility to achieve that strength by the increase in carbon content with some microalloying aiming to ensure necessary resistance to delayed fraction. Other approaches include consideration of strengthening with alloying. In particular, in the study by Arlazarov et al., synergetic effect of carbon and manganese on the martensite strength and strain hardening was detected and was then taken into account. As a result, YS = 1439 MPa, TS = 2150 MPa, and TE = 6.3 % were obtained for 0.38C–1.22Mn–0.23Si–0.10Cr–0.04Ti steel. On the other hand, the authors observed that due to essential increase in strain hardening, a low-carbon 0.15C–5Mn steel could achieve TS > 1700 MPa (Arlazarov et al. 2013).

Figure 7.4 shows that increase in Si content in 0.15C–1.8Mn–0.02Nb–0.15Mo steel results in increase in tensile strength of as-quenched and especially tempered martensite due to retardation of softening martensite at tempering (Johnson et al. 2013).

It is also known that grain refinement is the major remedy in improving the mechanical properties of martensitic steel. The size of martensite packets is directly related to the austenite grain size (Krauss 2005), so the data presented in Fig. 7.5 related to packet size, D, emphasize the essential impact of austenite grain refinement.

7.2.2   As-Rolled Martensite

Several producers supply hot-rolled sheet martensitic grades.

Processing of martensitic grades in hot-rolling mills requires fast cooling from the finishing rolling temperature immediately upon the exit from the mill down to temperatures below M_S to prevent or at least to minimize non-martensitic products of austenite transformation. The steels should be alloyed high enough to ensure

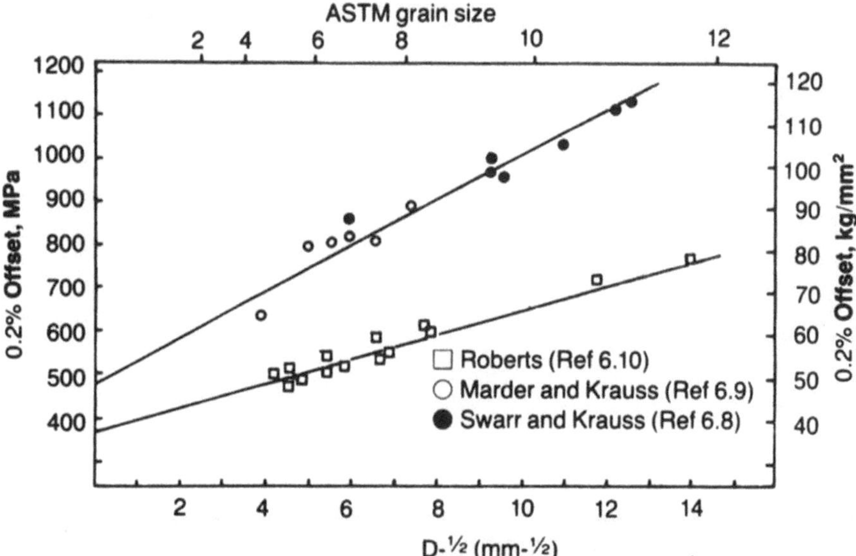

Fig. 7.5 Correlation between strength of lath martensite and packet size, D. The *upper curve* is for 0.2C martensite and the *lower curve* is for Fe–Mn carbon-free martensite (Swarr and Krauss 1976; Marder and Krauss 1970; Roberts 1970; Krauss 2005)

sufficiently high hardenability to prevent ferrite/pearlite transformation under cooling with real available rate (20–40 °C/s).

In particular, Salzgitter produces SZMS 1200 grade with TS above 1200 MPa, min YS = 900 MPa in thickness from 2 to 3 mm. For available cooling rate of relatively thick strips, the 0.18C–2.0Mn–0.6Cr–0.15Si composition is used to assure the necessary hardenability of steel (Saltzgitter Flachstahl GmbH catalog).

In general, chemical composition of as-hot-rolled martensitic grades depends on both cooling capability of hot-rolling mills and thickness of produced sheets. For example, ThyssenKrupp Steel produces as-hot-rolled martensite MS-W 1200 grade with TS > 1200 MPa and YS > 900 MPa using slightly higher alloying: 0.18 C with up to 2.0 Mn, 1.0 Cr, and 0.8 Si and various combinations of microalloying with Ti, B, and Nb.

7.2.3 Effect of Martensite Tempering

Tempering is a common practice to improve ductility or toughness of as-quenched martensite. During tempering or auto-tempering, a number of metallurgical effects take place: decrease of lattice distortion, release of residual stresses, and carbide precipitation. To some extent, these effects take place at tempering temperatures in the range of 150–200 °C but without any significant impact on tensile properties.

Since the tempering above 200 °C leads to a gradual decrease in strength, the martensitic grades are not usually tempered at temperatures higher than 180–200 °C.

7.3 Martensite Produced by Press Hardening

Press-hardening technology solves the long existing conflict of forming extremely high strength steels into complex shape without problems with cracking, excessive press forces, or spring back. Hot stamping as a process was developed in 1977, and with implementation of boron-containing (Al–Si coated) steels in 2000, more hot-stamped parts have been used in cars, and the number of annually produced parts has gone up to several hundred million parts with significant increase every year.

7.3.1 Basic Principles of Obtaining Martensite After Hot Stamping

The hot-stamping process currently exists in two main variants: direct and indirect hot stamping. In the direct hot-stamping process, a blank is heated up in the furnace, transferred to the press, and subsequently formed and quenched in the water-cooled closed tool as depicted in Fig. 7.6a. In the indirect hot-stamping process, a near net shape cold preformed part is subjected only to austenitization followed by

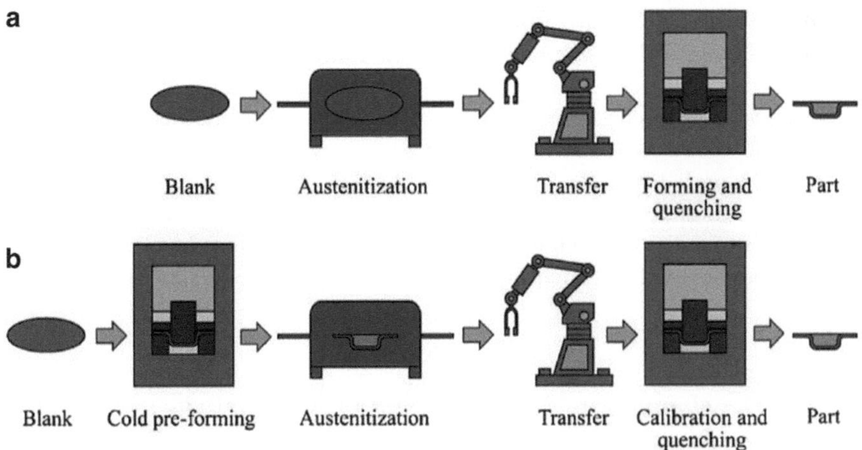

Fig. 7.6 Basic hot-stamping processes: (**a**) direct hot stamping. (**b**) Indirect hot stamping (Karbasian and Tekkaya 2010)

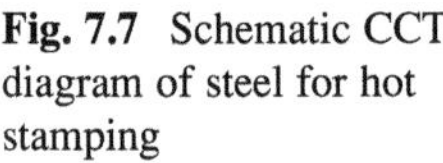

Fig. 7.7 Schematic CCT diagram of steel for hot stamping

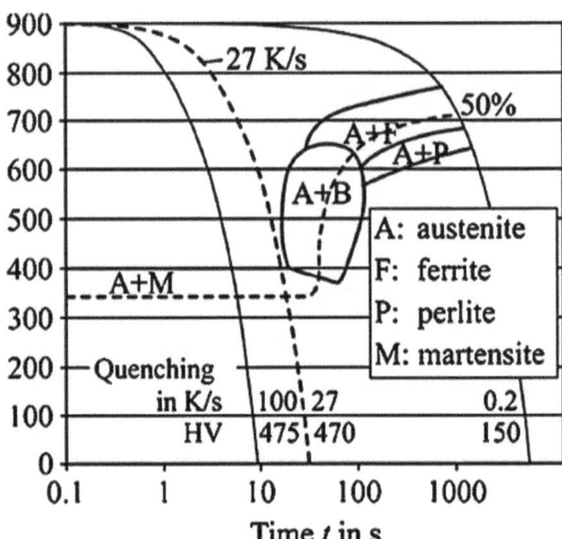

calibration operation and quenching in the press (Fig. 7.6b) (Karbasian and Tekkaya 2010).

22MnB5 steel is the most commonly used grade for hot stamping.

Supplied material (typically cold rolled and annealed or coated) has a ferrite–pearlite microstructure with a tensile strength of about 600 MPa. After the hot-stamping process, with complete martensite transformation taking place due to quenching in the water-cooled die, the component must have martensitic microstructure with strength of about 1500 MPa. To achieve such a microstructure, the blank is austenitized in the furnace at temperature above A_{c3}.

As shown in the CCT diagram in Fig. 7.7, the martensite start temperature of 22MnB5 steel is close to 375–400 °C (the martensite finish temperature is about 280 °C). The critical cooling rate to avoid the formation of softer phases (bainite, ferrite, and pearlite) is about 25 °C/s. Thus, the cooling rate in the die should exceed this critical rate to ensure full martensite transformation and the required final properties.

To make press-hardening technology a reliable process, numerous careful studies were performed, for example, by Naderi et al., focusing on calculations of the effects of variability of blank thickness and cooling conditions (cooling media) on critical cooling rate for martensite transformation (Naderi et al. 2008).

7.3.2 Development of Ultrahigh Strength Martensite for Press Hardening

As mentioned above, commercial martensitic grades for press hardening with TS ~1500 MPa are currently available. However, higher strength steels with TS > 1800 and 2000 MPa are required to lighten the parts to desirable levels and/or to increase crush resistance.

To increase the strength of martensitic steels for press hardening, various strategies are considered. They aim not only at increasing strength but also at ensuring robust manufacturability in steel mills and during press-hardening processing by the customers, as well as at better toughness and improved resistance to delayed fracture.

To reach TS $\geq$ 2000 MPa, which is significantly higher than for 22MnB5 grade, the carbon should be increased. To keep the necessary level of weldability, it is necessary to compensate higher carbon content by decrease in Mn content which, in particular, typically negatively affects hydrogen resistance. The presence of Si is considered as useful to improve the resistance to hydrogen embrittlement (Matsumoto et al. 2013). The beneficial effect of microstructural refinement on the strength and toughness of martensitic steels has been verified by many studies (Kubota et al. 2010). Nb should be considered as useful in this concept because grain refinement should be favorable for the resistance to delayed fracture.

Bain and Mohbacher are building their strategy on optimization of conventional alloy design based on 22MnB5 to reach TS $>$ 1800 MPa. They slightly increase carbon content to 0.25 % keeping the same 1.4 % Mn and 0.4 % Si, adding over 0.05 % Nb and 0.15 % Mo, and removing B + Ti.

Essential feature of the proposed improved press-hardenable steel is its fine grain microstructure. Additional useful role of microalloying by Nb can be related to prevention of grain growth during high temperature blank reheating, usually 950 °C, to compensate for temperature loss during transfer of the blank from the furnace to the die.

Since in press hardening the steel gauges are quite thin (usually $<$3 mm), in authors' opinion, cooling rates during die quenching are rather high, especially when the advanced press-hardening technology with increased die pressure is employed. This should allow for avoiding the formation of ferrite even at the absence of boron. Therefore, additions of Ti can be eliminated too. The bainite nose can be sufficiently delayed by adding more Mn, Mo, or Cr if lower cooling rates need to be tolerated.

The important component of the proposed strategy is the strengthening of austenite using additions of Mo. While boron blocks the nucleation of ferrite by segregating to austenite grain boundaries, Mo lowers the activity of carbon and slows its diffusion and consequently significantly delays ferrite, pearlite, and bainite transformations (Bian and Mohrbacher 2013).

7.3.3   *Modification of Press-Hardening Technology*

Several studies proposed to modify hot-stamping technology so as to reach the final mechanical properties of stamped parts close to the properties of the third

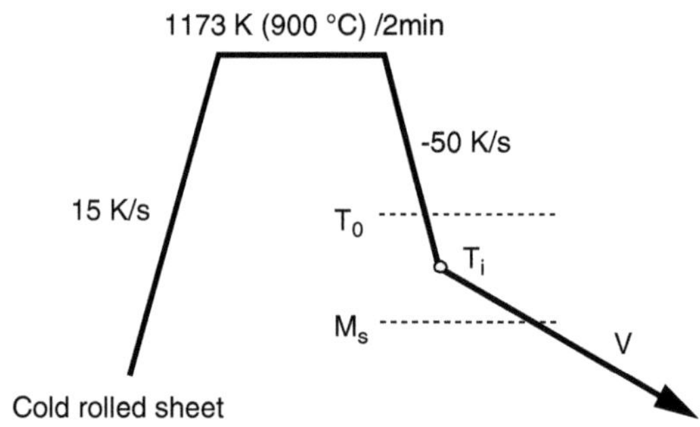

Fig. 7.8 Thermal cycle proposed for press-forming process to obtain carbide-free microstructure with retained austenite (Chen et al. 2014)

Generation steels, i.e., to combine ultrahigh strength of martensitic press-hardened parts with high ductility and toughness.

In particular, it was shown that medium Mn (4–5 % Mn) steels are not sensitive to cooling rate and can be fully transformed to martensite at air-cooling. This allows for hot stamping at lower temperatures because of substantially lower A_{c3} temperatures, and dies should be cooled only for productivity purposes.

Another quite novel modification to hot stamping is the attempt to combine it with the Q&P processing. Full austenitization and hot deformation to simulate hot stamping were performed with subsequent quenching to the temperature below Ms followed by isothermal holding at the same temperature. The resultant microstructure was refined and contained significant amounts of retained austenite that allowed for appreciable increase in elongation.

For example, Liu et al. proposed special 0.22C–1.58Mn–0.81Si–0.022Ti–0.0024B composition for such processing. After full austenitization for 5 min and hot deformation, the steel was quenched to 280, 300, and 320 °C, which are the temperature in the (M_S–M_F) range. This way, one-step Q&P process was realized. The best combinations of properties achieved with partitioning at 320 and 280 °C for 30 s were TS = 1510 MPa at TE = 14.8 % and TS = 1601 MPa at TE = 10.3 %, respectively (Liu et al. 2011).

Chen et al. suggested the so-called interrupted cooling that combined rapid cooling to temperature slightly above the Ms temperature (305–360 °C) with subsequent slow cooling to room temperature (Fig. 7.8). Using ~0.30C–1.6Si–2.3Mn–0.25Mo steel with up to 0.5Cr, the authors tried to obtain microstructure of TRIP type with bainitic ferrite matrix. By adjusting the composition and cooling interruption temperature, the authors managed to obtain TS > 1400 MPa at TE > 10 % (Chen et al. 2014)

One of the main concerns in hot stamping aimed at fully martensitic microstructure of the deformed part is to avoid austenite decomposition during cold plastic deformation. Therefore, steels considered for this application should have sufficiently stable austenite that could be deformed with minimum of deleterious decomposition products.

7.4 Susceptibility of Martensitic Grades to Delayed Fracture

The share of ultrahigh strength steel including martensitic grades is growing. However, there are some justified concerns that the susceptibility to delayed fracture (DF) controlled by hydrogen embrittlement (HE) increases with higher strength of steel. Generally, it is believed that the problem emerges for high-strength steel with a tensile strength of above 1000 MPa.

Since the strength of steel depends on both alloy composition and microstructure, it is difficult to independently distinguish the effects of strength itself from the effects of alloying and microstructure. In addition, different researchers have been using different test methods to evaluate the resistance to HE/DF, as well as high hydrogen content charged, often significantly higher than steel parts can experience after processing or in service.

In fact, strength has never showed a direct correlation with material behavior under hydrogen impact. Steels with the same strength level demonstrate different resistance to hydrogen embrittlement due to different combinations of microstructure and chemical composition that are believed to play the key role (Thiessen et al. 2011).

In particular, after three decades of producing lean chemistry martensite with strength of up to 1500 MPa, Inland Steel/Arcelor Mittal has received zero claims related to delayed fracture. The developed M1700 grade based on the same lean 0.45 % Mn matrix with only higher carbon content also passed very severe test of 600 h immersing the bent sample stressed to 85 % of TS into an acidic environment (0.1 N HCl) without crack appearance.

On the other hand, the study of martensite grades M1200 and M1400 containing 1.5 % Mn, 0.2–0.3 % Si, and 0.2–0.3 % Cr found them susceptible to hydrogen embrittlement, but it started from 4 and 1 weight ppm, respectively, that is in any way much higher than what could be expected (Lovicu et al. 2012).

The development of martensitic products with TS ~ 2000 MPa requires careful consideration of separate roles of chemical composition and microstructure on the resistance to DF/HE.

In accordance with data related mostly to HSLA, increased content of Mn in steels has a negative influence of susceptibility to hydrogen embrittlement (Hejazi et al. 2012)

Since HE involves intergranular fracture, boron should be considered useful in UHSS not only from the point of view of hardenability but probably for grain boundary strengthening. As shown by Nie et al. (2007), the increase in boron content from 0.0005 to 0.0016 % in medium carbon spring steel was accompanied by increase in delayed fracture strength.

According to Shiraga (1994), the positive effect of Ni is related to its increased concentration on the exterior surface that suppresses the permeation of diffusible hydrogen from aggressive solutions.

Effect of Cu additions was carefully studied by Toyoda et al. (Toyoda et al. 2008; Toyoda 2011). According to their data, steel with additions of Cu in the amount of 0.1–0.15 % demonstrated the same excellent DF resistance in both 1 N hydrochloric acid and saltwater as steel containing 0.5 % Cu.

It is known that additions of Al (1–1.5 %) solved the problem of DF of fully austenitic TWIP steels, although the mechanism of its effect is not fully understood.

The data related to effects of microalloying are rather inconsistent.

Effects of Nb and V additions are considered in the publication by Zhang et al. (2011). These authors showed that microalloying with Nb or Nb + V reduces the apparent diffusion coefficient of hydrogen in 0.6C–1.6Si–0.75Mn–0.35Cr (wt%) steel that results not only in smaller percentage of strength loss during testing of pre-charged specimens but also in some changes in DF character in favor of some portion of trans-granular and quasi-cleavage rapture instead of intergranular.

Meanwhile, the literature review shows that titanium or vanadium carbides can enhance or suppress DF depending on other conditions and primarily on micro-structure. For example, some publications demonstrate that VC increases the critical hydrogen content (threshold for DF), whereas other data show that it also increases the amount of absorbed hydrogen.

For example, Gladshtein et al. (1988) noted that additions of Ti (0.029–0.078 %), Nb (0.025–0.049 %), and boron (0.015–0.0044 %) to tempered 0.2C–2Cr–Ni–Mo steel with microstructure of lath martensite and bainite enhanced the resistance to crack initiation and propagation under hydrogenation. In the author's opinion, the positive role of Ti and Nb is the trapping capacity of their coarse particles, which decrease hydrogen saturation near the crack tip, whereas boron retards initiation and propagation of intergranular cracks.

In some cases, all microalloying elements (Ti, Nb, and V) are added as useful remedies against delayed fracture without distinguishing the role of Nb, as was done in case of steels for fasteners with TS above 1800 MPa (Kubota et al. 2010).

Significant improvement in DF resistance of low tempered (250 °C) martensitic 0.22C–0.5Mn–0.2Si–0.5Mo–1.8Cr (wt%) steel microalloyed with Nb, Ti, and B was noted by Glazkova et al. (1976). The authors suggested that microalloying elements play the dominating role due to both grain refinement and enhancement of trapping capacity by carbonitrides of Ti and Nb that significantly decreased the diffusivity of hydrogen.

Higher content of Si in steel can improve delayed fracture resistance because, on the one hand, Si inhibits/reduces carbide formation during tempering of martensite and, on the other hand, it reduces hydrogen diffusivity (Matsumoto et al. 2013).

Careful comparative study of the effects of 0.05 % V, 0.04 % Ti and, 0.03 % Nb on DF was conducted by Sergeeva et al. The authors evaluated the time before fracture of in situ hydrogen charged samples under stress, the content of occluded hydrogen after 1 h charging, and the desorption rate as an indicator of hydrogen trapping efficiency. It was found that 0.03 % Nb had the maximum effect on increasing time before fracture by two orders of magnitude of 0.31C–0.5Mn–0.25Si–0.15Mo–0.8Cr–1.5Ni (wt%) steel. This was explained by both increased toughness of steel and lower hydrogen consumption. Essential increase in time before fracture demonstrated by Ti additions was considered to be due to favorable compensating trapping role of Ti(C, N), as the total occlusion of H grew in the presence of Ti (Sergeeva et al. 1994).

Using Auger spectroscopy and investigating segregation at free surface that can simulate grain boundary segregation, the same authors established that positive effect of microalloying on delayed fracture was related also to the suppression of grain boundary segregations of P, Sb, C, and S. In terms of increasing the efficiency of this suppression and growing strength, the microalloying elements were ranked as V, Ti, and Nb (Drobyshevskaya et al. 1995).

Since the intergranular fracture between prior austenite grains is the dominant mechanism of delayed fracture, refining of prior austenite grains is important in improving grain boundary strength that was demonstrated, in particular, by Fuchigami et al. (2006).

Appropriateness of the above statements regarding the effects of alloying and microalloying was verified by lab research performed by Song et al. (2015), when Si and Nb were sequentially added to low-Mn–B–Ti steel with TS ~ 2000 MPa. The results of delayed fracture tests as well as sizes of prior austenite grains, defined using EBSD analysis, are presented in Table 7.4.

As shown in Fig. 7.9, the growing efficiency of alloying and microalloying additions was confirmed by measurements of diffusible hydrogen that decreased in the same order as the increase in time before delayed fracture.

Table 7.4 Basic mechanical properties and results of delayed fracture tests

Steel—ASTM-L	Time before cracking in hours during U-bend test. 0.1 N HCl, 85 % TS	Diffusible H (<300 °C) (ppm)	PAGS (μm)
Ti–B	<6 h	1.22	18.3
Ti–B–Si	137 h	0.89	10.3
Ti–B–Si-Nb	>600 h	0.60	6.4

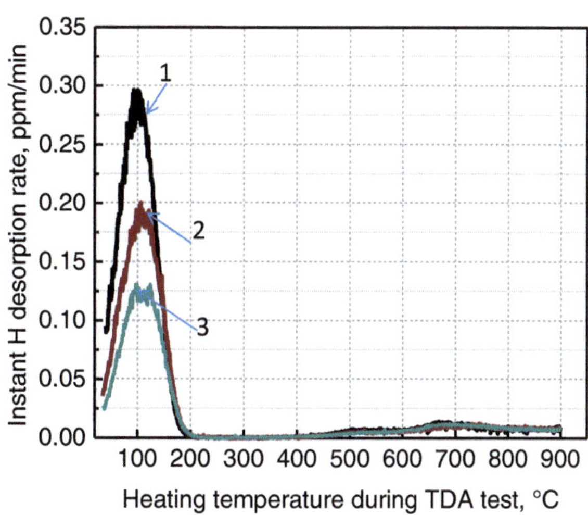

Fig. 7.9 Effect of alloying/microalloying elements on hydrogen desorption: 1-base Ti–B composition, 2-base plus Si, and 3-base plus Si plus Nb (original)

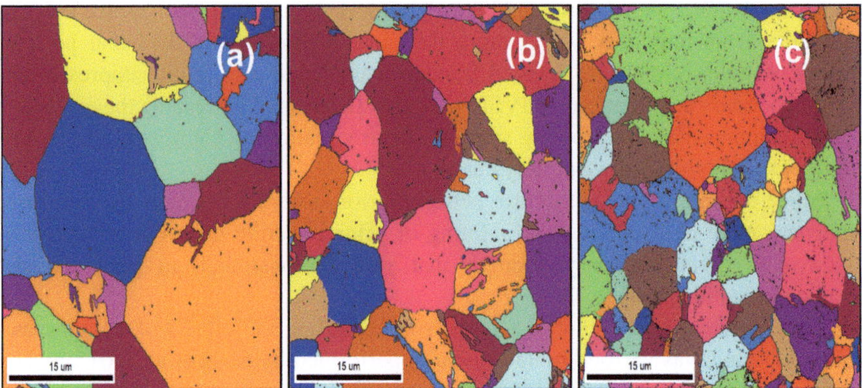

Fig. 7.10 EBSD analysis of PAGS: (**a**) base Ti–B composition, (**b**) base plus Si, (**c**) base plus Si plus Nb (original)

As was established by EBSD analysis, the added elements, besides their influence on trapping, also produce gradual refinement of prior austenite grain size (PAGS) (Fig. 7.10) with the smallest grain size detected for Nb microalloyed steel.

7.5   Summary

The use of martensitic steel with tensile strength of up to 2000 MPa and high YS/TS ratio has become extremely relevant to car body engineering, especially with respect to avoiding the intrusion by high speed impact loading during a crash.

Various types of manufacturing of martensitic steels by water quenching after annealing, by hot rolling, and by press hardening are available facilitating application of ultrahigh strength steels for safety critical car parts.

It was demonstrated that microstructure refinement being mainly related to austenite grain size prior to quenching is the key approach to improve combination of strength and resistance to delayed fracture.

References

ArcelorMittal. 2014. "Catalog."

Arlazarov, A., O Bouaziz, A. Hazotte, M. Gouné, and S. Allain. 2013. "Characterization and Modeling of Manganese Effect on Strength and Strain Hardening of Martensite Carbon Steels." *ISIJ International* 53 (6): 1076–80.

Bian, J., and H Mohrbacher. 2013. "Novel Alloying Design for Press Hardening Steels with Better Crash Performance." In *International Symposium: New Development in AHSS*, 251–63. Vail, CO, USA.

Chen, S., R. Rana, and C. Lahalie. 2014. "Study of TRIP-Aided Bainitic-Ferritic Steels Produced by Hot Press Forming." *Metallurgical and Materials Transactions* 45A (4): 2209–18.

Drobyshevskaya, I.S., A.I. Kovalev, T.K. Sergeeva, and D.A. Litvinenko. 1995. "Impurity Segregation, Temper Brittleness and Hydrogen Embrittlement of Steel Type 30KhNMA with Various Molybdenum Content." *Metal Science and Heat Treatment* 5: 21–24.

Fuchigami, H., H. Minami, and M. Nagumo. 2006. "Effect of Grain Size on the Susceptibility of Martensitic Steel to hydrogen-Related Failure." *Philosophical Magazine Letters* 86 (1): 21–29.

Gladshtein, L.I., V.M. Goritski, N.A. Evtushenko, and V.I. Sarrak. 1988. "Influence of Alloying Additives and Impurity Traces on Stress-Corrosion Cracking the High Strength Bolt Steels." *Rus. Metall.* 5: 176–81.

Glazkova, S.M., A.V. Pastoev, V.I. Sarrak, and G.A. Filippov. 1976. "Effect of Hydrogen on the Ductility and Fracture Steels 38 Cr." *Soviet Material Science* 12 (5): 478–80.

Hejazi, D., A.J. Hag, N. Yazdipour, and D.P. Dunne. 2012. "Effect of Manganese Content and Microstructure on the Susceptibility of X-70 Pipeline Steel to Hydron Cracking." *Material Science and Engineering A* 551: 40–49.

Johnson, J., H. J. Jun, N. Fonstein, and M. Enloe. 2013. "Effect of Silicon in as-Quenched and Quenched & Tempered Low Carbon Martensite." In . Vail, CO, USA.

Karbasian, H., and A.E. Tekkaya. 2010. "A Review of Hot Stamping." *Journal of Materials Processing Technology* 210: 2103–18.

Krauss, George'. 2005. *Steels: Processing, Structure and Performance*. TMS.

Kubota, M., S. Yoshida, T. Tauri, and H. Matsuda. 2010. "Steel with Excellent Delayed Fracture Resistance and Tensile Strength of 1801 MPa Class or More." US Patent 7,754,029 B2

Liu, H., X. Lu, X. Jin, H. Dong, and J. Shi. 2011. "Enhanced Mechanical Properties of a Hot Stamped Advanced High-Strength Steel Treated by Quenching and Partitioning Process." *Scripta Materialia* 64: 749–52.

Lovicu, G., M. Bottazi, F. D'Aiuto, and M. DeSanctis. 2012. "Hydrogen Embrittlement of Automotive Advanced High-Strength Steel." *Metallurgical and Materials Transactions A*.

Marder, A.R., and G. Krauss. 1970. "The Effect of Morphology on the Strength of Lath Martensite." In *Second International Conference on the Strength of Metals and Alloys, vol.III*, 822–23.

Matsumoto, Y., K. Takai, M. Ichiba, and T. Suzuki. 2013. "Reduction of Delayed Fracture Susceptibility O Tempered Martensitic Steel through Increased Si Content and Surface Softening." *ISIJ International* 53: 714–22.

Naderi, M., V. Uthaisengsuk, U. Prahl, and W Bleck. 2008. "A Numerical and Experimental Investigation into Hot Stamping of Boron Alloyed Heat Treated Steels." *Steel Research International* 79 (2): 77–84.

Nie, Y.H., W.J. Hui, W.-T. Fu, and Y.Q. Weng. 2007. "Effect of Boron on Delayed Fracture Resistance of Medium Carbon High Strength Spring Steel." *Journal of Iron and Steel Research, International* 14: 53–57,67.

Olsson, K., and J.-O. Sperle. 2006. "New Advanced Ultra-High Strength Steels for the Automotive Industry." *Auto Technology* 5: 46–49.

Roberts, M.J. 1970. "Effect of Transformation Substructure on the Strength and Toughness of Fe-Mn Alloys." *Metal Trans. A* 1: 3287–94.

Sergeeva, T.K., I.S. Drobyshevskaya, Litvinenko D.A., and V.N. Marchenko. 1994. "Resistance to Hydrogen Embrittlement of Low-Molybdenum Structural Steels Microalloyed with Carbonitride-Forming Elements." *Steel in USSR* 2: 75–79.

Shiraga, T. 1994. "Effect of Ni, Cu and Si on Delayed Fracture Properties of High Strength Steel with Tensile Strength of 1450 MPa." *CAMP-ISIJ* 7: 1646–47.

Song, R, N. Fonstein, N. Pottore, and H.J. Jun. 2015. "Effect of Nb on Delayed Fracture Resistance of Ultra-High Strength Martensitic Steels." In . China.

"SSAB Catalog." 2015.

Swarr, T.E., and G. Krauss. 1976. "The Effect of Structure on the Deformation of as-Quenched and Tempered Martensite in an Fe-0.2%C." *Metal. Trans. A* 7A: 41–48.

Thiessen, R.G., T Heller, K. Mraczek, A. Nitschke, and A Pichler. 2011. "Influence of Microstructure on the Susceptibility to Hydrogen Embrittlement." In *Steely Hydrogen Conference*. Gent, Belgium.

Toyoda, S. 2011. "Effect of Cu Addition on Hydrogen Absorption and Diffusion Properties of 1470 MPa Grade Thin-Walled Steel Tube under Atmosphere Corrosion." *ISIJ International* 51: 1416–1523.

Toyoda, S., Y. Ishiguro, Y. Kawabata, and K. Sakata. 2008. "Effect of Cu Addition on Delayed Fracture Resistance Welded Tube." *ISIJ International* 48: 640–48.

Zhang, C.L., Y.S. Liu, C. Jiang, and J.F. Xiao. 2011. "Effect of Niobium and Vanadium on Hydrogen-Induced Delayed Fracture in High Strength Spring Steel." *Journal of Iron and Steel Research, International* 18: 49–53.

Chapter 8
Candidates to AHSS of Third Generation: Steels with Carbide-Free Bainite

Contents

8.1 Introduction

The first group ("generation") of AHSS includes low-alloyed DP (Dual Phase), TRIP (Transformation-Induced Plasticity), CP (Complex Phase), and martensitic steels. Growing demands for weight saving and safety requirement motivate intensive elaborations of new concepts of automotive steels that can achieve higher formability simultaneously with higher strength compared to the existing Advanced High Strength Steels (AHSS).

Aiming at extended application of steels that allow for lightening car components, the current status of automotive materials is featured by increased requirements for steels with high and ultrahigh strength to be suitable for forming *complicated* parts. The sharp focus on car weight reduction makes car producers intend replacing grades of lower strength by stronger steels with the same formability as the predecessors. For this, the combinations of extremely high properties are required, for example, TS of 980 MPa at 21 % TE (the same elongation as for current 590 DP grade) or TS of 1180 MPa at TE > 14 % (the same as for 780 DP

© Springer International Publishing Switzerland 2015

N. Fonstein, *Advanced High Strength Sheet Steels*,
DOI 10.1007/978-3-319-19165-2_8

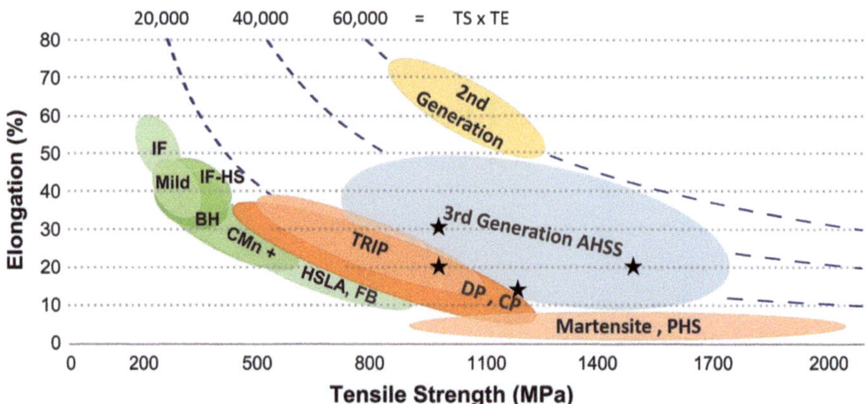

Fig. 8.1 Elongation–tensile strength balance diagram for the existing varieties of formable steels and the prospective "Third generation" grades with some targets marked

grade) or even higher as 980 MPa with 25–30 % TE or 1470–1500 MPa at 20 % elongation. New customer requirements not only include simultaneously higher strength and elongation but often specify flangeability (guaranteed values of hole expansion). These combinations cannot be achieved using existing types of micro-structures of DP or TRIP steels and calls for entirely new approaches.

The "second generation" of AHSS represented by high alloy, high Mn extremely ductile TWIP (*TW*inning-*I*nduced *P*lasticity) austenitic steels with ~1000 MPa tensile strength is described in Chap. 11. These steels delay with commercial applications.

The proposed "third" generation of AHSS (Matlock and Speer 2006) is supposed to be less alloyed (and less expensive) than the TWIP steels and to combine strength of 1000–1500 MPa with elongation of 20–30 %. To satisfy requirements of simul-taneously *H*igh *S*trength and *H*igh *F*ormability (HSHF), these steels should possess not only the extremely high elongation values that approach the area encircled in the diagram of Fig. 8.1 but, simultaneously, essentially high hole expansion (this "third axis" is not presented). Timely development of such steels to satisfy automotive demands is highly important for sustainability of steel in the future cars.

Therefore, the leading steel producers aim at all specified combinations of strength and formability of the "third generation" also keeping in mind the require-ments for increased yield strength and YS/TS. Several prospective candidates based on new microstructure concepts, which are approaching those goals, are presented in the current chapter and in Chaps. 9 and 10.

8.2 Steels with Carbide-Free Bainite or TRIP Steels with Bainitic Ferrite: General Concept

Advantages of bainitic microstructure constituent in multi-phase steels for improv-ing the hole expansion were convincingly shown and led to the development of complex phase steels (see Chap. 6). Improved flangeability is achieved due to smaller difference in hardness between microstructure constituents (Hasegawa et al. 2004). However, CP steels with TS = 800, 1000 MPa and 1180 MPa (the latter is still under development) have rather low elongation showing that they cannot meet the requirements for steels of third generation.

Intensive search for the ways to achieve substantially different combinations of properties appeared to be successful as kind of extrapolation of the advantages of TRIP steels. Traditional TRIP steels produced by cooling from intercritical region have very high strength and elongation at very high strain hardening that usually results in damaged edges during cutting or punching, which lowers the hole expansion. It was found though that preliminary full austenitization (with minimizing of ferrite fraction) and obtaining bainitic matrix with stable austenite that does not transform into martensite in the beginning of deformation can lead to a new type of mechanical behavior of steels.

It was shown that microstructure consisting of carbide-free lath bainite and inter-lath retained austenite can ensure the properties of HSHF steels. Steels with carbide-free bainite (CFB), as they are termed in Europe and USA, or TRIP steels with bainitic ferrite (TBF), as termed in Japan, are based on advantageous combinations of several positive ideas: the presence of substantial fraction of retained austenite that facilitates TRIP effect, elimination or minimizing the fraction of ferrite phase (to target at ferrite-free TRIP steel), and replacement of martensite with bainite thus lowering differences in hardness between phases as well as interface stresses responsible for initiation of cracking.[1]

The CFB steel concept has been intensively studied aiming at substantially higher balance of strength and ductility including higher yield strength at higher flangeability. It has been understood that, similar to TRIP steels, sufficiently high Si content is necessary not only to suppress carbide formation in bainite but also to simultaneously reach higher strength of ferritic bainite and thus to increase the energy absorption due to stronger ductile matrix and contribution to ductility from TRIP effect of retained austenite.

Comprehensive studies in this directions employing austempering from fully austenitic condition were performed by Sugimoto and coworkers beginning in 2000 (Sugimoto et al. 2000a, b, 2002a, b; Sugimoto and Murata 2007). These studies pointed at the advantages of ferrite-free bainite or martensite matrices that can be obtained depending on the position of isothermal holding temperature with respect to martensite transformation start temperature, M_S, as shown in Fig. 8.2.

Similar approach was elaborated in Europe.

In particular, steels with 0.2–0.3 % C, 2.5 % Mn, 1.5 % Si, and 0.8Cr were studied using annealing at 1080 °C and isothermal holding for 40 min at M_S, $M_S + 50$ °C, and $M_S - 50$ °C temperatures. Depending on the holding temperature, the microstructure was composed of fully carbide-free bainitic matrix, when

---

[1] From here on, the term CFB steels is used.

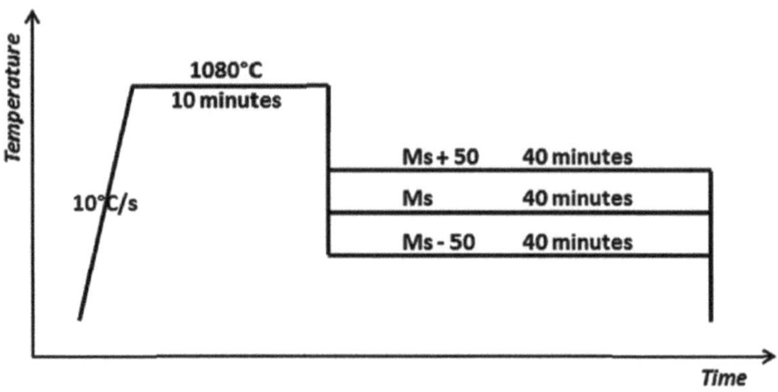

Fig. 8.2 Heat treatment applied to obtain CFB microstructure

Table 8.1 Example of mechanical properties of steels with CBF structure (Hell et al. 2011)

Steel	AT	YS (MPa)	UTS (MPa)	UE (%)	TE (%)	RA (%)	n
C2	$M_s - 50\ ^\circ C$	969	1290	6.6	15.1	46	0.28
	M_s	851	1300	8.6	17.9	41	0.31
	$M_s + 50\ ^\circ C$	835	1360	7.9	13.2	29	0.33
	$M_s - 50\ ^\circ C$	911	1570	7.9	15.2	37	0.24
C3	M_s	975	1490	7.6	15.4	39	0.25
	$M_s + 50\ ^\circ C$	797	1400	13.4	19.6	30	0.36

holding was above M_S, and of self-tempered martensite, when holding was below M_S.[2] Various amounts of retained austenite were obtained along with fresh martensite formed during final cooling by transformation of insufficiently stable austenite remaining at the end of the bainitic reaction (Hell et al. 2011). The resultant mechanical properties are exemplified in Table 8.1. Steel with 0.3 % C isothermally held at 450 °C and containing 21.5 % of retained austenite demonstrated YS = 797 MPa, TS = 1400 MPa with TE = 19.6 %. Fracture strain (reduction of area) up to 60 % suggested that higher values of hole expansion could be reached as well.

8.3 Fundamentals of CFB Steel Processing: Factors Affecting the Kinetics of Bainitic Reaction

Isothermal holding in bainite region should produce certain amount of bainite constituent. If the austempering cycle starts from an intercritical temperature, at least some initial portion of ferrite is unavoidable in the final microstructure

---

[2] Isothermal holding at temperatures below M_S is considered as "one-step Q&P" in the Q&P concept.

although the ferrite component is not desirable to obtain satisfactory high hole expansion because of nonuniform hardness of microstructure. To avoid ferrite transformation of austenite after annealing in supercritical region, the initial cooling to isothermal holding stage should be performed at a high rate. The holding time necessary to achieve a sufficient amount of bainite and to ensure carbon enrichment of the remaining austenite depends on the chemical composition of steel and the holding temperature. Fortunately, some continuous annealing furnaces allow for up to 6–10 min of isothermal holding (in the "overaging" zone) that makes it possible to reach the necessary carbon content and stability of the remaining austenite and to minimize the amount of fresh martensite in final cooling.

With regard to galvannealed/galvanized steels, steelmakers are facing three additional challenges in producing CFB steels. The first is related to limitations in annealing parameters: insufficiently fast initial cooling can induce the formation of undesirable ferrite. The second is related to insufficient time for isothermal/pseudo-isothermal holding and fairly high holding temperatures in bainite region (~460 °C in equalizing zone and Zn pot). This requires all means to accelerate bainitic reaction (lower carbon, minimum alloying). The third challenge is in achievement of necessary strength of bainite (>980 or 1180 MPa), which, as well as prevention of ferrite formation, requires entirely opposite approaches, i.e., higher carbon and heavier alloying.

To help finding a suitable compromise to meet these challenges, it is useful to carefully summarize the effects of steel chemical composition, initial microstructure, and annealing parameters on the kinetics of bainite transformation.

8.3.1   Effect of Temperature and Time of Isothermal Holding in Bainite Region

The relationship between the temperatures of isothermal holding and the rates of bainitic reaction is described by TTT diagrams.

As shown in the work by Hell, the increase in the temperature of isothermal holding of 2.5Mn–1.5Si–0.8Cr steels with 0.2 and 0.3 % C from $M_S - 50$ °C to M_S and then to $M_S + 50$ °C slowed the kinetics of bainite transformation (Hell et al. 2011). Similar trend was noted in numerous studies of bainitic reaction as, in particular, in (Alvarez et al. 2014) Fig. 8.3. It was pointed out that having isothermal holding shorter than 10 min is insufficient and low temperatures of bainite transformation within the range of $M_S + 10$ °C should be used.

Using   0.28C–1.96Mn–0.67Si–1.19Al–1.62Cr–0.34Ni–0.23Mo   steel,   Qian et al. intentionally employed isothermal holding at fairly low temperature of 320 °C (B_s of the steel was 486 °C) that required long holding, for 30–84 min, to obtain the optimal microstructure of fine bainitic laths and film-type retained austenite formed at low transformation temperature (Qian et al. 2012). With holding time increasing from 30 to 84 min, the tensile strength decreased slightly

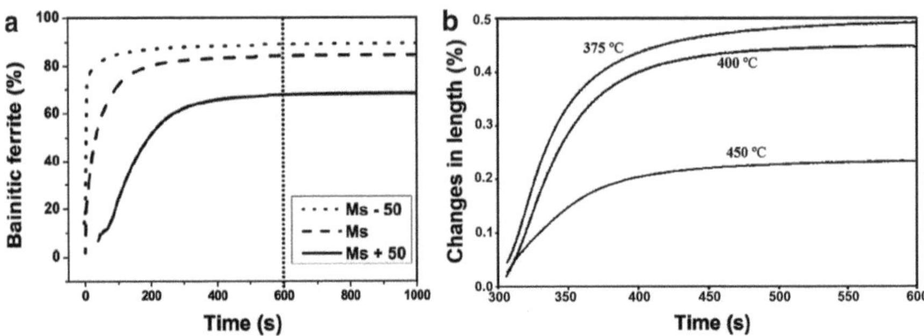

Fig. 8.3 Effect of isothermal holding temperature on the kinetics of bainite reaction: (**a**) 0.2C–2.5Mn–1.5Si–0.8Cr (Hell et al. 2011). (**b**) 0.24C–1.36Mn–0.55Si–0.85Al–2.1Cr steel, annealing at 900 °C, 100 s, cooling at 30 °C/s; $A_{C3} = 892$ °C (Alvarez et al. 2014)

from 1486 to 1432 MPa at the increase of elongation from 17 to 20 %, while the yield strength remained almost constant.

Misra and coauthors (Misra et al. 2012) also pointed out that the best combination of properties for 0.3C–1.76Si–1.57Mn–0.025Mo–0.144Cr steel was observed with isothermal holding at 350 °C for 30 min. They explained this fact by carbon partitioning from bainite to the remaining austenite that reached maximum under these conditions and lowered its M_S below room temperature, which consequently prevented the formation of fresh martensite during cooling. According to their observations, 1.7 % Si in steel did not prevent carbide formation at higher temperature or longer isothermal holding. The combination of YS = 1557 MPa, TS = 1656 MPa, and TE = 15.5 % was achieved.

Although the decrease in isothermal holding temperature complicates the annealing in coating lines, it is worth to note that preliminary (before Zn pot) overcooling with isothermal holding at temperatures as close to M_S as possible or even below M_S is favorable for the microstructure and properties of CFB steels. First, formed below M_S martensite drastically accelerates the kinetics of bainitic reaction (Kawata et al. 2010). Second, holding at lower temperature results in better bainite morphology, as well as higher volume fraction of retained austenite with the highest concentration of carbon (Sugimoto et al. 2007).

8.3.2 Effect of Alloying and Microalloying Elements

In general, the approach to the composition design for CFB steels is very similar to that for TRIP steels with a couple differences. First, there is almost no possibility to use Al since it strongly raises the A_{C3} temperature to the values that cannot be reached in industrial equipment when full austenitization is required. Second, higher alloying with austenite stabilizing elements like C, Mn, Cr, or even Cu or

Ni is to be used to prevent ferrite formation during cooling of austenite to the isothermal holding temperature.

The kinetics of bainite transformation is very important for processing CFB steels in view of limited overaging times achievable for the holding temperatures mostly around or below 400 °C in continuous annealing lines and around 460 °C in CGL.

With this regard, all the factors that can accelerate bainite transformation are beneficial for processing these steels.

Some pertinent data are presented in Chap. 6 because facilitation of bainite is also important to successfully manufacture complex phase steels.

Carbon Effect The important effect of carbon on hardenability of austenite is well known. Therefore, CFB steels have the carbon content increased to 0.2–0.3 % aiming to prevent ferrite formation during initial cooling.

On the other hand, carbon slows the bainite transformation by lowering the B_s temperature and hindering the growth of bainite (Ali and Bhadeshia 1989; Quidort and Brechet 2002), although the incubation period for ferrite formation is affected by carbon more significantly than that for bainite transformation (Xu et al. 2010), as was presented in Fig. 6.9 for low-carbon steels. As shown in CCT diagrams presented in Fig. 8.4, the same trend is observed when carbon content increases from ~0.2 to ~0.4 %.

Effects of Mn, Mo, and Cr Among other alloying elements, Mn, Mo, and Cr are the elements that have the strongest effects on the kinetics of bainite transformation (Quidort and Brechet 2002).

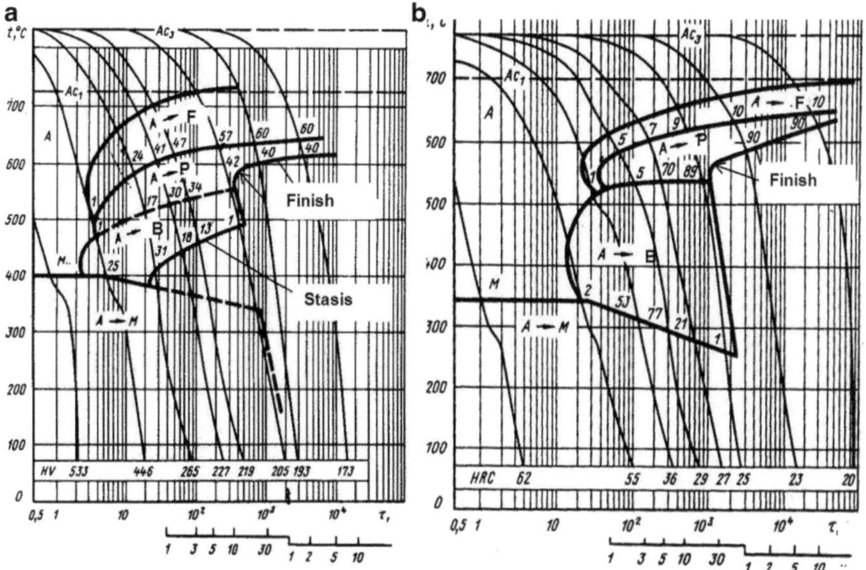

Fig. 8.4 Effect of carbon on CCT diagrams for 0.23C–0.4Si–1.53Mn (**a**) and 0.42C–0.27Si–1.42Mn (**b**)

On one hand, Mn as an austenite stabilizer slows the decomposition of austenite that helps preventing ferrite formation during cooling. However, higher Mn content lowers both A_{r3} and B_s temperatures that strongly delays the austenite-to-bainite transformation by solute drag. In addition, Mn segregation to prior austenite grain boundaries restrains the nucleation of bainitic ferrite. The measured enrichment factors of Mn segregation vary from 1.5 to 2.77. Thus, the TTT curves are shifted to longer times: when Mn content in the steel with 0.38 % C is increased from 1.73 to 3.11 %, the start and finish of bainite formation during isothermal holding at 450 °C or 400 °C increases from 2–3 s to 70 s and from 15 to 300 s, respectively.

Decrease in the austenitization temperature (the authors ensured the constancy of austenite grain size using the same preheating temperature) before fast cooling to isothermal holding appeared to shorten the time needed for the initiation of transformation by reducing nonequilibrium segregation of Mn to austenite grain boundaries (Liu and Zhang 1990).

As a result of those segregations, at the same holding time, the amount of austenite transformed to bainite decreases with increasing Mn content (Fig. 8.5a) (Bhadeshia 2001).

Effect of Mo and Cr during continuous cooling was studied by numerous researchers. The conclusion was that both substantially retard ferrite and especially pearlite formation.

Quidorf investigated the effect of Cr additions to base 0.2C–1.5Mn steel on bainite transformation. The retardation of isothermal transformation with increase in Cr content was very substantial. As shown in Fig. 6.10, the bainite transformation can be completely avoided in steel with addition of 3 % Cr (Quidort and Brechet 2001). The same very strong role of Cr is confirmed by data presented by Bhadeshia, as shown in Fig. 8.5b.

It was found that Mo slightly retards the isothermal phase transformation kinetics at a temperature above 675 °C but affects greatly the phase transformation at lower temperatures, when the higher the Mo content was, the more significant the retardation effect was observed.

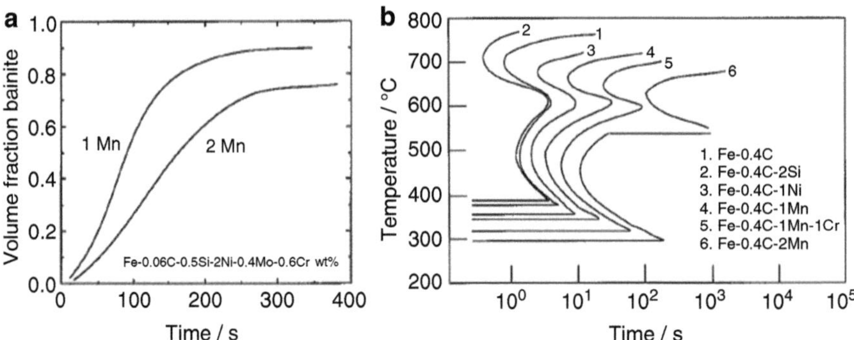

Fig. 8.5 Effect of Mn on the volume fraction of bainite (**a**) and comparative effects of various alloying elements on TTT diagram of steel with 0.4 % C (**b**) (Bhadeshia 2001)

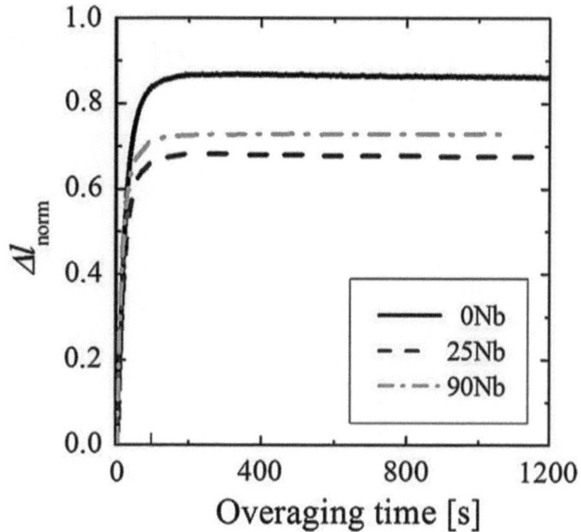

Fig. 8.6 Effect of Nb on normalized length of dilatometer samples during overaging at 400 °C (Hausmann et al. 2013b)

Effect of Nb, Boron, and B + Nb Additions   Steels with 0.05C–1.5 Mn containing 0.054 and 0.1 % Nb were studied. To completely dissolve Nb, steels were heated to 1250 °C, which produced the austenite grain size of 200 μm. Steels were then rapidly quenched to 600–450 °C to obtain TTT diagrams (Dacker et al. 2012). In both steels, the transformation started almost immediately after quenching. Addition of 0.1 % Nb induced a little delay in austenite decomposition at 600 °C but slightly accelerated the bainite transformation at 550 °C or lower temperatures. The same effect of Nb additions on the parameters of bainitic reaction was found by Hausmann et al., Fig. 8.6 (Hausmann et al. 2013b). The width of bainite lath reduced with a decrease in holding temperature.

It was observed that 0.025 % Nb significantly refined austenite grain size. Because of that, more nucleation sites (triple junctions or grain boundaries of prior austenite grains) are available, which usually accelerates ferrite formation and leads to acceleration of bainite formation as well. Since there was no significant grain refinement at Nb content above 0.025 %, quite similar transformation behavior was observed for all studied concentrations of Nb (including 0.045 and 0.090 %).

As shown in a number of studies, B retards the bainite transformation.

According to Song and De Cooman (2013), the addition of boron increased the incubation time, i.e., the time necessary to form a detectable amount of bainite, and slowed the overall transformation kinetics, more noticeably in the high temperature range of bainite transformation. The results for 0.15C–1.96Mn–0.3Si–0.3Cr–0.3Mo steels without B, with additions of boron, and with additions of B + Ti, when the presence of Ti enhanced the efficiency of boron by reducing the loss of soluble B to BN precipitation, are presented in Fig. 8.7.

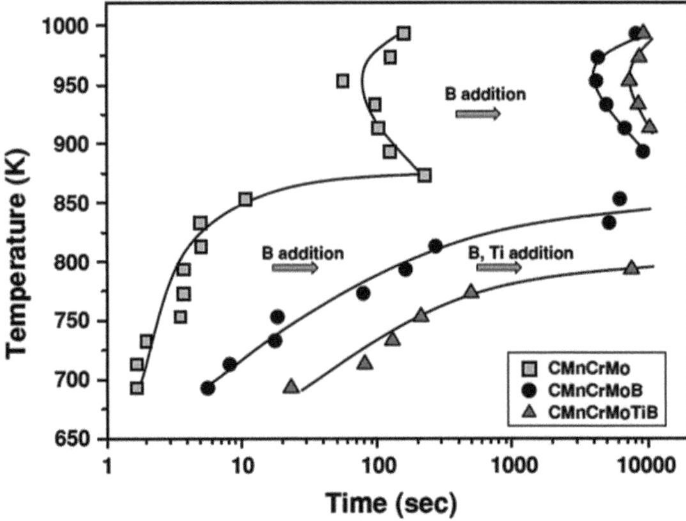

Fig. 8.7 Effects of B and B + Ti on TTT diagrams for 5 % bainite transformation (Song and De Cooman 2013)

Combined B + Nb additions significantly retarded the bainite transformation so that the transformation was not complete. Therefore, a large portion of "fresh" martensite should be present in the final microstructure. The increase in austenite grain size retards bainitic transformation kinetics in B-bearing steel even further.

In the study by Zhu et al. (2011), the mechanism of this type of behavior was considered in detail. In B-bearing steel, coarse $Fe_{23}(BC)_6$ carbides precipitate at prior austenite grain boundaries due to the segregation of both B and C. In B + Nb steel, the precipitation of these $Fe_{23}(BC)_6$ carbides is suppressed owing to decrease in carbon concentration at austenite grain boundaries as a result of lower C diffusivity due to solute Nb and the presence of Nb(C,N) clusters. Consequently, more free B atoms or very fine B precipitates cover the austenite grain boundaries so that the number of nucleation sites for bainitic ferrite is greatly reduced. Therefore, not only the nucleation kinetics for bainite formation in B + Nb steel is significantly retarded, but also the growth kinetics is slowed down since fewer groups of laths can develop.

Effects of Si and Al It is believed that the formation of cementite is the most critical way to consume carbon in front of bainitic ferrite, and this is an essential step to precede further bainite transformation. Therefore, the addition of Si and Al, which retard cementite formation, results in slowing the overall bainite transformation (Quidort and Brechet 2001). Sufficiently high additions of Si and Al can completely cease the formation of cementite resulting in formation of the so-called Carbide-Free Bainite.

As shown in the study of the combined effects of Mn and Si (Liu and Zhang 1990), Si enhances Mn segregation to austenite grain boundaries and inhibits the precipitation of Fe_3C, thereby increasing the carbon content in the untransformed austenite and consequently decreasing the nucleation of bainitic ferrite.

Quidort studied the effect of Si on isothermal bainite transformation in 0.5C–5Ni–Si steel. As was presented in Fig. 6.13, Si has strong retarding effect on formation of bainite (Quidort and Brechet 2002).

Aluminum, which suppresses carbide formation, was also found to stabilize austenite by increasing the concentration of carbon in it (Sugimoto and Murata 2007).

If it is possible (based on annealing parameters) to partially replace silicon with aluminum, then Al can be efficient in retention of austenite in CFB steels that can in turn lead to increase in uniform elongation. It has been also observed that the addition of aluminum tends to refine the bainitic lath microstructure, thus accelerating the formation of bainitic ferrite (DeMeyer et al. 1999; Hojo et al. 2008). This is one of the most important features potentially useful for producing CFB steels.

8.3.3 Grain Size Effect

As shown in several studies, particularly in (Godet et al. 2003), Fig. 5.4, the bainite transformation delayed in coarse grain steel. This is quite evident because bainitic ferrite mainly nucleates at prior austenite grain boundaries, and the increase in austenite grain size reduces the number of bainite nucleation sites.

Dacker et al. also demonstrated the effects of grain size on the kinetics of bainitic reaction, when grain size in 0.053C–0.292Si–1.495Mn steel was changed by varying the preheating temperatures (Dacker et al. 2012).

Alloying and especially microalloying usually refine austenite grain size that, in turn, should influence the bainite transformation kinetics. However, the transformation kinetics also strongly depends on the amount of austenite stabilizing elements. For example, comparison of two steels with different grain size and different alloying (C, Mn, etc.) revealed that the transformation kinetics, estimated by the time between the beginning and the end of transformation, in more alloyed steel albeit with smaller austenite grain size was much slower, being controlled by the dominating role of austenite stabilizers (Dacker et al. 2012).

Matsuzaki and Bhadeshia also found that the austenite grain size can induce opposite effects on the isothermal bainite transformation depending on steel chemical composition. In high C steel (0.96C–0.21Si–0.38Mn–1.26Cr), the refinement of austenite grain structure led to acceleration of transformation when the overall reaction is controlled by *slow growth* rate. In low C steel (in particular, 0.12C–2.03Si–2.96Mn), with the composition considered close to AHSS chemistries, the transformation is dominated by *rapid growth* from a limited number of *nucleation* sites. Therefore, austenite grain size refinement reduced the total volume of

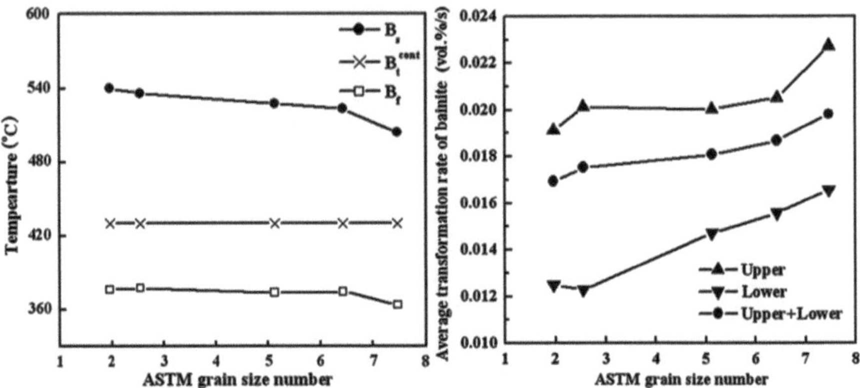

Fig. 8.8 The effects of austenite grain size on bainite transformation temperatures (*left*) and transformation kinetics (*right*) (Lee et al. 2008)

transformed bainite nuclei and hence retarded the overall reaction (Matsuzaki and Bhadeshia 1999).

Lee et al. studied the effect of austenite grain size on the bainitic transformation kinetics during continuous cooling of low-medium steel with 0.39C–0.65Mn–0.24Si–1.6Ni–0.67Cr–0.15Mo. It was found that the nucleation rate is much faster than the growth rate, which was similar rather to the high C steel in Matsuzaki's work. Figure 8.8 shows that refinement of austenite grain is slightly-lowering the B_s and B_f temperature enhancing the transformation kinetics (Lee et al. 2008).

8.4   Factors Determining the Strength of CFB Steels

CFB steels target tensile strength of 980, 1180 MPa and higher. Bearing in mind that the contribution of fresh martensite to strength is undesirable, strengthening of these steels is preferable mostly by the strength of bainite. The latter can be controlled by chemical composition of steels and the parameters of transformation that determine the morphology of bainite.

8.4.1   *Effect of Alloying and Microalloying Elements*

Alloying or microalloying additions to CFB steels affect the strength by changes in the morphology of bainite and dislocation density, as well as in solid solution strengthening of bainitic ferrite plates, distribution and morphology of cementite, and grain size.

Effects of Mn, Cr, and Mo Hausmann et al. developed the 0.2C–1 % Si-based steel with the combination of Mn, Cr, and Mo (within the sum of 2.6–2.8 %), which after full austenitization, cooling, and holding at 375 °C for 300 s demonstrated YS = 950 MPa, TS ~ 1200 MPa at TE = 12 % (Hausmann et al. 2013a).

The European "Ductaform: New AHSS Bainitic steels" research program with a long list of participants of the consortia coordinated by F. Caballero (CENIM-CSIC) considered the development of compositions and processing parameters to obtain bainitic microstructure after annealing and/or hot-dip galvanizing. The compositions under consideration contained 0.15–0.25 % C, ~1.5% Si, 1.5–2.5 % Mn, and 0.6–1.5Cr.

Coiling temperature was chosen to be below 600 °C to prevent possible inter-granular oxidation of silicon but was targeted at ferrite–pearlite microstructure of hot-rolled bands to enable cold reduction of 50 %.

It was shown that the 0.17C–1.47Mn–1.49Si–0.74Cr composition allowed for TS > 1200 MPa at TE > 15 % and high reduction of area, RA ~50 % indicative of increased HE, after processing using annealing cycle with direct cooling to isothermal holding in bainite area at 400 °C for 6 min.

It is necessary to note that so far, the compositions under consideration have not allowed for industrially obtaining bainitic microstructure using HDG cycles. More realistic is to match short pseudo-isothermal holding in bainite area using 0.20C–1.0Mn–1.5Si composition, but the complete transformation requires at least 60–70 s in this case as well.

Looking at various bainite structures that are featured with high strength, it is necessary also to mention SOLAM 1100 grade developed by ArcelorMittal with composition of 0.18C–1.7Mn–1Cr–0.5Ni + (Mo,Nb,B,Ti) that has been designed for obtaining UTS > 1100 MPa after cooling of forged parts. CCT diagram of this steel shows B_s around 600 °C in a very wide range of very slow cooling rates that starts lowering at cooling rates above 1 °C/s and reaches 500 °C at cooling rate above 10 °C/s.

In the bainite transformation temperature range 600–450 °C, the hardness of bainite increases with concentration of Mo. Addition of up to 0.5 % Mo leads to increase in hardness of about 50 HV (180 MPa). It was shown that the mean Kernel average misorientation as a function of holding temperature increases with addition of molybdenum. This can be related to higher dislocation density and/or finer sub-grain structure in the Mo-bearing steels.

An example of UHSS with bainitic matrix was presented by Qian et al. (2012). These authors showed that 0.28C–1.96Mn–0.67Si–1.19Al–1.62Cr–0.34Ni–0.23Mo steel (B_s = 486 and M_S = 283 °C) after austenitization at 920 °C and isothermal holding at 320 °C demonstrated TS ranging from 1486 to 1437 MPa and elongation from 17 to 20 %, respectively (depending on holding time) with YS close to 1200 MPa.

Effects of Si and Al Additions of Si and Al are very effective in suppressing the precipitation of cementite. The optimal microstructure consists of mainly fine carbide-free bainitic ferrite and the film-like retained austenite located between the bainitic ferrite laths and exhibits the excellent combination of strength, ductility, and toughness.

Based on thermodynamic and kinetics models, a set of four hot-rolled carbide-free bainitic steels with 0.3 % C with various combinations of 1.5–1.8 % Si, 1.5–2.25 % Mn, 0.17–1.47 % Cr, 0.25 % Mo, and (in one composition) 1.5 % Ni was designed. Steels were manufactured using a thermo-mechanical processing comprising hot rolling and two-step cooling: initial accelerated cooling at 50 °C/s followed by air cooling from 500–600 °C. This allowed achieving TS from 1500 to 1800 MPa and TE > 15 %. Reduction in carbon content to 0.2 % was proposed to improve in-use properties including weldability (Caballero et al. 2009).

Another prospective development was described by Sharma et al. (2011). These authors also used two relatively high carbon steels 0.47C–1.22Si–1.07Mn–0.7Cr and 0.3C–1.76Si–1.57Mn–0.144Cr and applied different isothermal holding temperatures and times. With isothermal holding at 350 °C, the steels demonstrated TS = 1550–1650 MPa at very high reduction of area, which was explained by the presence of carbide-free bainite as the dominant microstructure constituent. The reduction of area as a parameter of post-necking deformation is in good correlation with hole expansion (Sugimoto et al. 2000b).

Despite the fact that with additions of Al, certain amount of initial ferrite in the microstructure of CFB steel becomes unavoidable due to very high A_{C3} temperature; Sugimoto et al. pointed out some advantages of partial replacement of Si with Al, e.g., in 0.2C–0.98Si–1.48Mn–0.47Al steel, for improving galvanization and resistance to hydrogen embrittlement (Sugimoto et al. 2007).

Effect of Microalloying According to the European BainHard project report (Dacker et al. 2012), Nb-microalloyed CBF steels are generally harder than the reference Nb-free steels. However, no significant difference between 0.05Nb and 0.1Nb steels was found after processing at a given holding temperature. For Nb-free steels, the hardness increases with decreasing holding temperature, whereas higher hardness of Nb-added steels is almost the same regardless of the transformation temperatures. The latter observation can be related to the overlapping with precipitation hardening. The amount of Nb precipitates at lower holding temperatures should be smaller, so that hardness should decrease. But this decrease is probably compensated by hardening due to refinement of the bainite microstructure at lower holding temperatures.

Grain size is extremely important in determining the matrix properties, as confirmed by the study of Nb effect in 0.2 % C–1.5 % Si–1.5 % Mn steel with bainitic matrix. Addition of 0.05 % Nb raises both yield and tensile strength of CFB steels due to microstructure refinement, and Nb precipitates and improves stretch flangeability due to uniformly refined microstructure of the matrix. The total elongation is also improved (Sugimoto and Murata 2007).

A detailed work was performed by Hausmann et al. (2013b). The authors compared the effects of adding 0.0, 0.025, 0.045, and 0.09 % Nb to 0.17C–0.8Si–2.6 (Mn + Cr + Mo) steel. Steels were austenitized at 900 °C for 60 s, cooled at 50 °C/s to the temperature of isothermal holding (overaging) ranging from 350 to

475 °C, held at these temperatures for 30–600 s, and finally cooled to room temperature at 50 °C/s. It was found that the addition of 0.025 % Nb refined prior austenite grains from 12 μm to ~6 μm. Higher Nb content did not induce any further grain refinement. It was concluded that 0.025 % Nb was sufficient to achieve the advantages of microalloying with Nb.

Grain refinement resulted in the formation of ~20 % ferrite during cooling to isothermal holding at the cooling rate employed by these authors. The morphology of bainite and retained austenite with addition of Nb also changed: the lath-type bainite in Nb-free steel was replaced by globular bainite in Nb-added steels. This also altered the morphology of retained austenite from lath-type to blocky-type. The dominating factor, though, was that the reduced amount of cementite found in Nb-microalloyed steels resulted in more carbon available for stabilization of retained austenite and led to its higher volume fraction. Finally, adding of Nb appeared to be beneficial for production of sheet steel with TS above 980 MPa and higher uniform and total elongation.

As opposed to the conclusions made by Sugimoto et al., in the described work, Nb was not found to affect hole expansion of the developed steel, which demonstrated the tendency of decreasing HE with the overaging temperature increasing from 350 to 475 °C regardless of Nb content.

8.4.2   Bainite Morphology

If the initial cooling rate after full austenitization is high enough to avoid ferrite formation, the low-carbon steels undergo bainite transformation during isothermal holding in bainite region. Bainite transformation start (B_s) temperature, like F_s and M_S, depends on chemical composition of steel. Several empirical equations listed in Chap. 6 show the effects of alloying elements on the B_s temperature. As demonstrated by Steven and Haynes (1956) and Bodnar et al. (1989), the B_s temperature in low-alloyed steels is most strongly lowered by carbon, followed by Mn.

The morphology of bainite is determined by both temperature of transformation and chemical composition. The existing classification of bainite types is not quite suitable for CFB steels because it is based on cementite containing microstructures (Bramfitt and Speer 1990; Ohtani et al. 1990). According to Ohtani et al., BI is bainite formed at high temperatures without precipitation of carbides. BII bainite is formed in the intermediate temperature range with cementite precipitating between bainitic ferrite laths (upper bainite). BIII bainite is formed at low temperatures with fine cementite aligning specific plane within bainitic ferrite (lower bainite). Bramfitt and Speer classified the morphology of bainite according to the morphology of ferrite and the type of secondary phases such as cementite, martensite, and austenite.

Probably, the most suitable classification of bainite in CFB steels was recently proposed by Zajac et al. (2005) based on misorientation distribution measured by EBSD to distinguish the morphologies of lower, upper, and granular bainite depending on the misorientation angle, as shown in Fig. 8.9.

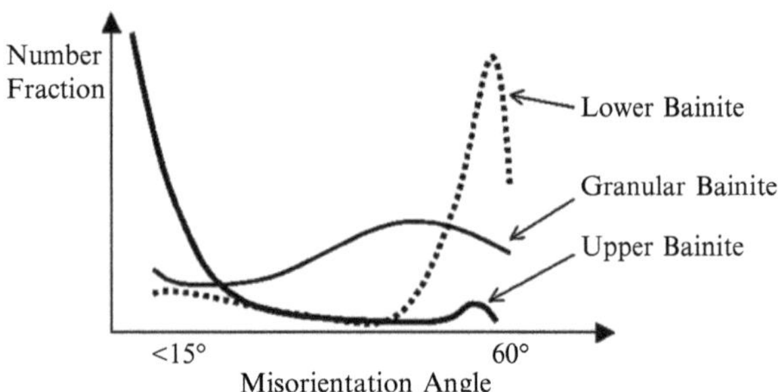

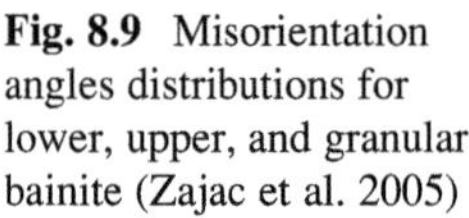
Fig. 8.9 Misorientation angles distributions for lower, upper, and granular bainite (Zajac et al. 2005)

The correlation between the strength of bainite and the B_s temperature as the main factor affecting morphology of bainite and its strength was presented in Chap. 6, where gradual increase in strength with lowering the temperature of bainite formation was shown (DeArdo et al. 2009).

Since important expectations for carbide-free bainitic ferrite steels concern better stretch flangeability, it is worth considering the effects of bainite morphology on strength and hole expansion of CFB steels. It was shown that lowering the temperature of isothermal holding resulted in higher energy of crack propagation and higher hole expansion of CP steels (Fonstein et al. 2011). Hausmann et al. came to the same conclusion in their study of CFB steels (Hausmann et al. 2013a).

Zhu et al. investigated the effects of B and Nb on bainite morphology. After isothermal holding at 600 °C, they observed granular morphology of bainite in base steel, whereas in B and B + Nb-alloyed steels bainite had lath morphology. Larger packets and longer laths were found in B + Nb-added steel. Based on EBSD analysis, the bainite was identified as upper bainite by measuring grain boundary misorientation angles. It is important to note that those investigations were performed after extremely high austenitization temperature to ensure Nb dissolution (1100 °C) at extra-low carbon content (0.04 %) (Zhu et al. 2011).

According to Song and De Cooman, the addition of boron to 0.15C–2Mn–0.3Si–0.3Cr–0.3Mo steel influenced the morphology of bainite formed at higher holding temperatures, such as 500 °C (Song and De Cooman 2013). The bainite microstructure consisted of bainitic ferrite matrix and M/A constituent. In boron-added steel, the M/A constituent had granular morphology, whereas it was of elongated type in B-free steel.

Hell et al. used EBSD analysis and detected only minor differences between the microstructures obtained at different isothermal holding temperature (Hell et al. 2011). This contradiction can probably be explained by rather low studied temperatures in quite narrow range (from $M_S - 50$ °C to $M_S + 50$ °C). All their samples exhibited crystallographic characteristics pertinent to lower bainite, when all misorientation angle distributions showed high peaks at 60 °C inherited from the bainite transformation.

8.4.3 Fresh Martensite

During short holding in bainite region, the amount of carbon rejected from bainite to the remaining austenite is not high enough, and the low carbon portion of austenite transforms into hard and relatively brittle fresh martensite that can deteriorate the total ductility of a CFB steel and especially its hole expansion.

One suggested remedy to compensate for this setback is the post-tempering of annealed steel. The other should include all possible ways to facilitate possible self-tempering of fresh martensite and its higher toughness.

8.5 Effects of Parameters of Retained Austenite on Ductility and Localized Fracture

The contribution of retained austenite to the ductility of bainitic steels is still not completely clear. In general, there are two possible situations. One is when metastable austenite transforms into martensite during deformation resulting in transformation-induced plasticity. To exploit this situation is the main purpose of microstructure optimization in what was originally called TRIP steels with bainitic ferrite. The other situation is when stable austenite with higher concentrations of austenite stabilizing elements (e.g., Mn, C) and smaller particle size remains untransformed during deformation. Some authors conclude that this tiny film-like austenite is useless and that this portion of RA does not really contribute to the TRIP effect. However, a potential role of duplex structure with strong ductile (fcc) constituent, reaching the total volume fraction up to 15 %, requires detailed studies regardless of its involvement in martensite transformation.

In particular, Wang et al. (2012) carried out TEM study of transformation of retained austenite into martensite and found out that the dislocation density in bainite became lower than it had been before deformation. The authors attributed this phenomenon to continuous dislocation absorption by retained austenite (DARA) that enhanced ductility of carbide-free bainitic steels. In authors' opinion, this effect brings hard bainite into "softened" or "non-work hardened" condition resulting in more harmonized deformation of bainite–austenite mixture.

More controversial is the role of retained austenite in flangeability. Whereas the strain-induced austenite-to-martensite transformation (TRIP effect) is well known to improve the ductility in tension, it is not clear whether this is good for such localized deformation as hole expansion. For example, the TRIP effect was found harmful for hole expansion of high-Mn steel due to formation of hard martensite during austenite–martensite transformation that facilitated crack propagation (De Cooman et al. 2012). Besides, the initial high strain hardening enhances damage of cut or punched edge, thus initiating a crack in the process of hole expansion test. This led to important conclusion that it is necessary to find the ways to balance high strain hardening typical for TRIP steels and fracture strain (reduction of area) with the latter related to flangeability (hole expansion) of CFB steels.

According to Hell et al., increase in isothermal holding temperature leads to coarser sizes of MA blocks with carbon content in austenite insufficient to stabilize all of the remaining austenite at room temperature (Hell et al. 2011). As a result, the volume fraction of fresh martensite increases at the expense of the volume fraction of retained austenite.

As shown in Table 8.1, which is based on the data obtained by Hell, raising the temperature of isothermal holding from $M_S - 50\,°C$ to M_S and further to $M_S + 50\,°C$ substantially increases strain hardening (lowers YS/TS ratio) that is accompanied by decrease in fracture strain (reduction of area) that correlates with hole expansion.

It was noted that more stable higher carbon retained austenite could improve the hole expansion ratio by contributing to ductility up to post-necking stage without damaging the zone of localized deformation by transformed martensite near the hole at initial lower strains (Sugimoto et al. 2000b). The general conclusion was that the smaller the amount of austenite and the higher its carbon content, the better the balance of formability of steel including elongation and hole expansion (Fig. 8.10) (Sugimoto et al. 2002a, b).

Some publications by Sugimoto were summarized in Chap. 6. These data demonstrate the importance of small volume fraction of retained austenite that should contain maximum possible amount of carbon and should preferably present in the form of inter-lath films. This can be interpreted as the necessity to ensure very high stability of austenite to prevent its strain-induced transformation at early

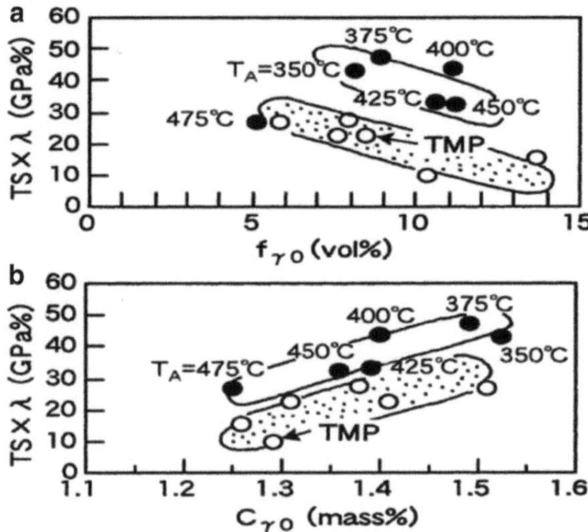

Fig. 8.10 Correlation between strength–stretch flangeability balance (TS × λ) and (**a**) initial volume fraction ($f_{\gamma 0}$) and (**b**) initial carbon concentration ($C_{\gamma 0}$) in retained austenite for fully austenitized TBF steels (*solid dots*) and steels austempered after intercritical annealing (*open dots*) (Sugimoto et al. 2000b)

deformation stages during hole expanding but to facilitate its contribution to localized ductility by the TRIP effect at the latter stages of the hole expansion process.

In these publications, the importance of obtaining a strong matrix with moderate ratio of strength of retained austenite to that of matrix strength is also emphasized, as this contributes to higher ability of CFB microstructure to bear higher internal stresses without inter-phase cracking.

8.6 Summary

CFB (or TBF) steels are expected to be superior to DP steels with respect to the combination of strength and uniform and total elongation and to surpass TRIP steels with polygonal ferrite in terms of edge formability (flangeability, hole expansion). It is the fact though that, depending on steel composition and primarily on time and temperature available for isothermal holding, these steels can contain certain amounts of fresh martensite, which can be formed during final cooling as a result of incomplete bainite reaction and which ideally should be tempered by post-annealing.

References

Ali, A., and H.K.D.H. Bhadeshia. 1989. "Growth Rate Data on Bainite in Alloy Steels." *Materials Science and Technology* 5 (April): 398–402.

Alvarez, D., J. Ferreno, and J.M. Artimez. 2014. "Development of Low Silicon Carbide-Free Bainitic Steels for Automotive Industry." In *TRA*, VII – 1–7. Paris.

Bhadeshia, H.K.D.H. 2001. *Bainite in Steels*.

Bodnar, R., T. Ohhashi, and R. Jaffee. 1989. "Effects of Mn, Si, and Purity on the Design of 3.5NiCrMoV, 1CrMoV, and 2.25Cr-1Mo Bainitic Alloy Steels." *Metallurgical and Materials Transactions A* 20 (8): 1445–1460.

Bramfitt, B.L., and J.G. Speer. 1990. "A Perspective on the Morphology of Bainite." *Metallurgical and Materials Transactions A* 21 (4): 817–829.

Caballero, F.G., M.J. Santofimia, C. García-Mateo, J. Chao, and C. García de Andrés. 2009. "Theoretical Design and Advanced Microstructure in Super High Strength Steels." *Materials & Design* 30 (6): 2077–2083.

Dacker et al., C.-A. 2012. *Bainitic Hardenability – Effective Use of Expensive and Strategically Sensitive Alloying Elements in High Strength Steels (BainHard)*. European Commission, Directorate-General for Research and Innovation.

DeArdo, A.J., M.J. Hua, K.G. Cho, and C.I. Garcia. 2009. "On Strength of Microalloyed Steels: An Interpretive Review." *Materials Science and Technology* 25 (9): 1074–1082.

De Cooman, B.C., O. Kwon, and K.G. Chin. 2012. "State-of-the-Knowledge on TWIP Steel." *Materials Science and Technology* 28: 513–527.

DeMeyer, M., D. Vanderschueren, and B.C. De Cooman. 1999. "The Influence of the Substitution of Si by Al on the Properties of Cold rolled C-Mn-Si TRP Steels." *ISIJ International* 39 (8): 813–832.

Fonstein, N., H.J. Jun, G. Huang, and et al. 2011. "Effect of Bainite on Mechanical Properties of Multi-Phase Ferrite-Bainite-Martensite Steels." In *MST'11*.

Godet, S., C. Georges, and P.J. Jacques. 2003. "On the Austenite Retention in Low Alloy Steels." In *TMS-ISS Conf.*, 523–536. Chicago.

Hasegawa, K., K. Kawamura, T. Urabe, and Y. Hosoya. 2004. "Effects of Microstructure on Stretch-Flange-Formability of 980 MPa Grade Cold-Rolled Ultra-High Strength Steel Sheets." *ISIJ International* 44 (3): 603–609.

Hausmann, K., D. Krizan, A. Pichler, and E. Werner. 2013a. "TRIP-Aided Bainitic-Ferritic Sheet Steel: A Critical Assessment of Alloy Design and Heat Treatment." In *MS&T'13*, 209–218. Montreal, Canada.

Hausmann, K., D. Krizan, K. Spiradek-Hahn, A. Pichler, and E. Werner. 2013b. "The Influence of Nb on Transformation Behavior and Mechanical Properties of TRIP-Assisted Bainitic-Ferritic Sheet Steels." *Material Science and Engineering A* 588: 142–150.

Hell, Jean-Christophe, Moukrane Dehmas, Sébastien Allain, Juscelino Mendes Prado, Alain Hazotte, and Jean-Philippe Chateau. 2011. "Microstructure – Properties Relationships in Carbide-Free Bainitic Steels." *ISIJ International* 51 (10): 1724–1732.

Hojo, T., K.-I. Sugimoto, Y. Mulai, and S. Ikeda. 2008. "Effect of Aluminium on Delayed Fracture Properties of Ultra-High Strength Low-Alloy TRIP-Aided Steels." *ISIJ International* 48 (6): 824–829.

Kawata, H., K. Hayashi, N. Sugiura, and et al. 2010. "Effect of Martensite in Initial Structure on Bainite Transformation." *Materials Science Forum* 638–642: 3307–3312.

Lee, Seok-Jae, June-Soo Park, and Young-Kook Lee. 2008. "Effect of Austenite Grain Size on the Transformation Kinetics of Upper and Lower Bainite in a Low-Alloy Steel." *Scripta Materialia* 59 (1): 87–90.

Liu, S.K., and J. Zhang. 1990. "The Influence of the Si and Mn Concentrations on the Kinetics of the Bainite Transformation in Fe-C-Si-Mn Alloys." *Metallurgical Transactions* 21A (6): 1517–1525.

Matlock, D., and J. Speer. 2006. "Design Consideration for the next Generation of Advanced High Strength Sheet Steels." In, 774–781. Korea.

Matsuzaki, A., and H.K.D.H. Bhadeshia. 1999. "Effect of Austenite Grain Size and Bainite Morphology on Overall Kinetics of Bainite Transformation in Steels." *Materials Science and Technology* 15 (5): 518–522.

Misra, A., S. Sharma, S. Sangal, A. Upadhyaya, and K. Mandal. 2012. "Critical Isothermal Temperature and Optimal Mechanical Behavior of High Si-Containing Bainitic Steels." *Material Science and Engineering A* 558: 725–729.

Ohtani, H., S. Okaguchi, Y. Fujishiro, and Y. Ohmori. 1990. "Morphology and Properties of Low-Carbon Bainite." *Metallurgical Transactions A* 21 (3): 877–888.

Qian, Lihe, Qian Zhou, Fucheng Zhang, Jiangying Meng, Ming Zhang, and Yuan Tian. 2012. "Microstructure and Mechanical Properties of a Low Carbon Carbide-Free Bainitic Steel Co-Alloyed with Al and Si." *Materials & Design* 39 (0): 264–268.

Quidort, D., and Y. Brechet. 2002. "The Role of Carbon on the Kinetics of Bainite Transformation in Steels." *Scripta Materialia* 47 (3): 151–156.

Quidort, D., and Y.J.M. Brechet. 2001. "Isothermal Growth Kinetics of Bainite in 0.5% C Steels." *Acta Materialia* 49 (20): 4161–4170.

Sharma, S., S. Sangal, and K. Mondal. 2011. "Development of New High-Strength Carbide-Free Bainitic Steels." *Metallurgical and Materials Transactions A* 42A (12): 3921–3931.

Song, T., and B.C. De Cooman. 2013. "Effect of Boron on the Isothermal Bainite Transformation." *Metallurgical and Materials Transactions A* 44 (4): 1686–1705.

Steven, W., and A.G. Haynes. 1956. "The Temperature of Formation of Martensite and Bainite in Low-Alloy Steels." *Journal of the Iron and Steel Institute* 183 (8): 349–359.

Sugimoto, K.-I., T. Iida, J. Sakaguchi, and T. Kashima. 2000a. "Retained Austenite Characteristics and Tensile Properties in a TRIP Type Bainitic Sheet Steel." *ISIJ International* 40 (9): 902–908.

Sugimoto, K.I., and M. Murata. 2007. "Applications of Niobium to Automotive Ultra High-Strength TRIP-Aided Steels with Bainitic Ferrite And/or Martensite Matrix." In *MS&T2007*, 1:15–26. Detroit, MI

Sugimoto, K.-I., J. Sakaguchi, T. Ida, and T. Kashima. 2000b. "Stretch-Flangeability of a High-Strength TRIP Type Bainitic Sheet Steel." *ISIJ International* 40 (9): 920–926.

Sugimoto, K.I., J. Tsuruta, and S.-M. Song. 2007. "Fatigue Strength of Ultra-High Strength TRIP-Aided Steels with Bainitic Ferrite Matrix." In *ICM'10*. Busan, S. Korea.

Sugimoto, Koh-ichi, Akinobu Kanda, Ryo Kikuchi, Shun-ichi Hashimoto, Takahiro Kashima, and Shushi Ikeda. 2002a. "Ductility and Formability of Newly Developed High Strength Low Alloy TRIP-Aided Sheet Steels with Annealed Martensite Matrix." *ISIJ International* 42 (8): 910–915.

Sugimoto, Koh-ichi, Kiyotaka Nakano, Sung-Moo Song, and Takahiro Kashima. 2002b. "Retained Austenite Characteristics and Stretch-Flangeability of High-Strength Low-Alloy TRIP Type Bainitic Sheet Steels." *ISIJ International* 42 (4): 450–455.

Wang, Y., K. Zhang, Z. Guo, N. Chen, and Y. Rong. 2012. "A New Effect of Retained Austenite on Ductility Enhancement in High Strength Bainitic Steel." *Materials Science and Engineering: A* 552 (0): 288–294.

Xu, Feng-yun, Yong-wei Wang, Bing-zhe Bai, and Hong-sheng Fang. 2010. "CCT Curves of Low-Carbon Mn-Si Steels and Development of Water-Cooled Bainitic Steels." *Journal of Iron and Steel Research, International* 17 (3): 46–50.

Zajac, S., V. Schwinn, and K.-H. Tacke. 2005. "Characterisation and Quantification of Complex Bainitic Microstructures in High and Ultra-High Strength Linepipe Steels." *Materials Science Forum* 500–501: 387–394.

Zhu, Kangying, Carla Oberbillig, Céline Musik, Didier Loison, and Thierry Iung. 2011. "Effect of B and B + Nb on the Bainitic Transformation in Low Carbon Steels." *Materials Science and Engineering: A* 528 (12): 4222–4231.

Chapter 9
Candidates for the Third Generation: Medium Mn Steels

Contents

9.1 Introduction

Steels with increased but still relatively moderate, "medium" amount of Mn (4–10 %) are seriously considered as potential candidates to achieve the performance of the third generation AHSS.

First detailed studies of steels with Mn in this "medium" concentration range (compared to Mn content in TWIP steels) were carried out by R. Miller (1972). Aiming to obtain ultrafine grain steels, Miller used 0.1C–5Mn steel, and by heavy cold reduction of 60–85 % combined with long soaking at a temperature in the lower range part of intercritical region, he managed to achieve the grain size of 0.3–1.1 μm. In addition to anticipated high yield strength, the material exhibited total elongation of about 30 % and significant amount of retained austenite of variable stability under deformation that depended on the extent of microstructure refinement.

© Springer International Publishing Switzerland 2015

N. Fonstein, *Advanced High Strength Sheet Steels*,

DOI 10.1007/978-3-319-19165-2_9

Later, M. Merwin repeated some of the Miller's experiments aiming to demonstrate a potential of steels with higher Mn to achieve the combination of strength and ductility intermediate between traditional TRIP and TWIP steels (Merwin 2006, 2007, 2008). Using steels with 0.1C and 5–7 % Mn, the author simulated different batch annealing cycles, which reproduced cold- and hot-temperature spots of a coil. The results showed that TS ~ 1100 MPa at total elongation of 30 % could be reached with the (TS × Uniform elongation) product as high as 25,000 MPa × %. The steels contained 20–30 % of retained austenite with the volume fraction and stability depending on temperature and duration of intercritical annealing. These works revealed very strong dependence of properties on the annealing temperature and the absence of direct correlation between ductility and volume fraction of retained austenite.

In contrast to ultrafine sub-micron grain single-phase materials that can exhibit high strength but very limited strain hardening (Morrison 1966), these "med Mn steels" demonstrated high strain hardening induced by TRIP effect due to the transformation of retained austenite.

The *combinations* of properties demonstrated by M. Merwin evoked huge interest for this steel concept, and numerous publications appeared with the intent to improve the properties even further, to widen the processing windows, to explain the nature of extraordinary high fraction of retained austenite and its stability, to understand methods for work hardening control, etc.

Some of the references to the pertinent publications are summarized in Table 9.1 based on data of Matlock and Speer 2006; Furukawa et al. 1994; You 2004; and others.

Table 9.1 Publications focused on Medium Mn concept

Composition	Initial condition	Temp. range of annealing (°C)	Time	Refs.
0.11C–5.7Mn[a]	CR	520–640	1–16 h	Matlock and Speer (2006), Miller (1972)
(0.01–0.4)C–5Mn–(0.006–2)Si	HR	700–900	1 h	Miller (1972), Furukawa (1989)
0.12C–5.1Mn	HR	650–700	20 min–26 h	Furukawa (1989)
(0.01–0.4)C–5Mn	HR	625–675	3 h	Huang et al. (1994)
0.15C–4 and 6.5Mn–0.5Si	HR, CR	620–675	1–12 h	Furukawa et al. (1994)
0.1C–(5–7)Mn–0.12Si	HR	540–730	~16 h	You (2004)
0.12C–(5,6)Mn–0.5Si–1.3Al	CR	660–840	2 min	Merwin (2006)
(0.2–0.4)C–(5.7)Mn	Forged	650	6 h	Suh et al. (2010a, b)
0.15C–6Mn–1.5Si	CR	640–680	3 min	Shi et al. (2010), Lee et al. (2008)

HR hot rolled, *CR* cold rolled
[a]Only medium Mn composition highlighted

Numerous studies of medium Mn steels also aimed at optimization of the composition of steels to obtain better combinations of properties, investigate the effects of microstructure and annealing parameters on mechanical properties and the conditions leading to the observed serrations in stress–strain curves.

9.2 Parameters Affecting the Properties of Mn Steels

Considering medium Mn steels as prospective candidates to approach the properties of 3rd generation AHSS, presented below the summary of known data concerning the critical factors that determine the properties of these steels, as well as explanations of experimental observations.

9.2.1 Effects of Temperatures and Time of Intercritical Annealing

The known data related to heat treatment of medium Mn steels are obtained on steels of different compositions with different initial microstructure, of hot and cold gauges, processed in different temperature ranges for different soaking times as shown in Table 9.1. However, all the studies show the same trends that allow for understanding the observed effects of the main annealing parameters.

Starting from the pioneering work by Furukawa (1989), researchers emphasize very strong dependence of the properties of Mn steel on the annealing temperature. For example, maximum TS × UE value after annealing at 675 °C for 1 h was observed for all steels with 5 % Mn regardless of their carbon content.

The effect of annealing temperature (AT) on mechanical properties of 0.16C–3.95Mn–0.5Si–0.05Nb cold-rolled steel annealed at various temperatures for 180 s and cooled at 10 °C/s is illustrated in Fig. 9.1. The volume fraction of retained

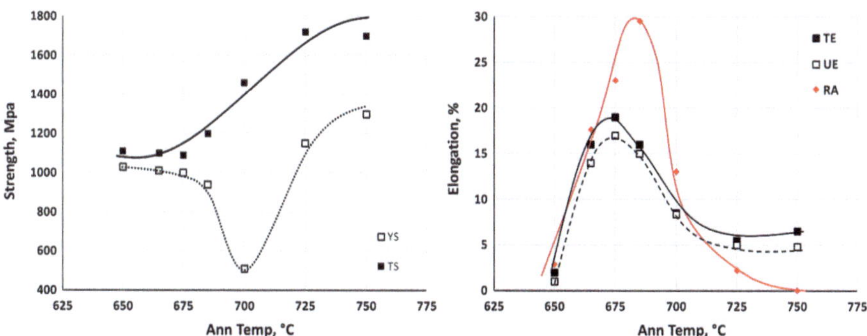

Fig. 9.1 Effects of annealing temperature on tensile properties and volume fraction of retained austenite: (**a**) yield/tensile strength and (**b**) uniform/total elongation and volume of retained austenite (Jun et al. 2011)

austenite measured by XRD is also shown. Both strength and ductility appear to be very sensitive to the annealing temperature. Tensile strength varies from 1000 to 1800 MPa but especially sharply changes in the range of temperatures between 675 and 700 °C. Yield strength is relatively constant (around 1000 MPa) for temperatures of up to 685 °C, then sharply drops at 700 °C, and subsequently grows as the annealing temperature is increased to 725 and 750 °C. Total elongation has a peak of 19 % at the annealing temperature of 675 °C and is about 16 % in the range of 665–685 °C with TS being 1100–1200 MPa. It should be noted that the uniform elongation, which is a more important estimate of formability, is about 15 % in this temperature range. The amount of austenite in ferrite–austenite–martensite microstructure increases up to ~30 % at the annealing temperature of 685 °C and then decreases with increasing annealing temperature. The best combination of TS = 1084 MPa at TE = 19 % was demonstrated at 675 °C (Jun et al. 2011).

The observations of microstructure revealed non-recrystallized ferrite, martensite, and cementite after cold rolling and annealing at 650 °C (Fig. 9.2a). The latter two constituents are formed due to tempering of martensite present in the initial hot-rolled band. This "recovery annealed" microstructure produces very high yield strength and low ductility. Since A_{C1} of this steel is 560 °C, some portion of austenite has been formed and later transformed to fresh martensite upon cooling.

After annealing at 665 and 675 °C, the microstructure consists of duplex-like mixture of partially non-recrystallized ferrite and fine solid M–A phase (<1 μm, Fig. 9.2b). Magnetic and XRD techniques detected substantial amounts of retained austenite. However, the volume fraction of M–A determined by the image analysis is higher than the volume fraction of only RA obtained using XRD, which indicates

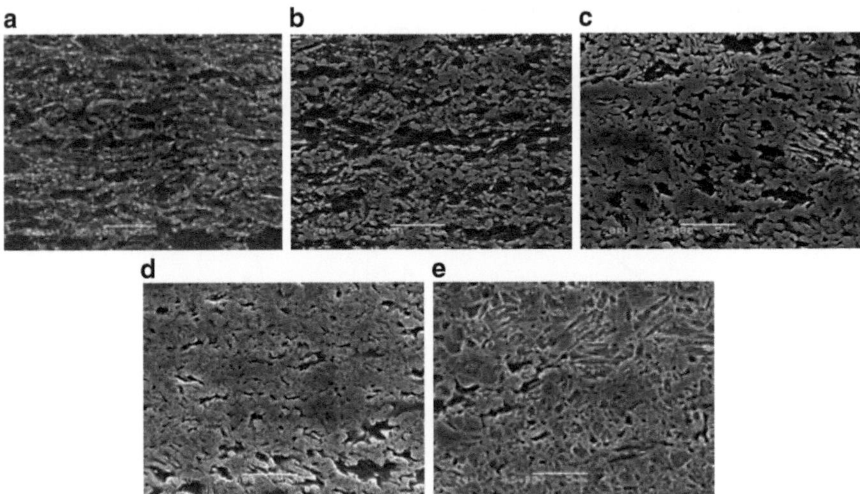

Fig. 9.2 SEM micrographs showing microstructure of samples of 0.15–4Mn steel annealed at various temperatures: AT = (**a**) 650, (**b**) 675, (**c**) 700, (**d**) 725, and (**e**) 750 °C

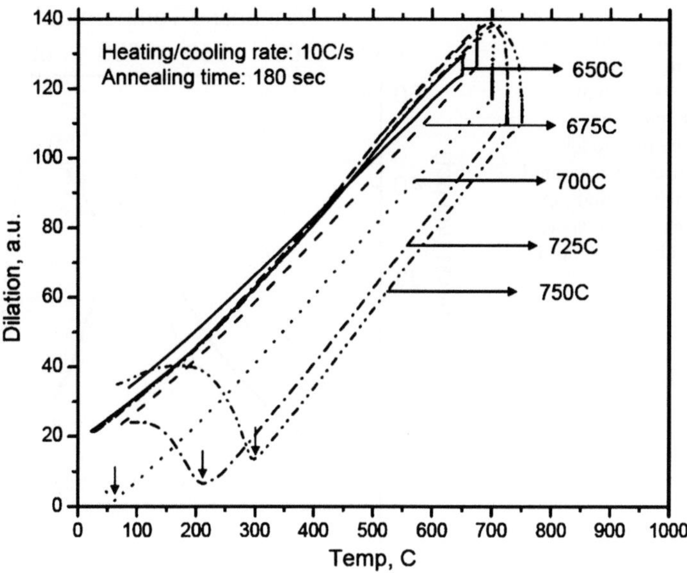

Fig. 9.3 Effect of annealing temperature on dilatation: annealing at AT = 650 and 675 °C—no martensite transformation is observed; annealing at AT = 700, 725, 750 °C—martensite transformation during cooling

a fact of partial transformation of the formed austenite to martensite (up to ~15 %) during cooling. After annealing at temperatures above 700 °C, the microstructure was typical for dual-phase steels, i.e., a mixture of ferrite and mostly martensite (Fig. 9.2c–e), which results in higher tensile strength and lower yield/tensile strength ratio.

These conclusions based on microstructure observations were consistent with the dilatometer data obtained for austenite transformation upon cooling from different initial annealing temperatures. As shown in Fig. 9.3, the dilatation curves for the cooling rate of 10 °C/s after annealing at 650 and 675 °C do not show, within the accuracy of the technique, any type of transformation of the formed austenite above the room temperature that confirms the existence of significant amount of retained austenite. At higher annealing temperatures, bigger volume of austenite with lower carbon content transforms to martensite during cooling, thus lowering the fraction of retained austenite.

The demonstrated very strong dependence of the properties and of the amount of retained austenite on annealing temperature is in a very good agreement with calculations and experimental results reported by Zhao et al. (2014), as well as by Gibbs et al. (2011), regardless of content of carbon or manganese, duration of annealing time, or upstream processing (Fig. 9.4).

In particular, as observed by Furukawa, the increase in annealing time, e.g., from 5 min to 1 h lowered the temperature of the "peak" properties. Ductility increased slightly due to expected higher amount of retained austenite. However, sharp variations in TS–TE balance with annealing temperature did not change (Furukawa 1989).

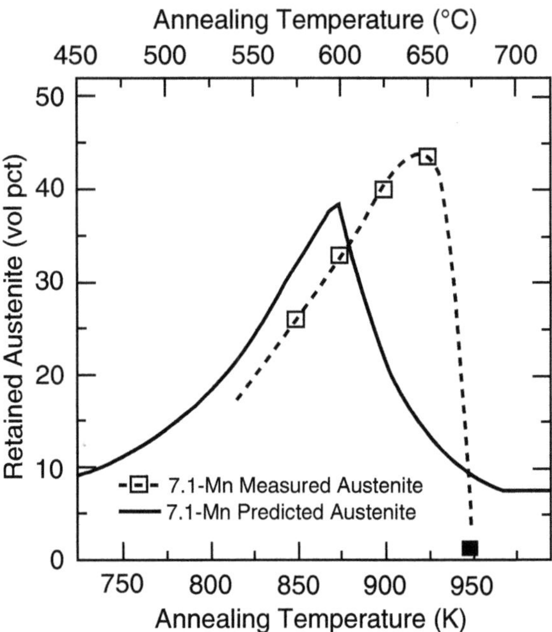

Fig. 9.4 Austenite fraction as a function of annealing temperature for 0.1C–7.1Mn–0.12Si steel after 1 week of annealing, as predicted by the model (*solid line*) and as measured by neutron diffraction (*open symbol*) and XRD (*shaded symbol*) (Gibbs et al. 2011)

9.2.2 Effect of Volume Fraction of Retained Austenite and Its Stability on the Properties of Medium Mn Steels

The first question related to strong dependence of mechanical properties on annealing temperature is whether it can be explained by the amount of retained austenite.

While studying the effects of annealing time ranging from 1 to 30 h on the properties of 0.09C–4.6Mn steel, Arlazarov et al. indicated a trend of increasing UE with increasing volume fraction of retained austenite, Fig. 9.5 (Arlazarov et al. 2012b).

At the same time, the majority of published data reveal no consistent direct correlation between elongation (uniform or total) and RA. Instead, the correlation was found between the volume fraction of retained austenite and the product of strength and ductility, TS × TE (MPa × %), when the unstable portion of austenite compensates the lowering elongation by increasing strength due to higher volume fraction of fresh martensite formed during cooling. In particular, as shown in Fig. 9.1, maximum total elongation is observed after annealing at 675 °C. At higher annealing temperatures up to 685 °C, the total elongation decreases in spite of growing volume fraction of retained austenite. Similar observations allow for assumption that ductility of medium Mn steels does not depend on the total amount of RA but rather on its stability, because unstable portions transform to martensite

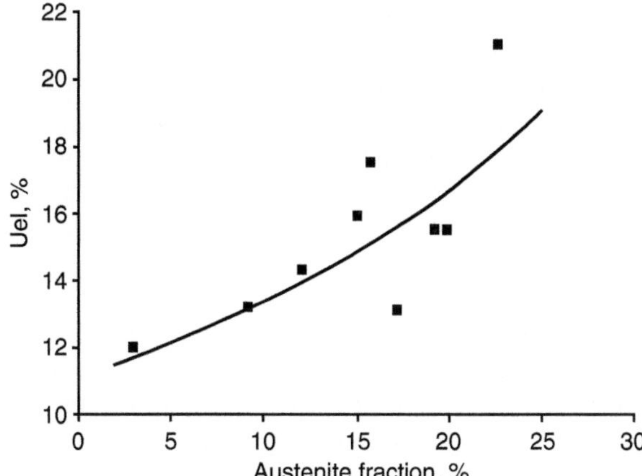

Fig. 9.5 Influence of retained austenite fraction on uniform elongation (Arlazarov et al. 2012b)

Table 9.2 Effect of annealing temperature on tensile properties and volume fraction of retained austenite in nondeformed and deformed areas (Jun et al. 2011)

						RA, % (nondeformed)		RA, % (def.)
AT (°C)	YS (MPa)	TS (MPa)	UE (%)	TE (%)	YPE (%)	Magnetic	XRD	Magnetic
650	1038	1105	0.6	1.1	0	4.2	3.0	na
665	1018	1094	14.2	15.9	14.2	22.7	18.0	11.0
675	996	1084	17.3	19.1	16.3	26.4	23.7	13.3
685	945	1187	15.1	15.9	14.2	32.8	29.6	11.6
700	528	1465	8.6	8.8	0	22.0	12.9	na
725	1145	1726	5.0	5.1	0	7.1	2.3	na
750	1307	1694	4.5	6.2	0	4.3	0.0	na

at early deformation stages, whereas very stable film-type RA does not contribute to TRIP effect at all (Jun and Fonstein 2008).

Table 9.2 presents the mechanical properties of 0.15C–4Mn–0.5Si–0.05Nb steel after annealing at different temperatures, as well as the amount of retained austenite in nondeformed and deformed (near fracture) areas measured by magnetic saturation and XRD. The results for temperature range of interest (665–685 °C) demonstrate the decrease in the amount of RA after tensile deformation that confirms the occurrence of the TRIP effect. On the other hand, different portions of RA transformed at the ultimate strain: 0.52, 0.50, and 0.65 after annealing at 665, 675, and 685 °C, respectively, indicating the maximum stability of retained austenite after annealing at 675 °C. These results are in agreement with the data obtained by Gibbs et al. who showed that the highest ductility is achieved under the conditions that create the highest stability of austenite.

Lee and De Cooman described the approach to select the annealing temperature for 0.3C–6.0Mn steel based on predictions of the amount of retained austenite after cooling to room temperature (Lee and De Cooman 2011). Their estimates were based on calculated chemical composition of austenite at each annealing temperature and its grain size under assumption of partial reverse transformation of martensite to austenite by nucleation and partial diffusion controlled growth, whereas the remaining martensite forms recrystallized ferrite during heating.

Using the modified equation

$$M_S(^\circ C) = 545 - 421 \text{ pct C} - 30.4 \text{ pcr Mn} \tag{9.1}$$

with C and Mn content computed by ThermoCalc, the authors evaluated the stability of the formed austenite using the M_S temperature position with respect to room temperature: $M_S < RT$ corresponded to fully stabilized austenite and $M_S > RT > M_F$ corresponded to complete transformation of austenite to martensite.

These calculations confirmed the existence of sharp maximum of the amount of retained austenite at certain annealing temperature, T_M (temperature of RA maximum) and distinguished three different types of tensile behavior depending on the austenite stability. At temperatures above T_M, tensile strength is high and ductility is limited due to significant amount of martensite fraction formed during cooling. Annealing at temperatures around T_M produces small amount of martensite and large amount of retained austenite that results in high strength at higher elongation due to the TRIP effect during tensile deformation and very high work hardening rate. The closer the annealing temperature was to T_M, the higher was the RA stability and the higher was the strain at which the TRIP effect was observed. At temperatures slightly below T_M, the microstructure was the mixture of UFG ferrite and stable austenite without any martensite transformation occurring during deformation that resulted in very limited work hardening, lower strength, and ductility.

These conclusions were supported by TEM analysis of fine structure at yielding of UFG 0.099C–7.09Mn–0.13Si steel (De Cooman et al. 2013) with comparison between two annealing temperatures, at ~T_M and above (600 and 650 °C). It was shown that at low intercritical temperature annealing (at 600 °C), UFG duplex microstructure consisting of ferrite and austenite was produced. Yielding of this structure occurred as initiation and propagation of the Luders bands due to plastic deformation of ferrite. Strain-induced transformation of austenite took place after considerable yield point elongation. After annealing at 650 °C, the microstructure consisted of ferrite, martensite, and austenite of lower stability. Yielding occurred at lower tensile stress by stress-induced transformation of austenite islands.

9.2.3 Effect of Manganese

Numerous studies of intercritically annealed medium Mn steels with high ductility achieved by utilizing the TRIP effect cover a wide range of Mn concentrations from 4 to 10 %.

It was shown that the higher the Mn content is, the higher the combination of strength and ductility can be obtained due to solid strengthening by Mn and the higher the volume fraction of retained austenite because of higher stability of austenite formed in two-phase region.

So far the known strength of med MN steel mostly does not exceed 1200 MPa, but both tensile strength and the balance with elongation (and therefore the product of TS × TE) are sensitive to Mn content.

In particular, for 0.16C–4Mn–0.5Si–0.05Nb steel, the best combination of properties was obtained as TS = 1100 MPa and TE = 19 % at ~30 % RA. For 0.16C–6.7Mn–1Si–1.5Al steel, TS = 1200 MPa and TE = 25 % at ~33 % RA were obtained (Jun et al. 2010).

On the other hand, the increase in Mn content lowers the A_{C3} temperature that calls for lower annealing temperatures in two-phase region, when the competition between austenite formation and recrystallization of ferrite becomes more pronounced.

9.3   Localized Deformation and Strain Hardening in Medium Mn Steels

Figure 9.6 presents a simplified schematic diagram of quite different strain hardening behaviors of Mn steel depending on the annealing temperature and therefore on the proportion between microstructure constituents (Lee and De Cooman 2011).

The sample of 0.3C–6Mn steel annealed at 620 °C (Fig. 9.6a) shows very limited work hardening due to the absence of strain-induced martensite transformation (TRIP effect). The mechanical behavior of samples annealed at 660 °C and 680 °C is shown in Fig. 9.6c. The sample annealed at 680 °C has high tensile strength but limited ductility mainly because of the presence of large volume fraction of martensite formed during cooling as a result of transformation of austenite with M_S above room temperature. On the other hand, the sample annealed at 660 °C has small amount of martensite but high volume fraction of retained austenite. Its strength is comparable with that of the sample annealed at 680 °C, but elongation is higher mainly due to the TRIP effect activated during tensile deformation. The graph also shows much higher work hardening rate at the beginning of deformation compared to that of the sample annealed at 640 °C. However, since most of martensite is formed at small strains, the ductility is lower than in case of the sample annealed at 640 °C.

The sample annealed at 640 °C (Fig. 9.6b) has a remarkably high work hardening rate due to the TRIP effect when large amount of retained austenite (80 %) transforms to martensite during tensile testing. The stability of austenite is the key factor here preventing its transformation to martensite on early stages of deformation.

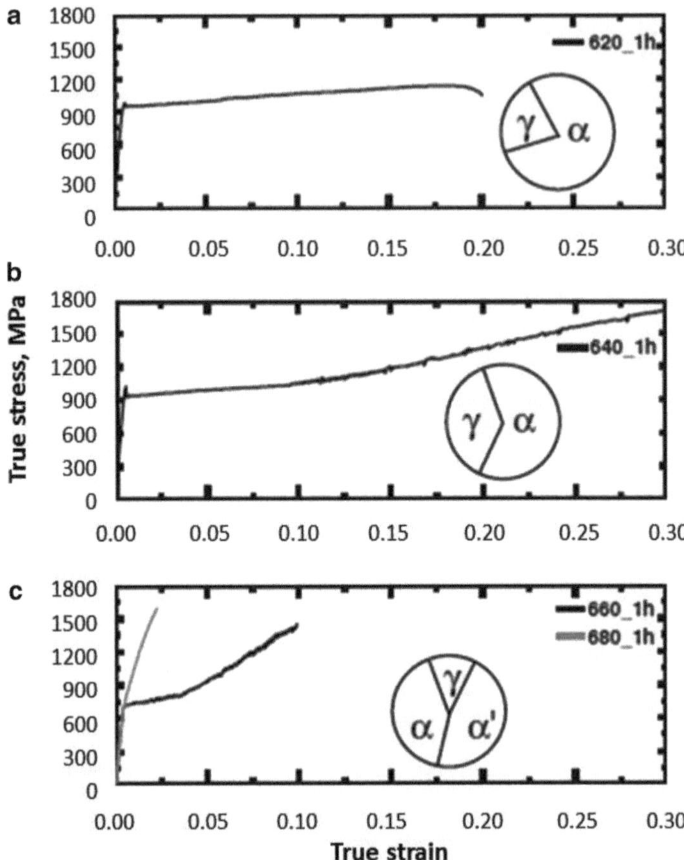

Fig. 9.6 True-stress—true-strain curves of 0.3C–6Mn steel samples annealed for 30 min at (**a**) 620 °C, (**b**) 640 °C, (**c**) 660 °C—*gray line* and 680 °C—*black line*. The pie chart indicates the phase content in the samples after intercritical annealing

The same conclusions were drawn by Lee et al. based on the study of 0.05C–6.15Mn–1.5Si steel annealed at different temperatures using high heating rate so that the *Austenite Reverted Transformation* (ART) of preliminary fully martensitic microstructure took place (Lee et al. 2011d). It was noted that stable austenite did not transform to martensite at all during tensile strain in case of low annealing temperature (640 °C), and without the TRIP effect, the highest total elongation was observed when deformation mostly occurred by propagation of Luders bands without work hardening. The observed high elongation was due to the presence of austenite as a second phase and because of beneficial strain partitioning between austenite islands and ferrite matrix, i.e., due to a duplex-type behavior of two coexisting ductile phases (Lee et al. 2011a, b, c).

After annealing at 660 °C, large yield point elongation due to several bands initiated at the early stages of tensile deformation, and subsequent transformation of

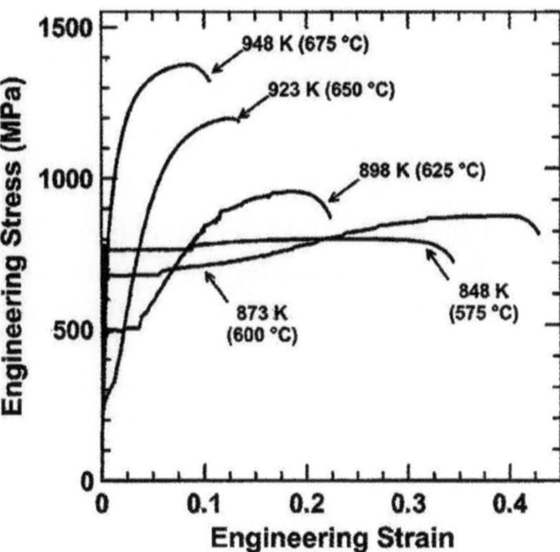

Fig. 9.7 Tensile engineering stress–strain curves for 0.1C–7.1Mn steel samples annealed at different temperatures for 1 week

austenite by TRIP mechanism resulted in high strain hardening rate and the best combination of strength and ductility.

Sun et al. also pointed at "abnormal work hardening rate" when substantially enhanced ductility was attributed to the propagation of Luders bands in ferrite matrix followed by TRIP effect of large fraction of austenite (Sun et al. 2012).

Very similar results were presented by Gibbs et al. who reported very high elongation demonstrated by 0.1C–7.1Mn steel annealed at 575 °C that had the highest yield strength and very significant YPE followed by limited strain hardening, Fig. 9.7 (Gibbs et al. 2011). This allows suggesting the limited contribution of strain-induced transformation of austenite to martensite (direct TRIP effect) that was confirmed by measurements of the fraction of RA as a function of strain. At slightly higher annealing temperature (600 °C), extended YPE was followed by work hardening due to TRIP effect accompanied by serrations in stress–strain curve but resulted in the highest observed elongation. Further increase in annealing temperature led to significant decrease in YS at the increase of work hardening at low strains, typical for DP steels.

Shi et al. proposed to distinguish three stages of work hardening behavior of med MN steels after optimal annealing (Shi et al. 2010). These authors used 0.2C–(5–7) Mn steels with initially martensitic microstructure, which after double annealing contained 34–44 % of retained austenite. Stage I is when the work hardening rate decreases with tensile strain. This is followed by increasing work hardening rate (Stage II) and then decreasing again (Stage III). On low strains (less than 0.05), the work hardening is contributed by the dislocation hardening in both ferrite and austenite. With increasing strain, the work hardening is determined by the austenite-to-martensite transformation (TRIP effect). At strains greater than ~0.15, the work hardening rate strongly decreases because of significant decrease in the fraction of retained austenite.

The main conclusion from the above works is that the work hardening behavior of medium Mn steels is controlled by stability of retained austenite that is in turn determined by the parameters of annealing (temperature and time), as well as by the chemical composition of steel.

9.4 Effect of Initial Microstructure and Austenite Reverted Transformation on the Evolution of Microstructure and Mechanical Properties of Medium Mn Steels

The upstream processes and corresponding microstructure prior to the final processing can affect the formation of austenite and its morphology, the stability of retained austenite, and the final mechanical properties.

Arlazarov et al. compared the mechanical properties after annealing of 0.1C–4.6Mn steels with two different initial microstructures: bainite–martensite mixture obtained after hot rolling and subjected to cold rolling and as-quenched (nondeformed) martensitic microstructure (Arlazarov et al. 2012c). Depending on temperature and time (up to 63 h) of annealing, different fractions of austenite were formed and ferrite underwent recrystallization to different extents, as shown in Fig. 9.8. Preliminary cold deformation accelerated the austenite formation during heating and strongly influences its morphology so that the resultant microstructure became fully polygonal.

Annealing of nondeformed martensite (double annealing) implies Austenite Reverted Transformation (ART) when austenite forms from martensite. The morphology of such austenite is either of the lath type or polygonal depending on the dominating ART mechanism: nucleation of the lath-type austenite along the laths of the initial martensite (Koo and Thomas 1977) or nucleation of polygonal austenite at high-angle grain boundaries of prior austenite formed during first full austenitization (Tokizane et al. 1982).

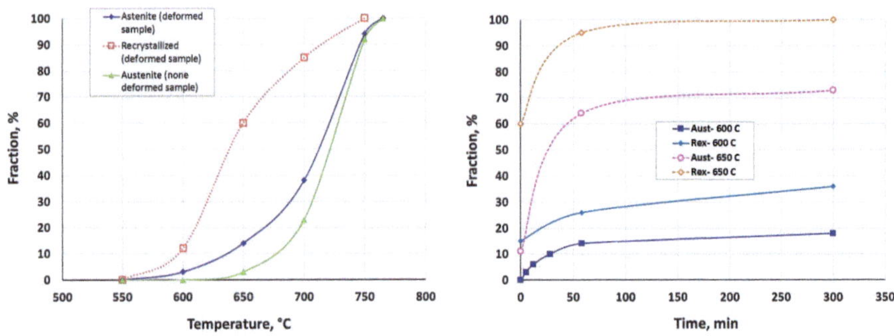

Fig. 9.8 Evolution of ferrite recrystallization and volume fraction of austenite with annealing temperature (**a**) and time of annealing (**b**) (Arlazarov et al. 2012c)

As shown in the study by Han and Lee, the reverted transformation depends on the heating rate (Han and Lee 2014). In 0.05C–(5–9)Mn alloys, ART occurred at heating rates higher than 15 °C/s without precipitation of cementite, whereas at low heating rate (below 15 °C/C), cementite first formed along interfaces and transformation from martensite to austenite occurred near the cementite particles by diffusive mechanism.

The type of initial structure and the realization of diffusionless reverted martensite-to-austenite transformation affect the morphology and stability of retained austenite. According to Han and Lee, diffusional transformation produced globular austenite of 200–250 nm with low dislocation density and Mn content higher than that in the initial tempered martensitic matrix. Diffusionless transformation produced lath-shaped austenite grains of bigger size (200–300 nm wide and 400–700 nm long) with high dislocation density and Mn content close to that in the initial martensite. Grounding on these observations, the authors came to the conclusion that most likely the "globular" austenite had higher stability than the lath austenite.

According to Cao et al., there is no recrystallization of ferrite accompanied ART during annealing of preliminary-quenched 0.2C–5Mn steel so that ultrafine lamellar ferrite–austenite microstructure could be observed even after long soaking. This can be attributed to Mn/C interface segregations (Cao et al. 2012).

As shown in Fig. 9.8, Arlazarov et al. found the interaction between recrystallization of ferrite and austenitization during annealing at the intercritical temperatures in the range of 650–750 °C, when the recrystallization of ferrite was hindered by the formation of austenite. Trying to obtain recrystallized microstructure with controlled amount of austenite, the authors concluded that a long soaking time can help to achieve the desired optimum. However, it was really a small positive effect of holding time (up to 5 h) during double annealing, whereas the duration of 63 h resulted in decreases of both strength and ductility.

The effects of the initial microstructure were evaluated using two steels with ~0.16 % C: 4Mn–0.5Si–0.05Nb and 0.16C–6.7Mn–1Si–1.48Al.

Since the as-rolled Mn steels had very high strength, portions of as-rolled materials were softened by batch annealing at various temperatures. The final annealing cycle included the heating to temperatures ranging from 650 to 750 °C at 10 °C/s, soaking at annealing temperatures (AT) for 180 s, and cooling to room temperature at cooling rate of 10 °C/s (Jun et al. 2010).

Microstructure of as-rolled steels and its evolution after batch annealing are displayed in Fig. 9.9. After hot rolling (Fig. 9.9a), the 4 % Mn steel had mixture microstructure with quasi-granular ferrite, martensite, and bainite. After batch annealing at 550 °C, the tempering of martensite is evident (Fig. 9.9b) and becomes more pronounced after annealing at 650 °C (Fig. 9.9c). Since the batch annealing temperature of 650 °C is above A_{C1}, a substantial part of martensite in as-rolled material is replaced with ferrite matrix containing coarse cementite particles.

As-rolled steel with 6.7 Mn has significantly more martensite and bainite (Fig. 9.9d) with martensite having clearly outlined lath substructure. After batch annealing at 550 °C, the tempering of martensite is clearly seen. Batch annealing,

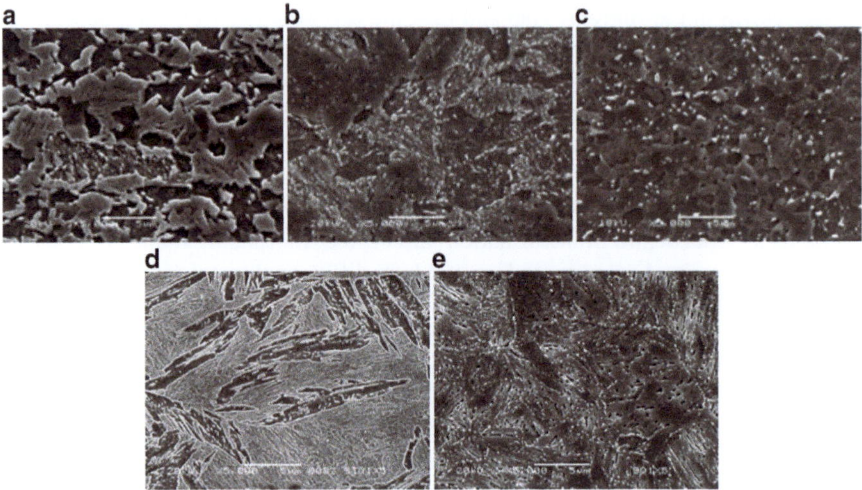

Fig. 9.9 Microstructure of as-rolled and batch-annealed steels with 4 and 6.7 % Mn: (**a**) 4Mn-as-hot rolled, (**b**) 4Mn-BAT550C, (**c**) 4Mn-BAT650C, (**d**) 6.7Mn-as-hot rolled, and (**e**) 6.7Mn-BAT550C

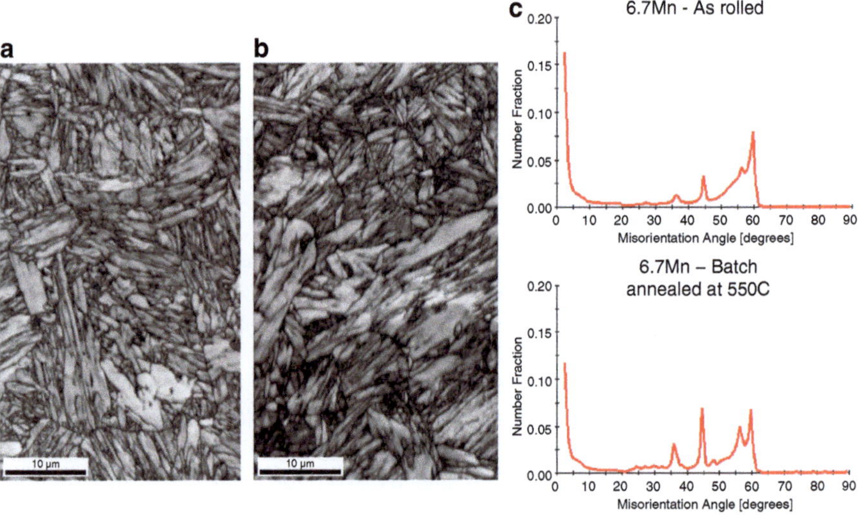

Fig. 9.10 EBSD images of 6.7 % Mn steel as-rolled (**a**) and (**b**) batch annealed at 550 °C, and misorientation distribution of microstructure elements in the same samples (**c**)

however, does not destroy the lath morphology of martensite, and cementite is precipitating along martensite laths (Fig. 9.9e).

EBSD analysis confirmed the presence of lath microstructure in as-rolled 6.7Mn steel (Fig. 9.10a) and its retaining after tempering during batch annealing at 550 °C

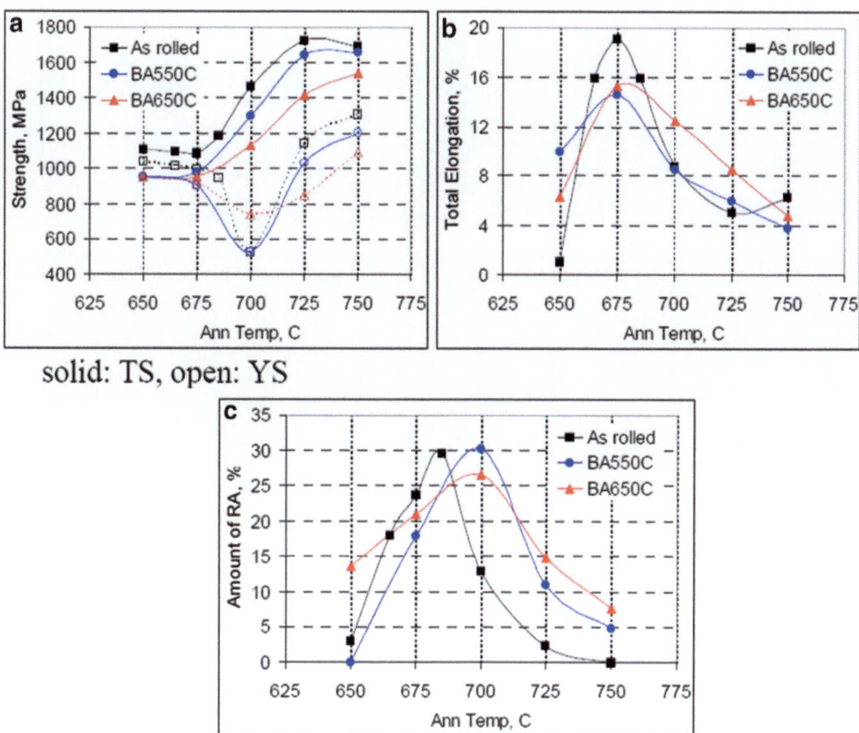

Fig. 9.11 Effect of initial microstructure and annealing temperature on tensile properties and the amount of retained austenite in 4Mn steel

(Fig. 9.10b). This fact is additionally proven by studying the misorientation distribution, Fig. 9.10c.

Preliminary batch annealing significantly softened the as-rolled specimens of 4Mn steel. The YS/TS ratio increased from 0.52 for as-rolled steel (low value typical for dual-phase structure) to 0.8–0.9 after batch annealing at 650 °C (typical high value for tempered martensite). Similar trend was observed in case of 6.7Mn steel after batch annealing at 400 and 550 °C (lower temperatures were used because of lower A_{C1} temperature).

Figure 9.11 shows that preliminary batch annealing of 4Mn steel hot bands reduced both strength and peak ductility after intercritical annealing. As presented earlier, the best combination of properties of the cold-rolled annealed steel (TS ~1100 MPa and TE ~19 %) was achieved using initial hot-rolled structure. Preliminary batch annealing of hot-rolled bands at both 550 and 650 °C deteriorated the combination of property to less than 1000 MPa and 16 % elongation. This degradation of properties by preliminary batch annealing of hot-rolled material is attributed to grain growth and austenitization kinetics during annealing, as presented in Fig. 9.12.

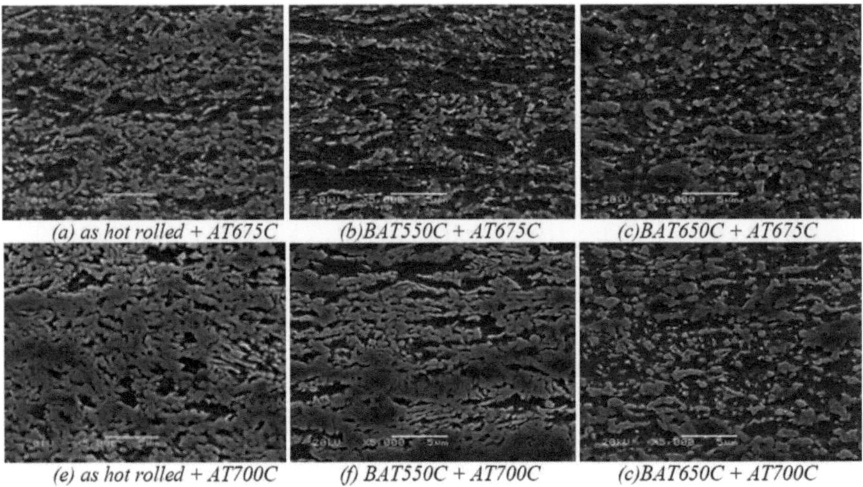

Fig. 9.12 SEM micrographs of 4Mn steel annealed at 675 and 700 °C with different initial microstructures (as-rolled and batch annealed at 650 °C)

Figure 9.12 clearly shows not only coarsening of microstructure but also the delay in austenitization during final intercritical annealing of preliminary (prior to cold rolling) batch-annealed steel. EDS mapping of sample batch annealed at 650 °C revealed particles enriched with Mn, as well as Mn depleted zones. During annealing, the C/Mn-rich particles transform to austenite at lower temperatures, whereas the Mn-depleted portions transform at higher temperature. After preliminary batch annealing, to obtain the same amount of austenite, a higher final annealing temperature is required than for the material based on as-hot-rolled initial structure. In addition, the maximum amount of retained austenite in cold-rolled annealed samples, which had been preliminary batch annealed after hot rolling, was observed after final annealing at higher temperature compared to "as-hot-rolled" microstructure without batch annealing before cold rolling and annealing (Fig. 9.11).

Maximum ductility does not exactly match the amount of retained austenite, since the stability of retained austenite and its contribution to the TRIP effect can change with grain size and C/Mn partitioning due to prior batch annealing (Jun et al. 2011). For example, the samples annealed at 700 °C after preliminary batch annealing of hot bands at 550 and 650 °C have very low ductility of 8–12 % TE, despite the presence of significant amount (25–30 %) of retained austenite. Similar fact was noted by Furukawa who compared properties of 0.1C–5Mn after annealing at 650–700 °C for 1 h (Furukawa 1989). Whereas the treatment at 700 °C produced a greater amount of austenite compared to that at 675 °C, this austenite appeared to be less stable against deformation than after annealing at 675 °C, which explains sharp decrease in ductility.

Quite a different picture was observed in the steel with 6.7 % Mn. The tensile properties and the amount of retained austenite after annealing of 6.7Mn steel with different initial microstructures are presented in Fig. 9.13. Tensile strength

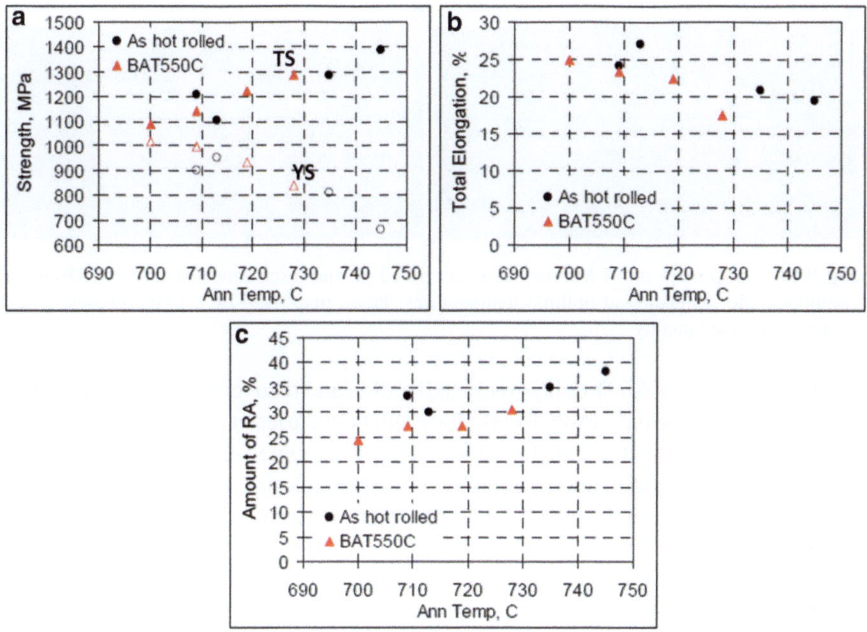

Fig. 9.13 Effect of the initial microstructure of 6.7Mn steel on tensile properties and the amount of retained austenite after intercritical annealing at different temperatures

increased with annealing temperature as bigger amount of hard phases had formed. Yield strength decreased since the microstructure of steel annealed at higher temperature contained austenite of lower stability, which, at least partially, transformed to martensite during cooling bringing about a dual-phase type of behavior with lower yield ratio (YS/TS). Significant ductility (25–17 %) of the 6.7Mn steel at such high tensile strength levels as 1100–1400 MPa was achieved due to TRIP of 25–40 % retained austenite (Fig. 9.13b, c).

Different responses of properties of preliminary batch-annealed microstructure vs. hot-rolled microstructure were observed at comparison of steels with different Mn content. The effect of preliminary batch annealing was significantly less pronounced for 6.7Mn steel than for 4Mn steel. This difference in behavior of the two steels can be explained through the effects of initial microstructure on the final microstructure after cold rolling and annealing. Figure 9.14 shows the effects of initial (prior to cold rolling) microstructure of 6.7 Mn steel, as-hot rolled (a) and following batch annealing at 550 °C (b), on the final microstructure after cold rolling and annealing at optimal temperatures. In both cases, the lath-type BCC matrix and retained austenite along lath boundaries were observed. As shown above, the initial microstructure of 6.7Mn steel, i.e., as-hot rolled or batch annealed, was not significantly different in terms of martensite morphology, when lath-type microstructure was preserved even after batch annealing of hot-rolled bands and cold rolling. Since

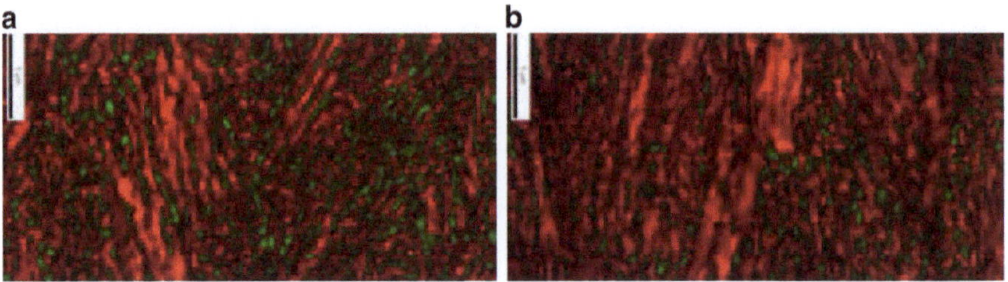

Fig. 9.14 EBSD maps of 6.7 Mn steel in as-hot rolled (**a**) and batch annealed at 550 °C (**b**) initial conditions after annealing at optimal temperatures: Phase map (*red*: BCC-based phases, *green*: austenite) + image quality

Mn in the amount of 6.7 % delays recrystallization and recovery of martensite during annealing, austenite preferably forms along lath boundaries of preexisting martensite through a kind of reverted transformation. It can be assumed that preservation of lath ferrite/martensite initial microstructures is responsible for small changes in final microstructure and properties of 6.7 Mn after preliminary batch annealing of hot-rolled bands.

Comparison between the two steels and their responses to preliminary batch annealing emphasizes the important role of grain size and morphology of austenite phase that determine its stability necessary for pronounced TRIP effect in med Mn steels.

9.5 Effect of Alloying and Microalloying on Mechanical Behavior of Medium Mn Steels

There is very limited knowledge directly related to the effects of alloying and microalloying elements on mechanical properties of Mn steels.

As was evident from the data presented above, one of the biggest anticipated manufacturability problems of medium Mn steels is very narrow processing window in terms of the annealing temperature range for achieving optimal combinations of strength and ductility. Besides, at 4 % Mn and higher, this temperature range is shifted downwards close to 670–700 °C, which is quite different from the typical temperatures for processing of commercial steels at continuous annealing or hot galvanizing lines. This should create additional problems of transitions. In this regard, it is assumed that changes in chemical composition can affect the annealing temperature range and the achievable combination of properties by altering the proportions of phases in microstructure.

In particular, some researchers consider potential effects of Al additions that should shift the annealing temperature range upwards also facilitating the recrystallization which is often incomplete at the temperatures corresponding to the maximum volume fraction of retained austenite. Additions of Al are also considered as the means to expand the processing window as they induce significant widening of the A_{C1}–A_{C3} temperature range.

In particular, Suh et al. studied the effects of additions of 1.1 % (Low-Al) or 3.1 % Al (High-Al) to steels containing 0.12C–0.5Si and 4.6 or 5.8 % Mn, respectively (Suh et al. 2010a, b). Short holding for 2 min was chosen to simulate heat treatment of cold-rolled sheet steels in CAL. The intercritical temperatures of 660, 720, and 780 °C were used for Low-Al and 720, 780, and 840 °C for High-Al compositions, respectively. Hot rolling was finished at 800 °C which falls into two-phase rolling temperature range for High-Al composition and resulted in banded ferrite–martensite microstructure with the segregation of Mn in the bands of initial austenite (6.9 % Mn in martensite bands and 5.5 % Mn in ferrite) that affected the further evolution of microstructure.

While after cold rolling the Low-Al steel underwent only recovery at all annealing temperatures employed, the recrystallized grains were observed in the High-Al steel under all annealing conditions. Accordingly, higher strength of non-recrystallized ferrite led to higher strength of Low-Al steel under all annealing conditions. Thus, additions of 3 % Al appeared to be effective in terms of promoting the recrystallization of ferrite and some widening of processing window. The Low-Al steel contained 6 % of retained austenite after annealing at 660 °C. This volume fraction increased to 28 % for 720 °C and then dropped to 6 % after annealing at 780 °C. In the High-Al composition, it was possible to preserve ~26–31 % of retained austenite after annealing at 720–780 °C, but this fraction dropped to 18 % after annealing at 840 °C. This drop was in agreement with the dilatometry study showing martensite transformation due to increase in M_S temperature after annealing at higher temperatures, which induced a decrease in carbon content in austenite and coarsening of austenite grains.

The authors pointed out that additions of Al raise the stability of retained austenite when they compared the 1.1Al steel annealed at 720 °C with 3.1 % Al steel annealed at 780 °C (Fig. 9.15).

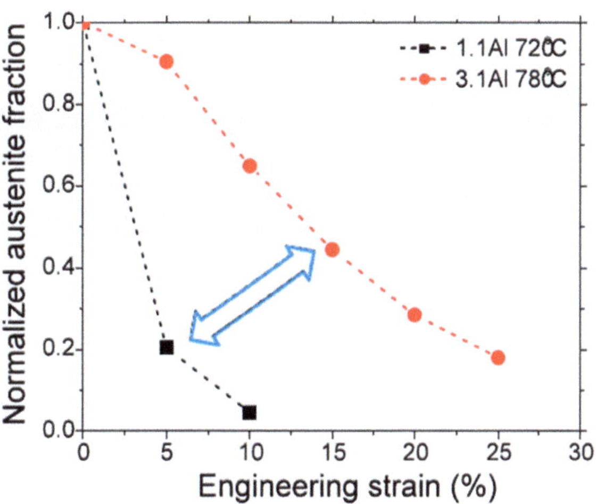

Fig. 9.15 The plot of normalized austenite fraction against engineering strain for 0.12C–Mn steels with 1.1 % Al and 3.1 % Al

Stability of austenite within relatively wide range of annealing temperatures can be explained by slower growth of austenite volume fraction with temperature due to higher Al content.

As to direct effects of alloying and microalloying on the properties of Mn steels, the study by Oh et al. showed that additions of 0.06 % Nb to 0.15C–6Mn–2.5Al steel increased the strength by almost 200 MPa compared to that of 6Mn–3Al steel without Nb. As a result, TS ~ 1200 MPa was achieved without any loss in total elongation (~30–35 %) (Oh et al. 2010).

According to the aforementioned study by Furukawa, additions of 2 % Si to 0.1C–5Mn steel improved the maximum TS × UE value from 23,000 MPa × % to ~28,000 MPa × %.

9.6   On the Stability of Retained Austenite in Medium Mn TRIP Steels

All researchers note the role of the TRIP effect in medium Mn steels. However, neither extraordinarily high amount of retained austenite in these steels nor its stability has been given consistent explanations.

In commercially produced TRIP steels that contain Si and/or Al to prevent carbide formation, the retained austenite at the end of bainite reaction accumulates high amount of carbon (>1 wt.%) that shifts M_S below room temperature. In some publications on medium Mn steels, the austenite stability is explained by Mn partitioning between ferrite and austenite during long soaking in the intercritical temperature range (Miller 1972; Furukawa et al. 1994). However, the only processing to realize a prolonged heat treatment is batch annealing. With this type of processing, the variations in mechanical properties induced by typical differences in "cold" and "hot" spot temperatures within a coil and hence by the differences in local Mn concentrations can be substantial (Merwin 2007). For this reason, short annealing in continuous lines is preferable from the viewpoint of uniformity of mechanical within a coil.

At the same time, the analysis of the published data leads to the conclusion that the existence of significant amount of retained austenite in medium Mn steels cannot be explained by chemical composition of gamma phase in the intercritical region. In particular, a potential partitioning of carbon is evidently insufficient regardless of holding time. For example, the M_S temperature of 0.11C–0.5 Si steel calculated under the assumption of complete partitioning of carbon to austenite is 169 °C for steel with 4.6 % Mn and 1.1 % Al annealed at 720 °C and is 156 °C for steel with 5.8 % Mn and 3.1 % Al annealed at 780 °C, whereas the retained austenite volume fractions of 28 and 31 % were observed, respectively, in these two steels (Suh et al. 2010a, b). Insignificant redistribution of Mn due to very short (2 min) annealing time was also expected.

In 0.05C–6.15Mn–1.5Si steel annealed for 180 s at 640 and 700 °C, 10 and 50 % of austenite were observed, respectively. Assuming that gamma phase was enriched with carbon to 0.5 and 0.1 % at these temperatures, the calculated M_S temperatures should have been 160 °C and 287 °C, respectively. In fact, as was established by dilatometry, the sample annealed at 700 °C showed the martensite transformation at 30 °C, whereas in all samples annealed at temperatures below 700 °C, no martensite transformation was observed with measurements performed down to −150 °C (Lee et al. 2008).

Another important aspect is the significant variation in the stability of retained austenite. For example, only 35 % of initial retained austenite in 0.12C–4.6Mn–3.1 % Al steel transformed into strain-induced martensite at tensile strain of 10 %, but 95 % of initial RA transformed to martensite at the same tensile strain in 5.8Mn–1.1 % Al steel (Suh et al. 2010a, b).

As mentioned above, strong dependence of mechanical properties of medium Mn steels on the annealing parameters is also explained by different stability of retained austenite that makes it important to consider all of the potentially affecting constituents.

Effect of annealing time has several different aspects. First, there are changes in carbon and Mn partitioning in austenite due to changes in the amount of the formed austenite, on the one hand, and due to more time available for diffusion, on the other. Second, it is the austenite grain growth.

Some experimental facts indicating significant role of the annealing time are summarized below.

Variations in annealing time within short duration (2–6 min) showed slight increase in the RA amount in 0.15C–6.06Mn–0.4Si–2.44Al–0.068Nb steel with time of annealing at 700 and 720 °C increasing from 2 to 4 min. This was accompanied by some increase in total elongation (Oh et al. 2010).

The effect of annealing time on the maximum amount of retained austenite was shown in the study by Huang and Furukawa (Huang et al. 1994). The peak of the volume fraction of retained austenite fraction during annealing of 0.12C–5.10Mn steel shifted to shorter holding times with increasing annealing temperature that allows assuming the favorable conditions for carbon and manganese partitioning. At the same time, longer annealing was accompanied by deteriorating stability of austenite that became apparent in dilatation curves showing martensite transformation above room temperature (Fig. 9.16) probably induced by grain growth.

As shown by Sun et al., longer holding during ART annealing increases the austenite stability and shifts the maximum work hardening rate to higher strains, which is indicative of slower decomposition of RA (Sun et al. 2012).

According to Arlazarov et al., with holding time increasing from 1 to 30 h during annealing of 0.092C–4.6Mn steel, the amount of RA increases from 15 to 22 % in the first 7 h. Beyond that, TE decreases and the volume fraction of RA drops down to 3 % after 30 h of holding (Arlazarov et al. 2012b).

The overall effect of annealing time should be determined by the balance between increasing concentrations of C and Mn due to enhanced diffusion and anticipated austenite grain growth. The role of grain growth during longer annealing will be separately analyzed below.

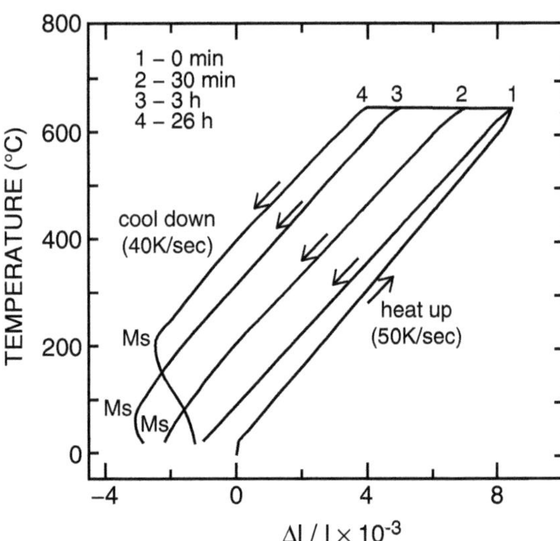

Fig. 9.16 Dilatation curves of 0.12C–5.1Mn steel after heating to 650 °C and soaking for various durations

9.6.1 Effect of Carbon Content

According to Thermo-Calc computations, the partitioning of carbon to austenite in Mn steels at the intercritical temperatures is substantial and can reach 0.5–0.6 %.

Small lab heats with 4 % Mn and carbon content ranging from ~0.15 to −0.6 % were produced to simulate austenite fraction formed at various intercritical temperatures. No Mn partitioning was assumed. Dilatation curves after annealing at 850 °C (full austenitization) are presented in Fig. 9.17. It is seen that all compositions underwent massive martensite transformation during cooling after full austenitization without any sign of incomplete transformation. Thus, the increase in carbon content up to ~0.6 %, at least in 4 % Mn steels, cannot sufficiently stabilize austenite and prevent its decomposition at room temperature.

9.6.2 Effect of Mn

As mentioned above, some authors explain the stability of retained austenite in medium Mn steels by Mn partitioning to γ-phase during annealing in the two-phase region. De Moor et al. considered the role of Mn partitioning in stabilization of austenite during long intercritical annealing keeping in mind that some essential increase in Mn content can really be expected based on Thermo-Calc predictions (De Moor et al. 2011), whereas high density of dislocations and the grain boundaries should facilitate Mn diffusion (De Moor et al. 2014). De Cooman et al. found enrichment of austenite with Mn even after short time annealing when in 6 % Mn

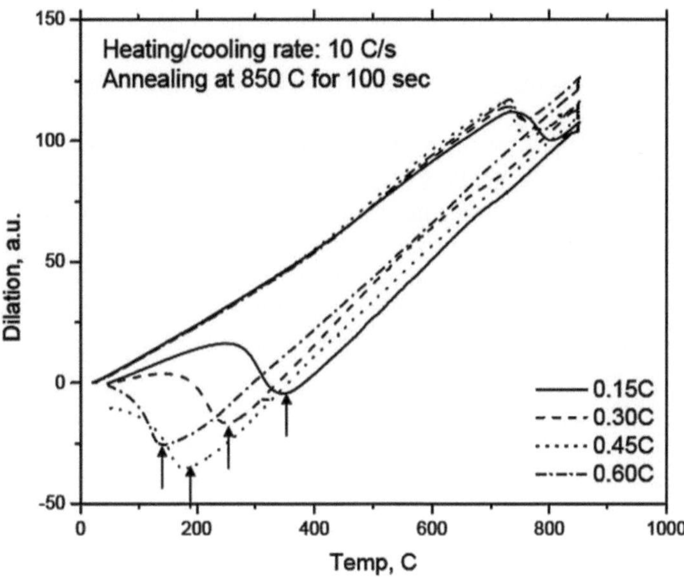

Fig. 9.17 Dilatation curves showing martensite transformation during cooling (-10 °C/s) from full austenitization in model alloys (0.15–0.6)C–4Mn

Table 9.3 Chemical composition (at.%) detected by Atom Probe at specific location of 0.15C–4Mn steel sample annealed at 675 °C

Location	Iron	Carbon	Manganese
Ferrite	97.13	0.02	2.15
Austenite	91.88	0.31	6.11
Interface ($<$5 nm)	95.92	0.81	9.93

steel, the concentration of Mn in austenite reached 11.6 % after annealing for 180 s (Lee et al. 2011a).

It can be assumed that relatively strong partitioning of Mn is facilitated by small size of austenite grains (short diffusion paths) and by the initial presence of carbides enriched with Mn. High Mn content of up to 11 % in cementite undissolved during annealing at 650 and 675 °C was in fact detected, but TEM–EDS analyses did not reveal any significant Mn partitioning within the volume of austenite grains after 180 s of annealing of 0.15C–4Mn–0.5Si–0.05Nb steel. The atom probe analysis (Table 9.3) detected significant, close to 10 %, growth of Mn concentration near ferrite/austenite interface induced by preferred nucleation of austenite in the regions near cementite particles and preferential segregation of Mn near austenite–ferrite interfaces. Thus, the interface segregation of Mn appeared to be essential even after short intercritical annealing and can contribute to stabilization of austenite boundary.

As predicted by Thermo-Calc ortho-equilibrium calculations, to suppress martensite transformation in 0.15C–4Mn steels below room temperature, the concentration of Mn in austenite must be about 10–11 %, which is significantly higher than average Mn content determined by atom probe (~6 %) and even higher than thermodynamically possible maximum partitioning at 675 °C (~8 %).

Computations by Lee and De Cooman also showed that the possible partitioning of C and Mn in austenite during intercritical annealing of steel even with 0.3C and 10 Mn is not sufficient to decrease the Ms temperature below room temperature so that to avoid the formation of athermal martensite (Lee and De Cooman 2014).

It should be concluded that enrichment of austenite with Mn cannot be the only explanation of significant RA stability in the studied steels.

9.6.3   Effect of Austenite Grain Size

As concluded above, the enrichment austenite phase with carbon and manganese during annealing in the two-phase region cannot be sufficient to effectively stabilize austenite, i.e., to shift the M_S temperature from calculated 150–200 °C to subzero temperatures.

Numerous publications point out that medium Mn steels with the best combinations of properties are featured by ultrafine grain microstructure, which is in agreement with micrographs presented in Fig. 9.2.

The nature of very small grain size in medium Mn steels is still being debated. It is known that Mn exerts a solute drag effect during recrystallization, thus retarding grain boundary migration. This effect can be stronger with higher Mn content. De Cooman et al. suggested that grain refinement takes place, in particular, due to the reversely transformed martensite that has been present in the initial microstructure (Lee et al. 2008).

The perception that very fine austenite grain size can significantly contribute to austenite stability is based on a lower probability of martensite nucleation within "small volume" of austenite and lower driving force for austenite-to-martensite transformation with decreasing austenite size (Olson and Cohen 1976). The effect of austenite grain size on temperature of martensite transformation was confirmed experimentally (Bleck et al. 1998) and is reflected, in particular, in the modified corrected equation for the M_S temperature (Lee and De Cooman 2011):

$$M_S(°C) = 545 - 423 \, \text{pctC} - 30.4 \, \text{pct Mn} - 60.5 V_\gamma^{-1/3} \qquad (9.2)$$

with V_γ being the average austenite grain volume in μm^3.

To study the effect of grain size on stability of retained austenite, a set of experiments were performed by Jun et al. (2011) that included isothermal annealing at 675 °C for various durations ranging from 3 min to 26 h. As shown in Fig. 9.18, the total amount of M–A constituent slightly grows with time. The grain size of

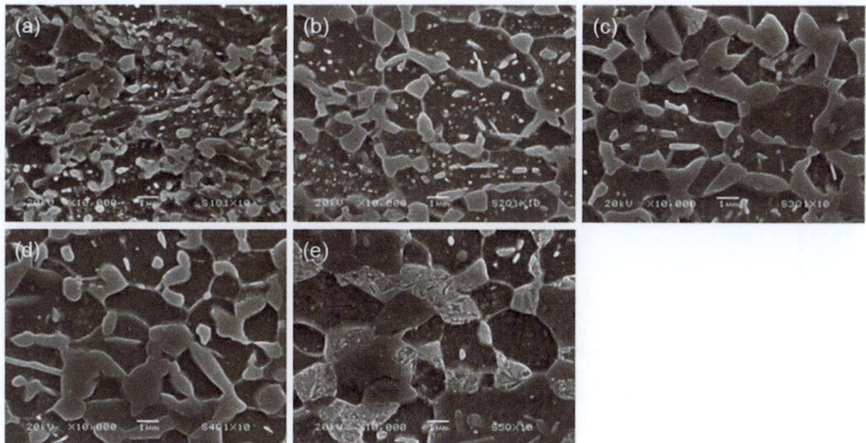

Fig. 9.18 SEM micrographs showing the microstructure after annealing at 675 °C for: (**a**) 3 min, (**b**) 1 h, (**c**) 3 h, (**d**) 6 h, (**e**) 26 h

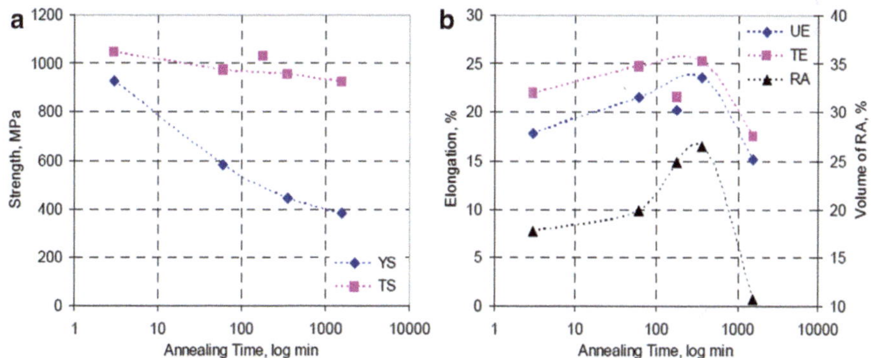

Fig. 9.19 Tensile properties and volume fraction of RA as a function of annealing time

austenite and/or martensite is initially very fine (<0.5 µm) and becomes coarser as the holding time increases. Mn-rich carbides gradually dissolve in austenite and disappear. Recrystallization of deformed ferrite is complete after 1 h of annealing.

Figure 9.19 describes the evolution of tensile properties and volume fraction of RA with the time of annealing. The amount of retained austenite gradually grows with time increasing to 6 h and then suddenly decreases. Both tensile and yield strengths decrease with time of annealing at 675 °C due to completion of recrystallization and grain growth. Serrations of flow stress and significant yield point elongation resulting in oscillations of instantaneous n values have been identified for samples annealed for 3 min–6 h, as shown in Fig. 9.20. Serration behavior gradually weakens and disappears with holding time. Instantaneous n value of

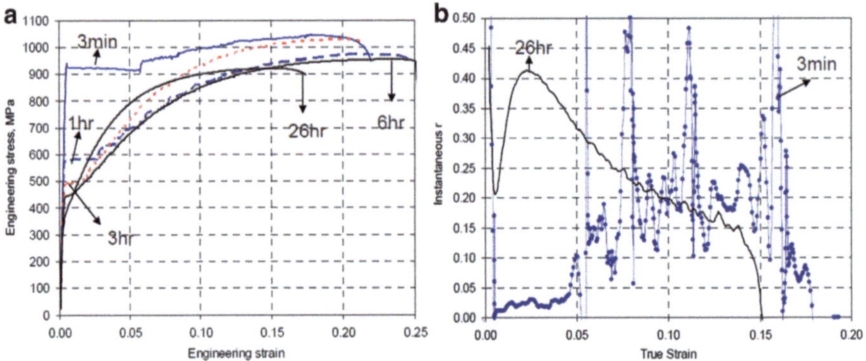

Fig. 9.20 Engineering stress–strain curves (**a**) and instantaneous n values (**b**)

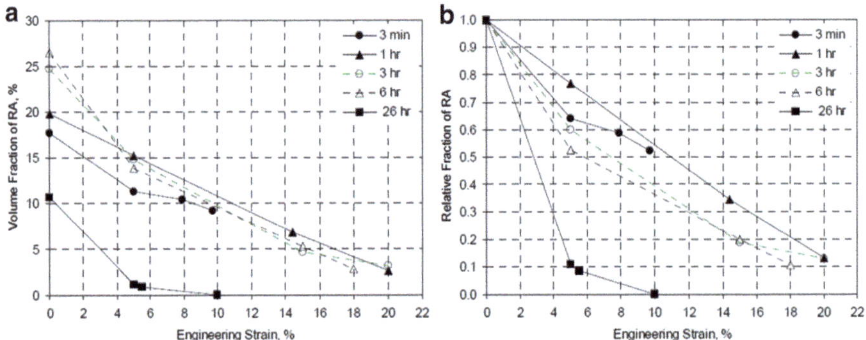

Fig. 9.21 Volume fraction of RA (**a**) and relative fraction of remained RA (**b**) as a function of engineering strain

sample annealed for 26 h showed no oscillations and demonstrated typical dual-phase trend due to a large amount of martensite.

The amount of RA initially grows with holding time increasing up to 6 h probably due to progressive Mn partitioning but decreases substantially thereafter. Much smaller amount of RA (~10 %) was detected after 26 h annealing because of lower stability of coarse grain austenite that transformed mostly to martensite during cooling.

Elongation is relatively constant with holding time even though the amount of RA increases up to 6 h from 17.7 to 26.5 %. This can be explained by decreasing RA stability manifested by the increasing portion of austenite transformed during tensile deformation (Fig. 9.21). Whereas samples annealed for 3 min and 1 h show gradual decrease in the amount of RA under tensile deformation, rapid decrease in the amount of RA is seen in cases of annealing for 3, 6, and 26 h. As a result of lower stability of RA after longer holding, the level of total elongation after annealing for 3 min and 1 h is the same as that after annealing for 3 and 6 h in spite of smaller amount of RA in the former cases.

Although the data related to grain sizes was not presented, it can be suggested that the drop in ductility and decreased fraction of retained austenite after 7 h of holding in the study by Arlazarov et al. conducted using pre-quenched 0.1C–4.6Mn steel (Arlazarov et al. 2012a) could also be related to austenite grain growth.

Since one can expect Mn partitioning increasing with annealing time, the austenite grain growth appears to have the prevailing effect on RA stability. This general statement is also quantitatively confirmed in the study by Lee et al. 2011b, c). Using 0.05C–6.15Mn–1.4Si steel, the authors compared the austenite stability in cold-rolled samples annealed at 680 °C for 180 s and 24 h. They found that after short annealing (180 s), the Mn content in austenite as a result of partitioning had reached about 7.3 %. In contrast, after long annealing (24 h), no retained austenite was detected, and martensite formed during cooling contained a little more than 8 % Mn (equilibrium solubility of Mn in austenite of this steel at 680 °C is ~8.5 %). This was a clear indication that Mn partitioning to the intercritical austenite was not sufficient to stabilize austenite at room temperature. The dominating factor here was that after 180 s annealing, the average austenite grain size was ~250 nm, whereas it increased sixfold to ~1.5 μm after 24 h annealing.

As a kind of compromise, the general conclusion is that high stability of austenite and enormously high amount of retained austenite in medium Mn steels are the combined result of fine austenite grain size and certain partitioning of carbon and manganese.

9.7 Summary

Combination of high tensile strength and ductility of medium Mn TRIP-assisted steels (e.g., TS = 1200 MPa at 25 % TE) motivates numerous researchers to find the ways to their practical utilization as the new third generation of steel.

Optimization of property balance of these steels, as well as widening of the processing windows, requires selection of appropriate alloying, initial microstructure, and annealing parameters. No published data related to their flangeability and hole expansion are available, but under certain conditions, the high YS/TS ratio that typically correlates with high HE values can be achieved. Tendencies to alloy of medium Mn steel with Al to widen the processing window should be also considered as an efficient method to enhance the resistance to delayed fracture, which is critical for ultrahigh strength steels.

As mentioned in other chapters, the "air hardenability" of medium Mn low-carbon steels stimulates studies aiming at applications of these steels for hot stamping (press hardening) that produces martensite microstructure or for Q&P processing.

References

Arlazarov, A., M. Gouné, O. Bouaziz, A. Hazotte, and F. Kegel. 2012a. "Effect of Intercritical Annealing Time on Microstructure and Mechanical Behavior of Advanced Medium Mn Steels." In *Materials Science Forum*, 706–709:2693–98.

Arlazarov, A., M. Gouné, O. Bouaziz, A. Hazotte, G. Petitgand, and P. Barges. 2012b. "Evolution of Microstructure and Mechanical Properties of Medium Mn Steels during Double Annealing." *Materials Science and Engineering: A* 542 (0): 31–39.

Arlazarov, A., A. Hazzotte, O. Bouaziz, and et al. 2012c. "Characterization of Microstructure Formation and Mechanical Behavior of an Advanced Medium Mn Steel." *Material Science and Technology, MS&T*, 1124–31.

Bleck, W., K. Hulka, and K. Papamentellous. 1998. "Effect of Niobium on the Mechanical Properties of TRIP Steels." *Material Science Forum* 284–286: 327–34.

Cao, W.Q., C. Wang, C.Y. Wang, J. Shi, and M.Q. Wang. 2012. "Microstructure and Mechanical Properties of the Third Generation Automobile Steels Fabricated by ART-Annealing." *Science China*, 1–9.

De Cooman, B.C., P.J. Gibbs, S. Lee, and D.K. Matlock. 2013. "Transmission Electron Microscopy Analysis of Ultrafine-Grained Medium Mn Transformation-Induced Plasticity Steel." *Metallurgical and Materials Transactions* 44A (6): 2563–72.

De Moor, E., S. Kang, J.G. Speer, and D.K. Matlock. 2014. "Manganese Diffusion in Third Generation Advanced High Strength Steels." In, *Key note lecture II*. Prague.

De Moor, E., D. Matlock, J. Speer, and M. Merwin. 2011. "Austenite Stabilization Through Manganese Enrichment." *Scripta Materialia* 64: 185–88.

Furukawa, T. 1989. "Dependence of Strength-Ductility Characteristics of Thermal History in Low Carbon, 5 Wt.% Mn Steels." *Material Science and Technology* 6 (5): 465–70.

Furukawa, T., H. Huang, and O. Matsamura. 1994. "Effect of Carbon Content on Mechanical Properties of 5%Mn Steels Exhibiting Transformation Induced Plasticity." *Material Science and Technology, MS&T*, 964–69.

Gibbs, P.J., E. De Moor, M.J. Merwin, B. Clausen, J.G. Speer, and D.K. Matlock. 2011. "Austenite Stability Effects on Tensile Behavior of Manganese-Enriched-Austenite Transformation-Induced Plasticity Steel." *Metallurgical and Materials Transactions A* 42 (12): 3691–3702.

Han, J., and Y.-K. Lee. 2014. "The Effect of the Heating Rate on the Reverse Transformation Mechanism and the Phase Stability of Reverted Austenite in Medium Mn Steels." *Acta Materialia* 67: 354–61.

Huang, H., O. Matsumura, and T. Furukawa. 1994. "Retained Austenite in Low Carbon, Manganese Steel after Intercritical Heat Treatment." *Material Science and Engineering, A* 10 (7): 621–26.

Jun, H.J., and N. Fonstein. 2008. "Microstructure and Tensile Properties of TRIP-Aided CR Sheet Steels: TRIP-Dual and Q&P." In, 155–61. Orlando, Florida.

Jun, H.J., O. Yakubovsky, and N. Fonstein. 2010. "Effect of Initial Microstructure and Parameters of Annealing of 4% and 6.7% Steels on the Evolution of Microstructure and Mechanical Properties." In . Houston, Texas.

Jun, H.J., O. Yakubovsky, and N.M. Fonstein. 2011. "On Stability of Retained Austenite in Medium Mn TRIP Steels." In *The 1st International Conference on High Manganese Steels*. Seoul, Korea.

Koo, J.Y., and G. Thomas. 1977. "Design of Duplex Fe/X/0.1C Steels for Improved Mechanical Properties." *Metallurgical Transactions A* 8 (3): 525–28.

Lee, S., and B.C. De Cooman. 2011. "On the Selection of the Optimal Intercritical Annealing Temperature for Medium Mn TRIP Steel." *Metallurgical and Materials Transactions A* 44 (11): 5018–24.

Lee, S., and B.C. De Cooman. 2014. "Tensile Behavior of Intercritically Annealed 10 Pct Mn Multi-Phase Steel." *Metallurgical and Materials Transactions A* 45A (2): 709–16.

Lee, S.-J., S. Lee, and B.C. De Cooman. 2011a. "Mn Partitioning during the Intercritical Annealing of Ultrafine-Grained 6% Mn Transformation-Induced Plasticity Steel." *Scripta Materialia* 64: 649–52.

Lee, S.-J., S. Lee, and B.C. De Cooman. 2011b. "Austenite Stability in Multi-Phase Ultrafine-Grained 6pct Mn Transformation-Induced Plasticity Steel." *Scripta Materialia* 64: 225–28.

Lee, S., K. Lee, and B.C. DeCooman. 2008. "Ultra Fine Grained 6wt% Manganese TRIP Steel." *Materials Science Forum* 654–656: 286–89.

Lee, S., S.-J. Lee, and B.C. De Cooman. 2011c. "Work Hardening Behavior of Ultrafine-Grained Mn Transformation-Induced Plasticity Steel." *Acta Materialia* 59: 7546–53.

Lee, S., S.-J. Lee, S. Santhosh Kumar, K. Lee, and B.C. De Cooman. 2011d. "Localized Deformation in Multiphase, Ultra-Fine-Grained 6 Pct Mn Transformation-Induced Plasticity Steel." *Metallurgical and Materials Transactions A* 42 (12): 3638–51.

Matlock, D.K., and J.G. Speer. 2006. "Design Consideration for the next Generation of Advanced High Strength Sheet Steels." In *The 3rd International Conference on Advanced Structural Steels,* 774–81. Gyeongju, Korea.

Merwin, M.J. 2006. *Method for Producing High Strength, High Ductility Manganese Steel Strip.*

———. 2007. "Low-Carbon Manganese TRIP Steels." In *Materials Science Forum,* 539–543:4327–32.

———. 2008. "Microstructure and Properties of Cold-Rolled and Annealed Lowcarbon Manganese TRIP Steels." *Iron & Steel Technology,* 66–86.

Miller, R.L. 1972. "Ultrafine-Grained Microstructures and Mechanical Properties of Alloy Steels." *Metallurgical Transactions* 3 (4): 905–12.

Morrison, W.B. 1966. "The Effect of Grain Size on Stress-Strain Relationship in Low Carbon Steels." *Trans of ASM* 59: 224–46.

Oh, C.-S., J. Kang, S.-J. Park, and S.-J. Kim. 2010. "Microstructure and Tensile Properties of Nb-Added High Manganese TRIP-Aided Steel Sheets." In *MS&T2010*. Houston, TX.

Olson, G.B., and M. Cohen. 1976. "A General Mechanism of Martensite Nucleation." *Metallurgical Transactions A* 7A: 1897–1904.

Shi, J., X. Sun, M. Wang, W. Hui, H. Dong, and W. Cao. 2010. "Enhanced Work Hardening Behavior and Mechanical Properties in Ultrafine Grained Steels with Large Fractioned Metastable Austenite." *Scripta Materialia* 63 (8): 815–18.

Suh, D.W., S.-J. Park, N.N. Han, and S.-J. Kim. 2010a. "Influence of Al on Microstructure and Mechanical Behavior of Cr Containing Transformation-Induced Plasticity Steel." *Metallurgical and Materials Transactions A* 41 (13): 3276–81.

Suh, D.-W., S.-J. Park, T.-H. Lee, C.-S. Oh, and S.-J. Kim. 2010b. "Influence of Al on the Microstructure Evolution and Mechanical Behavior of Low Carbon, Medium Manganese Transformation-Induced-Plasticity Steels." *Metallurgical and Materials Transactions A* 41A: 397–408.

Sun, R., W. Xu, C. Wang, J. Shi, H. Dong, and W. Cao. 2012. "Work Hardening Behavior of Ultrafine Grained Duplex Medium Mn Steels Processed by ART-Annealing." *Steel Research International* 83 (4): 316–21.

Tokizane, M., N. Matsumura, K. Tsuzaki, T. Maki, and I. Tamura. 1982. "Recrystallization and Formation of Austenite in Deformed Lath Martensite Structure of Low-Carbon Steels." *Metallurgical Transactions* 13 A: 1379–88.

You, J.S. 2004. "Effect of Reverse Transformation Treatment on the Formation of Retained Austenite and Mechanical Properties of C-Mn TRIP Steels." *Korean Journal of Materials Research* 14: 126–34.

Zhao, X., Y. Shen, L. Qiu, and X. Sun. 2014. "Effect of Intercritical Annealing Temperature on Mechanical Properties of Fe-7.9 Mn-0.14 Si-0.06Al-0.07C Steel." *Materials* 7: 7891–7906.

Chapter 10
Candidates for Third-Generation Steels: Q&P Processed Steels

Contents

10.1 Introduction

The necessity to simultaneously reduce the weight of cars by making thinner parts and to ensure high safety of passengers has made automakers to introduce new requirements for AHSS that include high YS values and unachievable for DP and TRIP steels' high YS/TS ratio at very high ductility parameters including hole expansion. The new heat treatment concept based on stabilizing austenite by carbon partitioning from pre-quenched martensite proposed by Speer and Streicher-Clarke

© Springer International Publishing Switzerland 2015

327

N. Fonstein, *Advanced High Strength Sheet Steels*,
DOI 10.1007/978-3-319-19165-2_10

(Speer et al. 2003; Streicher-Clarke et al. 2004) has become a very promising ground for development of steels with extremely high combination of mechanical properties. Partitioning of elements is unavoidable during any phase transformation in steels, but the authors named the new heat treatment "quenching and partitioning" to emphasize the key role of obtaining martensite by partial pre-quenching and subsequent partitioning of carbon between martensite and remaining untransformed austenite.

10.2   The Fundamentals of Q&P Process

Typical thermal cycle of Q&P process, Fig. 10.1, can begin from full or partial austenitization.

Initial martensite is obtained by quenching austenite or austenite plus ferrite mixture to the desired quench temperature (QT) below M_S when the carbon content in martensite at QT is equal to the bulk carbon content in austenite phase. Subsequent isothermal holding either at the initial quenching temperature (one-step Q&P) or at the elevated so-called partitioning temperature (two-step Q&P) is performed to induce gradual carbon enrichment of the remaining austenite. After this partitioning stage the steel is cooled to room temperature.

The main concept of Q&P process implies the partitioning of carbon from martensite to existing untransformed austenite by diffusion through martensite/austenite interface or, in case of quenching from two-phase region, partially through ferrite thus stabilizing the retained austenite at room temperature.

Stabilization of retained austenite is the key point in Q&P processing intended to achieve the desired combination of strength and ductility. J. Speer uses the

Fig. 10.1 Typical thermal cycle for Q&P process: AT is the annealing temperature, QT is the quenching temperature, PT is the partitioning temperature, and M_S and M_F are the martensite start and finish temperatures, respectively

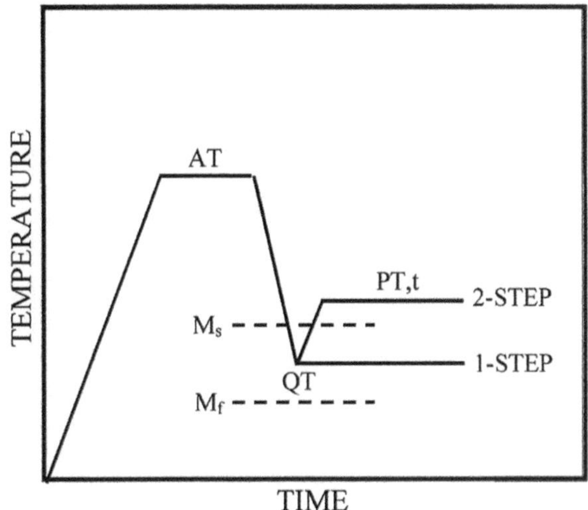

equation by Koistinen and Marburger to predict the austenite-to-martensite transformation in plain carbon steels at given quenching temperature QT (Koistinen and Marburger 1959)

$$f_\mathrm{m} = 1 - e^{-0.011(M_s - QT)} \tag{10.1}$$

where f_m is the fraction of austenite that transforms to martensite and M_S is the martensite start temperature. There are numerous equations that allow predicting the M_S temperatures from chemical composition of steels, summarized, in particular, by Krauss (2005).

The most common are

$$M_S \ (^\circ C) = 561 - 474\,C - 33\,Mn - 17\,Cr - 17\,Ni - 21\,Mo \tag{10.2}$$

proposed by Steven and Haynes (Steven and Haynes 1956), and

$$M_S \ (^\circ C) = 539 - 423C - 30.4Mn - 12.1Cr - 17.7Ni - 7.5Mo \tag{10.3}$$

proposed by Andrews (1965).

Recently, another equation

$$M_S = 539 - 423C - 30.4Mn - 7.5Si + 30Al \tag{10.4}$$

was proposed for CMnAl TRIP steel compositions (Imai et al. 1995).

A quench temperature selection methodology incorporates the calculated quenching temperature and carbon partitioning kinetics. The Koistinen–Marburger and the M_S temperature equations are used to predict the quenching temperature corresponding to the maximum amount of retained austenite that can be stable at room temperature. Selection of the initial quenching temperature between M_S and M_F is based on the target volume fraction of martensite calculated using K.-M. equation. Assuming an ideal carbon partitioning from martensite to austenite, where all other competing reactions are suppressed, the influence of partitioning kinetics is simulated by incorporating the local carbon content in austenite profile to predict an "optimum" quenching temperature, where the amount of retained austenite would be the maximal in the final microstructure.

The results can be presented in a graph exemplified in Fig. 10.2. In this figure, filled blue lozenges—martensite formed at quenching temperature (T_Q) that then will be a source of carbon partitioning (partitioned martensite PM); open blue squares—austenite retained at T_Q; yellow triangles—carbon content (wt%) of retained austenite (RA) considering complete partitioning; red crosses with orange line—martensite formed during second cooling step to room temperature (final fresh martensite—FM); black line—final retained austenite fraction at room temperature (RA) Carbon content in austenite, designated by yellow curve, at the end of partitioning is assessed from the balance of carbon, which is assumed to eventually

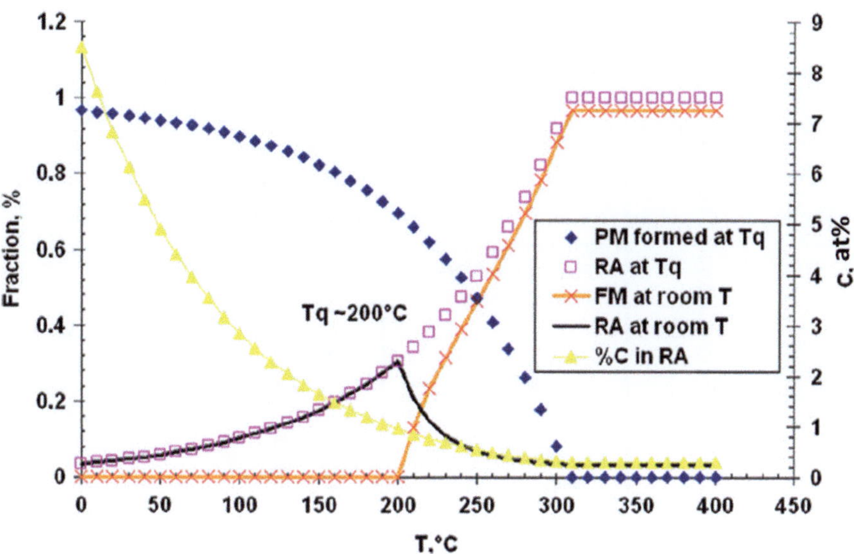

Fig. 10.2 Estimation of optimal quenching temperature according to the Speer et al.'s (2003) theory (Arlazarov et al. 2015)

become 0.015–0.02 % in martensite and all the remaining portion of carbon is in the retained austenite.

The model proposed by J. Speer is based on constrained carbon equilibrium (CCE) that assumes that diffusion of substitutional atoms is restricted and that the martensite/austenite interface allows only for diffusion of small interstitial atoms. Therefore, only carbon atoms are expected to diffuse from martensite to austenite. It is also assumed that carbide formation and carbon segregation to dislocations in martensite do not occur. Another assumption is that alternative reactions, such as carbide precipitation or bainite formation, are avoided and all carbon is partitioning from martensite to austenite, whereas iron and substitutional atoms cannot move.

Although the direct carbon partitioning from martensite to austenite is the basis for the Q&P concept, at least two coexisting or competing mechanisms of carbon partitioning to austenite should be considered: (1) partitioning of carbon to austenite from carbon supersaturated martensite (preferable concept) and (2) carbon enrichment of austenite, as in TRIP steel processing, due to formation of carbide-free bainite.

Consequently, the existing data should be divided into (1) those proving the mechanism of direct partitioning from martensite and (2) those that indicate prevailing phenomena of partitioning due to bainitic reaction, which is strongly accelerated by the initially formed martensite (Kawata et al. 2010).

Some data obtained using atom probe tomography (APT) can be considered as direct confirmation of carbon partitioning from martensite to austenite, as, for example, in the work by Toji et al. (2014). In similar experiments, high-carbon or/and alloyed steels are used to clearly separate carbon partitioning between

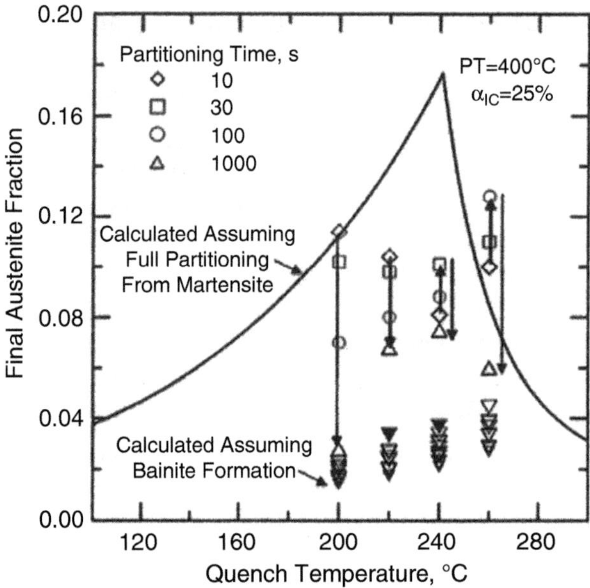

Fig. 10.3 Volume fractions of austenite calculated assuming two different possible mechanisms for austenite stabilization and determined experimentally (*open symbols* excluding *inverted triangles*) (Clarke et al. 2008)

martensite and austenite from bainitic transformation by suppressing the latter with heavy alloying. In the study by Toji, 0.59C (2.7 at.%C)–2Si–2.9Mn steel with M_F below room temperature was water quenched (~8 % of austenite was produced) and then tempered at 400 °C that gave convincing indication of enrichment of austenite with partitioned carbon up to 5–8 at.%.

Another direct confirmation of carbon transfer from martensite to austenite was obtained using in situ neutron diffraction measurements of carbon concentration in martensite and austenite during isothermal holding (partitioning) of pre-quenched 0.64C–4.57Mn–1.3Si steel (Speer et al. 2011). High Mn content allowed for retention of substantial amount of austenite at ambient temperature prior to heating, as well as for retardation of bainitic reaction.

Changes from initial **bct** martensite lattice to **bcc** or at least to low-carbon **bct** martensite lattice with low c/a ratio should be also considered as a direct proof of carbon transport from martensite to austenite with the former becoming progressively depleted with carbon (Edmonds et al. 2010).

Experiments performed, in particular, by Clarke et al. (2008) indicated that calculations of volume fractions of retained austenite under assumption of solely bainite formation during quenching and partitioning failed to explain the levels of carbon enrichment of austenite observed experimentally, Fig. 10.3. Partitioning of carbon to austenite from the initial martensite looked more consistent with experimentally observed austenite fraction than with one evaluated from formation of bainitic ferrite. On the other hand, the observed fraction of retained austenite was less than predicted by model that was explained by the loss of carbon atoms to

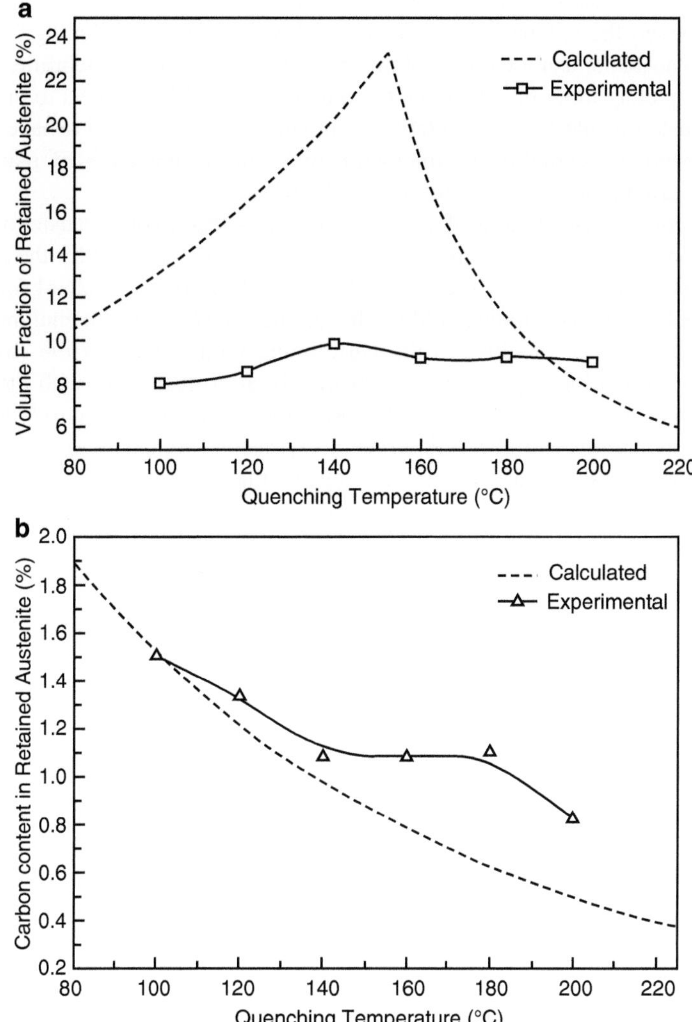

Fig. 10.4 Experimentally detected and calculated volume fraction of retained austenite (**a**) and its carbon content (**b**) as functions of QT; 0.2C–1.5Si–1.8Mn steel (Sun and Yu 2013)

competing processes of carbide precipitation, segregation of carbon atoms to dislocation, or some carbide-free bainite formation.

Significant inconsistences between calculations and experiments were noted by Sun and Yu who studied variations in volume fraction of retained austenite with quenching temperature. Experimentally detected carbon content in austenite was always higher than calculated when the difference was growing at increasing QT, Fig. 10.4.

Some implications of the partitioning model with respect to bainite transformation have been discussed by Speer et al. including the possibility of displacive bainite growth under carbon diffusion control (Speer et al. 2003). Santofimia et al. assessed the volume changes associated with carbon partitioning from martensite to austenite using theoretical models. Different interface mobilities ranging from completely immobile interface (assumed in early Q&P studies) to incoherent ferrite–austenite interface with relatively high mobility were found to have a strong influence on the evolution of microstructures during annealing of martensite–austenite grain assemblies (Santofimia et al. 2011a).

Theoretical considerations of partitioning process that involve transfer of carbon atoms from supersaturated martensite to austenite did not account for the extent of linear dilatation, observed experimentally in Q&P samples of 0.2C–1.5Si–2Mn–0.6Cr steel during partitioning holding. It was suggested that isothermal martensite and bainite form at or close to the partitioning temperature. While isothermal martensite formation began almost immediately after quenching with the rate of expansion decreasing with the time, the formation of bainite seemed to follow TTT diagram predictions with increasing rate of expansion (Somani et al. 2014).

The observed data make researchers suggest that at high partitioning temperatures the tempering of martensite competes with partitioning as well.

Due to combined kinetics of partitioning, tempering and bainite formation, as well as because of variability of austenite film/block sizes and carbon concentration profiles, the final fraction of retained austenite as a function of partitioning temperature and time can vary in a very complex manner.

10.3   Microstructure Evolution During Q&P Process

Q&P process aims at obtaining the final microstructure that consists of martensite and retained austenite. As mentioned earlier, cooling to a temperature between M_S and M_F should create the controllable fractions of martensite whereas subsequent thermal treatment intends to promote carbon transport into austenite thus stabilizing it at room temperature.

Details of microstructure evolution during Q&P processing cannot be described by metallography and are mostly supported by dilatometry or are based empirical on hypotheses (Edmonds et al. 2010). In analysis by Speer et al. (2011), all possibilities of microstructure evolution are considered. In addition to carbon partitioning into austenite, other processes pertinent to martensite tempering could occur during the "partitioning" step. These include carbon trapping at dislocations and interfaces in martensite, formation of carbides (both transitional carbides and/or cementite), and decomposition of austenite to bainite or other transformation products. The bottom line is that *various* competing processes *may* be possible.

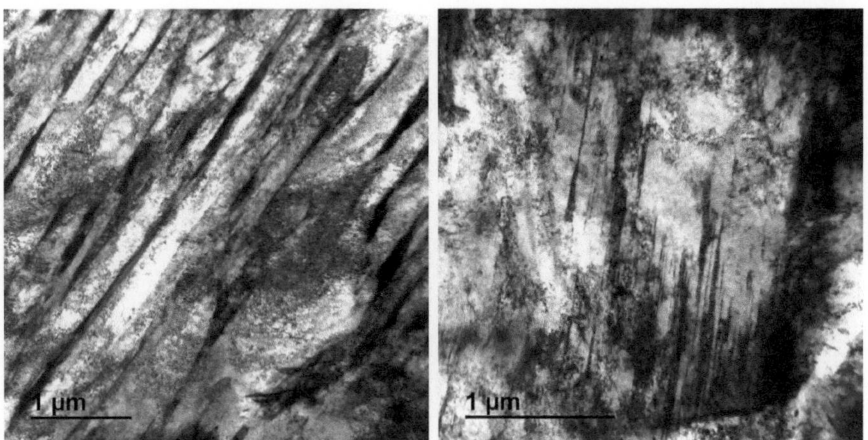

Fig. 10.5 Typical microstructure of Q&P steel, TEM

10.3.1   Morphology of Final Q&P Microstructure

The final microstructure of Q&P steel, as presented in Fig. 10.5, is the mixture of tempered martensite, lower bainite, retained austenite, and fresh martensite, formed during final cooling by decomposition of austenite portion of lower stability.

In some cases, TEM allows distinguishing lower bainite; however, its presence is mostly assessed by dilatometry data and simulating TTT diagrams conducted after pre-quenching. Nevertheless, lower bainite is a potential constituent of Q&P steels especially after quenching to relatively high temperature when the kinetics of bainite transformation is faster than at lower temperatures.

The important feature of Q&P treatment is a range of temperatures of initial martensite formation. As shown in Fig. 10.5, widely varying morphology of martensite can be observed.

Due to the short partitioning time, the carbon content in austenite grains is highly nonhomogeneous causing significant variability in stability and morphology of these grains. As shown in the study by Sun and Yu, three types of retained austenite can be observed in the microstructure of Q&P steels, Fig. 10.6. The first type is the ~100 nm wide films of retained austenite squeezed between martensite laths. The second type is the blocky retained austenite distributed in ferrite matrix. The third type is the ultrafine retained austenite films (20–30 nm thick) located between bainitic ferrite plates.

Numerous studies also note tempering of martensite during partitioning stage and the presence of fine carbides embedded in the final microstructure of Q&P processed steels.

Preventing the formation of carbides is an important aspect of the Q&P processing as these carbides reduce the amount of carbon available for enrichment of the remaining austenite. Transitional carbides which are more likely to form at

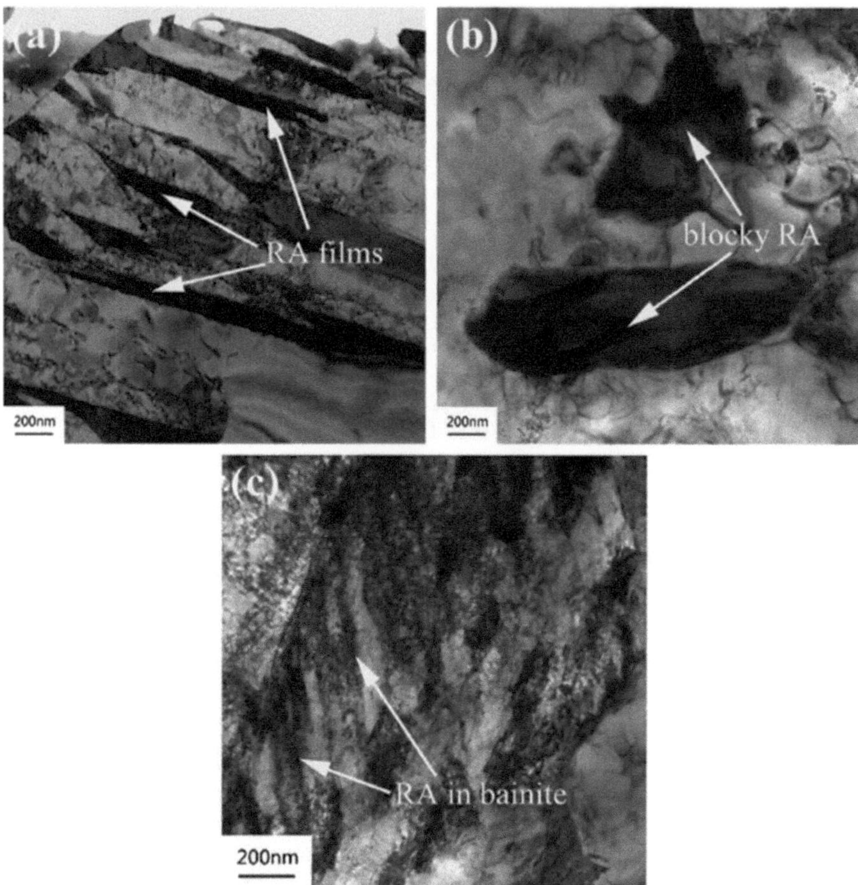

Fig. 10.6 TEM micrograph showing morphology and distribution of retained austenite in Q&P steel (Sun and Yu 2013)

low temperatures can redissolve at higher partitioning temperature and become a source of carbon for further enrichment of untransformed austenite provided the precipitation of cementite formation is prevented.

Unfortunately, there is lack of data related to quantification of another microstructure constituent, i.e., "fresh" martensite formed from insufficiently stabilized austenite during final cooling. The substructure and chemical composition of initially quenched and "fresh" martensite should vary because they form at different time and from parent austenite of different compositions. Wang et al. (2010) studied 0.21C–1.75Si–0.29Mn–1.03Cr–2.86Ni–0.31Mo–0.08V–0.049Nb steel and found "blocky" martensite, the structure of which was better revealed after 1 h tempering at 200 °C. After comparison with initial lath martensite that had also been tempered and was easy to etch, it was suggested that a "blocky" phase shown in Fig. 10.7 was "fresh" martensite formed from higher carbon austenite during final cooling.

Fig. 10.7 Microstructure of Q&P steel additionally tempered at 200 °C for 1 h to highlight the microstructure of the "blocky" martensite (Wang et al. 2010)

This approach allowed quantifying the volume fraction of fresh martensite and the distribution of its size (mostly less than 1 μm, with few particles over 3 μm). The authors were able to determine higher carbon content in "fresh" martensite as a result of insufficient partitioning to austenite.

Indirect approach to calculate the volume fraction of "fresh" martensite was suggested by (Paravicini Bagliaani et al. 2013). In the final microstructure of Q&P treated 0.28C–1.41Si–0.67Mn–1.49Cr–0.56Mo–Nb steel, they observed carbon-depleted martensite laths with some carbides inside that was interpreted as martensite *auto-tempered* before or during partitioning. In parallel, there were the unetched areas that could be retained austenite and/or untempered high-carbon martensite formed during final cooling. By deducting the amount of austenite detected by XRD from that unetched fraction, the volume fraction of untempered fresh martensite and its strong dependence on quenching temperature were determined, as presented in Fig. 10.8.

10.3.2   Effect of Initial Microstructure

According to the data summarized by Santofimia et al. (2011a), the initial microstructure before quenching can significantly affect the final microstructure and properties of Q&P processed steels.

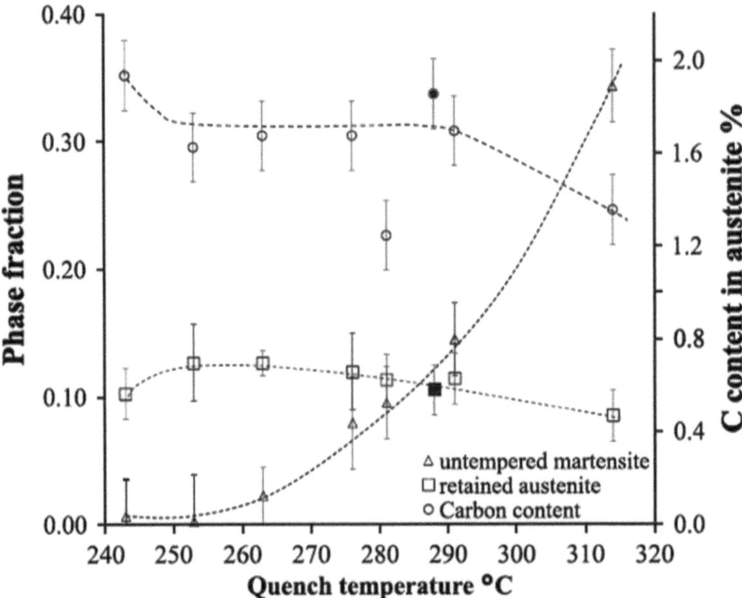

Fig. 10.8 Effect of quenching temperature on the volume fractions of retained austenite and untempered martensite and carbon content in retained austenite after Q&P treatment (Paravicini Bagliaani et al. 2013)

The initial martensite–ferrite mixture typical for high Mn steels (0.095C–3.5Mn–1.5Si) is converted by Q&P to microstructure consisting of martensite, *film-like ferrite*, and significant amount of retained austenite. This gives a very promising combination of strength and ductility: YS = 820 MPa, TS = 1630 MPa, UE = 10 %, and TE = 15 % or TS = 1415 MPa at TE = 23 %. However, the YS/TS ratio is about 0.5, which is rather typical for DP steel and can be perhaps explained by substantial portion of fresh martensite (Santofimia et al. 2010). The initial (after hot rolling) mixture of ferrite, bainite, and retained austenite in 0.19C–1.61Mn–0.35Si–1.1Al–0.09P (TRIP steel composition) steel resulted in *equiaxed ferrite* and martensite–retained austenite microstructure after partial austenitization and Q&P processing.

Zhang et al. investigated the role of pre-quenched microstructure after full austenitization of 0.18C–1.48Si–1.44Mn–0.15Al–0.025Nb steel at 900 °C on Q&P process including intercritical annealing at 850 °C for 3 min, quenching to 220 °C followed by partitioning at 400 °C for 5–500 s, and final quenching to room temperature.

It was found that the Q&Q–P cycle with double quenching produced fine martensite, which gave more nucleation sites for austenite leading to finer final microstructure. Film-like retained austenite was observed in both cases, whereas only Q&P samples contained some blocky retained austenite. Q&Q–P samples contained more retained austenite than the Q&P one.

While TS shows the general trend to decrease with partitioning time and TE tends to grow, Q&Q–P processing route demonstrated combinations of substantially higher TS and TE (for example, 1000 MPa at 30 % after 100 s partitioning) as well as higher n-value to the extended strains that was explained by larger fraction of more stable retained austenite and sustained TRIP effect at higher strains (Zhang et al. 2011).

10.3.3   Effect of Initial Cooling Rate

The rate of cooling during pre-quenching is important because it is necessary to prevent the formation of new ferrite in cooling from intercritical region and to prevent Widmanstatten ferrite or lower bainite in case of cooling after full austenitization.

CCT diagrams of steels allow predicting the critical cooling rate when uncontrolled transformations can take place. The majority of steels in question are alloyed high enough to be cooled at 50 °C/s without decomposition of initial austenite after full austenitization.

During cooling from intercritical region, the generation of ferrite is facilitated by preexisting ferrite interfaces that enable further epitaxial nucleation and growth of ferrite. Santofimia et al. considered in detail how the generation of new ferrite and the changes in carbon concentration profile within austenite grains can affect partitioning of carbon from martensite to austenite after pre-quenching. Commercial 0.19C–1.61Mn–0.35Si–1.10Al–0.09P steel with high content of ferrite forming aluminum and phosphorous was intercritically annealed at 900 °C for 600 s and then quenched at the cooling rate of 50 °C/s to 125, 150, and 175 °C with subsequent partitioning at 250 and 350 °C for various durations. The resultant microstructure mostly contained tempered martensite and ferrite with the latter included epitaxial ferrite formed during first cooling. This epitaxial ferrite, in authors' opinion, plays a role in the partitioning step as it induces carbon concentration gradients in austenite grains so that carbon concentration is higher in the vicinity of austenite grain boundaries and is lower in austenite grain interior.

It was observed that martensite formed during initial quenching was located inside of austenite grains, whereas austenite located at grain boundaries remained untransformed evidently due to lower M_S temperature.

The authors point out that excessive epitaxial ferrite can reduce the amount of martensite formed during first quenching and therefore could be detrimental for steel strength, for the amount of retained austenite, and for its carbon content. Therefore, cooling rate should be sufficiently high as well in spite of higher carbon content in the initial austenite after intercritical annealing.

Carbon concentration heterogeneity should lead to faster partitioning that can be useful for lines with short partitioning step, so the optimal amount of epitaxial ferrite can be controlled by combination of annealing temperature in the intercritical range and cooling rate (Santofimia et al. 2010).

10.3.4 Effect of Quenching Temperature

The quenching temperature (QT) determines the fraction of martensite that is supposed to provide carbon for partitioning to the adjacent austenite during the isothermal holding following quenching.

Quenching to temperature slightly above the M_F temperature leads to generation of appreciable amount of martensite, which leaves a small portion of austenite available for carbon enrichment and final retention. On the contrary, quenching temperature close to the M_S temperature results in a small fraction of martensite, so that carbon available for partitioning might not be sufficient to stabilize the untransformed austenite. These extreme conditions imply determination of the optimal quenching temperature to reach maximum volume fraction of retained austenite.

As mentioned above, calculation of the optimum quenching temperature recommended by Speer is based on prediction of the optimal fraction of retained austenite (see Fig. 10.2). This model assumes that there is full carbon partitioning from martensite to austenite at the end of partitioning step and austenite fraction does not change during partitioning (i.e., martensite/austenite interface is fixed and any other transformations such as bainite formation or carbide precipitation are suppressed). On the other hand, the formation of bainite is feasible, even in the case of partitioning temperatures below M_S, since the presence of previously formed martensite promotes generation of bainite practically without an incubation period (Kawata et al. 2010).

Later, Santofimia et al. (2008) proposed a model for carbon diffusion during partitioning step assuming migration of the martensite/austenite interface.

In fact, the choice of the optimum quenching temperature, QT, is, to a large extent, governed by the anticipated final strength of steel. For example, for (0.3–0.6)C–(1.5–2.5)Si–(1.5–2.0)Mn steel, Seto and Matsuda noted that the target strength close to ~1500 MPa at the highest elongation can be achieved using quenching to slightly below the M_S temperature (Seto and Matsuda 2013).

Even at a fair approximation of calculated optimum QT, the amount of carbon that actually partitions to the austenite as well as the final microstructure and properties depend not only on the partitioning temperature and time but also on the kinetics of carbon partitioning. As a result, the calculated predicted volume fraction of retained austenite as a function of QT is much less sensitive to QT than could be expected based on the assumption of full carbon partitioning. Experimental results suggest relatively wide "processing window" to obtain the desired amount of RA. For example, QT ranging from 200 to 270 °C resulted in approximately the same fractions of retained austenite in the final microstructure (Clarke et al. 2009).

Other researchers including Sun and Yu came to similar conclusion as was shown in Fig. 10.4.

Santofimia et al. also demonstrated experimental data describing the variations of volume fraction of retained austenite with quenching temperature (Fig. 10.9). Wide maximum of ~40 °C around QT = 230 °C was observed. Carbon content in

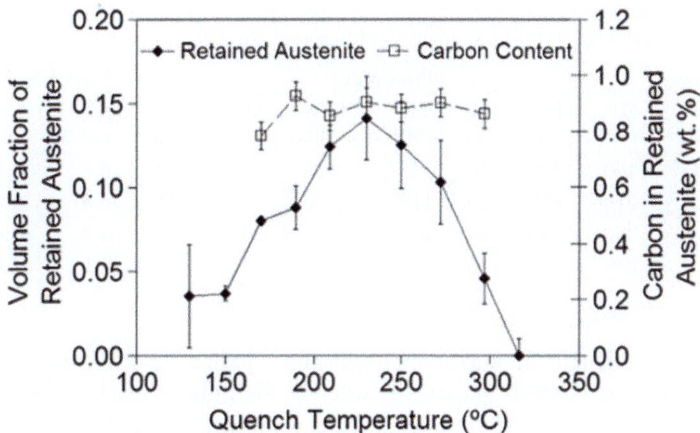

Fig. 10.9 Volume fraction of retained austenite and corresponding carbon content after full austentization, quenching to different temperatures, and partitioning at 400 °C for 100 s, 0.2C–2.5Mn–1.47Ni–1Cr–1.5Si steel (Santofimia et al. 2011b)

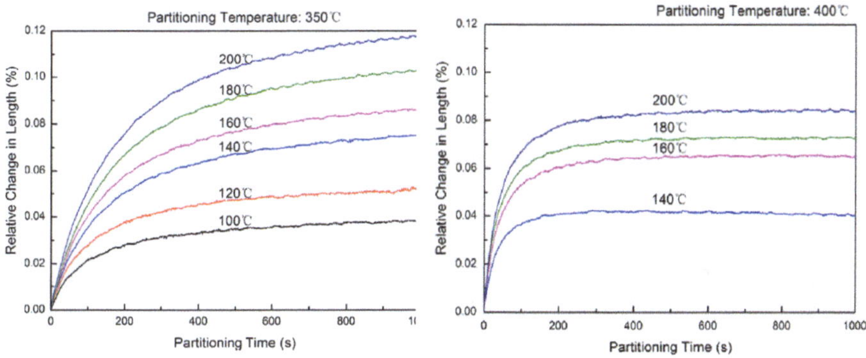

Fig. 10.10 Dilatation curves of 0.2C–1.5Si–1.9Mn during partitioning at 350 °C and 400 °C for 1000 s after quenching to various temperatures; 0.2C–1.5Si–1.8Mn steel

austenite ranging from 0.8 to 1 % appeared to be rather insensitive to quenching temperature, but it was high enough to lower the M_S temperature below the room temperature for given chemistry (Santofimia et al. 2011b).

Quenching temperature determines not only the amount of the formed martensite but naturally the fraction of untransformed austenite. Dilatometry studies show that all dilatation curves in specimens pre-quenched to different temperatures display isothermal expansion, proportional to the amount of untransformed austenite. In all cases, the amount of residual austenite tends to stabilize after a certain period of time (Fig. 10.10).

Table 10.1 Processing parameters, mechanical properties, and austenite characterization for the fully austenitized 0.3C–3Mn–1.6Si grade (De Moor et al. 2011a)

Processing					Properties				Austenite	
ST (°C)	St (s)	QT (°C)	PT (°C)	Pt (s)	YS (MPa)	UTS (MPa)	UE (%)	TE (%)	$f_{\gamma R}$ (vol%)	C content (wt%)
840	120	190	450	10	1257	1398	5	10	6.1	1.04
		210	450	10	1197	1356	8	12	5.8	1.02
		230	450	10	1050	1342	10	14	4.2	1.01
		250	450	10	981	1359	11	15	7.1	1.01

The larger dilatation at maximum expansion is indicative of a higher propensity for the formation of fresh martensite, whereas the time of reaching stable values increases with QT, i.e., with smaller fraction of initially formed martensite.

Similar results were obtained by Clarke et al. (2008) when the increase in QT of 0.19C–1.59Mn–1.63Si steel required longer partitioning time at constant 400 °C to achieve maximum volume fraction of austenite.

As to effects of QT on mechanical properties, the most sensitive to changes in quenching temperature are YS and TS, as they are related to the amount of martensite. Uniform elongation shows some tendency to depend on the total fraction of retained austenite, but rather on its portion that contributes to TRIP effect (i.e., its stability).

De Moor studied the effect of quenching temperature with partitioning at 400 and 450 °C using 0.2C–3Mn (or 5 Mn)–1.6Si steel after annealing in the intercritical region or full austenitization (De Moor et al. 2011a). The parameters of processing for conditions of full austenitization and partitioning at 450 °C are presented in Table 10.1.

In this example the partitioning time was short (10 s). Higher quenching temperature resulted in smaller amount of martensite and higher ductility. YS and TS values were high and tended to decrease (mostly YS) with increasing quenching temperature due to smaller fraction of martensite. Therefore, elongation increased with quenching temperature. No direct correlation between total elongation and the amount of retained austenite was found. Carbon content in austenite remained practically constant and independent of quenching temperature.

Higher quenching temperature and hence larger amount of untransformed austenite lead, at given partitioning time, to some decrease in the amount of retained austenite and in significant increase of amount of fresh martensite, formed during final cooling.

In particular, 0.2C–0.9Si–Nb steel with 2.8 % of (Mn + Cr + Mo) was chosen for a commercial Q&P steel grade with tensile strength of 1180 MPa (Hausmann et al. 2013b). For this purpose, cold-rolled material was fully austenized, quenched to the respective quenching temperature, and isothermally held at 400 °C. Microstructure studies revealed a general trend of transition from lath-like fine bainite to coarse globular bainite as the quenching temperature was increased. The microstructure after quenching to well below the M_S temperature clearly differed from that after quenching to higher temperatures. Finely dispersed retained austenite grains and much finer lath-like martensitic matrix was produced by quenching to temperatures well below the martensite start temperature, whereas the microstructures after quenching closer to M_S comprised coarse retained austenite/martensite islands at globular bainite.

While in all cases the tensile strength was within narrow range between 1200 and 1260 MPa, yield strength, elongation, and hole expansion strain showed significant variability. In particular, the n-values varied in wide range, from $n_{4-6} = 0.07$ to $n_{4-6} = 0.17$, indicating higher strain hardening potential for higher quenching temperatures. Increase in quenching temperature was accompanied by increase in elongation and decrease in yield strength and hole expansion due to evidently higher portion of fresh martensite.

Undercooling below the M_S temperature and therefore the amount of initial (eventually tempered) martensite play a minor role compared to the large differences in mechanical properties of microstructure constituents that depend on quenching temperatures. Higher quenching temperature leads to higher uniform and total elongations, as well as to higher n-values due to increase in the amount of retained austenite (and hence stronger TRIP effect) and softer matrix. Furthermore, with higher quenching temperatures the yield strength decreases reproducing DP steel behavior. This is because the difference in hardness between soft matrix and hard retained austenite/martensite islands increases as bigger volume of fresh martensite becomes harder and the matrix becomes softer with increasing QT. The effect of hardness differences between micro-constituents was particularly pronounced with respect to hole expansion strain.

Hole expansion is significantly higher (~32 %) for QT = 350 °C than for QT = 360 °C (~15 %), while for both $T_Q = 310$ °C and $T_Q = 340$ °C the hole expansion strain of about 40 %. Since the amount of retained austenite, obtained for QT = 360 °C, was even slightly higher than for QT = 350 °C and the average size of retained austenite/martensite islands was smaller, it can be assumed that fresh martensite plays the dominating role. More fresh high-carbon martensite was present in the final microstructure after QT = 360 °C, which in combination with softening matrix resulted in tremendous deterioration of edge formability as indicated by low hole expansion strain.

With respect to edge formability, the studied composition demonstrated great potential at quenching temperatures of 310–350 °C combined with partitioning at 400 °C. This allows for exceeding tensile strength of 1180 MPa, while achieving high YS/TS ratios and hole expansion higher than 30 % along with appropriate total elongation (Fig. 10.11).

No studies of the effect of holding time at quenching temperature can yet be found in the literature. It can be suggested that if this time is too long (>10 s) then martensite can auto-temper, especially with high M_S temperature of steel. This should lower the amount of retained austenite and reduce both YS and TS.

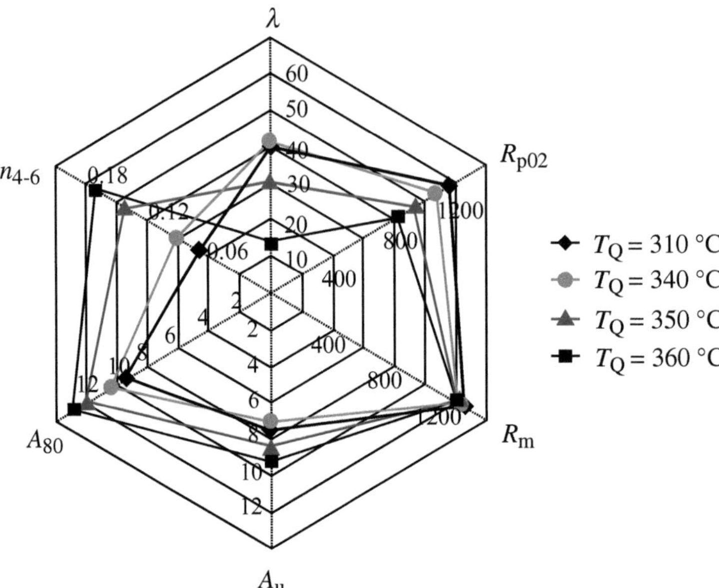

Fig. 10.11 Mechanical properties of Q&P grades produced at industrial annealing line with overaging/partitioning temperature of 400 °C and various quenching temperatures (Hausmann et al. 2013a)

10.3.5 Bainitic Reaction During Q&P Process

Using 0.3C–2.49Mn steel ($M_S = 330$ °C), Kawata et al. found that the bainitic transformation in austenite–martensite mixture is faster than that starting from single-phase austenite (Kawata et al. 2010). The authors suggest that the boundary between martensite and austenite serves as a nucleation site for bainitic ferrite. As shown in Fig. 10.12, the bainitic reaction is substantially accelerated even at temperature as low as 320 °C and its rate grows with volume fraction of martensite increasing to 0.22 and 0.40.

Fast bainitic reaction after formation of martensite was indicated also by Seto and Matsuda (2013). As shown in Fig. 10.13, isothermal holding after pre-quenching is accompanied by formation of substantial amount of bainite. The authors confirmed that bainitic transformation was strongly accelerated by the presence of martensite and this tendency was enhanced with higher carbon content in steel.

On the other hand, some researchers assume that small amounts of bainite proportional to the fraction of untransformed austenite are formed mostly in the beginning of partitioning (Sun et al. 2013). Further enrichment of remaining austenite with carbon lowers the B_s temperature which should significantly suppress the bainitic reaction so that eventually B_s can drop below the partitioning temperature and bainitic reaction can stop completely.

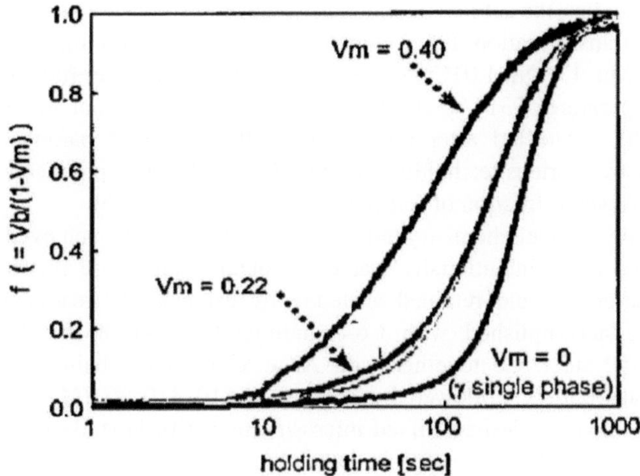

Fig. 10.12 Effect of the volume fraction of martensite in the initial microstructure on the rate of bainitic transformation at 320 °C Kawata (Kawata et al. 2010)

Fig. 10.13 Dilatation curve for 0.6C–2Si–1.5Mn steel annealed at 900 °C for 120 s, pre-quenched to 200 °C, and held at 400 °C for 300 s (Seto and Matsuda 2013)

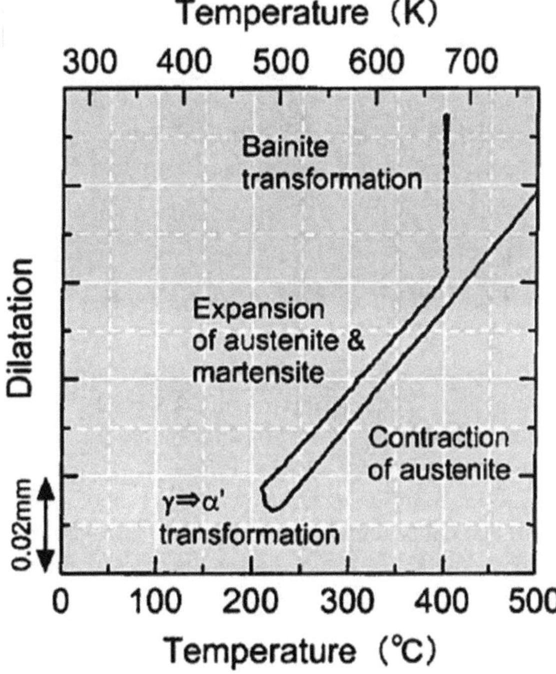

In some studies, the authors intentionally used high alloying to delay and suppress the bainitic transformation. It was done in particular by Santofimia et al. who used 0.204C–2.5Mn–1.47Ni–1.01Cr–1.50Si composition which after full austenitization had M_S temperature of 314 °C and $B_s = 320$ °C so that isothermal holding at 350 °C during 4000 s detected after cooling a small amount of bainite, which was surrounded by martensite; holding at 400 °C for 16,000 s produced martensitic matrix with a small fraction of ferrite (Santofimia et al. 2011b).

Depending on steel chemistry, an alternative effect can be achieved. For example, Samanta et al. intentionally wanted to obtain multiphase microstructure of bainite, martensite, and retained austenite by using Q&P treatment. This was successfully accomplished with Co-containing 0.32C–1.78Mn–0.64Si–1.75Al–1.2Co–0.035P steel. Co together with Al accelerates the bainitic reaction. By annealing at 1050 °C followed by cooling to 210–310 °C ($M_S = 355$ °C), the authors obtained the desired mixed microstructure. EBSD study showed random distribution of bainite and martensite laths. After short partitioning at 375 °C (3 s), the samples quenched to 290 °C exhibited both blocky and film-like austenite while only interlath films of RA were found in the sample quenched to 210 °C (Samanta et al. 2013).

At the same time, the study by Santofimia et al. showed that for the majority of temperature–time combinations, the amounts of retained austenite after Q&P were considerably higher and emerge at much shorter times than those obtained by direct cooling and holding in the bainitic transformation region. Performed dilatometry indicated that, if bainite is formed during partitioning in the Q&P process, its volume fraction was too small to justify the observed amounts of retained austenite. Therefore, it was assumed that Q&P process is more efficient for stabilization of austenite than bainitic isothermal treatment used to produce CFB steels.

10.3.6   Effect of Partitioning Time

The increase in partitioning time tends to reduce TS of steel due to softening (tempering) of martensite fraction. YS grows with partitioning time increasing up to certain duration (30–100 s depending on partitioning temperature), which is typical for tempering martensite.

During partitioning, the volume fraction of retained austenite initially decreases and then either remains approximately constant or grows with partitioning times. Carbon content in retained austenite initially grows quite rapidly, but this growth slows down at longer partitioning times.

In 0.19C–1.59Mn–1.63Si steel, intercritically annealed at 820 °C and quenched to QT = 220–280 °C, the austenite fraction was increasing during partitioning at 400 °C for up to 30 s, but it decreased already after 10 s of partitioning after quenching to QT = 200 °C. Carbon content in austenite under all these conditions was increasing with partitioning time of up to 100 s (Streicher-Clarke et al. 2004).

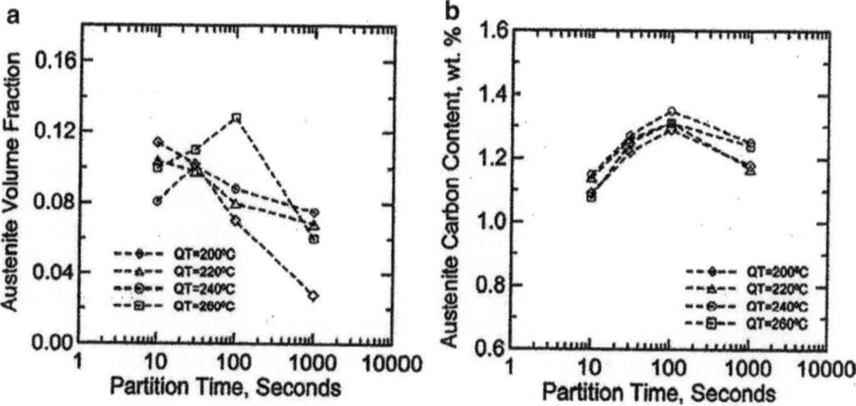

Fig. 10.14 Effect of partitioning time on the final volume fraction of austenite (**a**) and measured carbon concentration in retained austenite (**b**). 0.19C–1.59Mn–1.63Si steel; intercritical annealing at 820 °C, partitioning time of 400 s after quenching to various temperatures, 25–28 % of ferrite (Clarke et al. 2008)

As discussed above, the attainment of maximum retained austenite at certain partitioning time can be predicted by calculations of martensite formation and kinetics of carbon partitioning in the binary Fe–0.2C system. As shown in the study by Santofimia et al. (2011a), the calculations confirm the possibility of existence of two peaks, as was experimentally observed by Matlock et al. (2003) during annealing simulations. While the first peak of austenite fraction indicated fast stabilization of austenite due to rapid kinetics of carbon transport from martensite to untransformed austenite, the second peak after longer partitioning time was explained by re-dissolution of transitional carbides formed at lower temperatures provided that cementite formation is suppressed.

According to the data presented in Fig. 10.14 (Clarke et al. 2008), the optimal partitioning time at 400 °C strongly depends on the quenching temperature (i.e., evidently on the initial volume fraction of martensite) and increases from 10 to 100 s with QT increasing from 200 to 260 °C. Beyond this peak, sharp decrease in volume fraction of retained austenite with increasing partitioning time was due to carbide/cementite formation. On the other hand, regardless of QT, the maximum carbon content in austenite was always detected after 100 s of partitioning.

Using a typical TRIP 0.2C–1.5Si–1.9Mn steel composition, Sun et al. studied the effect of partitioning time on final properties and microstructure after intercritical annealing at 800 °C for 150 s, quenching to 160 °C for 5 s, and partitioning at 400 °C for 10–1000 s. The Q&P treatment produced excellent combination of strength and ductility with TS being between 990 and 1100 MPa and TE ranging from 25.9 % to 29.3 %. The general trend was that TS decreased with partitioning time, whereas elongation changed mostly in the same way as austenite stability (its carbon content).

Optimal partitioning time depends on both partitioning temperature and steel composition. The fact of optimal combination of properties after partitioning for about 100 s was noted by Tariq and Baloch who used 0.37C–1.24Si–0.86Mn–1.18Cr steel ($M_S = 328$ °C) and obtained TS = 2000 MPa and TE = 11.5 % with one-step Q&P processing. The Q&P cycle included full austenitization at 920 °C for 1200 s and quenching to 200 °C with further partitioning at the same temperature for 10, 100, and 1000 s to allow migration of carbon from supersaturated martensite to adjacent austenite (Tariq and Baloch 2014). While TS and YS showed the trend to decrease with partitioning time, elongation and amount of retained austenite demonstrated maximum after partitioning for 100 s and carbon content in austenite increased to 100 s and remained stable thereafter.

During partitioning process at 400 °C, the volume fraction of retained austenite showed a decrease between 10 and 50 s and remained approximately constant (0.09–0.10) thereafter. Carbon content in austenite rapidly increased from 1.08 to 1.28 % within the same period of time between 10 and 50 s and very slowly increased up to 1.32 % afterwards (Sun et al. 2013).

Samanta et al. also studied the effect of partitioning time on microstructure and in particular on bainite formation in 0.32C–1.78Mn–0.64Si–1.75Al–1.2Co–0.035P steel. It was established that for all quenching temperatures the carbon content in retained austenite increased with partitioning time, sharply in the beginning and slowly after 20 s, Fig. 10.15. Measured bainite fraction was almost constant after

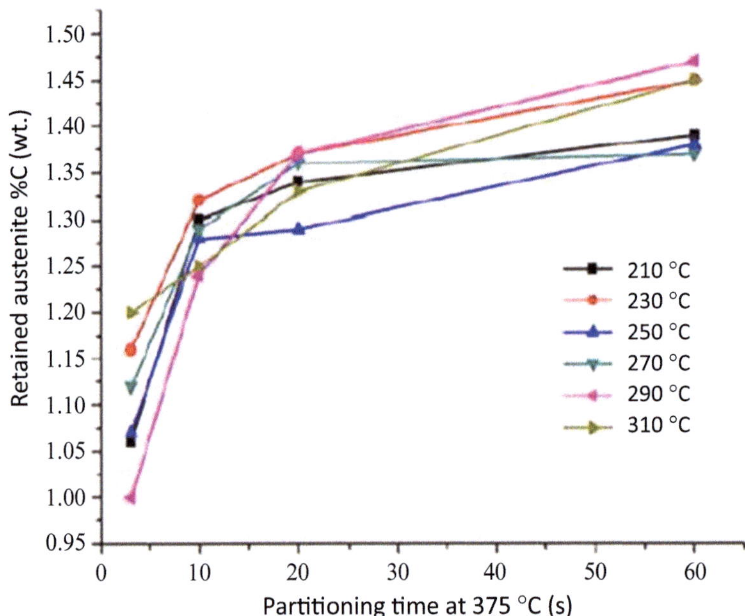

Fig. 10.15 Effect of quenching temperature and time of partitioning at 375 °C on carbon content in austenite (Samanta et al. 2013)

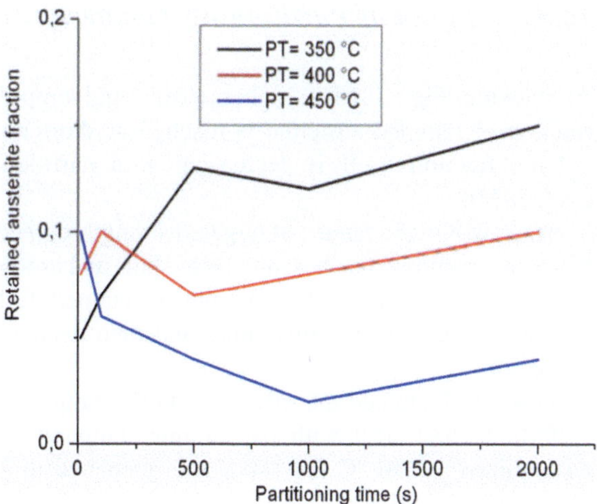

Fig. 10.16 Volume fraction of retained austenite after annealing at 900 °C for 600 s, quenching to 275 °C, and partitioning at 350, 400, and 450 °C (Santofimia et al. 2011a, b)

20 s of partitioning that could indicate that bainitic reaction took place only during first 20 s. However, since the most significant increase in austenite carbon content was observed also within initial 20 s of partitioning, authors concluded that carbon partitioned to austenite from both martensite and bainite.

Santofimia et al. conducted very comprehensive study that also included evaluation of the effect of partitioning time using abovementioned high alloyed 0.204C–2.5Mn–1.47Ni–1.01Cr–1.50Si steel that allowed for suppressing bainitic reaction. Some of the results are presented in Fig. 10.16.

During partitioning at 350 °C, the volume fraction of austenite continuously grows with partitioning time up to a maximum of around 0.15 after partitioning for 2000 s. The average carbon content in retained austenite (0.75–0.95 %) reaches maximum after partitioning for 100 s. During partitioning at 400 °C, the trend is similar: the volume fraction of RA reaches maximum of 0.10 after 100 s and remains approximately constant thereafter.

During partitioning at 450 °C, the maximum RA volume fraction of ~0.12 is observed after only 3 s. RA volume fraction gradual decreases with the time of the partitioning until 500 s and then remains approximately constant.

It is clear that the maximum in the volume fraction of retained austenite is reached at shorter times for higher partitioning temperatures, which is in agreement with the expected kinetics of carbon partitioning. Therefore, the selection of partitioning time should depend on partitioning temperature. It should be also noted that sufficient partitioning time is necessary to homogenize partitioned carbon content within the austenite grains. However, very long partitioning can lead to formation of transitional carbides and deterioration of mechanical properties.

10.3.7   Effect of Partitioning Temperature

As shown in Fig. 10.16, at higher partitioning temperatures, the maximum volume fraction of retained austenite is reached at shorter times; however, the austenite volume fraction itself is decreasing with partitioning temperature (Santofimia et al. 2011b).

Thus, within the ranges of targeted strength properties, the partitioning time and temperature should be chosen to be as short and as low as possible to avoid carbide/cementite precipitation. The amount of retained austenite and its carbon concentration increased with temperature and then decreased probably due to carbides' formation.

Zhao et al. studied the effect of Q&P treatment of low-carbon 0.17C–1.41Si–1.48Mn–0.25Al steel with partitioning temperature below M_S. Steel was fully austenitized at 910 °C for 180 s and quenched to 245 °C for 10 s. Then it was reheated and held at 350 °C for 120 s before final quenching. Dilatometry measurements revealed M_S, A_{c1}, and A_{c3} to be 390, 710, and 880 °C, respectively.

Microstructure obtained after this heat treatment consisted of martensite laths with high dislocation density and thin film-like retained austenite. Blocky shaped retained austenite was not present in the microstructure. The volume fraction of retained austenite was 0.08. Steel exhibited tensile strength of 1050 MPa and elongation of about 25 %. Although the initial M_S temperature of this steel was above the partitioning temperature, carbon enrichment of the initial austenite gradually lowered its M_S so that partitioning at 350 °C occurred above new M_S temperature stabilizing the remaining austenite (Zhao et al. 2008).

TS decreased with increasing partitioning temperature due to softening of martensite. YS increased with partitioning temperature, as typically observed during tempering of martensite up to certain limits depending on steel chemistry. Elongation increased with partitioning temperature despite of slight decrease in retained austenite fraction for temperatures above 420 °C, probably due to dominant role of material softening.

Zhang et al. studied the impact of partitioning temperature in Q&P cycle using cold-rolled 0.19C–1.53Mn–1.52Si–0.14Al–0.048Nb steel. Samples were fully austenitized at 900 °C for 180 s, quenched to 260 °C, and partitioned during 120 s at 350, 390, 420, and 450 °C, before being water quenched.

Roughening of martensite was observed with increasing partitioning temperature. YS decreased slightly with temperature increasing above 420 °C due to significant softening of tempered martensite. TS dropped sharply at the beginning and then remained stable between 390 °C and 420 °C. For temperatures above 420 °C, YS sharply decreased along with TS. Retained austenite volume fraction remained stable and then decreased sharply for temperatures higher than 420 °C, whereas elongation only slightly decreased with increasing partitioning temperature (Zhang et al. 2013).

Few data are available related to partitioning at temperatures close to 450–460 °C that can be relevant to galvanizing process. Using cold-rolled

0.29C–3Mn–1.4Si steel, Arlazarov et al. applied full austenitization and quenching to 200 °C for 2 s followed by partitioning at 460 °C in comparison with 400 °C. Authors concluded that partitioning temperature has significant effect on structure and properties, whereas the role of partitioning time is secondary.

During partitioning at 400 °C an improvement of uniform and total elongation was observed due to the enhanced stability of retained austenite showed by higher C content. Strength of the steel seemed to be insensitive to the time increase, as the fractions of constituents stayed the same. In the case of partitioning at 460 °C, TS did not change as the portion of martensite plus retained austenite did not vary; YS decreased slightly due to the higher amount of RA and higher degree of martensite partitioning. Elongation was not affected by the partitioning time, probably due to some balance of changes in the increased amount of RA fraction (from 0.16 to 0.22) and its some lower stability (decrease in carbon content from 1.05 to 0.97 %).

As was observed, MA islands contained fresh martensite in the center and retained austenite at the edges that reflects the non-homogeneity of austenite grains. Carbon enrichment sufficient to stabilize austenite at room temperature occurred only at the edges of the austenite grain. Consequently, the grain interior was not sufficiently stabilized and transformed into martensite during final quenching.

This phenomenon was also observed in the study by Santofimia et al. (2009).

Using 0.256–1.2Si–1.48Mn–1.51Ni–0.053Nb hot-rolled steel, Zhou et al. also studied partitioning at 450 °C. Steel was fully austenitized at 930 °C for 600 s, quenched to 290 °C, and partitioned at 450 °C for 30 s. The volume fraction of 0.112 of retained austenite with 1.25 % of carbon was detected. Steel exhibited YS of 1040 MPa, TS of 1220 MPa, and 18 % total elongation after partitioning at these high temperature and short time. The effects of partitioning temperature on RA amount, YS, and TS are illustrated in Fig. 10.17.

Thus, higher partitioning temperature results in martensite softening accompanied by decreasing the strength and increasing elongation. Depending on the chemical composition, there exists a temperature range, within which the combination of strength and elongation can be optimized.

10.4 Effect of Alloying and Microalloying Elements

So far, no study was attempted dedicated specifically on optimization of chemical composition for Q&P process. The majority of research work is focusing on TRIP steel-type chemical composition assuming that the main contribution to the enrichment of austenite with carbon comes from suppressing the carbide formation in bainite. Since bainitic reaction does not play the dominant role in Q&P processing, as well as in evolution of microstructure and properties of Q&P processed steels, the effects of alloy composition deserve special attention. Some data are available with regard to the effects of chemical composition, but only a few studies considered the intentional variation of any element. More often, the work employs several compositions simultaneously, like in the study by Maheswari et al. (2014) who used

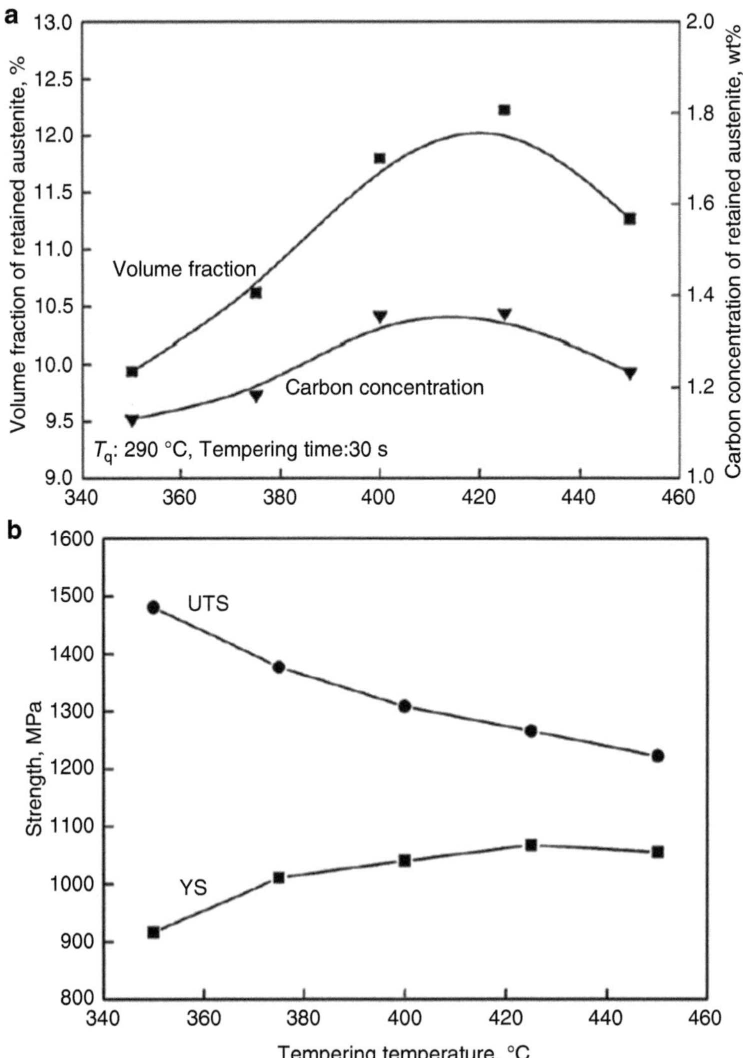

Fig. 10.17 Effect of partitioning temperature (partitioning time 30 s) on the volume of retained austenite and mechanical properties (Zhou et al. 2011)

three heats that simultaneously differed in concentrations of carbon, manganese, silicon, and aluminum.

Carbon The biggest effect on final microstructure and properties is observed with higher carbon content in steel, which increases the strength of martensite and austenite stability.

Carbon content in steel also affects the M_S temperature and the morphology of martensite, with the latter being mostly of the lath type at carbon content below

Table 10.2 Chemical compositions, mechanical properties, and characteristics of retained austenite (Seto and Matsuda 2013)

	Chemical composition, mass.%			M_S	TS	TE	UE	HE	V_γ	C_γ
	C	Si	Mn	(°C)	(MPa)	(%)	(%)	(%)	(%)	(%)
A1	0.3	2.5	2.0	305	1476	17	11	40	12	0.88
B1	0.4	2.0	2.0	277	1477	22	14	18	21	1.00
C1	0.6	2.0	1.5	212	1474	35	26	12	27	1.25

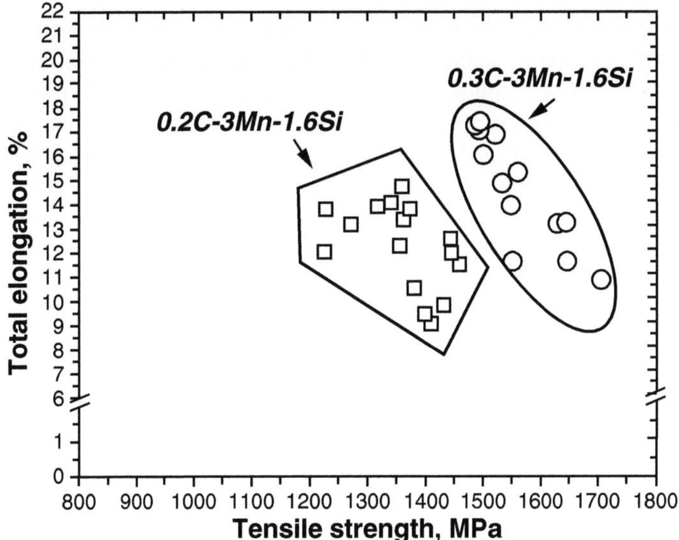

Fig. 10.18 Tensile properties obtained for the two grades after full austenitization (De Moor et al. 2012b)

0.3–0.5 % C and of the plate type at higher carbon level. In case of modification of QT and partitioning time aiming at the same level of tensile strength, steel with highest carbon content demonstrated the highest ductility not only due to increased volume fraction of retained austenite but rather due to higher carbon concentration in it, Table 10.2 (Seto and Matsuda 2013).

When studying 3Mn–1Si steels with 0.2 and 0.3 % C, De Moor et al. (2012b) showed that after full austenitization, quenching to QT = 250 °C, and partitioning at 450 °C for 10 s, the steel with 0.3 % C had tensile strength of 1500 MPa and total elongation of 17 %, while TS of only 1350 MPa and TE of 15 % were obtained for steel with 0.2 %. All combinations of properties at different QT and partitioning times at 400 and 450 °C depending on carbon content in steel are summarized in Fig. 10.18.

Manganese, Chromium, Molybdenum   Additions of Mn, Cr, or Mo are necessary to ensure necessary stability of the initial austenite so that to prevent its decomposition into ferrite, pearlite, or bainite during initial fast cooling/quenching. Therefore, the content of these elements is determined by the existing limitations in cooling capacity and by the target strength.

All these elements also retard bainitic reaction and can be intentionally added in higher amounts, sometimes together with ~1 % Ni, to consider partitioning of carbon to austenite from martensite avoiding the competing reactions. On the other hand, if carbide-free bainite is also considered to make efficient contribution to carbon enrichment of austenite, the alloying content should be reduced as much as possible (Santofimia et al. 2011a).

Mn additions strongly decrease the M_S temperature, but steels with high Mn are featured with banded structure and Mn segregations. De Moor et al. studied the effect of additions of 3 and 5 % Mn to 0.3C–1.6Si steel (De Moor et al. 2012b). Promising combinations of tensile strength and elongation were obtained for steels with 0.2 and 0.3 %C and 3 %Mn (see above). Steel with 5 % Mn quenched after full austenitization revealed untempered martensite after partitioning at 400 °C regardless of time, which consequently led to very low ductility with most samples failing at strains lower than 2 %. The intercritical annealing did not ensure full dissolution of cementite that resulted in rather low fraction of retained austenite.

In contrast, very promising combinations of strength and elongation were found in the study by Seo et al. who used 0.21C–4.0Mn–1.61Si–1.0Cr steel fully austenitized at 850 °C, quenched to 150–270 °C, and partitioned at 450 °C for 300 s. For example, YS ranged from 1050 to 1160 MPa with TE from 15 to 18 % and TS increased from 1400 to 1900 MPa with decrease in quenching temperature. Maximum amount of retained austenite was observed at QT = 210 °C and its highest stability was detected after QT = 150–190 °C (Seo et al. 2015).

Effect of Mo was investigated by De Moor et al. (2009) using 0.24C–1.6Mn steels with 1.41Al or 1.49Si. Significantly higher retained austenite fractions were observed in the Mo-added grade with maximum at 450 °C (close to 460 °C used in galvanizing cycle). Mo enhanced tensile strength, while its effect on ductility was less evident.

Silicon, Aluminum, Phosphorus   As was mentioned above, inhibition of carbide formation is critical for Q&P process to prevent consumption of carbon that is to diffuse into austenite. It is known that the formation of undesirable cementite can be inhibited by adding Si, Al, P, or their combination, which also retard tempering reactions.

Most of the Q&P studies employ Si-bearing grades with 1–2.5 % Si because of the above ability of Si to retard the precipitation of cementite during tempering or during bainitic transformation resulting in carbide-free bainite. At the same time, silicon appears to stabilize the transitional epsilon carbide, which consequently endures for longer time and to higher temperatures (Edmonds et al. 2006).

Seo et al. reported synergetic effect of Si and Cr additions in increasing retained austenite fraction during Q&P processing (Seo et al. 2015).

Table 10.3 Comparison of tensile properties and amount of retained austenite after Q&P processing of two steels with different Si content (Santofimia et al. 2010)

Steel	Partitioning time (s)	V_γ	YS (MPa)	TS (MPa)	UE (%)	TE (%)
HSi	10	0.08 ± 0.01	820	1630	10	15
	100	0.09 ± 0.01	520	1415	16	23
LSI	10	0.06 ± 0.01	720	1500	11	18
	100	0.12 ± 0.01	550	1320	12	20

Aluminum additions have also been explored because of the same ability and they have been proven to be efficient in enriching the retained austenite in TRIP steels (Santofimia et al. 2009). Utilization of Al in steels for Q&P processing is limited because Al has such a strong effect on the A_{c3} temperature that only the intercritical initial heating is possible. Al also does not lead to any strengthening, as opposed to Si.

As described in Table 10.3, which compares steels with different Si- and Al-alloyed 0.195C–3.5Mn–1.54Si–0.006Al (HSi) and 0.198C–3.5Mn–0.45Si–0.22Al (LSi) compositions, high Si shows significant advantages as it allows reaching combination of higher strength and ductility apparently due to bigger amount of retained austenite (Santofimia et al. 2010).

De Moor et al. carried out a systematic study of the Q&P process using dilatometry and differential scanning calorimetry (DSC). They verified that carbon diffuses from martensite laths into austenite without formation of transitional carbides. Comparing steels with Al and Si alloying they, in particular, concluded that Al may be less efficient in retarding precipitation of transitional carbide, which explains smaller volume fractions of retained austenite in CMnSiAl steel compared to CMnSi steel with the same carbon level (De Moor et al. 2009).

Niobium There is no known study where the effect of Nb on properties of Q&P steels was considered in comparison with steels without Nb. However, few papers can be found describing very high combinations of properties achieved in Nb-containing Q&P processed steels (Hausmann et al. 2013a; Fonstein et al. 2013).

In particular, Zhang et al. (2013) noted that the additions of Nb enhanced strength without deteriorating ductility.

10.5 On Stability of Retained Austenite in Q&P Process

Several studies reported the trend that the increase in RA is accompanied by increasing total elongation but not much so with UE, which reflects the contribution of TRIP effect. As shown above, the experiments with TRIP and CFB steels there often indicate no direct correlation between volume fraction of retained austenite and ductility of steel. This is even more evident for Q&P steels. The role of stability of the formed austenite looks more important than that of its volume fraction.

Among various factors affecting stability of retained austenite, the main three are carbon content, austenite morphology, and grain size.

As shown above, the carbon content in austenite in Q&P steels depends on partitioning parameters and was found to reach ~1 % and above. Apparently, the carbon content in austenite is not uniform throughout the bulk of steel or even across a single austenite grain so that the portions with lower carbon transform into "fresh" martensite during final cooling.

Stabilization of austenite with carbon during partitioning step comprises two important stages. The first is carbon diffusion from martensite and the second is homogenization of carbon concentration within austenite grains. Generally, carbon depletion of martensite is the fastest phenomenon. Li et al. calculated the time required for carbon to diffuse from martensite to austenite and the time for homogenization inside austenite grains for 0.41C–1.27Si–1.30Mn–1.01Ni–0.56Cr steel partitioned at 300 °C. The calculations showed that less than 1 s was needed for martensite to be depleted, whereas carbon homogenization inside austenite grains took 192 s. The smaller the austenite grain size, the faster the homogenization of carbon concentration (Li et al. 2010).

Besides, the stability of retained austenite depends on its morphology and therefore on the locations where it has been formed during microstructure evolution. In numerous studies the coexistence of two different morphologies of retained austenite, blocky and film-like, was observed with the film-like austenite being more stable. Sun et al. found three types of retained austenite in 0.2C–1.5Si–1.9Mn steel, including austenite films of about 100 nm in width between the martensite laths, blocky austenite embedded into ferritic matrix, and ultrafine austenite films (20–30 nm thick) between the plates of ferritic bainite, Fig. 10.19 (Sun and Yu 2013).

These different kinds of retained austenite have different stability and different carbon content, and transform into martensite at different strains during deformation. In particular, according to Xie et al. (2014), the blocky type RA starts to transform to martensite after 2 % tensile strain while the transformation of the film-like austenite begins only at strains over 12 %.

Xiong et al. (2013) studied 0.22C–1.4Si–1.8Mn steel after Q&P processing with QT = 280 °C and partitioning at 350 °C for 10 s using synchrotron X-ray diffraction and TEM. Interestingly enough, their observations revealed that the blocky austenite had higher carbon content (1.14 %) compared to that in film-like austenite (0.64 %) located between martensite laths. The M_S temperatures were estimated to be −8.4 °C and 203 °C for high- and low-carbon austenite, respectively. However, "high C" RA appeared to be less resistant to martensite transformation. The blocky austenite started to transform to twinned martensite at tensile strain of 2 % and at 12 % strain all blocky austenite vanished although numerous film-like austenite grains were still present indicating that carbon content is not always a dominating factor.

Several explanations of higher stability of film-like austenite can be recalled. In particular, since lath martensite surrounding film-like austenite has higher yield strength than ferrite around blocky austenite, the martensite transformation requiring volume expansion can be suppressed.

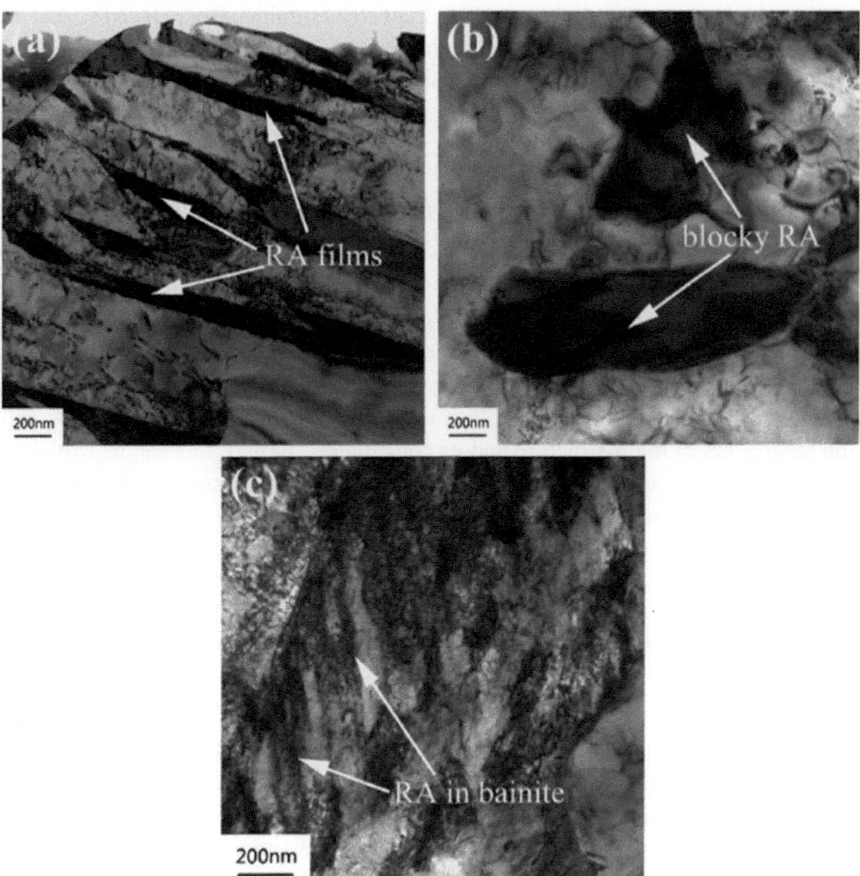

Fig. 10.19 TEM micrographs showing different morphology of RA: (**a**) film-like austenite in martensite laths; (**b**) blocky RA; (**c**) ultrafine austenite films trapped in carbide-free bainite (Sun and Yu 2013)

The other factor is the size effect. It is generally observed that coarse austenite grains are less stable, as discussed earlier in Chap. 5 (Olson and Cohen 1976). As was observed experimentally, the effect of austenite grain size manifests itself, in particular, through the M_S temperature decreasing with austenite grain refinement, as described by Lee and Lee (2005), Yang and Bhadeshia (2009).

Influence of grain size on austenite stability in Q&P processed steels was studied and optimal range of austenite size of 0.01 to 1 μm was established. RA grains coarser than 1 μm transform into martensite at very low strains, while very fine grains smaller than 0.01 μm are too stable and do not undergo strain-induced transformation (Samek et al. 2006).

De Knijf et al. pointed at the negative role of fresh martensite that indirectly affects the austenite stability. With fresh martensite, the amount of deformation that can be accommodated in the tempered martensite is significantly reduced so that the strain distribution in microstructure can change considerably. This constraining effect is decreasing austenite stability against transformation (De Knijf et al. 2014).

10.6   Relationship Between Microstructure and Properties of Q&P Steels

The majority of publications related to Q&P treatment are focusing on the partitioning mechanisms, on the effects of processing conditions, and on the evolution of retained austenite and its carbon content. There is lack of data regarding the correlation of possible combinations of mechanical properties with the parameters of microstructure and its constituents. In part, this can be explained by difficulties to even qualitatively distinguish between tempered martensite, bainite, and fresh martensite.

10.6.1   Combination of Strength and Ductility

The publications quoted above contain some data indicating combinations of very high tensile properties of Q&P processed steels: up to TS ~ 1500 MPa at TE ~20 %, TS ~ 1000 MPa at TE ~30 %, etc.

Jun and Fonstein (2008) compared the combination of properties of steel with fixed chemical composition after processing per CFB and Q&P thermal cycles. As was observed, although strength and elongation at the same tensile strength were approximately the same, the Q&P processing led to significantly higher YS (for example, from ~750 to 1100 MPa at TS ~ 1250 MPa). Consequently, higher YS/TS ratio was reached that is presently considered as a serious advantage. High YS/TS implies also relatively weak strain hardening favorable for higher hole expansion.

Decrease in tensile strength with partitioning time is attributed to softening of martensite. The yield strength clearly increases at the preliminary stage of partitioning due to tempering of the initial martensite. However, YS and YS/TS ratio are most sensitive to the volume fraction of fresh martensite formed during final cooling and the presence of fresh martensite can significantly reduce YS and YS/TS ratio to a level typical for DP steels.

Effect of quenching temperature was discussed above. The data in Fig. 10.20 quantitatively illustrate the effect of the volume fraction of untempered martensite.

Total (TE) and uniform elongation (UE) tend to increase with the amount of retained austenite but more important with increasing stability of retained austenite, i.e., due to TRIP effect.

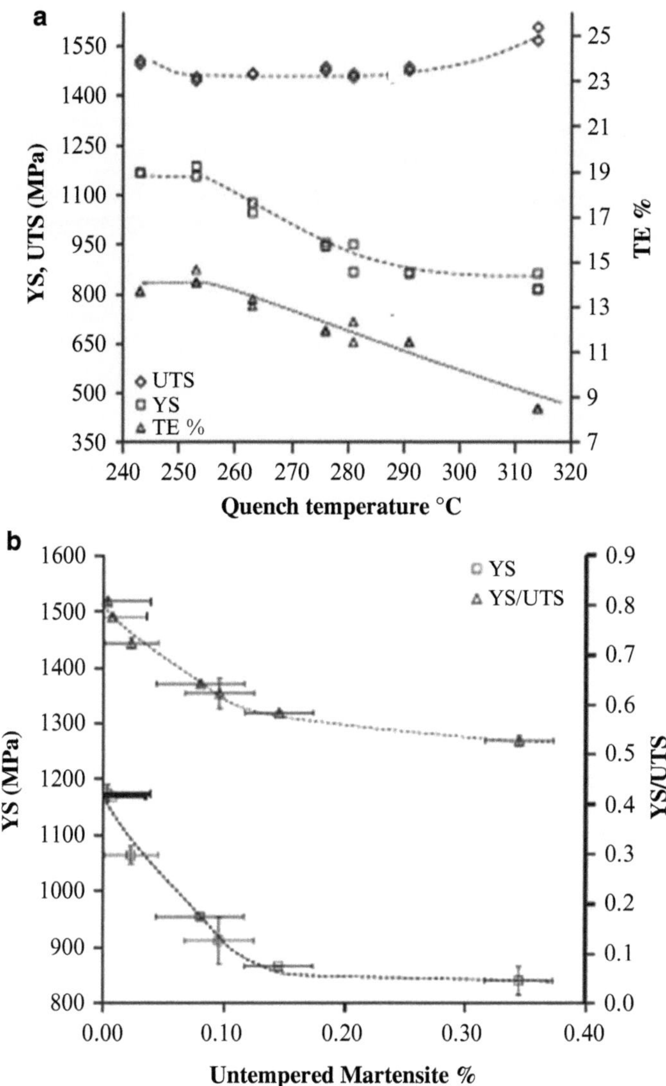

Fig. 10.20 Effect of quenching temperature (**a**) and volume fraction of untempered martensite (**b**) on tensile properties of Q&P steel (Paravicini Bagliaani et al. 2013)

10.6.2   Strain Hardening Rate

In general, Q&P steels with typical volume fraction of retained austenite of 0.05–0.10 have high YS/TS ratio due to significant amount of tempered martensite in microstructure. Therefore, strain hardening rate of these steels is relatively low, especially at small strains. When transformation of retained austenite

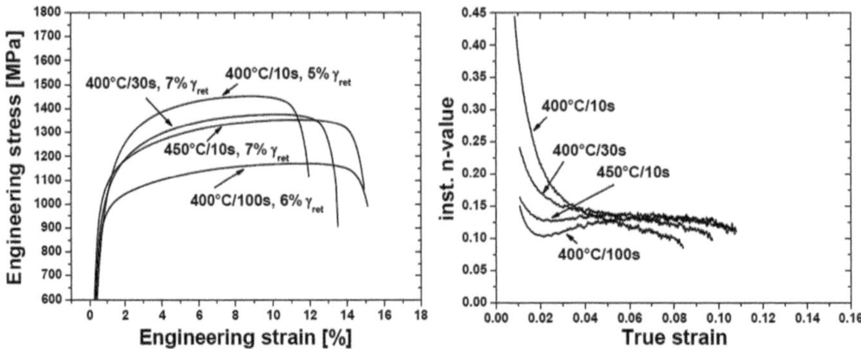

Fig. 10.21 Stress–strain curves (*left*) and instantaneous n_i values (*right*) for fully austenitized 0.2C–3Mn–1.6Si steel quenched at 250 °C (De Moor et al. 2011b)

starts (at larger strain), strain hardening increases due to the TRIP phenomenon. With longer partitioning time when retained austenite receives more higher carbon and becomes more stable, higher strain hardening can be observed (Fig. 10.21).

As shown by De Moor et al., the instantaneous strain hardening rate of Q&P steels strongly depends on the applied partitioning conditions. Lower partitioning temperature (PT) leads to instantaneous *n*-values continuously decreasing with strain, similarly to strain hardening behavior of dual-phase (DP) steels, whereas partitioning at higher PT for the given time raises the strain hardening rate significantly. After initial increase, *n*-values remain high up to considerable strains resulting in strain hardening behavior similar to that observed for austempered TRIP grades. Significant contribution of the TRIP effect to the increase of strain hardening rate with strain was observed at sufficient mechanical stability of the retained austenite as assessed by tensile testing (De Moor et al. 2012b).

DP-type strain hardening behavior can be observed for Q&P steels of different compositions when fairly large amounts of "fresh" martensite are formed during final cooling. Zhang et al. pointed at the correlation between the $n_i(\varepsilon)$ dependence and the final cooling rate: with the same parameters of partitioning the final air cooling led to extended strain range where high strain hardening rate was still detected (Zhang et al. 2013).

10.6.3   Hole Expansion

The microstructure of Q&P processed steels containing tempered martensite, low bainite, retained austenite and fresh martensite in various proportions possesses very serious advantage as the difference in hardness between microstructure constituents is not very significant so that the stress distribution upon loading is relatively uniform.

The improved behavior during edge cutting, flangeability, and other localized type deformation is expected due to higher YS/TS ratio of Q&P steels, which has been confirmed experimentally in some publications.

Detailed study by De Moor et al. (2012a) compared a few groups of two steels of typical compositions with different types of microstructure—dual-phase, quenched and tempered (Q&T), and Q&P after quenching from full and partial austenitization. As shown in Fig. 10.22, at the same or higher strength, Q&P steels demonstrate significantly higher hole expansion than DP or Q&T steels.

In the study by De Moor et al., the drilling/reaming of holes was used that resulted in elevated HE values, but it should not have changed the general trend. The combinations of target properties, TS > 1180 MPa at TE > 14 % and HE > 50 % (measured on punched hole samples), can be achieved for C–Mn–Si–Nb–Mo steel after supercritical annealing, 2 s pre-quenching at 300–350 °C, and isothermal holding at 460 °C within short partitioning times (~60 s) pertinent to galvanizing lines equipped with equalizing zones (Fonstein et al. 2013).

The negative correlation of HER with n-values and the effects of strain hardening on edge damage are shown in numerous publications, e.g., in Karelova et al. (2009). It can be suggested that the difference in behavior between, for example, TRIP and Q&P steels, both containing retained austenite can be in part explained by the difference in the morphology of retained austenite in TRIP and in Q&P steels. The key role can be attributed to different stability of retained austenite under deformation when the TRIP effect and potential damage of edge by localized deformation during punching is delayed in Q&P steels whose strain hardening at small strains is low.

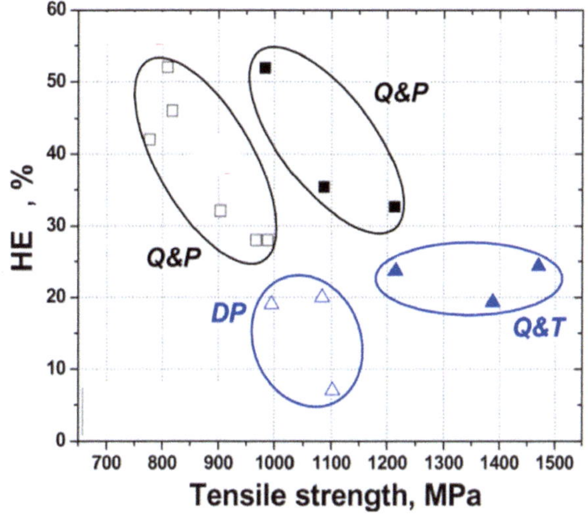

Fig. 10.22 Measured hole expansion ratio (HER) as a function of tensile strength for intercritically annealed C–Mn–Al–Si–P steel (*open symbols*) and fully austenitized C–Mn–Si steel (*solid symbols*). Holes were prepared by drilling and reaming in all cases (De Moor et al. 2012a)

10.7   Modifications of Q&P Thermal Cycles

Implementation of Q&P treatment implies upgrade of the existing annealing equipment that requires not only substantial capital investment but also the modifications to the layouts of annealing lines to incorporate some kind of induction reheating and furnace for isothermal partitioning in accordance with the cycle depicted in Fig. 10.1.

10.7.1   Alternative Designs of Q&P Process

To simplify the layout of various sections for Q&P processing, Sun and Yu (2013) suggested alternative designs that incorporate GI step before or after quenching and partitioning steps. In particular, quenching to QT temperature is carried after galvanizing at 460 °C with holding in Zn pot for 3 s followed by heating to partitioning temperature, as shown in Fig. 10.23a. In the alternative design, the strip after QT is heated to galvanizing temperature, which is followed by lowering the temperature to the partitioning stage, as depicted in Fig. 10.23b. In both cases, only one reheating is required instead of two in regular cycle.

For a valid confirmation of viability of proposed cycles, the authors used QT = 160 °C with holding for 5 s and partitioning at 350 °C for 100 s. Using 0.2C–1.5Si–1.9Mn steel, they achieved TS of 1018 and 1014 MPa at TE = 24 and 26 % for routes (a) and (b), with the volume fraction of retained austenite of 0.099 and 0.097, respectively.

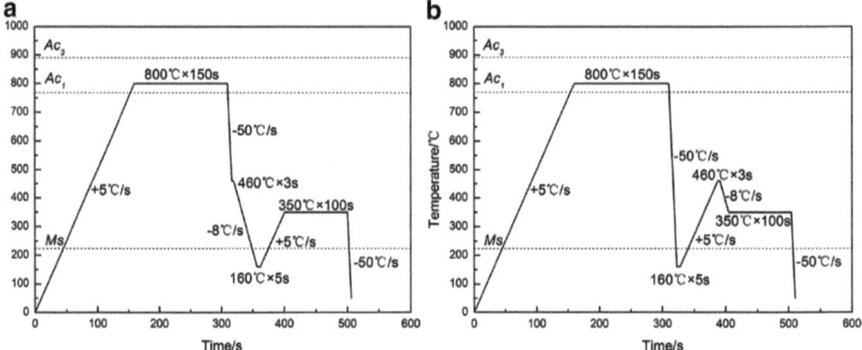

Fig. 10.23 Schematic illustration of heat treatment cycle applied to (**a**) galvanizing before Q&P process; (**b**) galvanizing and partitioning after quenching (Sun and Yu 2013)

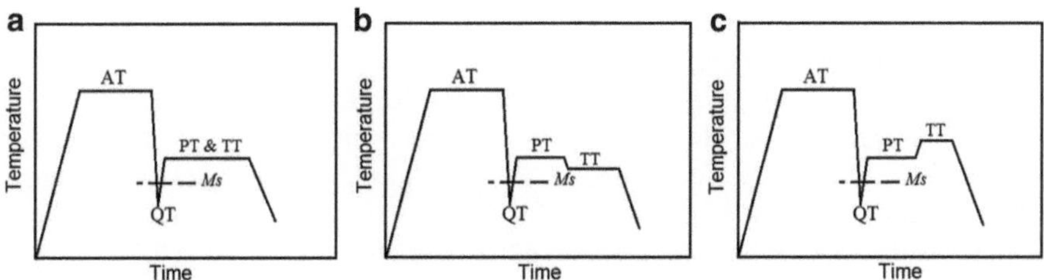

Fig. 10.24 Diagram of proposed Q–P–T process (Hsu 2010; Hsu and Jin 2011)

10.7.2   Post-tempering of Q&P Processed Steels

As mentioned above, the potential combination of high YS, TS, and ductility steels after Q&P processing can be destroyed by high-carbon "fresh" martensite formed during final cooling when the M_S temperature of a part of remaining austenite is above the room temperature. In view of that, several studies considered the so-called Q–P–T processing, when Quenching and Partitioning steps are followed by low-temperature tempering (Tan et al. 2014a, b), e.g., at 280 °C for 2 h, as shown in Fig. 10.24.

The term Q–P–T was proposed by Hsu for medium-alloy high-carbon steels, like 0.40C–2Mn–2Si–Cr, 0.5C–1.5Mn–1.5Si, or 0.488C–1.2Mn–1.2Si–0.9Ni–0.02Nb, targeting at very high strength of 1500–2000 MPa. Later Hsu et al. applied this concept to lower carbon of 0.15–0.2C–1.5Mn–1.5Si steels with small additions of Mo, Nb, or V. In case of Nb addition, the formation of complex Nb-containing carbides is assumed that provides additional strengthening of steel by very fine precipitates. Tensile properties, TS and TE, as well as the RA content were not very high, but the authors emphasized the increase in toughness (Hsu et al. 2013).

At the same time, additional low-temperature tempering, which would not destroy the retained austenite, could be beneficial for Q&P steels as it helps softening of "fresh" high-carbon brittle martensite.

10.7.3   Application of Q&P Approach for Hot Stamping

Quite novel modification of conventional Q&P processing is the application of this concept to hot stamping as was presented in Sect. 7.2.3. Currently hot stamping of ultrahigh-strength safety part is aiming at final martensite structure. The attempt of application of Q&P process to hot stamping is focusing on higher combination of strength and elongation obtained. Quenching to a temperature below M_S followed by isothermal holding at the same temperature (one-step Q&P) was performed after full austenitization and hot deformation that simulated hot stamping. The resulted microstructure was refined and contained significant amounts of retained austenite leading to substantial increase in elongation.

For example, Liu et al. proposed special 0.22C–1.58Mn–0.81Si–0.022Ti–0.0024B steel for such processing. After full austenitization for 5 min and hot deformation, the steel was quenched to 280, 300, and 320 °C in the (M_S–M_F) range. The best combinations of properties achieved after partitioning at 320 and 280 °C for 30 s were TS = 1510 MPa at TE = 14.8 % and TS = 1601 MPa at TE = 10.3 %, respectively (Liu et al. 2011).

Significantly more alloyed 0.19C–1.55Si–1.53Mn–0.95Ni–1.01Cr–0.45Mo–1.01Cu–0.033Ti–0.0027B steel was suggested for "dynamic" one-step Q&P process where initial fast cooling (20 °C/s) from full austenitization (and assumed hot stamping) followed by relatively slow oil or air cooling from M_S temperature allowed for TS = 1520 MPa at TE = 11 % (Zhang et al. 2014).

10.7.4   Q&P During Hot Rolling

The proposed Q&P processing in a hot rolling mill (Fig. 10.25) can be considered as one-step Q&P treatment. Thomas et al. (2008) studied the possibility to achieve the advantages of Q&P process with 0.19C–1.6Mn–1.6Si steel by using rapid cooling of hot-rolled strip and its pseudo-isothermal holding (really, slow cooling) during coiling below the M_S temperature (230–260 °C). Initial simulations revealed significant volume fractions of retained austenite (0.04–0.05) and high strength YS = 720–760 MPa, TS = 1105–1110 MPa at UE = 8–9 %.

Using 0.21C–1.67Si–1.65Si–0.2V steel, Tan et al. compared several Q&P cycles after hot rolling and ultrafast cooling at 200 °C/s to the M_S–M_F range (270–280 °C)

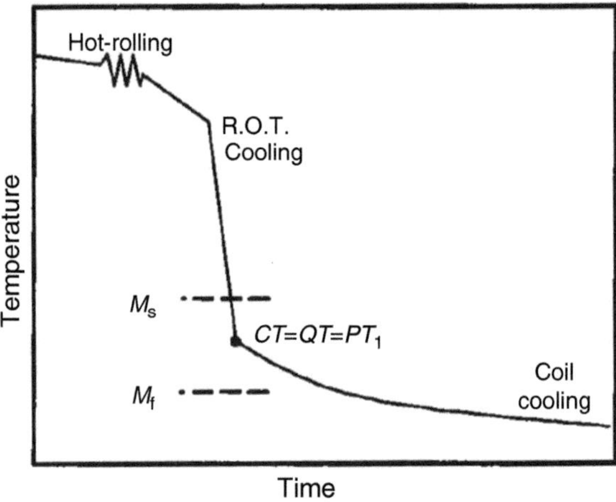

Fig. 10.25 Potential combination of hot strip rolling and Q&P (Thomas et al. 2008)

followed by air cooling. They showed that air cooling after pre-quenching allowed for realization of one-step "dynamic" Q&P process with obtaining RA volume fractions above 0.10 and the combination of TS = 1600 MPa at TE = 17 % (ASTM-T) (Tan et al. 2014a, b; Xu et al. 2014).

10.7.5 Q&P Process with Quenching at Room Temperature

Control of quenching temperature during Q&P process requires very sophisticated cooling equipment. In this context, very interesting studies were performed by Yi et al. The authors used high Al 0.4C–0.5Mn–0.2Si–0.5Cr–3.5Al steel with M_F temperature below the ambient temperature (based on dilatometry data, the M_S, A_{c1}, and A_{c3} temperatures of this alloy were 110 °C, 740 °C, and 1073 °C, respectively).

The Q&P treatment included intercritical annealing at 800 °C for 10 min, followed by water quenching to the room temperature and partitioning at 300 or 350 °C for 2, 10, and 30 min. The resultant microstructures consisted of retained austenite up to 10–12 % and martensite. Achieved combination of properties for the given chemistry was not very high—TS of 1200 MPa at TE = 10–12 % that should be improved by correction of chemical composition. The main conclusion was that conventional water quenching to room temperature followed by partitioning in overaging section can be implemented using traditional CAL, as described in Fig. 10.26.

Fig. 10.26 Schematic of Q&P treatment proposed for quenching at room temperature (Yi et al. 2013)

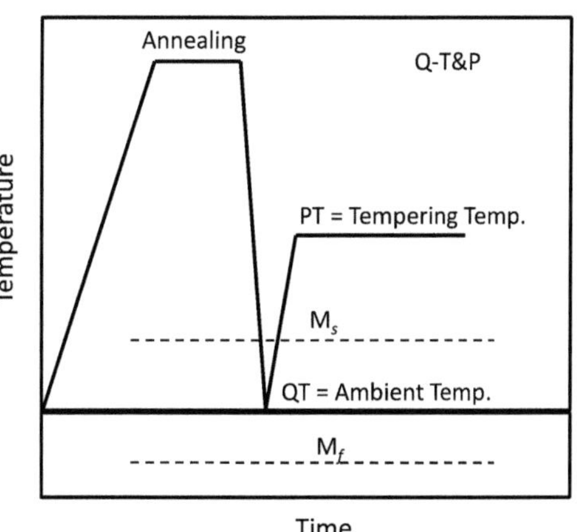

10.8 Summary

Quenching and partitioning is a complex process, which involves several parameters that should be optimized in order to obtain the desired microstructure and mechanical properties. There is no universal combination of these parameters to achieve the best properties of steel, but some general trends should be taken into account when considering the adjustments of the Q&P cycle for specific chemical composition.

The chemical composition of steel for Q&P processing should contain alloying elements that inhibit carbides formation (Si, Al) and stabilize austenite (Mn, Ni, Mo) to reduce bainite formation window and prolong bainite incubation time.

Microstructure refinement is very beneficial for Q&P processing. It was shown that microalloying with Nb can help refining the microstructure and allow for higher strength without loss of ductility. Similar effect could be obtained by pre-quenching the steel prior to Q&P treatment.

In 2009, Baosteel commissioned the Q&P processing line that currently produces the 1000 MPa steel grade (Wang and Feng 2010). It is also known that a number of successful industrial trials of Q&P steels with TS of 980, 1180, and 1470 MPa have been performed elsewhere to expect their commercialization in short future, and even limited quantities of these grades have been produced commercially.

The most important conclusion is that Q&P processed steels demonstrate some combinations of properties that match expectations from third generation of steel.

References

Andrews, K.W. 1965. "Empirical Formulae for Calculation of Some Transformation Temperatures." *Iron and Steel Institute Journal* 203 (Part 7): 721–27.

Arlazarov, A., O. Bouaziz, J.P. Masse, and F. Kegel. 2015. "Characterization and Modeling of Mechanical Behavior of Quenching and Partitioning Steels." *Material Science and Engineering A* 620: 293–300.

Clarke, A.J., J.G. Speer, D.K. Matlock, F.C. Rizzo, D.V. Edmonds, and M.J. Santofimia. 2009. "Influence of Carbon Partitioning Kinetics on Final Austenite Fraction during Quenching and Partitioning." *Scripta Materialia* 61: 149–52.

Clarke, A.J., J.G. Speer, M.K. Miller, R.E. Hackenberg, and D.V. Edmonds. 2008. "Carbon Partitioning to Austenite from Martensite or Bainite during the Quench and Partition (Q&P) Process. A Critical Assessment." *Acta Materialia* 56 (1): 16–22.

De Knijf, D., R. Petrov, C. Fojer, and L.A.I. Kestens. 2014. "Effect of Fresh Martensite on the Stability of Retained Austenite in Quenching and Partitioning Steel." *Material Science and Engineering A* 615: 107–15.

De Moor, E., D.K. Matlock, J. Speer, and et al. 2012a. *Comparison of Hole Expansion Properties of Quench & Partitioned, Quenched & Tempered and Austempered Steels*. SAE International 2012-01-0530.

De Moor, E., J.G. Speer, D.K. Matlock, J.-H. Kwak, and S.-B. Lee. 2011a. "Effect of Carbon and Manganese on the Quenching and Partitioning Response of CMnSi Steels." *ISIJ International* 51 (1): 137–44.

————. 2012b. "Quenching and Partitioning of CMnSi Steels Containing Elevated Manganese Levels." *Steel Research International* 83 (4): 322–27. doi:10.1002/srin.201100318.

De Moor, E., J.G. Speer, D.K. Matlock, J. Penning, and C. Fojer. 2009. "Effect of Si, Al and Mo Alloying on Tensile Properties Obtained by Quenching and Partitioning." In *Material Science and Technology*, 1554–63. Pittsburgh.

De Moor, E., J. Speer, J.-H. Kwak, S.-B. Lee, and D.K. Matlock. 2011b. "Quenching and Partitioning of CMnSi Steels Containing Elevated Manganese Levels." In *HMnS – 2011*, B – 34, 1–9. Seoul, South Korea.

Edmonds, D., D. Matlock, and J. Speer. 2010. "Development in High Strength Steels with Duplex Microstructures of Bainite or Martensite with Retained Austenite: Progress with Quenching and Partitioning (Q&P) Heat Treatment." In 229–41. Beijing.

Edmonds, D.V., K. He, F.C. Rizzo, B.C. De Cooman, D.K. Matlock, and J.G. Speer. 2006. "Quenching and Partitioning Martensite – A Novel Steel Heat Treatment." *Material Science and Engineering A* 438–440: 25–34.

Fonstein, N., H.J. Jun, O. Yakubovsky, R. Song, and N. Pottore. 2013. "Evolution of Advanced High Strength Steels (AHSS) to Meet Automotive Challenges." In Vail, CO, USA.

Hausmann, K., D. Krizan, A. Pichler, and E. Werner. 2013a. "TRIP-Aided Bainitic-Ferritic Sheet Steel: A Critical Assessment of Alloy Design and Heat Treatment." In *MS&T'13*, 209–18. Montreal, Canada.

Hausmann, K., D. Krizan, K. Spiradek-Hahn, A. Pichler, and E. Werner. 2013b. "The Influence of Nb on Transformation Behavior and Mechanical Properties of TRIP-Assisted Bainitic-Ferritic Sheet Steels." *Material Science and Engineering A* 588: 142–50.

Hsu, T.Y. 2010. "Ultra-High Strength Steel Treated by Using Quenching-Partitioning-Tempering Process." In 64–70. Beijing.

Hsu, T.Y., and X. Jin. 2011. "Ultra-High Strength Treated by Using Quenching-Partitioning-Tempering Process." In *Advanced Steels: The Recent Scenario in Steelscience and Technology*, 67–73. Springer.

Hsu, T.Y., X.J. Jin, and Y.H. Rong. 2013. "Strengthening and Toughening Mechanisms of Quenching-Partitioning-Tempering (Q-P-T) Steels." *Journal of Alloys and Compounds* 575: 5568–71.

Imai, N., N. Komatsubara, and K. Kunishige. 1995. "Effect of Alloying Element and Microstructure on Mechanical Properties of Low-Alloy TRIP-Steels." *CAMP-ISIJ* 8: 572–75.

Jun, H.J., and N. Fonstein. 2008. "Microstructure and Tensile Properties of TRIP-Aided CR Sheet Steels: TRIP-Dual and Q&P." In 155–61. Orlando, Florida.

Karelova, A., E. Werner, and T. Hebesberger. 2009. "Hole Expansion of Dual-Phase and Complex-Phase AHS Steels – Effect of Edge Condition." *Steel Research International* 80: 71–77.

Kawata, H., K. Hayashi, N. Sugiura, N. Yoshinaga, and M. Takahashi. 2010. "Effect of Martensite in Initial Structure on Bainite Transformation." *Materials Science Forum* 638–642 (Pt. 4, THERMEC 2009): 3307–12. doi:10.4028/www.scientific.net/MSF.

Koistinen, D.P., and P.E. Marburger. 1959. "A General Equation Prescribing the Extent of the Austenite-Martensite Transformation in Pure Iron-Carbon Alloys and Plain Carbon Steels." *Acta Metall* 7: 59.

Krauss, George'. 2005. *Steels: Processing, Structure and Performance*. TMS.

Lee, S.-J., and Y.-K. Lee. 2005. "Effect of Austenite Grain Size in Martensitic Transformation of a Low Alloy Steel." *Material Science Forum* 475–479: 3169–72.

Li, H.Y., X.W. Lu, W.J. Li, and X.J. Jin. 2010. "Microstructure and Mechanical Properties of an Ultrahigh-Strength 40SiMnNiCr Steel during the One-Step Quenching and Partitioning Process." *Metallurgical and Materials Transactions* 41 (5): 1284–1300.

Liu, H., X. Lu, X. Jin, H. Dong, and J. Shi. 2011. "Enhanced Mechanical Properties of a Hot Stamped Advanced High-Strength Steel Treated by Quenching and Partitioning Process." *Scripta Materialia* 64: 749–52.

Maheswari, N., S.G. Chowdhury, K.C.H. Kumar, and S. Sankaran. 2014. "Influence of Alloying Elements on the Microstructure Evolution and Mechanical properties in Quenched and Partitioned Steels." *Material Science and Engineering A* 600: 12–20.

Matlock, D.K., V.E. Brautigam, and J.G. Speer. 2003. "Application of the Quenching and Partitioning (Q&P) Process to a Medium Carbon High Si Microalloyed Bar Steel." In *Materials Science Forum*, 1089–94.

Olson, G.B., and M. Cohen. 1976. "A General Mechanism of Martensite Nucleation." *Metallurgical Transactions A* 7A: 1897–1904.

Paravicini Bagliaani, E., M.J. Santofimia, L. Zhao, J. Sietsma, and E. Anelli. 2013. "Microstructure, Tensile and Toughness Properties after Quenching and Partitioning Treatment of a Medium-Carbon Steel." *Material Science and Engineering A* 559: 486–95.

Samanta, S., S. Das, D. Chakrabarti, I. Samajdar, S. Singh, and A. Haldar. 2013. "Development of Multiphase Microstructure with Bainite, Martensite and Retained Austenite Un a Co-Containing Steel through Quenching and Partitioning (Q&P) Treatment." *Metallurgical and Material Transactions A* 44 (13): 5653–64.

Samek, L., E. De Moor, J. Penning, and B.C. De Cooman. 2006. "Influence of Alloying Elements on the Microstructure Evolution and Mechanical Properties in Quenched and Partitioned Steels." *Metallurgical and Materials Transactions* 37 (1): 109–24.

Santofimia, M.J., T. Nguyen-Minh, and et al. 2010. "New Low Carbon Q&P Steels Containing Film-like Intercritical Ferrite." *Material Science and Engineering A* 527: 6429–39.

Santofimia, M.J., I. Zhao, and J. Sietsma. 2011a. "Overview of Mechanisms Involved during the Quenching and Partitioning Process in Steels." *Metallurgical and Materials Transactions A* 42A (12): 3620–26.

Santofimia, M.J., L. Zhao, R. Petrov, C. Kwakernaak, W.G. Sloof, and J. Sietsma. 2011b. "Microstructure Development during the Quenching and Partitioning Process in a Newly Designed Low-Carbon Steel." *Acta Materialia* 59: 6059–68.

Santofimia, M.J., L. Zhao, and J. Sietsma. 2008. "Model for Interaction between Interface Migration and Carbon Diffusion during Annealing of Martensite-Austenite Microstructure in Steels." *Scripta Materialia* 59 (2): 159–62.

Santofimia, M.J., L. Zhao, and J. Sietsma. 2009. "Microstructural Evolution of a Low-Carbon Steel during Application of Quenching and Partitioning Heat Treatments after Partial Austenitization." *Metallurgical and Materials Transactions A* 40 A (1): 46–56.

Seo, E.J., L. Cho, and B.C. De Cooman. 2015. "Application of Quenching and Partitioning Processing to Medium Mn Steels." *Metallurgical and Materials Transactions* 46 (1): 27–31.

Seto, K., and H. Matsuda. 2013. "Application of Nanoengineering to Research and Development and Production of High Strength Sheets." *Material Science and Technology* 29 (10): 1158–65.

Somani, M.C., D.A. Porter, L.P. Karialainen, and R.D.K. Misra. 2014. "On Various Aspects of Decomposition of Austenite in a High Silicon Steel During Quenching and Partitioning." *Metallurgical and Materials Transactions A* 45A (3): 1247–57.

Speer, J.G., E. De Moor, K.O. Findley, D.K. Matlock, B.C. De Cooman, and D.V. Edmonds. 2011. "Analysis of Microstructure Evolution in Quenching and Partitioning Automotive Sheet Steels." *Metallurgical and Materials Transactions A* 42 A (12): 3591–3601.

Speer, J., D.K. Matlock, Cooman De, and J.G. Schroth. 2003. "Carbon Partitioning into Austenite after Martensite Transformation." *Acta Materialia* 51 (9): 2611–22.

Steven, W., and A.G. Haynes. 1956. "The Temperature of Formation of Martensite and Bainite in Low-Alloy Steels." *Journal of the Iron and Steel Institute* 183 (8): 349–59.

Streicher-Clarke, A., J. Speer, D. Matlock, and B.C. De Cooman. 2004. "Quenching and Partitioning Response of a Si-Aided TRIP Sheet Steels." In 51–62. Winter Park, CO.

Sun, J., and H. Yu. 2013. "Microstructure Development and Mechanical Properties of Quenching and Partitioning (Q&P) Steels and an Incorporation of Hot-Dipping Galvanization During Q&P Process." *Material Science and Engineering, A* 586 (1): 100–107.

Sun, J., H. Yu, S. Wang, and Y. Fan. 2013. "Study of Microstructure Evolution, Microstructure-Mechanical Properties Correlation and Collaborative Deformation-Transformation Behavior of Quenching and Partitioning (Q&P) Steel." *Material Science and Engineering A* 585: 132–38.

Tan, X., Y. Xu, X. Yang, and Z. Liu. 2014. "Effect of Partitioning Procedure on Microstructure and Mechanical Properties of a Hot-Rolled Directly Quenched and Partitioned Steel." *Material Science and Engineering A* 594: 149–60.

Tan, Z.-l., K.-k. Wang, G.-h. Gao, X.-l. Gui, and B.-z. Bai. 2014. "Mechanical Properties of Steels Treated by Q-P-T Process Incorporating Carbide-Free Bainite/martensite Multiphase Microstructure." *Journal of Iron and Steel Research, International* 21 (2): 191–96.

Tariq, F., and R.A. Baloch. 2014. "One-Step Quenching and Partitioning Heat Treatment of Medium Carbon Low-Ally Steel." *JMEPEG* 23: 1726–39.

Thomas, G.A., J.G. Speer, and D.K. Matlock. 2008. "Considerations in the Application of the 'Quenching and Partitioning' Concept in Hot Rolled AHSS Production." In 227–39. Orlando, Florida.

Toji, Y., H. Matsuda, M. Herbig, P. Choi, and D. Raabe. 2014. "Atom-Scale Analysis of Carbon Partitioning between Martensite and Austenite by Atom Probe Tomography and Correlative Transmission Electron Microscopy." *Acta Materiala* 65: 215–28.

Wang, C.Y., J. Shi, W.Q. Cao, and H. Dong. 2010. "Characterization of Microstructure Obtained by Quenching and Partitioning Process in Low Alloy Martensitic Steel." *Materials Science and Engineering: A* 527 (15): 3442–49.

Wang, L., and W. Feng. 2010. "Development and Application of Q&P Sheet Steels." In 242–45. Beijing.

Xie, Z.J., Y.Q. Ren, W.H. Zhou, J.R. Yang, C.J. Shang, and R.D.K. Misra. 2014. "Stability of Retained Austenite in Multi-Phase Microstructure during Austempering and Its Effect on the Ductility of a Low Carbon Steel." *Material Science and Engineering A* 603: 69–75.

Xiong, X.C., B. Chen, M.X. Huang, J.F. Wang, and L. Wang. 2013. "The Effect of Morphology on the Stability of Retained Austenite in a Quenched and Partitioned Steel." *Scripta Materialia* 68: 321–24.

Xu, Y., X. Tan, X. Yang, and Z. Hu. 2014. "Microstructure Evolution and Mechanical Properties of a Hot-Rolled Directly Quenched and Partitioned Steel Containing Proeutectoid Ferrite." *Material Science and Engineering A* 607: 460–75.

Yang, H.-S., and H.K D.H. Bhadeshia. 2009. "Austenite Grain Size and the Martensite-Start Temperature." *Scripta Materialia* 60: 493–95.

Yi, H.L., P. Chen, Z.Y. Hou, and N. Hong. 2013. "A Novel Design: Partitioning Achieved by Quenching and Tempering (Q-T&P) in an Aluminium Added Low-Density Steel." *Scripta Materialia* 68 (6): 370–74.

Zhang, F., H. Song, M. Cheng, and X. Li. 2014. "Microstructure Development and Mechanical Properties of a Hot Stamped Low-Carbon Advanced High Strength Steel Treated by a Novel Dynamic Carbon Partitioning Process." In *ICHSU 2014*. Chongqin, China.

Zhang, J., H. Ding, R.D.K. Misra, and C. Wang. 2011. "Enhanced Stability of Retained Austenite and Consequent Work Hardening Rate Through Pre-Quenching prior to Quenching and Partitioning in a Q-P Microalloyed Steel." *Material Science and Engineering, A* 611: 252–56.

Zhang, J., H. Ding, C. Wang, J. Zhao, and T. Ding. 2013. "Work Hardening Behavior of a Low Carbon Nb-Microalloyed Si-Mn Quenching and Partitioning Steel with Different Cooling Styles after Partitioning." *Material Science and Engineering A* 585: 132–38.

Zhao, C., D. Tang, H.-T. Jiang, and S.-S. Jhao. 2008. "Process Simulation and Microstructure Analysis of Low Carbon Si-Mn Quenched and Partitioned Steel." *Journal of Iron and Steel Research, International* 15 (4): 82–85.

Zhou, S., K. Zhang, Y. Wang, J.F. Gu, and Y.H. Rong. 2011. "High Strength-Elongation Product of Nb-Microalloyed Low-Carbon Steel by a Novel Quenching-Partitioning-Tempering Process." *Material Science and Engineering A* 528 (27): 8006–12.

Chapter 11
Austenitic Steels with TWIP Effect

Contents

11.1   Introduction

The twinning-induced plasticity (TWIP)-aided steels as prospective auto-body materials became known over a decade ago when the POSCO patent of 1991 and the work by Frommeyer and Grassel et al. (Grassel et al. 2000; Frommeyer et al. 2003; Bouaziz and Guelton 2001) pointed at their potential unique combination of strength and ductility, Fig. 11.1. The term "TWIP steels" was first proposed by Frommeyer et al.

Initial studies of TWIP steels including those by European, Korean, and Japanese researchers led to attempts to industrially produce steels with ~22–28 % Mn; however, these attempts faced some difficulties of manufacturing. Nevertheless, the demonstrated combinations of tensile characteristics and ductility (~900–1000 MPa at ~50–60 %) were adopted as benchmarks for the so-called "Second generation" of AHSS quoted in all "banana diagrams." Mn with its specific weight being less than that of iron and additions of Al make these materials lighter than conventional steels. A subgroup of carbon-free TWIP steels with slightly lower strength was

© Springer International Publishing Switzerland 2015   369
N. Fonstein, *Advanced High Strength Sheet Steels*,
DOI 10.1007/978-3-319-19165-2_11

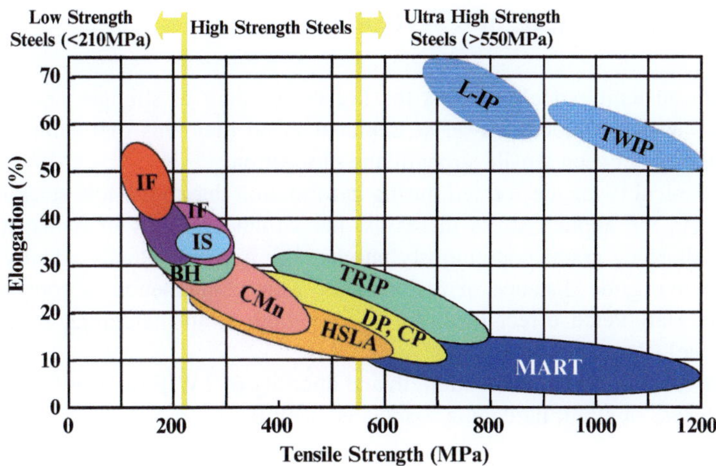

Fig. 11.1 Relationship of tensile strength and elongation of various automotive steels

termed L-IP (Light-Induced Plasticity) steels, Light steels with Induced Plasticity. The L-IP FeMnAlSi concept was studied and developed mostly by Posco, MPIE, and Saltzgitter. Carbon-based concepts of X-IP (Extra-Induced Plasticity) FeMnC TWIP steels were studied mainly by ArcelorMittal.

The implementation of both groups of steels, besides manufacturing challenges, has faced some problems with applications that include low yield strength of these steels, lower Young modulus, propensity to delayed fracture/hydrogen embrittlement, and high cost.

As a result of dedicated studies and intensive research, the problem of sensitivity to hydrogen embrittlement seems to have been solved, and the ways to reduce Mn content in these steels down to ~14–18 %, as well as to significantly lower their price, were found (Kwon et al. 2010; Kwon 2011). Modification of steel chemistry resulted in some changes in their properties, which, nevertheless, remain the highest with regard to ductility range, ever achieved so far, motivating carmakers to look for possibilities to use TWIP steel in vehicles. First coils of TWIP-980 steel produced at POSCO commercial line moved TWIP steels into the stage of industrial testing and commercialization. TWIP steel with TS>1180 is marked as being under development. Commercialization of TWIP-980 steel and development of 1180 grade were also announced by Ansteel (Jinyu et al. 2013); however all of this progress is currently occurring mostly in Asia.

The austenitic high-Mn steels can be considered as prospective candidates for AHSS. However, although these steels have been developed earlier, their readiness for commercial production is far behind that of other AHSS. Therefore, it is reasonable to discuss these steels together with the group of candidates for the "3rd generation" AHSS.

This chapter is based on numerous publications and very comprehensive reviews on the matter (Bouaziz et al. 2011b; Chen et al. 2013; De Cooman et al. 2011, 2012; Neu 2013).

11.2   Fundamentals of TWIP steels

High-Mn austenitic steels possess the highest product of strength and ductility, normally above 50,000 MPa·%, as depicted in all diagrams that summarize the relationships between tensile strength and elongation.

Mechanical twins are formed during deformation due to the low stacking-fault energy. As the applied stress increases, the volume fraction of twins increases steadily dividing grains continuously into smaller fragments that gradually reduce the effective glide distance of dislocations. This phenomenon is considered as dynamic Hall–Petch effect resulting in very high strain hardening observed in TWIP steel (Remy 1978).

Extra-high combination of strength and ductility of TWIP steels is explained by the high rate of strain hardening associated with the phenomenon of deformation twinning.

11.2.1   The Role of SFE

Stacking-Fault Energy (SFE) is the key factor that controls the mechanical properties of the high-Mn alloys and plays essential role in the occurrence of twinning phenomena and the TWIP effect. Some authors indicated that the SFE required for the TWIP effect should be within the range 20–50 mJ/m^2, but it is still unclear why this minimum value is essential for occurrence of the strain-induced twinning. The minimum SFE value could be related to suppression of athermal $\gamma \rightarrow \varepsilon$ martensitic transformation, as described in the schematic diagram by Remy (Fig. 11.2).

In some studies, stable and fully austenitic microstructures featured by TWIP properties were reported to have the SFE in the range of 20–30 mJ/m^2 or higher.

Fig. 11.2 Effect of SFE on deformation mode (Remy 1978)

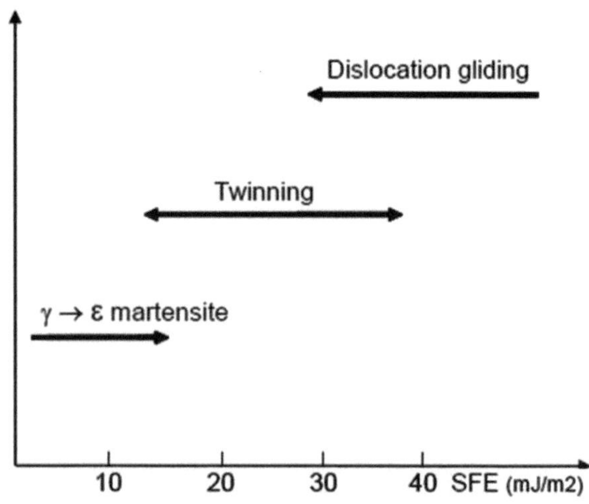

Carbon additions are desirable to lower SFE, but the carbon content is limited by formation of M_3C carbides. According to Yakubtsov et al. (1999), additions of up to ~1 % carbon to Fe–22Mn alloy reduce SFE from about 30 mJ/m^2 to approximately 22 mJ/m^2.

Frommeyer et al. (2003) indicated that SFE higher than 25 mJ/m^2 resulted in twinning of stable γ-phase, whereas SFE lower than 16 mJ/m^2 led to formation of ε-phase, which is in agreement with Remy's schematic diagram. Allain et al. (2004b) gave much narrower range. According to their data, SFE should be at least 19 mJ/m^2 to induce mechanical twinning, whereas SFE less than 10 mJ/m^2 results in ε-phase formation. Dumay et al. (2008) mentioned that with SFE below 18 mJ/m^2, twins tend to disappear and are replaced with ε-platelets; SFE of about 20 mJ/m^2 is necessary for the best hardening rate.

The uncertainty related to the value of SFE required to ignite deformation by twinning mechanism has two explanations. First, low SFE is apparently a necessary condition to initiate twinning but is not sufficient, and activation of dislocation gliding should be hampered. Second, it is worth noting that the experimental measurement of SFE with good accuracy is very cumbersome, and most of the data found in the literature are based on thermodynamic computations and not on direct measurements with TEM.

Among direct measurements of SFE using TEM, the values of 15 mJ/m^2 for Fe20Mn1.2C TWIP steel (Idrisi et al. 2010), 13 ± 3 mJ/m^2 for Fe18Mn0.6C TWIP steel and 30 ± 10 mJ/m^2 for Fe18Mn1.5Al TWIP steel (Kim et al. 2011b) should be mentioned.

With all other conditions kept constant, lower SFE leads to higher density of strain-induced twins, which act as barriers to dislocation glide, and ultimately results in higher strain hardening. This was observed, in particular, in studying Fe18Mn0.6C and Fe18Mn0.6C1.5Al TWIP steels (Kim et al. 2011b).

11.2.2 Deformation Mechanisms of TWIP Steels

Deformation of TWIP steels as low SFE fcc alloys occurs by a competition between dislocation slip, deformation twinning, and austenite-to-martensitic transformation. In the most promising TWIP steels for automotive applications, the austenite-to-martensite transformation is largely suppressed (*which is particularly crucial in view of manufacturability as well*); dislocation slip occurs during early stages of deformation with deformation twinning becoming active after a threshold level of strain has been reached. The latter is of the order of 5 %, although it can vary with SFE, grain size, temperature, etc.

Thus, the dominating deformation mode in TWIP steels is the dislocation glide. Strain-induced twins formed due to low stacking-fault energy gradually reduce the effective dislocation slip distance. This leads to the so-called "Dynamic Hall–Petch effect" illustrated schematically in Fig. 11.3 and is considered to be cause for very high strain hardening of TWIP steels. An alternative model for the stress–strain

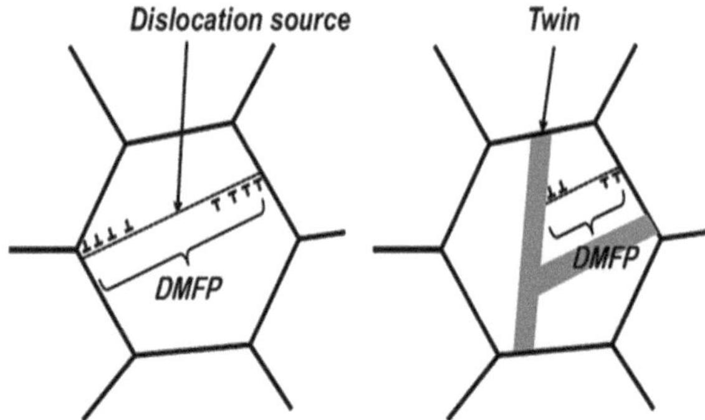

Fig. 11.3 Dynamic Hall–Petch effect as strain-hardening mechanism in TWIP steels (DMFP - Dislocation Mean Free Path)

behavior of austenitic TWIP steels was proposed by Sevillano. The suggested explanation emphasized the key contribution of austenite reinforcement by thin (nano-size) twins both to the macroscopic strain hardening and to the development of forward and backward stresses in a twin and in matrix, respectively (Gil Sevillano 2009).

Low SFE is critical for increasing the dislocation density while maintaining relatively long homogeneous slip distances. Exceptional ductility and strength are explained by multistage strain hardening at first, planar dislocation slip is activated as commonly observed in low SFE alloys. With additional strain, wavy slip is promoted leading to dislocation configurations that are progressively refined as the applied stress increases. The nucleation of deformation twins requires the critical dislocation density coupled with relatively long distances available for homogeneous slip. The activation of deformation twinning raises strain hardening of the alloy even further and therefore increases its ductility. Finally, the secondary twin system can be activated triggering intensive twin intersections that further strain harden the alloy (El-Danaf et al. 1999).

The transformation to martensite is also controlled by SFE, so twinning ends with the formation of martensite that depends on chemical composition and temperature, both of which influence SFE. As SFE decreases, the active deformation mechanism changes from dislocation slip to twinning and then to martensitic transformation that first produces ε-martensite and then α'-martensite (Grassel et al. 2000; Allain et al. 2004b). Since SFE increases with temperature, the same order of active deformation mechanisms is expected as temperature is decreased. Allain et al. studied Fe–22Mn–0.6C TWIP steel at different temperatures ($-196\,^{\circ}$C, 20 $^{\circ}$C, and 420 $^{\circ}$C). Martensitic transformation took place when SFE was below 18 mJ/m^2, and twinning was active for SFE between 12 and 35 mJ/m^2, with 20 mJ/m^2 being the most commonly reported value of the intrinsic SFE at room temperature (Allain et al. 2004b).

Park et al. studied the dependence of dislocation glide and mechanical twinning on SFE in fully austenitic TWIP steels by comparing Fe–22Mn–0.6C containing 3 % and 6 % Al and having SFE of respectively 20 and 50 mJ/m^2. Using optical microscopy and TEM, the authors studied the materials deformed in tension to different strains and noted the domination of planar glide of dislocations prior to the onset of mechanical twinning regardless of SFE and/or Al content. They also pointed out that planar glide was dominant over wider strain ranges with higher SFE, as manifested by increase in critical strain for the initiation of mechanical twinning with increasing SFE (Park et al. 2010).

At low strains, the dislocation density increases and the grain boundaries can serve as particularly efficient source of isolated stacking faults. Therefore, the deformation mechanism of TWIP steel at low strains is considered to be planar slip and the formation of wide stacking faults. The onset of twinning requires multiple slip within deformed grains. At significant strain (~20 %), higher dislocation density between deformation twins clearly shows that twin boundaries act as strong barriers for dislocation movement. Twins are very thin and there seems to be a continuous nucleation of new deformation twins of increasingly smaller size.

The nucleation of twins in TWIP steel may not be a homogeneous process. Instead, the nucleation stage of deformation twinning is closely related to prior dislocation activity, as the process always occurs after some amount of dislocations have been generated and dislocations in different slip systems have begun to interact. Twins are initiated within special dislocation configurations created by these interactions generally resulting in multilayer stacking faults which can act as twin nuclei.

11.2.3 Stress–Strain Response and Localization of Deformation

TWIP steels tend to have yield strength close to 250 MPa when C content is low (<0.1 %) or 400–600 MPa, when carbon is 0.6 %. According to Bouaziz et al., tensile strength of TWIP steels increases by 187 MPa per 0.1%C in the range of 0.2–0.6 wt% C (Bouaziz et al. 2011b). Microalloyed grades with TS close to 900–1000 MPa exhibit high strain hardening and high ductility reaching elongation of 60 %. These features imply considerable energy absorption defined by area under stress–strain curve.

Jerky serrated flow during tensile test of FeMnC-type TWIP steels is typically observed, which is explained by the formation of Portevin-Le Chatelier (PLC) bands. Strain is localized in these bands, and the local strain rate is much greater in those bands than in other parts of material. The underlying phenomenon is related to the classic dynamic strain aging (DSA) mechanism based on dynamic interaction between mobile dislocations and diffusing solute atoms (Bouaziz et al. 2011b).

The overview of the published stress–strain characteristics of high-Mn steels clearly differentiates between FeMnC and low-carbon FeMnSiAl TWIP steel concepts by the absence of jerky flow in carbon-free austenitic steels. Moreover, it is seen that neither serrations nor DSA are observed in FeMnC TWIP steels with 0.2–0.4%C.

Room temperature DSA occurs in most commonly studied carbon-alloyed TWIP steels such as Fe–22Mn–0.6C and Fe–18Mn–0.6C. DSA-related type serrations are explained by the presence of C–Mn complexes (De Cooman et al. 2011). According to Hong et al., DSA occurs because the reorientation of C atoms of Mn–C complexes obstructs the movement of twinning partial dislocations. Although the character of serration in steels with 0.6%C and 18 or 22 Mn was similar, the number of bands and the corresponding fluctuations of strain-hardening rate increased due to enhancement of DSA with higher Mn content (Hong et al. 2014). Since the same phenomenon was also observed in FeNiC TWIP/TRIP steels, the presence of Mn–C dipoles may not be a sole explanation and may not be the only mechanism to consider.

Detailed DSA studies were carried out by Chen et al. (2007) and Kim et al. (2009) who analyzed the properties of PLC bands. These authors reported that the band velocity decreases with increasing strain, but the strain rate within the bands was 15–100 times higher than the externally applied strain rate.

Alloying with Al weakens DSA (Kim et al. 2009) that can be explained by prevailing effect of increasing SFE that suppresses twinning. For example, serrations in stress–strain curves were already visible at the early stages of plastic deformation of Fe–18Mn–0.6C TWIP steel, while in case of Fe18Mn0.6C–1.5Al steel, the serrations began to appear only after considerable strain (Kim et al. 2011b). Similar effect is expected from additions of nitrogen less than 0.3 mass-%, when SFE also increases (De Cooman et al. 2011).

DSA is an undesirable phenomenon for automotive materials because it can lead to strain localization that deteriorates ductility and can in principle impede press forming. However, in case of Fe–22Mn–0.6C TWIP steel, neither the occurrence of PLC bands in uniaxial tensile testing nor poor press-forming performance was reported (Allain et al. 2008). This is probably because the occurrence of DSA-induced surface defects depends on the applied strain and strain rate, as well as on the stress state during deformation.

The effects of strain rate on stress–strain responses of Fe–22Mn–0.6C steel with grain size of 3 μm are shown in Fig. 11.4. It is seen that serrated flow occurs at higher strains, stresses, and strain rates.

11.2.3.1 Effect of Deformation Temperature on Mechanical Behavior of TWIP Steels

Twinning is a dominating deformation mechanism of TWIP steels at room temperature due to low SFE. Since SFE depends on temperature, the mechanical properties and deformation mode in steels with SFE-controlled deformation mechanisms are expected to strongly depend on deformation temperature.

Earlier work by Allain et al. using Fe–22Mn–0.6C steel also indicated that deformation mechanism and microstructure change with deformation temperature. In particular, the transformation to ε-martensite occurred at SFE < 18 mJ/m^2, while

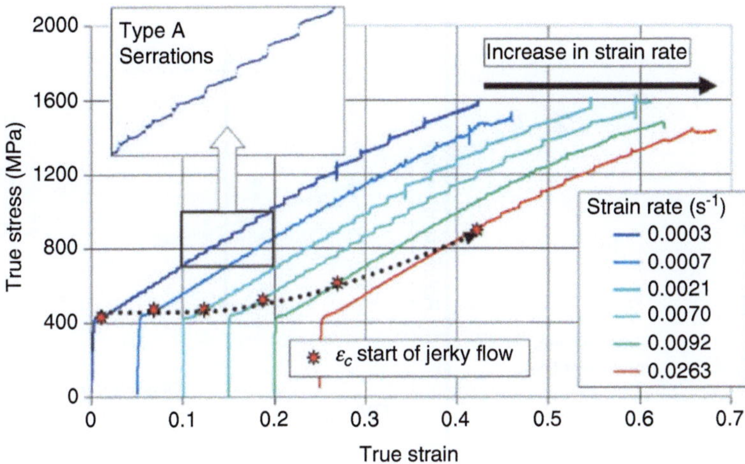

Fig. 11.4 Room temperature tensile response of 22Mn–0.6C TWIP steel at different strain rates. Note: strain offset is pointed out for clarity (Allain et al. 2008)

mechanical twinning was observed with 12 mJ/m^2 < SFE < 35 mJ/m^2 (Allain et al. 2004b).

According to Remy (1978), transformation into ε-martensitic occurs at relatively low temperatures, deformation twinning occurs at intermediate temperatures, and dislocation glide prevails at relatively high temperatures. Dynamic strain aging is facilitated at higher deformation temperatures due to increasing diffusivity of solute atoms.

The temperature dependences of dynamic strain aging and deformation twinning are opposite, and the temperature ranges for ε-martensitic transformation and deformation twinning are also different. Thus, deformation temperature controls the dominance of one of the three deformation modes.

As noted above, the TWIP mechanism responsible for high strain-hardening rate and high elongation is activated at critical strain, which is higher than the plastic yield onset in the studied steels and which is attained by dislocation glide. The work performed by Allain et al. was focusing on the thermally activated dislocation and intense dislocation gliding glide as the main contributors to very high elongation reported for TWIP steels, as the final volume fractions of twins remained below 0.1 (Allain et al. 2010).

As displayed in Fig. 11.5, the general trend is that both the yield and the ultimate tensile strengths decrease with increasing deformation temperature, whereas the uniform elongation demonstrates maxima near the room temperature and within certain range above 200 °C (Scott et al. 2005). SFE was found to decrease with decreasing temperature, and lower SFE promotes deformation twinning and inhibits dislocation glide which nevertheless remains the main deformation mode. Deformation twins formed during plastic deformation can act as obstacles for

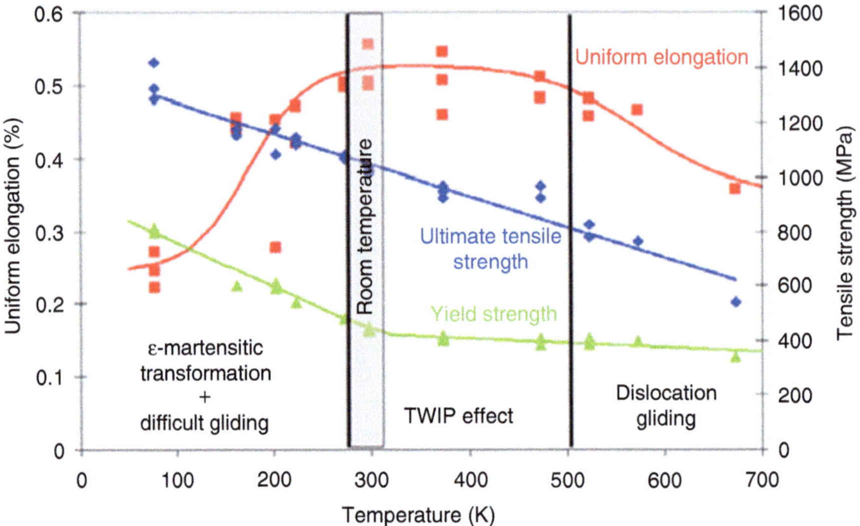

Fig. 11.5 Results of the tensile tests (UE, TS, and 0.2 % YS) carried out at different temperatures on laboratory Fe–22Mn–0.6C cold strips (grain size 3 μm). The activated deformation mechanisms were confirmed by TEM observation (Scott et al. 2005)

dislocation motion resulting in high strain-hardening rate so that both high elongation and tensile strength can be obtained at relatively low temperatures.

At temperatures above ~220 °C (500 K), the SFE is high enough to suppress twinning, and the only active deformation mechanism is planar dislocation glide. Since hardening effects at this temperature are limited to interactions between dislocations, the strain hardening is lowered. This causes a decrease in uniform elongation.

According to Liu et al., the temperature controls the competition between generation and annihilation of dislocations and influences the twinning parameters, i.e., the volume fraction of shear bands, the triggering (initial) strain, and the driving force. In the temperature range of 298–673 K, fracture strain and strain-hardening exponent of Fe–22Mn–0.6C TWIP steel decrease with temperature due to the weakening twinning-induced plasticity of austenite (Liu et al. 2015).

Koyama et al. studied the effects of deformation temperature on deformation modes and work hardening behavior of Fe–17Mn–0.6C and Fe–17Mn–0.8C TWIP steels. The variation in deformation temperature revealed the dominating effect of specific deformation mode on work hardening rate and ductility of steel. The contribution of deformation mode to strain hardening was found to decrease in the following order: ε-martensitic transformation > deformation twinning > dynamic strain aging. In particular, the ε-martensite transformation, active in Fe–17Mn–0.6C up to 294 K, made this steel almost brittle at 173 K, despite strain-hardening rate of approximately 4 GPa. In Fe–17Mn–0.9C steel, deformation twinning produced strain-hardening rates of approximately 3 GPa at 173 K and of 2.3–3 GPa at 373 K. Dominant deformation by twinning and DSA observed at 293 K led to uniform elongation of the latter steel exceeding 70 % with tensile strength higher than 1100 MPa (Koyama et al. 2011b).

11.2.3.2   Effect of Strain Rate on Mechanical Behavior of TWIP Steels

Response of TWIP steels to high strain rate has two aspects: stress–strain state during high strain rate cold forming and behavior in collision.

Like the temperature of deformation, strain rate also affects mechanical behavior of TWIP steels. Strength and elongation of TWIP steels depend on the alloy system, whereas the deformation mechanism varies with strain rate.

In carbon-free Fe30Mn3Si3Al alloys, the strain rate range for thermally activated plastic yield strength was lower compared to that for Fe22Mn0.6C steels, as the former alloys contain high amount of silicon that significantly affects dislocation mobility (Allain et al. 2010).

Intensive twin formation at high strain rate was reported by Frommeyer et al. (2003) who studied Fe–15Mn–3Al–3Si TWIP steel in the strain rate range from 10^{-3} to 10^3 s^{-1}. Under these conditions, the tensile strength remained practically constant, whereas the yield strength increased with strain rate. Extremely high ductility and energy absorption were observed because of the dominant role of intensive mechanical twinning due to favorable SFE and phase stability, both of which are controlled by concentrations of Mn and Al.

In the study of microstructure and mechanical properties of low-carbon high-manganese Fe–18.1Mn–3.15Si–3.12Al–0.03C TRIP/TWIP steel, during tensile deformation at room temperature, the inverse effect of strain rate on strength of steel was observed, which is usually inherent for FeMnC TWIP concept due to the DSA phenomenon. In fact, strength and ductility decreased with strain rate increasing within the quasi-static range between 1.67×10^{-4} and 1.67×10^{-1} s^{-1}. At the same time, in the dynamic tensile strain rate range of 10–1000 s^{-1}, strength and ductility increased significantly with increasing strain rate indicating normal sensitivity to strain rate. At strain rates below 10 s^{-1}, the martensite transformation and deformation twinning occurred during tensile test and adiabatic temperature rise softening the martensitic matrix (Chen et al. 2013). This negative strain rate sensitivity can to some extent explain poor post-neck elongation of FeMnC steels.

11.2.4   Strain Hardening of TWIP Steels

Superior properties of TWIP steels are to a large extent attributable to their sustained high work hardening rate. Strain hardening of TWIP steels can be seen to increase steadily up to strains of approximately 0.25 and then it becomes constant on the level of about 0.5.

However, despite several decades of research, the real origins of high work hardening rate in these steels are still not fully understood.

For strains below the threshold value for twinning, strain hardening is induced by interactions between dislocations. Slip–twin interactions are responsible for hardening at strains beyond the threshold until the volume fraction of twins reaches saturation. As a result of both dislocation slip and deformation twinning, these alloys exhibit increased ductility in comparison to other high strength steels.

The most common explanation for high strain hardening and high tensile strength refers to the refinement of microstructure with formation of very fine twins that serve as obstacles for dislocation motion (Remy 1978).

Several other mechanisms have been proposed to explain the observed strain hardening in TWIP steels in addition to deformation twinning. One of the proposed mechanisms implies interactions of gliding dislocations with twin boundaries, which increases dislocation storage rate and reduces the dislocation mean free path (MFP) in grains where twins are present (Fig. 11.3). As mentioned above, this mechanism is often called the dynamic Hall–Petch effect (Remy 1978; Allain et al. 2004a). Twin boundaries can be considered as essentially grain boundaries with different resistance to slip transmission. The MFP represents the average distance of dislocation travel before they become immobile, and therefore, it is a key microstructure parameter related to strain hardening.

As mentioned above, in the study of 22Mn–0.6C steel, Gil Sevillano proposed a model for strain hardening, which is alternative to the models based exclusively on the dynamic Hall–Petch effect. He emphasized the key contribution to the macroscopic strain hardening from strengthening of austenite by thin nano-size deformation twins (Gil Sevillano 2009).

Based on the study of 18Mn–0.6C TWIP steel, Chen et al. concluded that high strain hardening of this alloy is due to dynamic strain aging (DSA) caused by interactions between Mn–C octahedral clusters and dislocations (Chen et al. 2007).

The ε-martensitic transformation, dynamic strain aging, and twinning have all been reported to increase the work hardening rate in carbon-containing TWIP steels. Koyama et al. studied the behavior of two 17Mn steels with 0.6 % C and 0.8 % C. The authors intentionally tested steels at different temperatures to distinguish the contributions from ε-martensitic transformation, deformation twinning, and dynamic strain aging. Based on the results, they concluded that the quantitative contributions of different mechanisms to work hardening rate could be presented in the following order: ε-martensitic transformation > deformation twinning > dynamic strain aging (Koyama et al. 2011a). Each of these mechanisms works within specific temperature ranges that also depend on strain rate.

As emphasized in the review by Bouaziz et al., among all possible mechanisms (twinning, pseudo-twinning, DSA, etc.) suitable to explain the strain hardening of TWIP alloys as the obstacles to dislocation glide, twinning should be preferentially taken into account (Bouaziz et al. 2011b). This conclusion is based on the facts that in the absence of twinning, the work hardening of TWIP steels is much lower.

Recently, Kim et al. also confirmed that twinning is the most important strain-hardening mechanism in TWIP steels. They estimated that the contribution from

DSA does not exceed 20 MPa (<3 %) of the total flow stress of an 18Mn–0.6C–1.5Al, which is negligibly small (Kim et al. 2009).

According to Barbier et al., who studied the microstructure evolution during tensile tests of fine grain 22Mn–0.6C TWIP steel, the formation of nano-size mechanical twins led to increase in strain-hardening rate at early stages of plastic deformation. At high strains, the development of pronounced (111) fiber in tension direction sustained mechanical twinning and maintained high strain-hardening rate (Barbier et al. 2009).

Yoo et al. suggested that continuously increasing strain-hardening rate in high-manganese 28Mn–9Al–0.8C austenitic steel that leads to exceptionally high elongation of ~100 % could be attributed to the initiation and intersection of micro-bands formed by geometrically necessary dislocations (Yoo et al. 2009).

On the other hand, comparison between Fe–Mn–C and Fe–Mn–Si–Al TWIP steels showed that steels with similar volume fraction of twins can exhibit quite different work hardening behavior (Idrisi et al. 2010). Grounding on these observations, Bouaziz et al. extended their physical model for work hardening of TWIP steels to include the effect of carbon. It was shown that carbon mainly controls the maximum number of dislocations piled up at twin boundary, which dictates the magnitude of back stresses and therefore the overall work hardening (Bouaziz et al. 2011a).

11.2.5   Effect of Grain Size on Mechanical Behavior of TWIP Steels

The microstructure of a TWIP steel is single phase austenite with rather coarse grains often containing wide annealing twins, as revealed by De Cooman et al. (2012).

For fcc metals and alloys with low SFE, such as TWIP steels, the information on grain size effects on strength and ductility is quite limited.

Figure 11.6 depicts the effect of grain refinement on stress–strain curves of Fe–22Mn–0.6C TWIP steel. As was shown, the yield strength in X-IP 1000 alloys closely followed the Hall–Petch relationship. In general, TS also increased with yield stress as the grain size decreased (Scott et al. 2005).

As was shown by Qian et al., compared to the coarse-grained steel, the fine-grained steel increases the critical strains of onset of serration, especially in the low strain-rate region, where a negative strain-rate sensitivity is observed. The findings, not always confirmed in other studies, suggest that the grain refinement of Fe–22Mn–0.6C TWIP steel suppresses the serrated flow, possibly due to the enhanced dynamic recovery associated with the decreased planar slip length in low stacking-fault-energy steel (Qian et al. 2013).

The available data suggest that grain refinement is accompanied by refinement of deformation twins that reduces ductility, as was shown for low-carbon

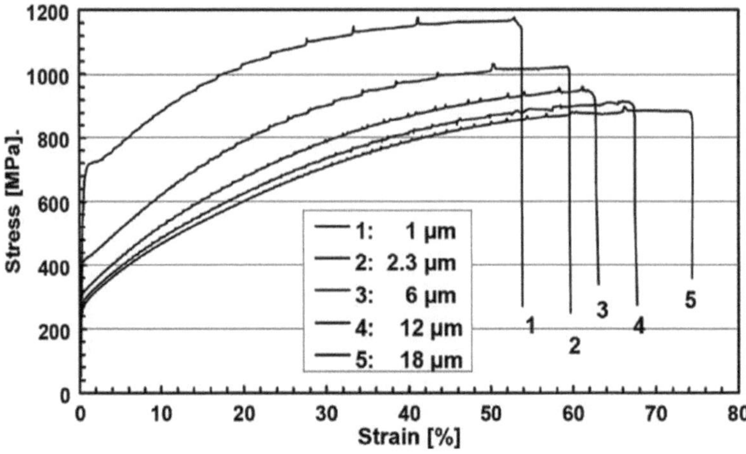

Fig. 11.6 Effect of grain size on tensile response of Fe–22Mn–0.6C (Scott et al. 2005)

Table 11.1 The tensile properties of TWIP steel with different grain sizes

Grain size (μm)	YS (MPa)	UTS (MPa)	YS/UTS	WH (MPa)	Elongation (%)	Uniform elongation (%)
2.1	572.0	825.0	0.69	253.0	47.2	39.8
4.7	404.4	723.8	0.56	319.4	68.4	56.5
11.4	311.9	648.1	0.48	336.2	75.9	63.1
23.7	208.6	575.4	0.36	366.8	82.4	70.0
72.6	123.1	519.4	0.24	396.3	85.0	76.3

Note: WH work hardening

Fe–31Mn–3Al–3Si and 24.8Mn–3.17Si–3.12Al–0.022C TWIP steels. The number of deformation twins increases with grain size leading to greater TWIP effect in coarse-grained material than in fine grain one. This can be attributed to grain size dependence of the critical stress for onset of deformation twinning, Table 11.1 (Chen et al. 2013). As described in the literature, twinning occurs only when the applied resolved shear stress reaches a certain critical value, the critical resolved shear stress (CRSS). Among other factors, CRSS is contributed by dislocation pile-ups at grain boundaries that act as stress concentrators for twin nucleation (Allain et al. 2004a).

However, fine-grained Fe–31Mn–3Al–3Si steel (grain size $d = 1.8$ μm) showed high strength with adequate ductility, which is different from alloys with medium to high SFE. This observation suggests that high ductility of these steels is related not only to twinning but also to weak dynamic recovery due to low SFE.

Fine grain size can completely inhibit the formation of twins and martensite in low SFE materials (El-Danaf et al. 1999). It was suggested that the initiation of twinning requires accumulation of critical dislocation density, i.e., the twins start to

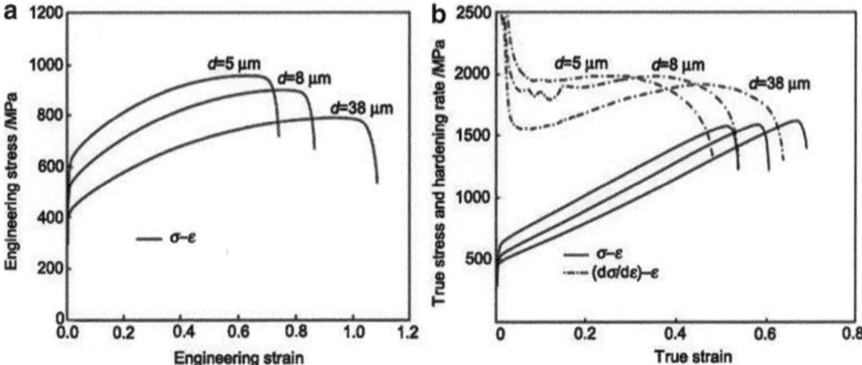

Fig. 11.7 Representative engineering stress–strain curves of Fe–28Mn–9Al–0.8C steel with three different grain sizes tested at 25 °C and the initial strain rate of 10^{-3} s^{-1} (**a**), and true stress–strain curves and the corresponding strain-hardening rate—true strain curve (**b**) (Yoo et al. 2009)

form after a certain (critical) plastic strain by dislocations glide has been attained. Table 11.1 shows the mechanical properties as a function of work hardening changed by grain size. The microstructural observations indicate that planar dislocation pile-ups are necessary for triggering mechanical twinning and that grain refinement suppresses mechanical twinning.

The same conclusion was confirmed experimentally in the study of early deformation stage using 31Mn–3Al–3Si steel (Dini and Ueji 2011).

Thus, yield strength (YS) and tensile strength (TS) increase with decreasing grain size. The trend for elongation is opposite. The product of TS and total elongation decreases with decreasing grain size in spite of the increase in strength indicating the dominating role of elongation as a tensile characteristic of steel.

The representative engineering stress–strain curves of 28Mn–9Al–0.8C steel with three different grain sizes are shown in Fig. 11.7a. The plot of strain-hardening rate superimposed on the true stress–strain curves is depicted in Fig. 11.7b. Strain-hardening rate of fine grain (5 μm) steel remains constant up to true strain of ∼0.3 and then decreases. By contrast, strain-hardening rate of coarse-grained (38 μm) steel is lower than that of fine grain steel at the initial stage of deformation, but it grows continuously with true strain increasing up to ∼0.5.

11.3   Effect of Alloying/Microalloying on Microstructure and Properties of TWIP Steels

Fe–22Mn–0.6C can be considered a baseline TWIP composition showing the desirable deformation twinning. However, this alloy has several disadvantages including very low yield strength in the annealed condition (around 250 MPa),

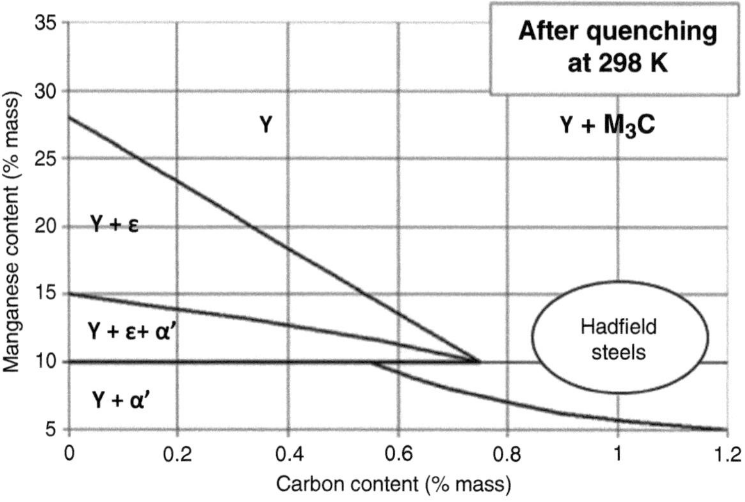

Fig. 11.8 Fe–Mn–C phase stability diagram at room temperature (Scott et al. 2005)

cementite precipitation during annealing, and susceptibility to hydrogen-induced delayed fracture. Consequently, the alloy design should consider the conditions for maintaining austenitic structure at all processing temperatures, prevention of martensite formation during cold working operations, optimization (increase) of yield strength, tensile strength and elongation at room temperature, prevention of carbide formation under normal processing conditions, and optimization of austenite stability/SFE so that twinning is activated at sufficiently high strains as well as the resistance to delayed fracture.

Review of TWIP steels published by Bouaziz et al. outlines, in particular, the history of evolution of high-manganese austenitic steels beginning from the development of Hadfield steel in 1868. This review also describes the criterion

$$wt\%Mn \ + \ 13wt\%C \geq \ 17$$

for concentrations of C and Mn required for stabilization of the fully austenitic microstructure (Bouaziz et al. 2011b).

Figure 11.8 shows the effects of additions of Mn and C on stabilization of the fcc austenite γ-phase.

To obtain fully austenitic microstructure, TWIP steels should contain sufficient amount of Mn. Other elements, e.g., C, Si, and/or Al, are also needed to ensure high strength and high uniform elongation associated with strain-induced twinning.

Depending on the alloy system, the carbon content in TWIP steels is either low (<0.05 wt%) or high, typically in the range of 0.5–1.0 wt%. Manganese stabilizes austenite. However, if its concentration is less than 15 wt%, α'-martensite can be formed, which aggravates ductility. If Mn content exceeds 30–32 wt%, brittle β-Mn

Table 11.2 Effects of alloying elements on the properties of high-manganese steels (Chen et al. 2013)

Effect	C	Mn	Si	B	Ti	N	Al
γ-Stabilizer	✓	✓				✓	✓
Solid solution strengthening austenite	✓		✓			✓	✓
ε-Martensite refinement			✓				✓
Hot ductility				✓	✓		

phase forms in microstructure. Thus, Mn content is normally in the range of 15–30 wt%.

The effects of alloying elements on the properties of high-manganese steels are illustrated in Table 11.2.

In Fe–Mn phase diagram, the Fe-rich side has an open γ loop. Therefore, within the range of 5–25 wt% Mn, α'-martensite at lower Mn content and ε-martensite at higher Mn content can be present in microstructure (De Cooman et al. 2012). Mn concentration of approximately 27 wt% is required to obtain stable austenite at the room temperature in carbon-free alloys. The $\gamma \rightarrow \varepsilon$ transformation temperatures decrease with increasing Mn content.

Computations of SFE for Fe–xMn–3Si–3Al system showed that SFE increases with manganese content resulting in change of deformation mechanism from TRIP-type to TWIP-type mode (Chen et al. 2013).

The manganese content does not seem to play any role in the thermally activated mechanisms, (e.g., dislocation gliding) as the two binary alloys Fe22Mn1C and Fe13Mn1.2C present similar activation volumes.

It is well known that carbon can improve the austenite stability and strengthen the steels. It also inhibits the formation of ε-martensite by increasing SFE. With high carbon (>0.6 wt%), martensite-free austenitic microstructures can be obtained at Mn content as low as 12 wt%, for example, in Fe–12Mn–1C Hadfield steel.

As shown by Allain et al., the higher the carbon content, the lower the activation energy for dislocation glide, which is an important mechanism that contributes to ductility of TWIP steels (Allain et al. 2010). Their analysis confirmed the major role of carbon in plastic deformation of such alloys, not only with respect to work hardening behavior but more importantly with respect to dislocation mobility.

From the view of the role of thermally activated dislocation gliding, aluminum content does not play significant roles. On the contrary, silicon seems to reduce the apparent activation volume (Allain et al. 2010).

Effect of alloying on SFE of steels with Fe22Mn0.6C is presented in Fig. 11.9 (Dumay et al. 2008).

Si and Al can be added to achieve stable and fully austenitic microstructure with SFE in the range of 15–30 mJ/m^2 as the critical precondition for twinning during deformation. In particular, the SFE is linearly raised with Al additions. For example, in 18Mn–0.6C steel, SFE increases from 13 mJ/m^2 to 30 mJ/m^2 with the addition of 1.5 % Al (Kim et al. 2011b).

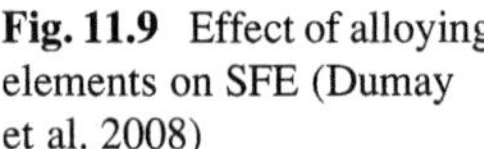

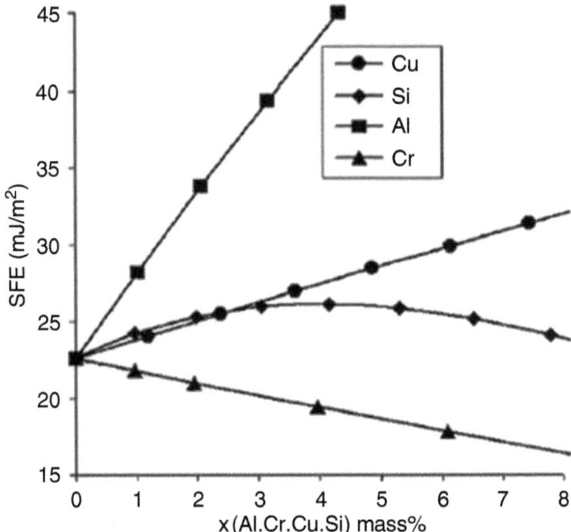

Fig. 11.9 Effect of alloying elements on SFE (Dumay et al. 2008)

While small additions of Al facilitated the TWIP effect, the formation of ε-martensite is effectively suppressed by addition of 1.5 wt% Al to 15Mn–0.6C TWIP steel.

Additions of 1.5 % Al to 18Mn–0.6C reduce tensile strength while nearly doubling the elongation by suppressing cementite precipitation during cooling after hot rolling and during annealing due to decrease in both activity and diffusivity of C in austenite (Jin and Lee 2012). Jin and Lee tried to summarize various aspects the effects of Al. According to their study, Al addition suppressed precipitation of cementite during cooling after hot rolling, linearly raised the stacking-fault energy with a constant slope of 7.8 mJ/m^2 per 1 wt% Al, increased the yield stress, and reduced area, uniform (e_u), and post-uniform (e_{pu}) elongations. In particular, as depicted in Fig. 11.10, although post-uniform strain e_{pu} of TWIP steel without Al was almost zero, it was improved up to 7 % by adding 2 wt% Al. Although both Al free and 2 % Al compositions have negative strain rate sensitivities, the authors explain the observed effect of Al by prevailing role of lower rates of strain hardening and dynamic strain aging (DSA) as Al reduces both the activity of carbon and its diffusivity in austenite.

According to Jin and Lee, nitrogen was found to suppress DSA and serrations, that is, the effects of nitrogen are analogous to those of Al. Similar observations were reported by De Coman et al. (De Cooman et al. 2011). With nitrogen content below 0.3 wt%, the SFE increases. Nitrogen is an efficient austenite strengthening element, e.g., adding nitrogen to Fe–16.5Mn alloy lowers the martensite start temperature and also reduces the volume fraction of ε-martensite.

Silicon improves strength by solid solution strengthening, refines ε-martensite plates, and increases fracture strength, although it does not improve ductility. Higher Si tends to stabilize austenite against its transformation into ε-martensite.

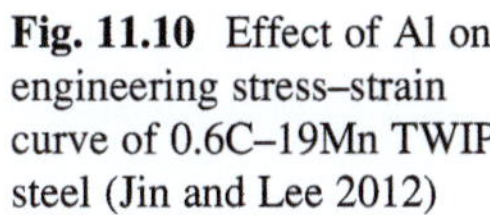

Fig. 11.10 Effect of Al on engineering stress–strain curve of 0.6C–19Mn TWIP steel (Jin and Lee 2012)

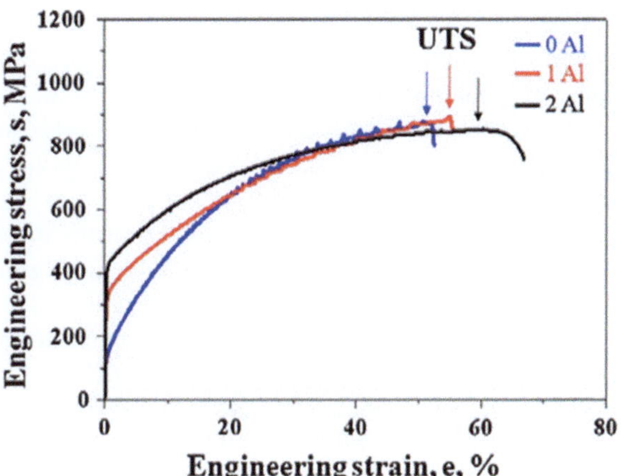

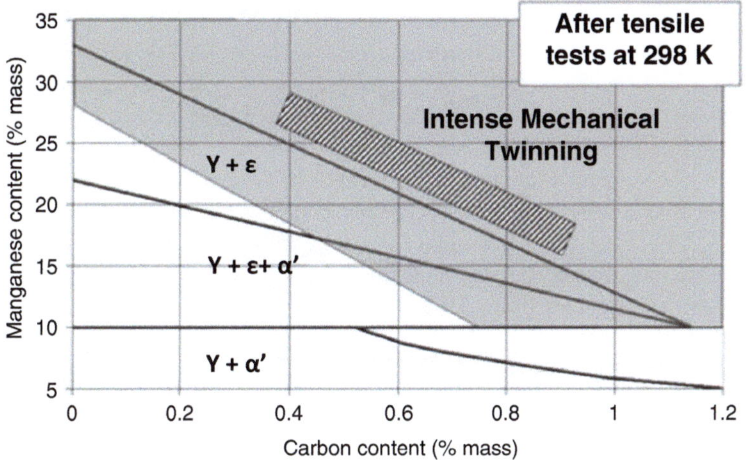

Fig. 11.11 Fe–Mn–C phase stability diagram after tensile testing at room temperature (Scott et al. 2005)

As shown in Fig. 11.9, Cu increases SFE and Cr decreases SFE of austenite.

As was emphasized above, the aim of optimization of TWIP steel composition is to promote deformation by twinning and suppress martensitic transformation.

Figure 11.11 describes, based on literature available in 2004, the estimated compositions that exhibit intensive mechanical twinning during tensile deformation. These compositions are fully austenitic and maintain the austenitic microstructure during deformation. This figure also shows the compositions in which strain-induced transformation of austenite to ε-martensite and α′-martensite can occur. For example, in 18Mn–3Al–3Si–0.04C composition, the transformation to ε-martensite occurs first, and then deformation twinning starts to dominate at higher strain. The formation of α′-martensite results in lower ductility and generally is undesirable. At lower temperatures, α'-martensite forms after ε-martensite. In accordance with Fig. 11.11, with lower C, more Mn is needed to prevent strain-induced transformation to martensite (Allain et al. 2004b).

It was shown that yield strength of TWIP steels can be significantly increased by grain refinement through controlled recrystallization of cold-rolled steels (Hamada et al. 2012).

Microalloying was considered as the means to increase the yield strength by combining the effects of precipitation hardening, grain refinement, and recrystallization control in cold-rolled steels (Scott et al. 2011).

The effects of Nb on mechanical twinning and tensile deformation behavior were studied using hot-rolled Fe–18Mn–0.6C–1.5Al TWIP steels without Nb and with 0.1 % Nb. Compared to Nb-free TWIP steel with fully recrystallized and equiaxed grains, the Nb-bearing TWIP steel showed non-recrystallized and elongated grains with well-dispersed NbC particles and high dislocation density. Higher yield and tensile strengths of Nb-added TWIP steel were mainly due to dislocation hardening of non-recrystallized grains rather than due to precipitation hardening by NbC particles or solid solution strengthening by Nb atoms. At the same time, uniform elongation and strain-hardening rate of Nb-added TWIP steel were lower due to inactive mechanical twinning, probably because the dislocations interfere with the movement and dissociation of dislocations, necessary for mechanical twinning (Kang et al. 2012) (Fig. 11.12).

Reyes-Calderón et al. compared the efficiency of B, V, and Nb additions to Fe–21.5Mn–1.5Al–1.5Si–0.4C steel in retarding grain growth within temperature range of 1100–1250 °C. It was shown that the strongest pinning effect within the entire temperature range was demonstrated by Nb resulting in the smallest grain size. Among three elements, boron was the weakest grain growth inhibitor (Reyes-Calderón et al. 2011).

The effects of precipitation hardening of FeMnC TWIP steel by vanadium carbides were studied in several works. High density of nano-size V_4C_3 carbides in Fe–21.6Mn–0.63C–0.87V TWIP steel was observed that resulted in 200 MPa increase in yield strength (Yen et al. 2012).

For decades, mainly two types of TWIP steel compositions have been extensively researched: FeMnC X-IP TWIP steels as Fe22Mn–0.6C (Allain 2004), Fe–18Mn–0.6C, Fe–18Mn–0.6C–1.5Al (Kim et al. 2008), and (L-IP) low-carbon Fe-(25–30) Mn–3Si–3Al (Grassel et al. 2000).

At the same time, reducing Mn content has become a very serious practical issue to cut the cost and facilitate manufacturing of TWIP steels. In the work by Kim et al., 18Mn–1.5Al–0.6C steel and Fe–12Mn–2Si–0.9C steel were tested at various temperatures (25–600 °C) and strain rates (10^{-4}/s–3×10^2/s) to study the temperature, strain rate, and Mn effects on mechanical properties. Strength and ductility of 18Mn TWIP steel increased with decreasing strain rate, but in case of lean (12Mn) TWIP steel, these properties decreased with decreasing the strain rate. However, the overall effect of strain rate on strength and ductility of studied steels was insignificant.

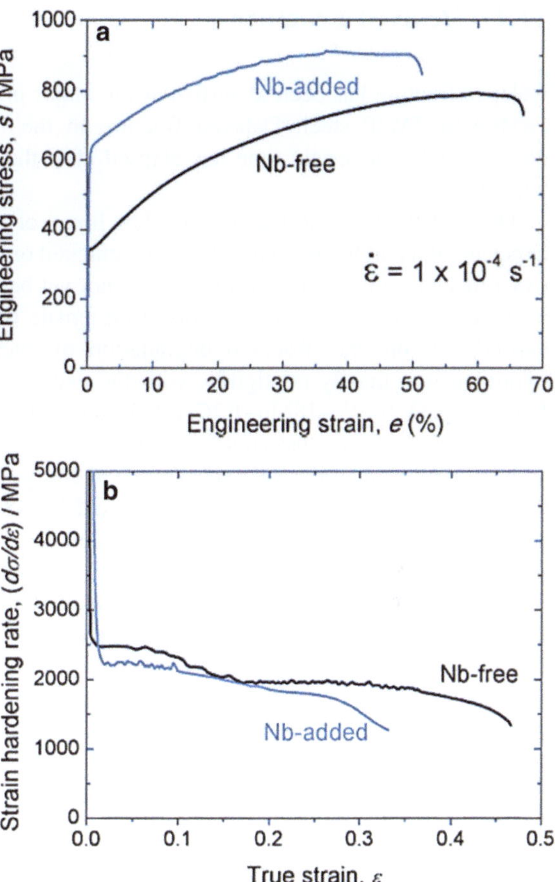

Fig. 11.12 Engineering stress–strain curves (**a**) and strain-hardening curves (**b**) for Nb-free and free-added (~0.1 %) 0.6C–18Mn–1.5Al TWIP steels (Kang et al. 2012)

Deformation mechanisms of lean Mn TWIP steels were also studied. Twin formation in 18Mn TWIP steel weakened with increasing strain rate as stacking-fault energy increased due to adiabatic heating. In 12Mn TWIP steel, however, twinning intensified at increasing strain rate. TS appeared to be directly related to the volume fraction of twinned grains.

Tensile strength of 18Mn TWIP steel was nearly 1000 MPa at elongation over 60 % when tested at the strain rate of $10^{-4}\ s^{-1}$. For 12Mn TWIP steel, TS was over 1100 MPa at elongation of nearly 60 % when tested at the strain rate of $10^{-1}\ s^{-1}$. Strain-hardening rate and the number of serration in both steels also decreased with increasing strain rate (Kim et al. 2011a).

It is noteworthy that the important advantage of FeMnAlSi concept of TWIP steels is their relatively low density that is in direct relation to their alloying. For example, the largest component of TWIP steels, Mn, is 5.5 % less dense than Fe. Typical alloying elements are even less dense than Fe: Al is 66 % and Si is 70 % less dense than Fe. As a consequence, the density of TWIP steels is lower than that of conventional steels and other AHSS.

11.4   Delayed Fracture

Delayed fracture has been identified as the major problem for Fe–22Mn–0.6C and FeMnAlSi TWIP steel. Delayed fracture in the form of deep edge cracks is observed when a certain time has elapsed after the part (for example, a cup) has been drawn.

Delayed fracture can happen in a few hours or days after forming and during exposure of a part to air even if it was not cracked or fractured during stamping. The exact mechanism of delayed fracture has not yet been identified.

Several studies of hydrogen impact on tensile properties of TWIP steels were carried out, but the observed degradation of ductility after pre-charging with hydrogen was mostly negligible, with the exception of findings in the study by Koyama et al. for Fe–18Mn–1.2C TWIP steel, where a significant loss of ductility during testing of pre-charged specimen was detected (Koyama et al. 2012).

Kim et al. studied the influence of $\gamma \rightarrow \alpha'$ and $\gamma \rightarrow \varepsilon$ martensitic transformations during tensile testing of Fe–18Mn–0.6C and Fe–18Mn–0.6C–1.5Al TWIP steels. These authors suggested that the observed delayed fracture was related to martensitic transformation in the presence of residual stresses and possibly hydrogen (Kim et al. 2008). It was established that as-annealed Al-alloyed TWIP steel remained free of martensite. Both TWIP steels contained martensite after cup drawing, but in Al-alloyed TWIP steel, the amount of martensite was slightly less. This could stem from higher SFE due to Al additions and the related limited nucleation of martensite phase, which otherwise could cause embrittlement of steel in the presence of small amounts of solute hydrogen.

In spite of numerous confirmations that Al-added TWIP steels should be immune to the problem, the mechanism of the effect of Al is not clear. According to Ryu, adding Al can enhance the resistance to cracking by lowering the chances not only for ε-martensite formation but also for mechanical twinning due to increasing SFE (Ryu 2012).

Dieudonne et al. also studied the mechanisms of improved resistance to delayed fracture due to additions of 1.7%Cu or 1.5%Al to Fe–18Mn–0.6C TWIP steel that improved both resistance to hydrogen embrittlement and corrosion behavior. It was shown that copper had no influence on corrosion rate of austenitic Fe–18Mn–0.6C steels, whereas the enrichment of oxide layer with metallic copper promoted hydrogen recombination and consequently reduces hydrogen absorption. The effect of aluminum on hydrogen uptake was somewhat different. The formation of oxide layer enriched with aluminum during corrosion increases the passivation ability of steel and reduces the corrosion rate. Therefore, decrease in hydrogen absorption during corrosion can be explained by beneficial effect of aluminum on the kinetics of corrosion. It was shown that the presence of phosphorus enhances the corrosion rate of Al-enriched alloys and promotes hydrogen absorption (Dieudonne et al. 2014).

Another mechanism of improving hydrogen resistance was studied by Milard et al. using Fe–18Mn–0.6C alloy with additions of 0.2 %V. Quite strong hydrogen trapping by vanadium carbides was observed. The authors did not detect any interactions between hydrogen and vanadium in solid solution, but significant interaction between hydrogen and VC precipitates was revealed when 5 wt ppm of hydrogen could be trapped in precipitates (Millard et al. 2012).

The phenomenon of delayed cracking of TWIP steels is very likely related to hydrogen impact, but it requires further fundamental analysis.

11.5 Summary

TWIP (Twinning Induced Plasticity) steels, among the known materials offered for automotive industry, possess the highest elongation of up to 50–60 %, although, so far, at strength not higher than 900–1100 MPa. At the same time, several issues such as low yield strength still present problems. Increase in yield strength by pre-strain (cold rolling/heavy temper rolling) is not efficient because it reduces the "remaining" ductility, and other ways such as microalloying and grain refinement should be intensively explored. Delayed fracture is suppressed by additions of Al, but the mechanism still lacks understanding.

Due to intensive studies in Europe, Korea, and China, as well as due to tremendous efforts of POSCO, the industrialization of TWIP steels began (Lee et al. 2014), but further technological research should focus on the alloy designs to reach higher values of strength—elongation product with leaner chemical composition, with more comfortable manufacturability in both steel and auto industries.

Physical metallurgy of TWIP steels also requires more fundamental research works to clarify the operation of twinning mechanism and the ways to control deformation, as well as the mechanism of delayed fracture.

References

Allain, S., O. Bouaziz, and J.-P. Chateau. 2010. "Thermally Activated Dislocation Dynamics in Austenitic FeMnC Steels at Low Homologous Temperature." *Scripta Materialia* 62 (7): 500–503.

Allain, S., J.-P. Chateau, and O. Bouaziz. 2004a. "A Physical Model of the Twinning Induced Plasticity Effect in a High Manganese Austenitic Steel." *Material Science and Engineering A* 387–389: 143–47.

Allain, S., J.-P. Chateau, O. Bouaziz, and S. Migot. 2004b. "Correlations between the Calculated Stacking Fault Energy and the Plasticity Mechanisms in Fe-Mn-C Alloys." *Material Science and Engineering A* 387–389: 158–62.

Allain, S., P. Cugy, C. Scott, and J.-P. Chateau. 2008. "The Influence of Plastic Instabilities on the Mechanical Properties of a High-Manganese Austenitic FeMnC Steel." *International Journal of Materials Research* 99 (7): 734–38.

Barbier, D., N. Gey, S. Allain, and N. Bozzolo. 2009. "Analysis of the Tensile Behavior of a TWIP Steel Based on the Texture and Microstructure Evolutions." *Material Science and Engineering A* 500: 196.

Bouaziz, O., S. Allain, M. Huang, and D. Barbier. 2011a. "Effects of Microstructural Length Scales and of Carbon Content on the Work Hardening of Twinning Induced Plasticity Steels." In *HMnS – 2011*. Seoul, South Korea.

Bouaziz, O., S. Allain, C. Scott, P. Cugy, and D. Barbier. 2011b. "High Manganese Austenitic Twinning Induced Plasticity Steels. A Review of the Microstructure Properties Relationships." *Current Opinion in Solid State and Material Science* 15 (4): 141–68.

Bouaziz, O., and N. Guelton. 2001. "Modelling of TWIP Effect on Work-Hardening." *Material Science and Engineering A* A319–321: 246–49.

Chen, L., H.-S. Kim, S.K. Kim, and B.C. De Cooman. 2007. "Localized Deformation due to Porteviin-LeChatelier Effect in 18Mn-0.6C TWIP Austenitic Steel." *ISIJ International* 47: 1804.

Chen, L., Y. Zhao, and X. Qin. 2013. "Some Aspects of High Manganese Twinning-Induced Plasticity (TWIP) steel. A Review." *Acta Metallurgica Sinica. English Letters* 26 (1): 1–15.

De Cooman, B.C., K.G. Chin, and J. Kim. 2011. "High Mn TWIP Steels for Automotive Application." In *New Trends and Developments in Automotive System Engineering*, 101–28. INTECH.

De Cooman, B.C., O. Kwon, and K.G. Chin. 2012. "State-of-the-Knowledge on TWIP Steel." *Material Science and Technology* 28: 513.

Dieudonne, T., L. Marchetti, M. Wery, and F. Miserque. 2014. "Role of Copper and Aluminium on the Corrosion Behavior of Austenitic Fe-Mn-C TWIP Steels in Aqueous Solutions and the Related Hydrogen Absorption." *Corrosion Science* 83: 234–44.

Dini, G., and R. Ueji. 2011. "Effect of Grain Size and Grain Orientation on Dislocations Structure in Tensile Strained TWIP Steel during Initial Stages of Deformation." In *HMnS – 2011*. Seoul, South Korea.

Dumay, A., J.-P. Chateau, S. Allain, S. Migot, and O. Bouaziz. 2008. "Influence of Addition Elements on the Stacking-Fault Energy and Mechanical Properties of an Austenitic Fe-Mn-C Steel." *Material Science and Engineering A* 483–484: 184–87.

El-Danaf, E., S.R. Kalidindi, and R.D. Doherty. 1999. "Influence of Grain Size and Stacking-Fault Energy on Deformation Twinning in Fcc Metals." *Metallurgical and Materials Transactions* 30: 1223–33.

Frommeyer, G., U. Brux, and P. Neumann. 2003. "Supra-Ductile and High-Strength Manganese – TRIP/TWIP Steels for High Energy Absorption Purposes." *ISIJ International* 43 (3): 438–46.

Grassel, O., G. Kriger, G. Frommeyer, and L. Meyer. 2000. "High Strength Fe-Mn-(Al.Si) TRIP/ TWIP Steel Development – Properties-Application." *International Journal of Plasticity* 16: 1391–1409.

Hamada, A.S., F.M. Haggag, and D.A. Porter. 2012. "Non-Destructive Determination of the Yield Strength and Flow Properties of High Manganese Twinning-Induced Plasticity Steels." *Material Science and Engineering A* 558: 766–70.

Hong, S., S.Y. Shin, J. Lee, and D.-H. Ahn. 2014. "Serration Phenomena Occurring during Tensile Tests of Three High-Manganese Twinning-Induced Plasticity (TWIP) Steels." *Metallurgical and Materials Transactions* 45 (2): 633–45.

Idrisi, H., K. Renard, L. Ryelandt, D. Schryvers, and P.J. Jacques. 2010. "On the Mechanism of Twin Formation in Fe-Mn-C TWIP Steels." *Acta Materialia* 58: 2364–2476.

Jin, J.E., and Y.K. Lee. 2012. "Effect of Al on Microstructure and Tensile Properties of C-Bearing High Mn TWIP Steel." *Acta Materialia* 60 (4): 1680–88.

Jinyu, G., L. Rendong, M. Jinsong, and T. Fuping. 2013. "Study on High-Strength Steel and Application for Automobile in Anstel." In *The 2nd Intern. Symposium on Automobile Steel*. Anshan, China.

Kang, S., J.-G. Jung, and Y.-K. Lee. 2012. "Effect of Niobium on Mechanical Twinning and Tensile Properties of a High Mn Twinning-Induced Plasticity Steel." *Material Transaction of Japan Institute of Metals* 53 (12): 2187–90.

Kim, J.-K., L. Chen, H.-S. Kim, and S.-K. Kim. 2009. "On the Tensile Behavior of High Manganese Twinning-Induced Plasticity Steels." *Metallurgical and Materials Transactions* 40 (13): 3147–58.

Kim, J., I. Choi, and Y.K. Lee. 2011a. "Twin Formation in Strain-Controlled Lean Mn TWIP Steels." In *HMnS – 2011*. Seoul, South Korea.

Kim, J., S.-J. Lee, and B.C. De Cooman. 2011b. "Effect of Al on the Stacking Fault Energy of the Fe-18Mn-0.6C Twinning-Induced Plasticity." *Scripta Materialia* 65 (4): 363–66.

Kim, Y., N. Kang, Y. Park, I. Choi, and G. Kim. 2008. "Effect of the Strain Induced Martensite Transformation on the Delayed Fracture for Al-Added TWIP Steel." *Journal of the Korean Institute of Metals and Materials* 46: 780–87.

Koyama, M., E. Anelli, T. Sawaguchi, and D. Raabe. 2012. "Hydrogen-Induced Cracking at Grain and Twin Boundaries in an Fe–Mn–C Austenitic Steel." *Scripta Materialia* 66 (7): 469–462.

Koyama, M., T. Sawaguchi, T. Lee, C.S. Lee, and K. Tsuzaki. 2011a. "Work Hardening Associated with E-Martensitic Transformation, Deformation Twinning and Dynamic Strain Ageing in Fe-17Mn-0.6C and Fe-17Mn-0.8C TWIP Steels." *Material Science and Engineering A* 528: 7310–16.

Koyama, M., T. Sawaguchi, and K. Tsuzaki. 2011b. "Effect of Deformation Temperature on Work Hardening Behavior in Fe-17Mn-0.6C and Fe-17Mn-0.8C Steels." In *HMnS – 2011*. Seoul, South Korea.

Kwon, O. 2011. "Development of High Performance High Manganese TWIP Steels in POSCO." In *HiMnS – 2011*. Seoul, Korea.

Kwon, O., K. Lee, G. Kim, and K.G. Chin. 2010. "New Trends in Advanced High Strength Development for Automotive Applications." *Material Science Forum* 638–642: 136–41.

Lee, K.Y., S.K. Kim, J.-H. Kwak, Y.R. Cho, and S.D. Choo. 2014. "Recent Development of Automotive Sheet Steels." In *Materials on Car Body Engineering*. Bad Nauheim.

Liu, F., W.J. Dan, and Zhang W.G. 2015. "Strain Hardening Model of Twinning Induced Plasticity Steel at Different Temperatures." *Materials and Design* 65: 737–42.

Millard, B., B. Remy, C. Scott, and A. Deschamps. 2012. "Hydrogen Trapping by VC Precipitates and Structural Defects in a High Strength Fe-Mn-C Steel Studied by Small-Angle Neutron Scattering." *Material Science and Engineering A* 536: 110–16.

Neu, R.W. 2013. "Performance and Characterization of TWIP Steels for Automotive Applications." *Materials Performance and Characterizations* 2 (1): 244–84.

Park, K.T., K.-G. Jin, S.H. Han, and S.W. Hwang. 2010. "Stacking Fault Energy and Plastic Deformation of Fully Austenitic Manganese Steels: Effect of Al Addition." *Material Science and Engineering A* 527: 3651–61.

Qian, L., P. Guo, J. Meng, and F. Zhang. 2013. "Unusual Grain-Size and Strain-Rate Effects on the Serrated Flow in FeMnC Twin-Induced Plasticity Steels." *Journal of Material Science* 48: 1669–74.

Remy, L. 1978. "Kinetics of F.c.c. Deformation Twinning and Its Relationship to Stress-Strain Behavior." *Acta Metallurgica* 26 (3): 443–51.

Reyes-Calderón, F., I. Mejía, C. Bedolla-Jacuinde, and J. Calvo. 2011. "Effect of Microalloying Elements (B, Nb, and V) on Solution Heat Treatment Microstructure of Fe-Mn-Al-Si-C TWIP Steels." In *HmnS-2011*. Seoul, South Korea.

Ryu, J.H. 2012. "Hydrogen Embrittlement in TRIP and TWIP Steels." Ph.D., Pohang, South Korea: Pohang University of Science and Technology.

Scott, C., N. Guelton, S. Allain, and M. Faral. 2005. "The Development of a New Fe-Mn-C Austenitic Steel for Automotive Applications." In *Material Science and Technology*, 2:127–38.

Scott, C., B. Remy, J.L. Collet, and A. Cael. 2011. "Precipitation Strengthening in High Manganese Austenitic TWIP Steels." *International Journal of Materials Research* 102 (5): 538–49.

Sevillano, G. 2009. "An Alternative Model for the Strain Hardening of FCC Alloys That Twin, Validated for Twinning-Induced Plasticity Steel." *Scripta Materialia* 60: 336.

Yakubtsov, I.A., A. Airapour, and Perovic. 1999. "Effect of Nitrogen on Stacking Fault Energy of FCC Iron-Based Alloys." *Acta Materialia* 47: 1271.

Yen, H.-W., M. Huang, C.P. Scott, and J.R. Yang. 2012. "Interactions between Deformation-Induced Defects and Carbides in a Vanadium-Containing TWP Steel." *Scripta Materialia* 66: 1018–23.

Yoo, J.D., S.W. Hwang, and K.T. Park. 2009. "Factors Influencing Tensile Behavior of a Fe-28Mn-9Al-0.8C Steel." *Material Science and Engineering A* 508: 234.

Index

© Springer International Publishing Switzerland 2015

N. Fonstein, *Advanced High Strength Sheet Steels*,

DOI 10.1007/978-3-319-19165-2